MyWorkBook

Christine Verity

Prealgebra and Introductory Algebra

Fourth Edition

Margaret L. Lial
American River College

Diana L. Hestwood
Minneapolis Community and Technical College

John Hornsby
University of New Orleans

Terry McGinnis

PEARSON

Boston Columbus Indianapolis New York San Francisco Upper Saddle River
Amsterdam Cape Town Dubai London Madrid Milan Munich Paris Montreal Toronto
Delhi Mexico City São Paulo Sydney Hong Kong Seoul Singapore Taipei Tokyo

Reproduced by Pearson from electronic files supplied by the author.

Publishing as Pearson, 75 Arlington Street, Boston, MA 02116.

ISBN-13: 978-0-321-85478-0
ISBN-10: 0-321-85478-0

1 2 3 4 5 6 OPM 17 16 15 14 13

www.pearsonhighered.com

PEARSON

CONTENTS

Name: Date:
Instructor: Section:

Chapter 1 INTRODUCTION TO ALGEBRA: INTEGERS

1.1 Place Value

Learning Objectives

1. Identify whole numbers.
2. Identify the place value of a digit through hundred-trillions.
3. Write a whole number in words or digits.

Key Terms

Use the vocabulary terms listed below to complete each statement in exercises 1–3.

place value system **digits** **whole numbers**

1. The ten ______________________ in our number system are 0, 1,2 3, 4, 5, 6, 7, 8, and 9.
2. The ___________________________ are 0, 1, 2, 3, 4, and so on.
3. A _________________________ is a number system in which the location, or place, where a digit is written gives it a different value.

Guided Examples

Review this example for Objective 1:

1. Identify the whole numbers in this list.

$67, -3,\ 200,\ 1.9,\ \frac{8}{9},\ 0,\ 0.333,\ 8\frac{3}{4},\ 4$

The whole numbers are 0, 4, 67, and 200.

Now Try:

1. Identify the whole numbers in this list.

$59, -6,\ 350,\ 9.7,\ \frac{9}{8},\ 0,\ 7\frac{1}{3},$

$0.999,\ 5$

____

Review this example for Objective 2:

2. Identify the place of each 5 in the number.
7,946,145,057

7,946,145,057
↑ ↑ tens place
thousands place

Now Try:

2. Identify the place of each 3 in the number.
9,854,310,327

____

Name: Date:
Instructor: Section:

Review these examples for Objective 3:

3.

a. Write 7,096,381 in words.

seven million, ninety-six thousand, three hundred eighty-one

b. Write 41,677,000,500,000 in words.

forty-one trillion, six hundred seventy-seven billion, five hundred thousand

4. Write each number using digits.

a. Six hundred forty-two thousand, ten

The answer is 642,010.

b. Eighty-nine billion, twenty-five thousand, six hundred

The answer is 89,000,025,600.

Now Try:

3.

a. Write 9,075,862 in words.

____

b. Write 54,800,543,700,100 in words.

____

4. Write each number using digits.

a. Seven hundred fifty-three thousand, six

____

b. Eleven billion, ten thousand, twenty-five

____

Objective 1 Identify whole numbers.

For extra help, see Example 1 on page 5 of your text and Section Lecture video for Section 1.1 and Exercise Solutions Clip 7.

Choose the whole numbers in each set of numbers.

1. $-\frac{3}{4}, 2, 1.398, -2\frac{1}{2}, -1.04, -2, \frac{1}{5}, -2.6428$ **1.** ____________

2. $-0.5,\ -1,\ 4,\ 3,\ -4.87,\ -22,\ \frac{4}{3},\ 1\frac{1}{2}$ **2.** ____________

3. $2.718, -1, 365, 22.4, -6.02, 1, -4\frac{2}{5}$ **3.** ____________

Objective 2 Identify the place value of a digit through hundred-trillions.

For extra help, see Example 2 on page 5 of your text and Section Lecture video for Section 1.1 and Exercise Solutions Clip 9 and 15.

Give the place value of the digit 6 in each number.

4. 639,111,192 **4.** ____________

Name: Date:
Instructor: Section:

5. 94,164,372,757 **5.** ________________

6. 6458 **6.** ________________

Objective 3 Write a whole number in words or digits

For extra help, see Examples 3–4 on pages 5–6 of your text and Section Lecture video for Section 1.1 and Exercise Solutions Clip 21, 27, 31, 35, and 39.

Write each number in words.

7. 59,504,806,873 **7.** ________________

8. 9,671,000,637 **8.** ________________

Write the number using digits.

9. Nine hundred eighty-seven million, three hundred thirty **9.** ________________

Name: Date:
Instructor: Section:

Chapter 1 INTRODUCTION TO ALGEBRA: INTEGERS

1.2 Introduction to Integers

Learning Objectives

1	Write positive and negative numbers used in everyday situations.
2	Graph numbers on a number line.
3	Use the < and > symbols to compare integers.
4	Find the absolute value of integers.

Key Terms

Use the vocabulary terms listed below to complete each statement in exercises 1–3.

number line **integers** **absolute value**

1. The ______________________________ of a number is its distance from 0 on the number line.

2. A __________________ is used to show how numbers relate to each other.

3. The whole numbers together with their opposites and 0 are called ____________________.

Guided Examples

Review these examples for Objective 1:

1. Write each negative number with a negative sign. Write each positive number in two ways.

a. The Dead Sea is 1299 feet below sea level.

The answer is –1299 ft.

b. The mountain had elevation of 17,000 feet.

The answer is +17,000 feet or 17,000 feet.

Now Try:

1. Write each negative number with a negative sign. Write each positive number in two ways.

a. A diver descends to a depth of 150 ft below sea level.

____

b. The temperature rose to 68°F.

____

Name: Date:
Instructor: Section:

Review this example for Objective 2:

2. Graph each number on the number line.

$-\frac{1}{2}, -3, -\frac{5}{2}, \frac{1}{4}, 1\frac{7}{8}, 3$

Draw a dot at the correct location for each number.

$-3, -\frac{5}{2}, -\frac{1}{2}, \frac{1}{4}, 1\frac{7}{8}, 3$

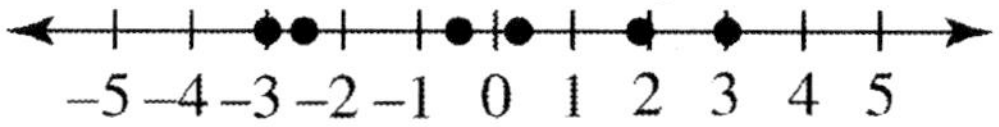

Now Try:

2. Graph each number on the number line.

$-3, -5, -1\frac{1}{2}, \frac{2}{3}, 3.5, 4$

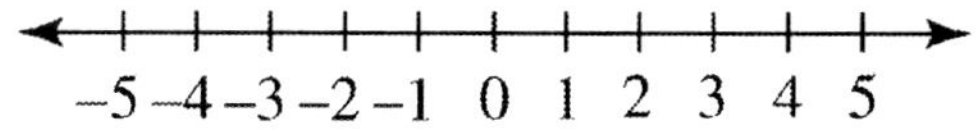

Review these examples for Objective 3:

3. Write < or > between each pair of integers to make a true statement.

a. 0 _____ 7

0 is to the left of 7 on the number line, so 0 is less than 7. Write 0 < 7.

b. –2 _____ –5

–2 is to the right of –5, so –2 is greater than –5. Write –2 > –5.

c. –6 _____ 7

–6 is to the left of 7, so –6 is less than 7. Write –6 < 7.

Now Try:

3. Write < or > between each pair of integers to make a true statement.

a. 0 _____ –8

____

b. –3 _____ –9

____

c. 11 _____ –4

____

Review these examples for Objective 4:

4. Find each absolute value.

a. $|3|$

The distance from 0 to 3 on the number line is 3 spaces. So, $|3| = 3$.

b. $|-3|$

The distance from 0 to –3 on the number line is 3 spaces. So, $|-3| = 3$.

c. $|0|$

$|0| = 0$ because the distance from 0 to 0 on the number line is 0 spaces.

Now Try:

4. Find each absolute value.

a. $|12|$

____

b. $|-13|$

____

c. $|0|$

____

Name: Date:
Instructor: Section:

Objective 1 Write positive and negative numbers used in everyday situations.

For extra help, see Example 1 on page 14 of your text and Section Lecture video for Section 1.2 and Exercise Solutions Clip 3 and 7.

Write a signed number for each of the following.

1. Following a defeat, an army retreats 46 kilometers. **1.** ________________

2. The company had a profit of \$830. **2.** ________________

3. A corporation has a shortfall of \$1.2 million **3.** ________________

Objective 2 Graph numbers on a number line.

For extra help, see Example 2 on page 14 of your text and Section Lecture video for Section 1.2 and Exercise Solutions Clip 11.

Graph each set of numbers on a number line.

4. $-2, -1, 0, 1, 2, 5$ **4.**

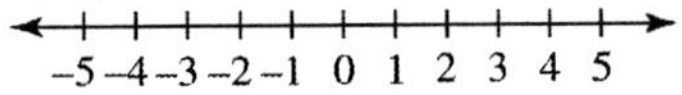

5. $-4.5, -1.5, -0.5, 0, 1.5, 2.5$ **5.**

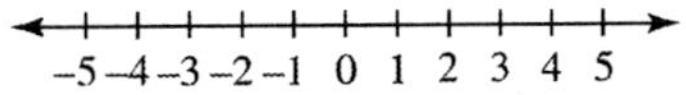

6. $4\frac{1}{2}, 1, -2\frac{3}{4}, -1\frac{1}{2}, -\frac{1}{4}$ **6.**

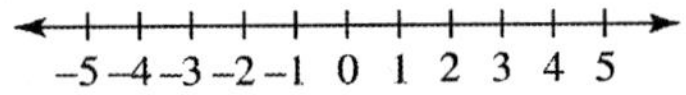

Objective 3 Use the $<$ and $>$ symbols to compare integers.

For extra help, see Example 3 on page 15 of your text and Section Lecture video for Section 1.2 and Exercise Solutions Clip 17, 23, and 31.

Write $<$ or $>$ in each blank to make a true statement.

7. -23 ____ -32 **7.** ________________

8. -6 ____ 0 **8.** ________________

9. -5 ____ -3 **9.** ________________

Name: Date:
Instructor: Section:

Objective 4 Find the absolute value of integers.

For extra help, see Example 4 on page 16 of your text and Section Lecture video for Section 1.2 and Exercise Solutions Clip 33, 35, 41, 43, and 44.

Find each absolute value.

10. $|-95|$ **10.** ________________

11. $|21|$ **11.** ________________

12. $|-788|$ **12.** ________________

Name: Date:
Instructor: Section:

Chapter 1 INTRODUCTION TO ALGEBRA: INTEGERS

1.3 Adding Integers

Learning Objectives
1 Add integers.
2 Identify properties of addition.

Key Terms

Use the vocabulary terms listed below to complete each statement in exercises 1–5.

addends **sum** **addition property of 0**

commutative property of addition

associative property of addition

1. By the________________________________, changing the order of the addends in an addition problem does not change the sum.

2. In addition, the numbers being added are called the ________________.

3. By the________________________________, changing the grouping of the addends in an addition problem does not change the sum.

4. The answer to an addition problem is called the ________________.

5. The ________________________ says that adding 0 to any number leaves the number unchanged.

Guided Examples

Review these examples for Objective 1:

1. Use a number line to find $(-3)+(-2)$.

Start at 0. Move 3 places to the left. Then move 2 places to the left.

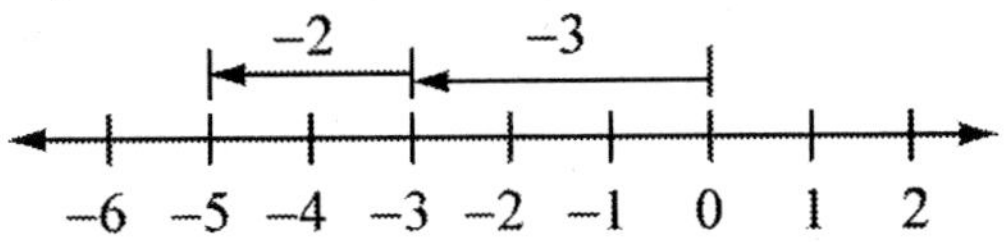

$(-3)+(-2)=-5$

Now Try:

1. Use a number line to find $(-4)+(-1)$.

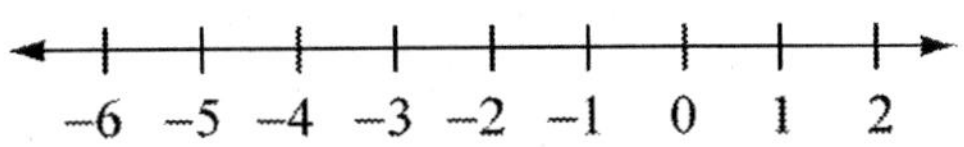

____

Name: Date:
Instructor: Section:

2. Add.

a. $-9+(-4)$

Step 1 Add the absolute values.
$|-9|=9$ and $|-4|=4$
Add 9 + 4 to get 13.
Step 2 Use the common sign as the sign of the sum. Both numbers are negative, so the sum is negative.
$-9+(-4)=-13$

b. $7+8$

Both numbers are positive, therefore the sum is positive.
$7+8=15$

3. Add.

a. $-7+4$

Step 1 $|-7|=7$ and $|4|=4$
Subtract $7-4$ to get 3.
Step 2 –7 has the greater absolute value and is negative, so the sum is also negative.
$-7+4=-3$

b. $-6+15$

Step 1 $|-6|=6$ and $|15|=15$
Subtract $15-6$ to get 9.
Step 2 15 has the greater absolute value and is positive, so the sum is also positive.
$-6+15=9$

4. A football team has to gain at least 10 yards during four plays in order to keep the ball. The football team lost 3 yards, gained 6 yards, lost 4 yards, and gained 9 yards. Did the team gain enough to keep the ball?

$$-3 \text{ yards} + 6 \text{ yards} + (-4 \text{ yards}) + 9 \text{ yards}$$
$$3 \text{ yards} \quad + (-4 \text{ yards}) + 9 \text{ yards}$$
$$-1 \text{ yards} \quad +9 \text{ yards}$$
$$8 \text{ yards}$$

The team gained 8 yards, which are not enough yards to keep the ball.

2. Add.

a. $(-25)+(-11)$

____

b. $19+21$

____

3. Add.

a. $-19+3$

____

b. $-4+25$

____

4. The temperature was –20°, then rose 32° and then dropped 15°. What was the new temperature?

____

Name: Date:
Instructor: Section:

Review these examples for Objective 2:

5. Rewrite each sum, using the commutative property of addition. Check that the sum is unchanged.

a. $78+62$

$$78+62=62+78$$
$$140 = 140$$

Both sums are 140, so the sum is unchanged.

b. $-24+(-53)$

$$-24+(-53)=-53+(-24)$$
$$-77 = -77$$

Both sums are –77, so the sum is unchanged.

6. In each addition problem, pick out the two addends that would be easiest to add. Write parentheses around those addends. Then find the sum.

a. $-10+10+7$

Group $-10+10$ because the sum is 0.

$$(-10+10)+7$$
$$0 \quad +7$$
$$7$$

b. $-35+16+4$

Group $16+4$ because the sum is 20, which is a multiple of 10.

$$-35+(16+4)$$
$$-35+ \quad 20$$
$$-15$$

Now Try:

5. Rewrite each sum, using the commutative property of addition. Check that the sum is unchanged.

a. $254+63$

____

b. $-44+(-77)$

____

6. In each addition problem, pick out the two addends that would be easiest to add. Write parentheses around those addends. Then find the sum.

a. $-20+19+(-19)$

____

b. $8+2+(-17)$

____

Objective 1 Add integers.

For extra help, see Examples 1–4 on pages 21–23 of your text and Section Lecture video for Section 1.3 and Exercise Solutions Clip 3, 5, 7, 13, 21, and 23.

Add by using a number line.

1. $-8+5$

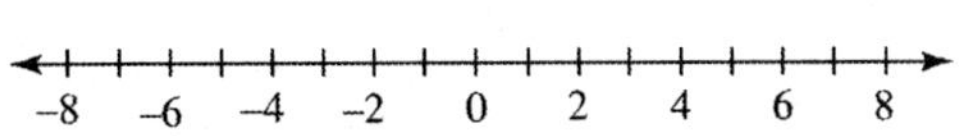

1. ____________

Name: Date:
Instructor: Section:

Add.

2. $7+(-22)$ **2.** ________________

Write an addition problem for the situation and find the sum

3. While playing a card game, Diane first gained 23 points, then lost 16 points, and finally gained 11 points. What was her final score? **3.** ________________

Objective 2 Identify properties of addition.

For extra help, see Examples 5–6 on pages 24–25 of your text and Section Lecture video for Section 1.3 and Exercise Solutions Clip 71.

Rewrite each sum using the commutative property of addition, and find the sum each way.

4. $-6+13$ **4.** ________________

In each addition problem, write parentheses around the two addends that would be easiest to add. Then find the sum.

5. $8+{}^{-}15+{}^{-}5$ **5.** ________________

6. ${}^{-}11+11+2$ **6.** ________________

Name: Date:
Instructor: Section:

Chapter 1 INTRODUCTION TO ALGEBRA: INTEGERS

1.4 Subtracting Integers

Learning Objectives
1. Find the opposite of an integer.
2. Subtract integers.
3. Combine adding and subtracting of integers.

Key Terms

Use the vocabulary terms listed below to complete each statement in exercises 1–2.

opposite **additive inverse**

1. The ____________________ of a number is the same distance from 0 on the number line as the original number, but located on the other side of 0.

2. The opposite of a number is its ______________________________.

Guided Examples

Review these examples for Objective 1:

1. Find the opposite (additive inverse) of each number. Show that the sum of the number and its opposite is 0.

a. 4

The opposite of 4 is –4 and $4+(-4)=0$.

b. –15

The opposite of –15 is 15 and $-15+15=0$.

c. 0

The opposite of 0 is 0 and $0+0=0$.

Now Try:

1. Find the opposite (additive inverse) of each number. Show that the sum of the number and its opposite is 0.

a. 25 ______________

b. –39 ______________

c. 0 ______________

Review these examples for Objective 2:

2. Make two pencil strokes to change each subtraction problem into an addition problem. Then find the sum.

a. $6-9$

Change 9 to –9. Change subtraction to addition.
$6-9=6+(-9)=-3$

Now Try:

2. Make two pencil strokes to change each subtraction problem into an addition problem. Then find the sum.

a. 13 – 15 ______________

b. $-8-(-5)$

Change –5 to 5. Change subtraction to addition.
$-8-(-5)=-8+5=-3$

c. $4-(-9)$

Change –9 to 9. Change subtraction to addition.
$4-(-9)=4+9=13$

d. $-3-12$

Change 12 to –12. Change subtraction to addition.
$-3-12=-3+(-12)=-15$

b. $-10-(-9)$

____

c. $12-(-11)$

____

d. $-8-16$

____

Review this example for Objective 3:

3. Simplify by completing all the calculations.
$-6-15-18+3$

Change all subtractions to adding the opposite. Change 15 to –15. Change 18 to –18. Then add from left to right.

$$\begin{aligned} &-6 - 15 - 18 + 3 \\ &-6+(-15)+(-18)+3 \\ &\quad -21 + (-18)+3 \\ &\qquad -39 \quad +3 \\ &\qquad\quad -36 \end{aligned}$$

Now Try:

3. Simplify by completing all the calculations.
$-9-14-21+6$

____

Objective 1 Find the opposite of an integer.

For extra help, see Example 1 on page 30 of your text and Section Lecture video for Section 1.4 and Exercise Solutions Clip 1, 3, and 4.

Find the opposite (additive inverse) of each number. Show that the sum of the number and its opposite is 0.

1. –11 **1.** ______________

____

2. 4 **2.** ______________

____

3. –5 **3.** ______________

____

Name: Date:
Instructor: Section:

Objective 2 Subtract integers.

For extra help, see Example 2 on page 31 of your text and Section Lecture video for Section 1.4 and Exercise Solutions Clip 9, 15, 17, 19, and 21.

Subtract by changing subtraction to addition.

4. $-2-(-11)$ **4.** ________________

5. $-13-1$ **5.** ________________

6. $12-(-5)$ **6.** ________________

Objective 3 Combine adding and subtracting of integers.

For extra help, see Example 3 on page 31 of your text and Section Lecture video for Section 1.4 and Exercise Solutions Clip 41.

Simplify.

7. $-2-16-(-18)$ **7.** ________________

8. $13-(-8)+(-7)-7$ **8.** ________________

9. $-14-(-3)-0+(-11)$ **9.** ________________

Name: Date:
Instructor: Section:

Chapter 1 INTRODUCTION TO ALGEBRA: INTEGERS

1.5 Problem Solving: Rounding and Estimating

Learning Objectives	
1	Locate the place to which a number is to be rounded.
2	Round integers.
3	Use front end rounding to estimate answers in addition and subtraction.

Key Terms

Use the vocabulary terms listed below to complete each statement in exercises 1−3.

rounding **estimate** **front end rounding**

1. ____________________________ is rounding to the highest possible place so that all the digits become zeros except the first one.

2. In order to find a number that is close to the original number, but easier to work with, use a process called ______________________.

3. ______________________ to find an answer close to the exact answer.

Guided Examples

Review these examples for Objective 1:

1. Locate and draw a line under the place to which each number is to be rounded. Then answer the question.

a. Round –49 to the nearest ten. Is –49 closer to –40 or –50?

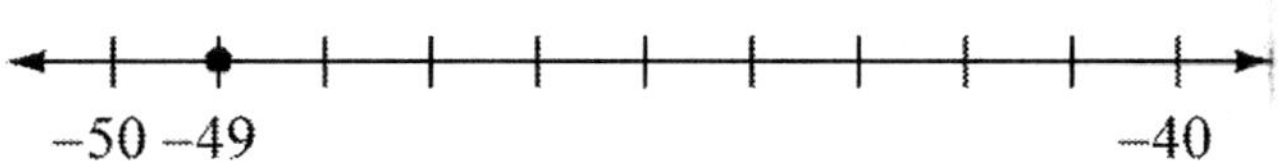

–49 is closer to –50.

b. Round $573 to the nearest hundred. Is $573 closer to $500 or $600?

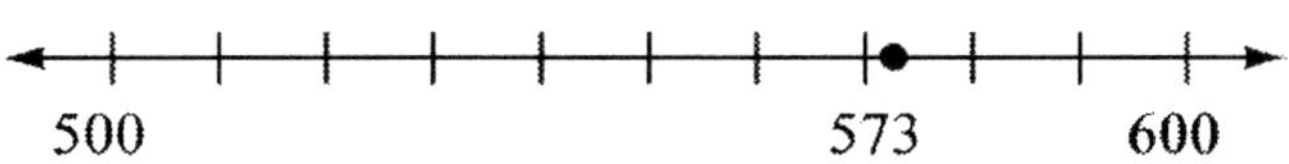

$573 is closer to $600.

c. Round –63,155 to the nearest thousand. Is –63,155 closer to –63,000 or –64,000?

–63,155 is closer to –63,000

Now Try:

1. Locate and draw a line under the place to which each number is to be rounded. Then answer the question.

a. Round –32 to the nearest ten. Is –32 closer to –30 or –40?

____

b. Round $642 to the nearest hundred. Is $642 closer to $600 or $700?

____

c. Round –76,542 to the nearest thousand. Is –76,542 closer to –76,000 or –77,000?

____

Name: Date:
Instructor: Section:

Review this example for Objective 2:

2. Round 541 to the nearest hundred.

Step 1 Locate the place to which the number is being rounded. Draw a line under that place.

541
hundred's place

Step 2 Because the next digit to the right of the underlined place is 4, which is 4 or less, do not change the digit in the underlined place.

541

Step 3 Change all digits to the right of the underlined place to zeros.

541 rounded to the nearest hundred is 500.

3. Round 67,899 to the nearest thousand.

Step 1 Locate the place to which the number is being rounded. Draw a line under that place.

67,899
thousand's place

Step 2 Because the next digit to the right of the underlined place is 8, which is 5 or more, add 1 to the underlined place.

67,899
change 7 to 8

Step 3 Change all digits to the right of the underlined place to zeros.

67,899 rounded to the nearest thousand is 68,000.

4.

a. Round –5623 to the nearest ten.

Step 1 –5623
ten's place

Step 2 The next digit to the right is 3, which is 4 or less.

–5623
Leave 2 as 2.

Step 3 –5623 rounds to –5620.

b. Round 37,968 to the nearest hundred.

Step 1 37,968

Now Try:

2. Round 732 to the nearest hundred.

____

3. Round 94,567 to the nearest thousand.

____

4.

a. Round –4738 to the nearest ten.

____

b. Round 56,978 to the nearest hundred.

____

Name: Date:
Instructor: Section:

Step 2 The next digit to the right of 9 is 6, which is 5 or more.

37,<u>9</u>68

Change 9 to 10; write 0 and regroup 1 into thousands place.

7 + regrouped 1 = 8

Step 3 37,968 rounds to 38,000.

5.

a. Round –86,732 to the nearest ten-thousand.

Step 1 –<u>8</u>6,732

Step 2 The next digit to the right is 6, which is 5 or more.

–<u>8</u>6,732

Change 8 to 9.

Step 3 –<u>8</u>6,732 rounds to –90,000.

b. Round 937,401,682 to the nearest million.

Step 1 93<u>7</u>,401,682

Step 2 The next digit is 4 or less.

93<u>7</u>,401,682

Leave 7 as 7.

Step 3 937,401,682 rounds to 937,000,000.

5.

a. Round –75,493 to the nearest ten-thousand.

____

b. Round 246,488,937 to the nearest million.

____

Review these examples for Objective 3:

6. Use front end rounding to round each number.

a. –324

Round to the highest possible place, that is, the leftmost digit. In this case the leftmost digit, 3, is in the hundreds place, so round to the nearest hundred.

For –<u>3</u>24, the next digit is 4 or less.

–<u>3</u>24 rounds to –300.

b. 95,602

The leftmost digit, 9, is in the ten-thousands place, so round to the nearest ten-thousand.

For <u>9</u>5,602, the next digit is 5 or more.

Change 9 to 10.

Regroup 1 into hundred-thousands place.

95,602 rounds to 100,000.

Now Try:

6. Use front end rounding to round each number.

a. –849

____

b. 974,312

____

Name: Date:
Instructor: Section:

7. Use front end rounding to estimate an answer. Then find the exact answer.
A checkbook had a balance of $912. A check was written for $394. What is the new balance?

Estimate: Use front end rounding to round $912 and $394.

$912 rounds to $900 $394 rounds to $400

Use the rounded number and subtract to estimate the new checkbook balance.

$900 – $400 = $500

Exact: Use the original numbers and subtract.

$912 – $394 = $518

7. Use front end rounding to estimate an answer. Then find the exact answer.
A monthly gross pay is $3207, with deductions of $795. Find the net pay after deductions.

Estimate ____________

Exact ____________

Objective 1 Locate the place to which a number is to be rounded.

For extra help, see Example 1 on page 34 of your text and Section Lecture video for Section 1.5 and Exercise Solutions Clip 5 and 13.

Locate the place to which the number is rounded by writing the appropriate digit.

1. $^{-}135$ Nearest ten **1.** ____________

2. $^{-}99{,}102$ Nearest hundred **2.** ____________

3. 645,371 Nearest ten-thousand **3.** ____________

Objective 2 Round integers.

For extra help, see Examples 2–5 on pages 35–37 of your text and Section Lecture video for Section 1.5 and Exercise Solutions Clip 5 and 13.

Round each number to the indicated place.

4. $^{-}6694$ to the nearest thousand **4.** ____________

5. 814,118 to the nearest ten thousand **5.** ____________

6. $^{-}4{,}697{,}134$ to the nearest million **6.** ____________

Name: Date:
Instructor: Section:

Objective 3 Use front end rounding to estimate answers in addition and subtraction.

For extra help, see Examples 6–7 on pages 38–39 of your text and Section Lecture video for Section 1.5 and Exercise Solutions Clip 41 and 51.

Use front end rounding to round the number.

7. The first printing run of a new book is 170,000 copies.

7. ________________

First use front end rounding to estimate each answer. Then find the exact answer.

8. $^{-}5172-3850$

8.
Estimate_____________

Exact ______________

First use front end rounding to estimate the answer to the application problem. Then find the exact answer.

9. A cannon is raised above the horizontal by an angle of 36°. The cannon is lowered 8°, then lowered another 17°, and finally raised 24°. What is the cannon's final elevation angle?

9.
Estimate_____________

Exact ______________

Name: Date:
Instructor: Section:

Chapter 1 INTRODUCTION TO ALGEBRA: INTEGERS

1.6 Multiplying Integers

Learning Objectives	
1	Used a raised dot or parentheses to express multiplication.
2	Multiply integers.
3	Identify properties of multiplication.
4	Estimate answers to application problems involving multiplication.

Key Terms

Use the vocabulary terms listed below to complete each statement in exercises 1–7.

factors **product** **multiplication property of 0**

multiplication property of 1 **commutative property of multiplication**

associative property of multiplication **distributive property**

1. By the____________________________, changing the order of the factors in a multiplication problem does not change the product.

2. In multiplication, the numbers being multiplied are called the ________________.

3. By the____________________________, changing the grouping of the factors in a multiplication problem does not change the product.

4. The ________________________ says that multiplying a number by 0 gives a product of 0.

5. The answer to a multiplication problem is called the ________________.

6. The ______________________________________ states that $a(b + c) = ab + bc$.

7. The ________________________ says that multiplying a number by 1 leaves the number unchanged.

Guided Examples

Review these examples for Objective 1:

1. Rewrite each multiplication in three different ways, using a dot or parentheses. Also identify the factor and the product.

a. 60×3

Rewrite it as $60 \cdot 3$ or $60(3)$ or $(60)(3)$.
The factors are 60 and 3. The product is 180.

Now Try:

1. Rewrite each multiplication in three different ways, using a dot or parentheses. Also identify the factor and the product.

a. 80×6

____

Name: Date:
Instructor: Section:

b. 5×90

Rewrite it as $5 \cdot 90$ or $5(90)$ or $(5)(90)$.
The factors are 5 and 90. The product is 450.

b. 8×15

____

Review these examples for Objective 2:

2. Multiply.

a. $-3 \cdot 9$

The factors have different signs, so the product is negative.

$-3 \cdot 9 = -27$

b. $-11(-7)$

The factors have the same sign, so the product is positive.

$-11(-7) = 77$

c. $8(-12)$

The factors have different signs, so the product is negative.

$8(-12) = -96$

3. Multiply.

a. $-7 \cdot (2 \cdot 4)$

Multiply inside the parentheses first.

$-7 \cdot (2 \cdot 4)$

$-7 \cdot 8$

-56

Note that the factors –7 and 8 have different signs, so the product is negative.

b. $-3 \cdot (-3) \cdot (-3)$

There is no work to do inside the parentheses, so multiply $-3 \cdot (-3)$ first. The factors have the same sign, so the product is positive.

$-3 \cdot (-3) \cdot (-3)$

$9 \cdot (-3)$

-27

Note that the factors 9 and –3 have different signs, so the product is negative.

Now Try:

2. Multiply.

a. $-6 \cdot 7$

____

b. $-4(-15)$

____

c. $9(-6)$

____

3. Multiply.

a. $-5 \cdot (7 \cdot 2)$

____

b. $-4 \cdot (-4) \cdot (-4)$

____

Name: Date:
Instructor: Section:

Review these examples for Objective 3:

4. Multiply. Then name the property illustrated by each example.

a. $(0)(-57)$

$(0)(-57)=0$
This illustrates the multiplication property of 0.

b. $763(1)$

$763(1)=1$
This illustrates the multiplication property of 1.

5. Show that the product is unchanged and name the property that is illustrated in each case.

a. $-5\cdot(-8)=-8\cdot(-5)$

$-5\cdot(-8)=-8\cdot(-5)$
$40 = 40$
This example illustrates the commutative property of multiplication.

b. $6\cdot(7\cdot3)=(6\cdot7)\cdot(3)$

$6\cdot(7\cdot3)=(6\cdot7)\cdot(3)$
$6\cdot21=42\cdot(3)$
$126=126$
This example illustrates the associative property of multiplication.

6. Rewrite each product using the distributive property. Show that the result is unchanged.

a. $5(6+9)$

$5(6+9)=5\cdot6+5\cdot9$
$5(15)=30+45$
$75=75$
Both results are 75.

b. $-3(-8+2)$

$-3(-8+2)=(-3)\cdot(-8)+(-3)\cdot(2)$
$-3(-6)=24+(-6)$
$18=18$
Both results are 18.

Now Try:

4. Multiply. Then name the property illustrated by each example.

a. $671(0)$

____

b. $1(-92)$

____

5. Show that the product is unchanged and name the property that is illustrated in each case.

a. $-9\cdot(-7)=-7\cdot(-9)$

____

b. $-11\cdot(3\cdot6)=(-11\cdot3)\cdot(6)$

____

6. Rewrite each product using the distributive property. Show that the result is unchanged.

a. $7(4+8)$

____

b. $-9(-5+3)$

____

Name: Date:
Instructor: Section:

Review this example for Objective 4:	**Now Try:**
7. Use front end rounding to estimate an answer. Then find the exact answer. A local cable company is losing about 1955 customers per month. How many customers are lost in a year?	**7.** Use front end rounding to estimate an answer. Then find the exact answer. Enrollment at a community college has decreased by 595 students for the last 3 semesters. What is the total decrease?
Estimate: Use front end rounding: –1955 round to –2000, and 12 months rounds to 10 months. $(-2000)\cdot(10) = -20,000$ customers	**Estimate** ____________
Exact: $(-1955)\cdot(12) = -23,460$ customers	**Exact** ____________

Objective 1 Used a raised dot or parentheses to express multiplication.

For extra help, see Example 1 on page 44 of your text and Section Lecture video for Section 1.6 and Exercise Solutions Clip 5 and 29.

Rewrite each multiplication, first using a dot and then using parentheses.

1. $^{-}3\times 6$ **1.** ____________

____

2. $12\times {}^{-}3$ **2.** ____________

____

3. 4×5 **3.** ____________

____

Objective 2 Multiply integers.

For extra help, see Examples 2–3 on page 46 of your text and Section Lecture video for Section 1.6 and Exercise Solutions Clip 11, 17, and 31.

Multiply.

4. $-7(-1)(-4)$ **4.** ____________

5. $-2\cdot 31$ **5.** ____________

6. $-5\cdot(-2)\cdot(6)$ **6.** ____________

Name: Date:
Instructor: Section:

Objective 3 Identify properties of multiplication.

For extra help, see Examples 4–6 on pages 47–49 of your text and Section Lecture video for Section 1.6 and Exercise Solutions Clip 41, 49, 51, and 53.

Fill in the blank to make a true statement.

7. $0 = 18(\_\_\_\_)$ **7.** ________________

Rewrite each multiplication, using the stated property. Show that the result is unchanged.

8. Commutative property **8.** ________________
$-22 \cdot 4$

9. Associative property **9.** ________________
$(3 \cdot (-2)) \cdot (-9)$

Objective 4 Estimate answers to application problems involving multiplication.

For extra help, see Example 7 on page 49 of your text and Section Lecture video for Section 1.6 and Exercise Solutions Clip 59.

First use front end rounding to estimate the answer to each application problem. Then find the exact answer.

10. An airplane on an approach path to an airport is losing altitude at the rate of 820 feet per minute. How much altitude does the plane lose in 18 minutes?

10.
Estimate______________
Exact ______________

11. A television is purchased for $375 down and twelve monthly payments of $55 each. A set of speakers costing a total of $280 is also part of the purchase. What is the total cost of the system?

11.
Estimate______________
Exact ______________

12. A tree grows from a seedling for 15 years at a rate of 6 feet per year. At the end of that time, the tree is felled with a cut 4 feet above the ground. Seven feet of the tip are trimmed off. How long is the resulting log?

12.
Estimate______________
Exact ______________

Name: Date:
Instructor: Section:

Chapter 1 INTRODUCTION TO ALGEBRA: INTEGERS

1.7 Dividing Integers

Learning Objectives
1. Divide integers.
2. Identify properties of division.
3. Combine multiplying and dividing of integers.
4. Estimate answers to application problems involving division.
5. Interpret remainders in division application problems.

Key Terms

Use the vocabulary terms listed below to complete each statement in exercises 1−3.

factors **product** **quotient**

1. Numbers that are being multiplied are called ______________________________.
2. The answer to a division problem is called the __________________________.
3. The answer to a multiplication problem is called the ________________________.

Guided Examples

Review these examples for Objective 1:

1. Divide.

a. $\frac{-30}{6}$

The numbers have different signs, so the quotient is negative.

$\frac{-30}{6}=-5$

b. $\frac{-27}{-9}$

The numbers have the same sign, so the quotient is positive.

$\frac{-27}{-9}=3$

c. $70\div(-5)$

The numbers have different signs, so the quotient is negative.

$70\div(-5)=-14$

Now Try:

1. Divide.

a. $\frac{-35}{7}$

____

b. $\frac{-40}{-10}$

____

c. $24\div(-8)$

____

Name: Date:
Instructor: Section:

Review these examples for Objective 2:

2. Divide. Then state the property illustrated by each example.

a. $\frac{-672}{-672}$

Any nonzero number divided by itself is 1.

$\frac{-672}{-672}=1$

b. $\frac{85}{1}$

Any number divided by 1 is the number.

$\frac{85}{1}=85$

c. $\frac{0}{23}$

Zero divided by any nonzero number is 0.

$\frac{0}{23}=0$

d. $\frac{67}{0}$

Division by 0 is undefined.

$\frac{67}{0}$ is undefined.

Now Try:

2. Divide. Then state the property illustrated by each example.

a. $\frac{-703}{-703}$

____

b. $\frac{93}{1}$

____

c. $\frac{0}{39}$

____

d. $\frac{51}{0}$

____

Review these examples for Objective 3:

3. Simplify.

a. $8(-12)\div(-4\cdot 2)$

$8(-12)\div(-4\cdot 2)$

$8(-12)\div(-8)$

$-96\div(-8)$

12

b. $-36\div(-3)(4)\div(-6)$

$-36\div(-3)(4)\div(-6)$

$12(4)\div(-6)$

$48\div(-6)$

-8

Now Try:

3. Simplify.

a. $5(-16)\div(-2\cdot 4)$

____

b. $-48\div(-4)(3)\div(-9)$

____

Name: Date:
Instructor: Section:

c. $-80 \div (-16) \div (-1)$

$$-80 \div (-16) \div (-1)$$
$$5 \div (-1)$$
$$-5$$

Now Try:

c. $-56 \div (-8) \div (-7)$

____

Review this example for Objective 4:

4. Use front end rounding to estimate an answer. Then find the exact answer.
Matthew lost 108 pounds last year. What was the average weight loss per month?

Estimate: Use front end rounding: –108 rounds to –100, and 12 months rounds to 10 months.
$-100 \div 10 = -10$ pounds each month

Exact:
$-108 \div 12 = -9$ pounds each month

Now Try:

4. Use front end rounding to estimate an answer. Then find the exact answer.
During the last year, John lost \$3708 in the stock market. What was the average loss each month?

____

Review these examples for Objective 5:

5. Divide; then interpret the remainder in each application.

a. A reading club has \$150 to buy children's books. If each book is \$9, how many can be purchased? How much is left over?

We solve using division.

$$\begin{array}{r} 16 \\ 9\overline{)150} \\ \underline{9} \\ 60 \\ \underline{54} \\ 6 \end{array}$$

The club can buy 16 books. There will be \$6 left over.

b. The college marching band has 118 members and will spend the night in a hotel. Each room can accommodate 4 people. How many rooms are needed?

We use division to solve the problem.

$$\begin{array}{r} 29 \\ 4\overline{)118} \\ \underline{8} \\ 38 \\ \underline{36} \\ 2 \end{array}$$

Now Try:

5. Divide; then interpret the remainder in each application.

a. John is serving 50 hot dogs at a party. If each package of rolls contains 8 rolls, how many packages of rolls should he buy?

____

b. In the lobby of a skyscraper, 150 people are waiting for elevators. If the capacity of each elevator is 18 people, how many elevator trips are needed?

____

Name: Date:
Instructor: Section:

If 29 rooms are rented, two members will have to sleep on a bus. So, 30 rooms must be rented. (One room will have only 2 band members.)

Objective 1 Divide integers.

For extra help, see Example 1 on page 54 of your text and Section Lecture video for Section 1.7 and Exercise Solutions Clip 9, 15, 23, and 29.

Divide.

1. $\frac{28}{-4}$

1. ________________

2. $\frac{-36}{-12}$

2. ________________

3. $\frac{-100}{10}$

3. ________________

Objective 2 Identify properties of division.

For extra help, see Example 2 on page 55 of your text and Section Lecture video for Section 1.7 and Exercise Solutions Clip 7, 17, and 27.

Complete the equation, then state the property represented by the equation.

4. $\frac{0}{14} =$

4. ________________

5. $\frac{-6}{-6} =$

5. ________________

6. $\frac{-48}{0} =$

6. ________________

Objective 3 Combine multiplying and dividing of integers.

For extra help, see Example 3 on page 56 of your text and Section Lecture video for Section 1.7 and Exercise Solutions Clip 33 and 39.

Simplify.

7. $-120 \div (-6) \div 10$

7. ________________

Name: Date:
Instructor: Section:

8. $28 \cdot (2 \cdot (-1)) \div -8$ **8.** ________________

9. $84 \div (-3) \cdot 3 \div 4$ **9.** ________________

Objective 4 Estimate answers to application problems involving division.

For extra help, see Example 4 on page 57 of your text and Section Lecture video for Section 1.7 and Exercise Solutions Clip 59.

Solve these application problems by using addition, subtraction, multiplication, or division. First estimate the answer using rounding; then find the exact answer.

10. A honeybee flies from its hive a distance of 52 meters to one flowering bush, and from there it flies a distance of 17 meters from there to the next bush. What is the bee's total distance of travel so far?

10.
Estimate ______________
Exact ______________

11. A car salesperson, bargaining with a customer over a car with a displayed price of \$24,800, agrees to knock \$2200 off that price. What is the new price at which the car is being offered?

11.
Estimate ______________
Exact ______________

12. A sound wave has a frequency of 468 cycles per second. This frequency is multiplied by 5 to produce a new sound. What is the new frequency?

12.
Estimate ______________
Exact ______________

Name: Date:
Instructor: Section:

Objective 5 Interpreting Remainders in Division Applications.

For extra help, see Example 5 on pages 57–58 of your text and Section Lecture video for Section 1.7.

Divide, then interpret the remainder in each application.

13. A group of 96 construction workers is to be transported on 7 trucks. The workers are to be divided as evenly as possible. How many workers will be transported on each truck?

13. _______________

14. A student plans to memorize 213 pages of notes in seven days. How many pages per day should the student memorize?

14. _______________

15. A graduation ceremony will be held in an auditorium with 1500 seats. If there are 468 students in the class, how many tickets will be allotted to each student?

15. _______________

Name: Date:
Instructor: Section:

Chapter 1 INTRODUCTION TO ALGEBRA: INTEGERS

1.8 Exponents and Order of Operations

Learning Objectives

1. Use exponents to write repeated factors.
2. Simplify expressions containing exponents.
3. Use the order of operations.
4. Simplify expressions with fraction bars.

Key Terms

Use the vocabulary terms listed below to complete each statement in exercises 1–2.

exponent **order of operations**

1. For problems or expressions with more than one operation, the ________________________ tells what to do first, second, and so on, to obtain the correct answer.

2. An ____________________ tells how many times a number is used as a factor in repeated multiplication.

Guided Examples

Review these examples for Objective 1:

1. Given the factored form, give the exponential form, simplified form, and how it is read.

a. $4 \cdot 4 \cdot 4$

Exponential form: 4^3
Simplified: 64
Read as: 4 cubed, or 4 to the third power

b. $(3)(3)$

Exponential form: 3^2
Simplified: 9
Read as: 3 squared, or 3 to the second power

c. 11

Exponential form: 11^1
Simplified: 11
Read as: 11 to the first power

Now Try:

1. Given the factored form, give the exponential form, simplified form, and how it is read.

a. $5 \cdot 5 \cdot 5 \cdot 5$

____

b. $(6)(6)$

____

c. 23

____

Name: Date:
Instructor: Section:

Review these examples for Objective 2:

2. Simplify.

a. $(-4)^2$

$(-4)^2 = (-4)(-4) = 16$

b. $(-3)^3$

$(-3)^3 = (-3)(-3)(-3)$
$= 9(-3)$
$= -27$

c. $(-4)^4$

$(-4)^4 = (-4)(-4)(-4)(-4) = 256$

d. $5^3(-2)^2$

$$5^3(-2)^2 = \underbrace{(5)(5)(5)}\underbrace{(-2)(-2)}$$
$$= \quad (125) \qquad (4)$$
$$= 500$$

Now Try:

2. Simplify.

a. $(-7)^2$ ______

b. $(-10)^3$ ______

c. $(-3)^4$ ______

d. $4^2(-3)^3$ ______

Review these examples for Objective 3:

3. Simplify.

a. $-9-(-7)+(-13)$

$-9-(-7)+(-13)$
$-9 \;+7\; +(-13)$
$-2 \;+(-13)$
-15

b. $-18 \div (-6)(5)$

$-18 \div (-6)(5)$
$(3)(5)$
15

4. Simplify $8+5(24-6)\div 9$.

$8+5(24-6)\div 9$
$8+\;5\;(18)\;\div 9$
$8\;+\;90\;\div 9$
$8+10$
18

Now Try:

3. Simplify.

a. $-11-(-4)+(-19)$ ______

b. $-27 \div (-9)(8)$ ______

4. Simplify $5+4(25-3)\div 11$. ______

Name: Date:
Instructor: Section:

5. Simplify.

a. $-9 \div (7-4) - 13$

$-9 \div (7-4) - 13$

$-9 \div (3) - 13$

$-3 - 13$

-16

b. $5 + 3(9-14) \cdot (18 \div 6)$

$5 + 3(9-14) \cdot (18 \div 6)$

$5 + 3(-5) \cdot (18 \div 6)$

$5 + 3(-5) \cdot (3)$

$5 + (-15) \cdot (3)$

$5 + (-45)$

-40

6. Simplify.

a. $6^2 - (-7)^2$

$6^2 - (-7)^2$

$6^2 - 49$

$36 - 49$

-13

b. $(-3)^4 - (5-7)^2(-9)$

$(-3)^4 - (5-7)^2(-9)$

$(-3)^4 - (-2)^2(-9)$

$81 - 4(-9)$

$81 - (-36)$

$81 + 36$

117

5. Simplify.

a. $-10 \div (8-6) - 19$

____

b. $6 + 4(11-15) \cdot (24 \div 8)$

____

6. Simplify.

a. $8^2 - (-9)^2$

____

b. $(-5)^3 - (6-9)^2(-8)$

____

Name: Date:
Instructor: Section:

Review this example for Objective 4:

7. Simplify $\dfrac{-9+6(7-10)}{5-6^2 \div 18}$.

First do work in the numerator.

$-9+6(7-10)$

$-9+\ 6(-3)$

$-9+(-18)$

$-27 \leftarrow$ Numerator

Now do the work in the denominator.

$5-6^2 \div 18$

$5-36 \div 18$

$5-2$

$3 \leftarrow$ Denominator

The last step is the division.

$\begin{matrix}\text{Numerator} \rightarrow \\ \text{Denominator} \rightarrow\end{matrix} \dfrac{-27}{3} = -9$

Now Try:

7. Simplify $\dfrac{-15+7(9-12)}{8-10^2 \div 25}$.

____

Objective 1 Use exponents to write repeated factors.

For extra help, see Example 1 on page 67 of your text and Section Lecture video for Section 1.8 and Exercise Solutions Clip 1, 5, and 7.

Rewrite each number in factored form as a number in exponential form; then state how the exponential form is read.

1. $12 \cdot 12 \cdot 12$ **1.** ______________

____

2. $6 \cdot 6 \cdot 6 \cdot 6 \cdot 6$ **2.** ______________

____

3. $7 \cdot 7 \cdot 7 \cdot 7 \cdot 7 \cdot 7 \cdot 7$ **3.** ______________

____

Objective 2 Simplify expressions containing exponents.

For extra help, see Example 2 on page 68 of your text and Section Lecture video for Section 1.8 and Exercise Solutions Clip 19, 25, and 27.

Simplify.

4. $(-1)^{91}$ **4.** ______________

Name: Date:
Instructor: Section:

5. $2^4 \cdot (-3)^2$ **5.** ________________

6. $(-2)^5 \cdot (-3)^2$ **6.** ________________

Objective 3 Use the order of operations.

For extra help, see Examples 3–6 on pages 68–71 of your text and Section Lecture video for Section 1.8 and Exercise Solutions Clip 45, 51, 55, and 67.

Simplify.

7. $3(-2+6)-(8-13)$ **7.** ________________

8. $3-(-6)\cdot(-1)^8$ **8.** ________________

9. $(-2)^3 \cdot (7-9)^2 \div 2$ **9.** ________________

Objective 4 Simplify expressions with fraction bars.

For extra help, see Example 7 on page 72 of your text and Section Lecture video for Section 1.8.

Simplify.

10. $\dfrac{-8+4^2-(-9)}{11-3-9}$ **10.** ________________

11. $\dfrac{-3\cdot 4^2-2(5+(-5))}{-3(8-13)\div -5}$ **11.** ________________

12. $\dfrac{4\cdot 2^2-8(7-2)}{7(6-8)\div 14}$ **12.** ________________

Name: Date:
Instructor: Section:

Chapter 2 UNDERSTANDING VARIABLES AND SOLVING EQUATIONS

2.1 Introduction to Variables

Learning Objectives
1. Identify variables, constants, and expressions.
2. Evaluate variable expressions for given replacement values.
3. Write properties of operations using variables.
4. Use exponents with variables.

Key Terms

Use the vocabulary terms listed below to complete each statement in exercises 1–5.

variable **constant** **expression**

evaluate the expression **coefficient**

1. An ____________________________ tells the rule for doing something.
2. A ______________________ is a letter that represents a number that varies or changes, depending on the situation.
3. To ____________________________, replace each variable with specific values and then follow the order of operations.
4. A ________________________ is a number that is added or subtracted in an expression.
5. The number part in a multiplication expression is the ______________________.

Guided Examples

Review this example for Objective 1:

1. Write an expression for this rule. Identify the variable and the constant.
   Order the class limit minus 8 lunches because some students will brown bag.

   Let c represent the variable and 8 is the constant.
   $c-8$

Now Try:

1. Write an expression for this rule. Identify the variable and the constant.
   Order the class limit plus 5 extra lunches because some students are football players.

   ____

Name: Date:
Instructor: Section:

Review these examples for Objective 2:

2. Use this rule (from Example 1) for ordering lunches: Order the class limit minus 8. The expression is $c - 8$.

a. Evaluate the expression when the class limit is 39.

Replace c with 39 and follow the rule.

$c-8$

$39-8$

31 Order 31 lunches.

b. Evaluate the expression when the class limit is 22.

Replace c with 22 and follow the rule.

$c-8$

$22-8$

14 Order 14 lunches.

3. The expression (rule) for finding the perimeter of a square shape is $4s$. Evaluate the expression when the length of a side of a square garden is 20 feet.

Replace s with 20 feet.

The total distance around the square garden is 80 feet.

4. Evaluate the expression $100+\frac{a}{2}$ when the age of the person is 36.

Replace a with 36, the age of the person. Then follow the rule using the order of operations.

$100+\frac{a}{2}$

$100+\frac{36}{2}$

$100+18$

118

The approximate systolic blood pressure is 118.

Now Try:

2. Use this rule (from Now Try 1) for ordering lunches: Order the class limit plus 8. The expression is $c + 5$.

a. Evaluate the expression when the class limit is 24.

____

b. Evaluate the expression when the class limit is 45.

____

3. The expression (rule) for finding the perimeter of a square shape is $4s$. Evaluate the expression when the length of a side of a square garden is 8 yards.

____

4. Evaluate the expression $100+\frac{a}{2}$ when the age of the person is 56.

____

Name: Date:
Instructor: Section:

5.

a. Find your average score if you bowl three games and your total score for all three games is 384.

Use the expression (rule) for finding your average score. Replace t with your total score of 384, and replace g with 3, the number of games.

$$\frac{t \rightarrow 384}{g \rightarrow 3}$$

Divide 384 by 3 to get 128. Your average score is 128.

b. Complete these tables to show how to evaluate each expression.

Value of x	Value of y	Expression (Rule) $x+y$
3	6	$3+6$ is 9
-7	4	__+__ is __
0	17	+ is

Value of x	Value of y	Expression (Rule) xy
3	4	$3 \cdot 4$ is 12
-8	3	__·__ is __
0	19	· is

The expression (rule) is to *add* the two variables. So the completed table is:

Value of x	Value of y	Expression (Rule) $x+y$
3	6	$3+6$ is 9
-7	4	$-7+4$ is -3
0	17	$0+17$ is 17

The expression (rule) is to *multiply* the two variables. So the completed table is:

Value of x	Value of y	Expression (Rule) xy
3	4	$3 \cdot 4$ is 12
-8	3	$-8 \cdot 3$ is -24
0	19	$0 \cdot 19$ is 0

5.

a. Find your average score if you bowl four games and your total score for all four games is 384.

____

b. Complete these tables to show how to evaluate each expression.

Value of x	Value of y	Expression (Rule) $x+y$
10	21	__+__ is __
-12	9	__+__ is __
93	0	+ is

Value of x	Value of y	Expression (Rule) xy
7	6	__·__ is __
-9	5	__·__ is __
31	0	· is

Name: Date:
Instructor: Section:

Review this example for Objective 3:

6. Use the variable b to state the property: When any number is added to 0, the sum is the number.

Use the letter b to represent any number.
$b+0=b$

Now Try:

6. Use the variable b to state the property: When any non-zero number is divided into 0, the quotient is zero.

____

Review these examples for Objective 4:

7. Rewrite each expression without exponents.

a. x^6

x^6 can be written as $x \cdot x \cdot x \cdot x \cdot x \cdot x$.

b. $13a^2b$

$13a^2b$ can be written as $13 \cdot a \cdot a \cdot b$.

c. $-10p^3q^4$

$-10p^3q^4$ can be written as $-10 \cdot p \cdot p \cdot p \cdot q \cdot q \cdot q \cdot q$.

8. Evaluate each expression.

a. x^3 when x is –5

x^3 means $x \cdot x \cdot x$ Replace each x with -5.

$$(-5)\cdot(-5)\cdot(-5)$$
$$25 \quad \cdot(-5)$$
$$-75$$

So x^3 becomes $(-5)^3$, which is $(-5)\cdot(-5)\cdot(-5)$, or –125.

b. x^2y^3 when x is –6 and y is –3

Replace x with –6 and replace y with –3.

x^2y^3 means $x \cdot x \cdot y \cdot y \cdot y$

$$(-6)\cdot(-6)\cdot(-3)\cdot(-3)\cdot(-3)$$
$$36 \quad \cdot(-3)\cdot(-3)\cdot(-3)$$
$$-108 \quad \cdot(-3)\cdot(-3)$$
$$324 \quad \cdot(-3)$$
$$-972$$

So x^2y^3 becomes $(-6)^2(-3)^3$, which is $(-6)\cdot(-6)\cdot(-3)\cdot(-3)\cdot(-3)$, or –972.

Now Try:

7. Rewrite each expression without exponents.

a. z^4

____

b. $14rs^2$

____

c. $-13m^3n^5$

____

8. Evaluate each expression.

a. z^2 when z is –9

____

b. x^4y^2 when x is –2 and y is –3

____

Name: Date:
Instructor: Section:

c. $-7a^2b$ when a is 4 and b is 6

Replace a with 4 and replace b with 6.

$-7a^2b$ means $-7 \cdot a \cdot a \cdot b$

$$-7 \cdot 4 \cdot 4 \cdot 6$$
$$-28 \cdot 4 \cdot 6$$
$$-112 \cdot 6$$
$$-672$$

So $-7a^2b$ becomes $-7(4)^2(6)$, which is $-7 \cdot 4 \cdot 4 \cdot 6$, or –672.

c. $-6c^3d$ when c is 2 and d is 5

____

Objective 1 Identify variables, constants, and expressions.

For extra help, see Example 1 on page 95 of your text and Section Lecture video for Section 2.1.

Identify the parts of each expression. Choose from **variable**, **constant**, *and* **coefficient**.

1. $-7 + h$

1. ____________

____

2. $-2w$

2. ____________

____

3. $9k + 1$

3. ____________

____

Objective 2 Evaluate variable expressions for given replacement values.

For extra help, see Examples 2–5 on pages 95–98 of your text and Section Lecture video for Section 2.1 and Exercise Solutions Clip 17 and 23.

Evaluate each expression.

4. The expression (rule) for the weight of an object (in Newtons) is $10m$, where m is the object's mass in kilograms. Evaluate the expression when
(a) the object has a mass of 14 kilograms.
(b) the object has a mass of 92 kilograms.

4. a. ____________

b. ____________

5. The expression (rule) for the total number of dollars spent on a car, when there is a \$2300 down payment and monthly payments of \$215, is $\$2300 + 215t$, where t is the number of monthly payments. Evaluate the expression when
(a) there are 36 monthly payments.
(b) there are 48 monthly payments.

5. a._______________

b._______________

6. The expression (rule) for the amount of insurance paid out on a policy with a \$2500 deductible is $\frac{8c}{10} - \$2500$, where c is total medical costs. Evaluate the expression when
(a) total medical costs are \$10,625.
(b) total medical costs are \$12,480.

6. a._______________

b._______________

Objective 3 Write properties of operations using variables.

For extra help, see Example 6 on page 99 of your text and Section Lecture video for Section 2.1.

State the property represented by each equation.

7. $h(a + b) = h(a) + h(b)$ **7.** _______________

8. $(1)(k) = k$ **8.** _______________

9. $x + 0 = x$ **9.** _______________

Name: Date:
Instructor: Section:

Objective 4 Use exponents with variables.

For extra help, see Examples 7–8 on page 100 of your text and Section Lecture video for Section 2.1 and Exercise Solutions Clip 35, 41, and 49.

Rewrite each expression without exponents.

10. $-7g^3h$ **10.** ________________

11. $u^3v^2w^2$ **11.** ________________

Evaluate the expression.

12. $-3k^2g^3$ when k is -2 and g is 5 **12.** ________________

Name: Date:
Instructor: Section:

Chapter 2 UNDERSTANDING VARIABLES AND SOLVING EQUATIONS

2.2 Simplifying Expressions

Learning Objectives

1. Combine like terms using the distributive property.
2. Simplify expressions.
3. Use the distributive property to multiply.

Key Terms

Use the vocabulary terms listed below to complete each statement in exercises 1–5.

simplify an expression **term** **constant term**

variable term **like terms**

1. ______________________________ are terms with exactly the same variable parts.
2. Each addend in an expression is called a ________________.
3. To ____________________________________, combine all the like terms.
4. A ______________________ has a coefficient multiplied by a variable part.
5. A ______________________ is a type of term that is just a number.

Guided Examples

Review these examples for Objective 1:

1. List the like terms in each expression. Then identify the coefficients of the like terms.

a. $-6w+(-6w^2)+4wz+w+(-6)$

The like terms are $-6w$ and w.
The coefficient of $-6w$ is -6, and the coefficient of w is understood to be 1.

b. $9ab^2+9a^2b+(-4a^2b)+3+(-9ab)$

The like terms are $9a^2b$ and $-4a^2b$.
The coefficient of $9a^2b$ is 9, and the coefficient of $-4a^2b$ is -4.

Now Try:

1. List the like terms in each expression. Then identify the coefficients of the like terms.

a. $-8a+(-8a^2)+9ab$
$+a+(-11)$

____

____

b. $21rs^2+21r^2s+(-5r^2s)$
$+11+(-33rs)$

____

____

c. $12xy+19+(-18x)+15y+(-7)$

The like terms are 19 and –7.
The like terms are constants (there are no variable parts).

c. $13pq+26+(-10p)$
$+15q+(-2)$

____

____

2. Combine like terms.

a. $7x+x+9x$

$$7x+x+9x$$
$$7x+1x+9x$$
$$(7+1+9)x$$
$$17x$$

b. $-5y^2-9y^2$

$$-5y^2-9y^2$$
$$-5y^2+(-9y^2)$$
$$[-5+(-9)]y^2$$
$$-14y^2$$

2. Combine like terms.

a. $12x+7x+x$

____

b. $-13z^2-26z^2$

____

Review these examples for Objective 2:

3. Simplify each expression by combining like terms.

a. $9ab+7b+8ab$

The like terms are $9ab$ and $8ab$. Use the commutative property first, then combine like terms.

$$9ab+7b+8ab$$
$$9ab+8ab+7b$$
$$(9+8)ab+7b$$
$$17ab \quad +7b$$

The simplified expression is $17ab + 7b$.

Now Try:

3. Simplify each expression by combining like terms.

a. $5pq+q+6pq$

____

Name: Date:
Instructor: Section:

b. $4c-9-c+11$

Write 1 as the coefficient of *c*. Change subtractions to adding the opposite, and then combine like terms.

$$4c \quad - \quad 9 \quad - \quad 1c \quad +11$$
$$4c+(-9)+(-1c)+11$$
$$4c+(-1c)+(-9)+11$$
$$[4+(-1)]c+(-9)+11$$
$$3c \quad + \quad 2$$

The simplified expression is $3c + 2$.

b. $8c-13-c+4$

____

4. Simplify.

a. $6(11x)$

Use the associative property.

$6(11x)$ can be written as $(6\cdot 11)\cdot x$

$66\cdot x$

$66x$

So, $6(11x)$ simplifies to $66x$.

b. $-7(5b)$

Use the associative property.

$-7(5b)$ can be written as $(-7\cdot 5)b$

$-35b$

So, $-7(5b)$ simplifies to $-35b$.

c. $-9(-3a^2)$

Use the associative property.

$-9(-3a^2)$ can be written as $[-9\cdot(-3)]a^2$

$27a^2$

So, $-9(-3a^2)$ simplifies to $27a^2$.

4. Simplify.

a. $9(12b)$

____

b. $-4(9x)$

____

c. $-8(-5y^2)$

____

Review these examples for Objective 3:

5. Simplify.

a. $7(4x+3)$

$7(4x+3)$ can be written as $7\cdot 4x+7\cdot 3$

$7\cdot 4\cdot x+21$

$28\cdot x+21$

$28x+21$

So, $7(4x+3)$ simplifies as $28x + 21$.

Now Try:

5. Simplify.

a. $6(5x+2)$

____

b. $-5(8b+4)$

$-5(8b+4)$ can be written as $-5\cdot 8b+(-5)\cdot 4$

$-5\cdot 8\cdot b+(-20)$

$-40\cdot b+(-20)$

$-40b+(-20)$

So, $-5(8b+4)$ simplifies as $-40b-20$.

c. $8(d-9)$

$8(d-9)$ can be written as $8[d+(-9)]$

$8\cdot d+8(-9)$

$8d+(-72)$

$8d-72$

So, $8(d-9)$ simplifies as $8d-72$.

b. $-9(6x+4)$

____

c. $28(x-4)$

____

6. Simplify: $9+4(x-5)$

Use the distributive property first.

$9+4(x-5)$

$9+4\cdot x-4\cdot 5$

$9+4x-20$

$4x+9-20$

$4x\quad -11$

The simplified expression is $4x-11$.

6. Simplify: $12+6(x-3)$

____

Objective 1 Combine like terms using the distributive property.

For extra help, see Examples 1–2 on pages 108–109 of your text and Section Lecture video for Section 2.2 and Exercise Solutions Clip 19.

Identify the like terms in each expression. Then identify the coefficients of the like terms.

1. $x^2+2x+(-5x^3)+(-3x)+8$

1. ______________

____

Simplify each expression.

2. $6d+d$

2. ______________

3. $c^3z^4-11c^3z^4-9c^3z^4$

3. ______________

Name: Date:
Instructor: Section:

Objective 2 Simplify expressions.

For extra help, see Examples 3–4 on pages 110–111 of your text and Section Lecture video for Section 2.2 and Exercise Solutions Clip 31 and 47.

Simplify each expression by combining like terms. Write each answer with the variables in alphabetical order and any constant term last.

4. $5b^4m+10-2b^4m-1$ **4.** ________

5. $-8s+6st-9t-11st+2s+4t+9-11t-1$ **5.** ________

Simplify by using the associative property of multiplication.

6. $-14(-a^3c^2t)$ **6.** ________

Objective 3 Use the distributive property to multiply.

For extra help, see Examples 5–6 on pages 112–113 of your text and Section Lecture video for Section 2.2 and Exercise Solutions Clip 53, 55, 57, and 67.

Use the distributive property to simplify each expression.

7. $-5(t+4)$ **7.** ________

8. $-8(3s-4)$ **8.** ________

Simplify the expression.

9. $-q+7(3q-2)+11$ **9.** ________

Name: Date:
Instructor: Section:

Chapter 2 UNDERSTANDING VARIABLES AND SOLVING EQUATIONS

2.3 Solving Equations Using Addition

Learning Objectives

1. Determine whether a given number is a solution of an equation.
2. Solve equations using the addition property of equality.
3. Simplify equations before using the addition property of equality.

Key Terms

Use the vocabulary terms listed below to complete each statement in exercises 1–5.

equation **solve an equation** **solution**

addition property of equality **check the solution**

1. A____________________ of an equation is a number that makes the statement true when it is substituted for the variable.

2. The ____________________________ states that adding the same quantity to both sides of an equation keeps the equation balanced.

3. An ________________ is a mathematical statement that contains an equals sign.

4. To ______________________________, find a number that can replace the variable and make the equation balance.

5. To ______________________________, go back to the original equation and replace the variable with the solution.

Guided Examples

Review this example for Objective 1:

1. Which of these numbers, 65, 45, or 85, is the solution of the equation $c-15=70$?

   Replace c with each of the numbers. The one that makes the equation balance is the solution.

   $65-15\neq 70$ Does not balance: 65 – 15 = 50.

   $45-15\neq 70$ Does not balance: 45 – 15 = 30.

   $85-15=70$ Balances: 85 – 15 = 70.

   The solution is 85 because when c is 85, the equation balances.

Now Try:

1. Which of these numbers, 25, 29, or 21, is the solution of the equation $25=c-4$?

   ____

Name: Date:
Instructor: Section:

Review these examples for Objective 2:

2. Solve each equation and check the solution.

a. $c+7=20$

Add –7 to each side.

$$\begin{aligned} c+7&=20 \\ \underline{-7}\ \ &\ \ \underline{-7} \\ c+0&=13 \\ c\ &=13 \end{aligned}$$

Because *c* balances with 13, the solution is 13.

Check the solution by replacing *c* with 13 in the original equation.

$$\begin{aligned} c+7&=20 \\ 13+7&=20 \\ 20\ &=20 \end{aligned}$$

The equation balances. Therefore, 13 is the correct solution.

b. $-7=x-4$

Change subtraction to adding the opposite. Then Add the opposite of –4, which is 4.

$$\begin{aligned} -7&=x-4 \\ -7&=x+(-4) \\ \underline{4}\ \ &\ \ \underline{\quad\quad 4} \\ -3&=x+\ 0 \\ -3&=\ x \end{aligned}$$

Because *x* balances with –3, the solution is –3.

Check the solution by replacing *c* with –3 in the original equation.

$$\begin{aligned} -7&=x-4 \\ -7&=-3-4 \\ -7&=-3+(-4) \\ -7&=\ -7 \end{aligned}$$

When *x* is replaced with –3, the equation balances, so –3 is the correct solution.

Now Try:

2. Solve each equation and check the solution.

a. $c+13=29$

____

b. $-10=x-8$

____

Name: Date:
Instructor: Section:

Review these examples for Objective 3:

3. Solve each equation and check the solution.

a. $y+9=5-8$

Simplify the right side first. Then get y by itself on the left side by adding the opposite of 9, which is –9.

$$\begin{aligned} y+9&=5-8\\ y+9&=5+(-8)\\ y+9&=\ \ -3\\ \underline{-9}\ \ &\ \ \underline{-9}\\ y&=-12 \end{aligned}$$

The solution is –12. Now check the solution.

Check Go back to the original equation and replace y with –12.

$$\begin{aligned} y+9&=5-8\\ -12+9&=5-8\\ -3&=5+(-8)\\ -3&=-3 \end{aligned}$$

So when y is replaced with –12, the equation balances, so –12 is the correct solution.

b. $-5+5=-7b-10+8b$

Combine like terms on each side first.

$$\begin{aligned} -5+5 &= -7b-10+8b\\ 0 &= -7b+(-10)+8b\\ 0 &= -7b+8b+(-10)\\ 0 &= 1b\quad +(-10)\\ \underline{10}\ &\ \ \underline{\qquad\qquad 10}\\ 10 &= b \end{aligned}$$

The solution is 10.

Check Go back to the original equation and replace b with 10.

$$\begin{aligned} -5+5&=-7b-10+8b\\ -5+5&=-7\cdot 10+(-10)+8\cdot 10\\ 0&=-70+(-10)+80\\ 0&=\quad -80\quad +80\\ 0&=\quad 0 \end{aligned}$$

When b is replaced with 10, the equation balances, so 10 is the correct solution.

Now Try:

3. Solve each equation and check the solution.

a. $x+10=14-15$

____

b. $-9+9=-15b-6+16b$

____

Name: Date:
Instructor: Section:

Objective 1 Determine whether a given number is a solution of an equation.

For extra help, see Example 1 on page 123 of your text and Section Lecture video for Section 2.3 and Exercise Solutions Clip 3.

In each list of numbers, find the one that is a solution of the given equation.

1. $r - 12 = 0$
$-8, -2, 12$

1. ________________

2. $p + 7 = 10$
$-7, 3, 7$

2. ________________

3. $6 - 2y = -8$
$-3, 5, 7$

3. ________________

Objective 2 Solve equations using the addition property of equality.

For extra help, see Example 2 on pages 125–126 of your text and Section Lecture video for Section 2.3 and Exercise Solutions Clip 11.

Solve each equation and check the solution.

4. $v - 72 = 32$

4. ________________

5. $-30 = 36 + n$

5. ________________

6. $65 = u + 74$

6. ________________

Name: Date:
Instructor: Section:

Objective 3 Simplify equations before using the addition property of equality.

For extra help, see Example 3 on page 126–127 of your text and Section Lecture video for Section 2.3 and Exercise Solutions Clip 39 and 43.

Simplify each side of the equation when possible. Then solve the equation. Check each solution.

7. $11a - 10a = -5 + 8$ **7.** ________________

8. $40 - 2u + 3u = 19$ **8.** ________________

9. $66n + 53 - 65n = 9$ **9.** ________________

Name: Date:
Instructor: Section:

Chapter 2 UNDERSTANDING VARIABLES AND SOLVING EQUATIONS

2.4 Solving Equations Using Division

Learning Objectives

1 Solve equations using the division property of equality.
2 Simplify equations before using the division property of equality.
3 Solve equations such as $-x = 5$.

Key Terms

Use the vocabulary terms listed below to complete each statement in exercises 1–2.

division property of equality

addition property of equality

1. The ______________________________ states that dividing both sides of an equation by the same nonzero number will keep it balanced.

2. When the same quantity is added to both sides of an equation, the ______________________________ is being applied.

Guided Examples

Review these examples for Objective 1:

1. Solve each equation and check each solution.

a. $5s = 35$

On the left side of the equation, the variable is multiplied by 5. To undo the multiplication, divide by 5.

$$5s = 35$$
$$\frac{5s}{5} = \frac{35}{5}$$
$$s = 7$$

So, 7 is the solution. We check the solution by replacing s with 7 in the original equation.

Check

$$5s = 35$$
$$5 \cdot 7 = 35$$
$$35 = 35$$

When s is replaced with 7, the equation balances, so 7 is the correct solution.

Now Try:

1. Solve each equation and check each solution.

a. $9s = 54$

____

b. $56 = -7w$

On the right side of the equation, the variable is multiplied by –7. To undo the multiplication, divide by –7.

$$56 = -7w$$
$$\frac{56}{-7} = \frac{-7w}{-7}$$
$$-8 = w$$

The solution is –8.

Check
We check the solution in the original equation.

$$56 = -7w$$
$$56 = -7 \cdot (-8)$$
$$56 = 56$$

When w is replaced with –8, the equation balances, so –8 is the correct solution.

b. $64 = -4w$

____

Review these examples for Objective 2:

2. Solve each equation and check each solution.

a. $6y - 10y = -20$

First, simplify the left side by combining like terms.

$$6y - 10y = -20$$
$$6y + (-10y) = -20$$
$$\frac{-4y}{-4} = \frac{-20}{-4}$$
$$y = 5$$

The solution is 5.

Check
Go back to the original equation and replace each y with 5.

$$6y - 10y = -20$$
$$6 \cdot 5 - 10 \cdot 5 = -20$$
$$30 - 50 = -20$$
$$30 + (-50) = -20$$
$$-20 = -20$$

When y is replaced with 5, the equation balances, so 5 is the correct solution.

Now Try:

2. Solve each equation and check each solution.

a. $5y - 14y = -36$

____

b. $13-21+8=h+6h$

First, combine like terms.

$$13-21+8=h+6h$$
$$13+(-21)+8=1h+6h$$
$$-8 \quad +8=7h$$
$$\frac{0}{7}=\frac{7h}{7}$$
$$0=h$$

The solution is 0.

Check Go back to the original equation and replace each h with 0.

$$13-21+8=h+6h$$
$$13+(-21)+8=0+6\cdot 0$$
$$-8 \quad +8=0+0$$
$$0=0$$

When h is replaced with 0, the equation balances, so 0 is the correct solution.

b. $24-43+19=h-9h$

____

Review this example for Objective 3:

3. Solve $-x=9$ and check the solution.

Start by writing in the coefficient of x.

$$-x=9$$
$$-1x=9$$
$$\frac{-1x}{-1}=\frac{9}{-1}$$
$$x=-9$$

The solution is –9.

Check Go back to the original equation and replace x with 0.

$$-x=9$$
$$-(-9)=9$$
$$9=9$$

When x is replaced with –9, the equation balances, so –9 is the correct solution.

Now Try:

3. Solve $-y=13$ and check the solution.

____

Name: Date:
Instructor: Section:

Objective 1 Solve equations using the division property of equality.

For extra help, see Example 1 on page 135 of your text and Section Lecture video for Section 2.4 and Exercise Solutions Clip 9.

Solve each equation and check.

1. $-3m = 21$ **1.** ________________

2. $-56 = -7b$ **2.** ________________

3. $15h = 420$ **3.** ________________

Objective 2 Simplify equations before using the division property of equality.

For extra help, see Example 2 on page 136 of your text and Section Lecture video for Section 2.4 and Exercise Solutions Clip 23 and 27.

Solve each equation and check.

4. $53 - 8 = -9d$ **4.** ________________

5. $35m - 33m = 100 - 34$ **5.** ________________

Name: Date:
Instructor: Section:

6. $140 - 116 = 20x - 12x$

6. ________________

Objective 3 Solve equations such as $-x = 5$.

For extra help, see Example 3 on page 137 of your text and Section Lecture video for Section 2.4 and Exercise Solutions Clip 39.

Solve each equation and check.

7. $-b = 19$

7. ________________

8. $-17 = -w$

8. ________________

9. $4 = -v$

9. ________________

Name: Date:
Instructor: Section:

Chapter 2 UNDERSTANDING VARIABLES AND SOLVING EQUATIONS

2.5 Solving Equations with Several Steps

Learning Objectives

1 Solve equations using the addition and division properties of equality.
2 Solve equations using the distributive, addition, and division properties.

Key Terms

Use the vocabulary terms listed below to complete each statement in exercises 1–3.

distributive property

division property of equality

addition property of equality

1. The ________________________________ states that if $a = b$, then $a + c = b + c$.

2. The ________________________________ states that $a(b + c) = ab + bc$.

3. The ________________________________ states that if $a = b$, then $\frac{a}{c} = \frac{b}{c}$ for $c \neq 0$.

Guided Examples

Review these examples for Objective 1:

1. Solve this equation and check the solution: $7m + 2 = 37$.

Step 1 Adding –2 to the left side will leave $7m$ by itself. To keep the balance, add –2 to the right side also.

$$7m + 2 = 37$$
$$\underline{\quad -2 \quad -2}$$
$$7m + 0 = 35$$
$$7m = 35$$

Step 2 Divide both sides by the coefficient of the variable term, 7.

$$\frac{7m}{7} = \frac{35}{7}$$
$$m = 5$$

Now Try:

1. Solve this equation and check the solution: $8n + 3 = 59$.

____

Step 3 Check the solution by going back to the original equation. Replace m with 5.

$$7m+2=37$$
$$7(5)+2=37$$
$$35+2=37$$
$$37=37$$

When m is replaced with 5, the equation balances, so 5 is the correct solution.

2. Solve this equation and check the solution: $5x-6=8x-18$.

To keep the variable on the left side, add the opposite of $8x$, which is $-8x$ to both sides.

$$\begin{aligned} 5x-6 &= \quad 8x-18 \\ \underline{-8x} \quad & \quad \underline{-8x \quad} \\ -3x-6 &= \quad 0\ -18 \\ -3x+(-6) &= 0+(-18) \\ \underline{\quad\quad 6} \quad & \quad \underline{\quad\quad 6} \\ -3x\ +\ 0 &= 0+(-12) \\ \frac{-3x}{-3} &= \frac{-12}{-3} \\ x &= 4 \end{aligned}$$

Check

$$5x-6=8x-18$$
$$5(4)-6=8(4)-18$$
$$20+(-6)=32+(-18)$$
$$14=14$$

When x is replaced with 4, the equation balances, so 4 is the correct solution.

2. Solve this equation and check the solution: $8x-15=10x+5$.

____

Review these examples for Objective 2:

3. Solve this equation and check the solution: $-12=6(y-2)$.

We can use the distributive property to simplify the right side of the equation. Then use the steps to solve for y.

$$-12=6(y-2)$$
$$-12=6\cdot y-6\cdot 2$$
$$-12=6y-12$$
$$-12=6y+(-12)$$
$$\underline{\ 12\ } \qquad \underline{\qquad\quad 12}$$
$$0=6y$$

Now Try:

3. Solve this equation and check the solution: $8(m-3)=-24$.

____

Name: Date:
Instructor: Section:

$$\frac{0}{6}=\frac{6y}{6}$$
$$0=y$$

The solution is 0.

Check Go back to the original equation and replace y with 0.

$$-12=6(y-2)$$
$$-12=6(0-2)$$
$$-12=6[0+(-2)]$$
$$-12=6(-2)$$
$$-12=-12$$

When y is replaced with 0, the equation balances, so 0 is the correct solution.

4. Solve this equation and check the solution: $3+6(n+5)=5+2n$.

Step 1 Use the distributive property on the left side.

$$3+6(n+5)=5+2n$$
$$3+6n+30=5+2n$$

Step 2 Combine like terms on the left side.

$$6n+33=5+2n$$

Step 3 Add $-2n$ to both sides.

$$\underline{-2n} \qquad \underline{-2n}$$
$$4n+33=5+0$$
$$4n+33=5$$

Step 3 To get $4n$ by itself, add -33 to both sides.

$$\underline{\quad -33} \quad \underline{-33}$$
$$4n+0=-28$$

Step 4 Divide both sides by 4, the coefficient of the variable term $4n$.

$$\frac{4n}{4}=\frac{-28}{4}$$
$$n=-7$$

Step 5 Check

$$3+6(n+5)=5+2n$$
$$3+6(-7+5)=5+2(-7)$$
$$3+6(-2)=5+(-14)$$
$$3+(-12)=-9$$
$$-9=-9$$

When n is replaced with -7, the equation balances, so -7 is the correct solution.

4. Solve this equation and check the solution: $5+8(m+2)=5m+6$.

____

Name: Date:
Instructor: Section:

Objective 1 Solve equations using the addition and division properties of equality.

For extra help, see Examples 1–2 on pages 142–143 of your text and Section Lecture video for Section 2.5 and Exercise Solutions Clip 1.

Solve each equation and then check.

1. $4m-19=-19$

1. ________________

2. $-14+8d=-24+3d$

2. ________________

3. $5q+10=19+8q$

3. ________________

Objective 2 Solve equations using the distributive, addition, and division properties.

For extra help, see Examples 3–4 on pages 144–145 of your text and Section Lecture video for Section 2.5 and Exercise Solutions Clip 25, 45, and 47.

Solve each equation and check.

4. $40=5(b+8)$

4. ________________

5. $t+28-10=4(t+6)-24$

5. ________________

6. $-87+3g=9(g-5)+8g$

6. ________________

Name: Date:
Instructor: Section:

Chapter 3 SOLVING APPLICATION PROBLEMS

3.1 Problem Solving: Perimeter

Learning Objectives

1 Use the formula for perimeter of a square to find the perimeter or the length of one side.
2 Use the formula for perimeter of a rectangle to find the perimeter, the length, or the width.
3 Find the perimeter of parallelograms, triangles, and irregular shapes.

Key Terms

Use the vocabulary terms listed below to complete each statement in exercises 1–6.

formula	**perimeter**	**square**
rectangle	**parallelogram**	**triangle**

1. A figure with exactly three sides is a ____________________.
2. The distance around the outside edges of a flat shape is the ____________.
3. A ____________________ is a rule for solving common types of problems.
4. A figure with four sides that are all the same length and meet to form right angles is a ________________.
5. A four-sided figure in which all sides meet to form right angles is a ________________.
6. A four-sided figure in which opposite sides are both parallel and equal in length is a ________________.

Guided Examples

Review these examples for Objective 1:

1. Find the perimeter of the square that measures 7 ft on each side.

   Use the formula for perimeter of a square, $P = 4s$. Replace s with 7 ft.

   $P = 4s$

   $P = 4 \cdot 7$ ft

   $P = 28$ ft

   The perimeter of the square is 28 ft.

Now Try:

1. Find the perimeter of the square that measures 15 in. on each side.

   ____

2. If the perimeter of a square is 120 ft, find the length of one side.

Use the formula for perimeter of a square, $P = 4s$. Replace P with 120 ft.

$$P = 4s$$
$$120 \text{ ft} = 4s$$
$$\frac{120 \text{ ft}}{4} = \frac{4s}{4}$$
$$30 \text{ ft} = s$$

The length of one side of the square is 30 ft.

2. If the perimeter of a square is 500 cm, find the length of one side.

____

Review these examples for Objective 2:

3. Find the perimeter of a rectangle with length 32 m and width 17 m.

The replace l with 32 m and w with 17 m.

$$P = 2l + 2w$$
$$P = 2 \cdot 32 \text{ m} + 2 \cdot 17 \text{ m}$$
$$P = 64 \text{ m} + 34 \text{ m}$$
$$P = 98 \text{ m}$$

The perimeter of the rectangle is 98 m.

Check
To check the solution, add the lengths of the four sides.

$$P = 32 \text{ m} + 32 \text{ m} + 17 \text{ m} + 17 \text{ m}$$
$$P = 98 \text{ m}$$

4. If the perimeter of a rectangle is 32 ft and the width is 4 ft, find the length.

Use the formula $P = 2l + 2w$. Replace P with 32 ft and w with 4 ft.

$$P = 2l + 2w$$
$$32 \text{ ft} = 2l + 2 \cdot 4 \text{ ft}$$
$$32 \text{ ft} = 2l + 8 \text{ ft}$$
$$\underline{-8 \text{ ft}} \qquad \underline{-8 \text{ ft}}$$
$$24 \text{ ft} = 2l$$
$$\frac{24 \text{ ft}}{2} = \frac{2l}{2}$$
$$12 \text{ ft} = l$$

The length is 12 ft.

Now Try:

3. Find the perimeter of a rectangle with length 54 ft and width 38 ft.

____

4. If the perimeter of a rectangle is 48 in. and the width is 6 in., find the length.

____

Name: Date:
Instructor: Section:

Review these examples for Objective 3:

5. Find the perimeter of the parallelogram with adjacent sides 15 m and 7 m.

To find the perimeter, add the lengths of the sides.

$P = 15 \text{ m} + 15 \text{ m} + 7 \text{ m} + 7 \text{ m}$

$P = 44 \text{ m}$

6. Find the perimeter of the triangle with sides 17 m, 8 m, and 15 m.

To find the perimeter, add the lengths of the sides.

$P = 17 \text{ m} + 8 \text{ m} + 15 \text{ m}$

$P = 40 \text{ m}$

7. Find the perimeter of the figure.

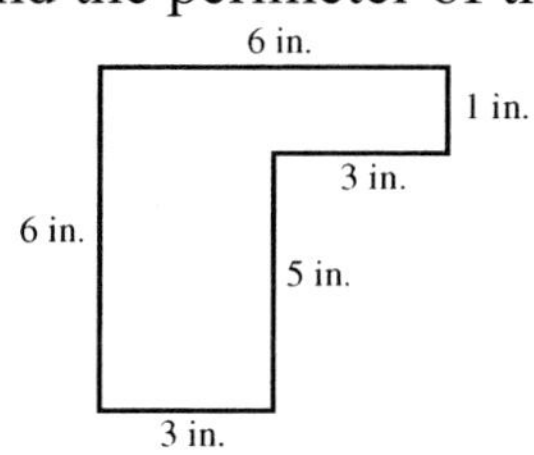

Find the perimeter by adding the lengths of the sides.

$P = 6 \text{ in.} + 1 \text{ in.} + 3 \text{ in.} + 5 \text{ in.} + 3 \text{ in.} + 6 \text{ in.}$

$P = 24 \text{ in.}$

Now Try:

5. Find the perimeter of the parallelogram with adjacent sides 25 m and 12 m.

____

6. Find the perimeter of the triangle with sides 23 mm, 29 mm, and 18 mm.

____

7. Find the perimeter of the figure.

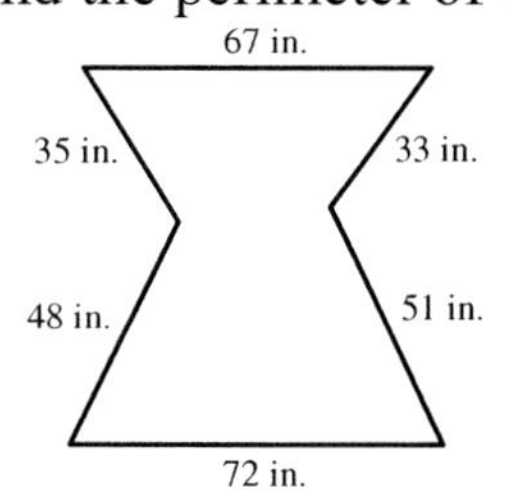

____

Objective 1 Use the formula for perimeter of a square to find the perimeter or the length of one side.

For extra help, see Examples 1–2 on page 165 of your text and Section Lecture video for Section 3.1 and Exercise Solutions Clip 5.

Find the perimeter of each square, using the appropriate formula.

1.

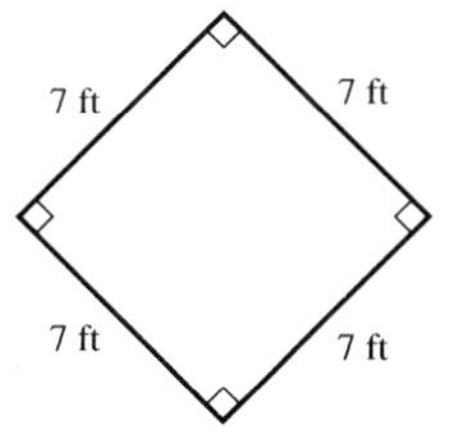

1. ____________

2.

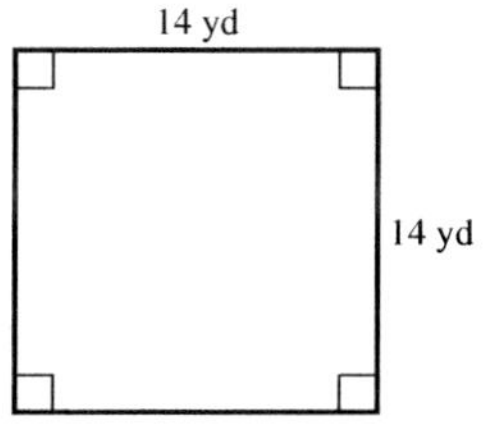

2. ____________

Name: Date:
Instructor: Section:

For the given perimeter of the square, find the length of one side using the appropriate formula.

3. The perimeter is 76 miles. **3.** ______________

Objective 2 Use the formula for perimeter of a rectangle to find the perimeter, the length, or the width.

For extra help, see Examples 3–4 on pages 166–167 of your text and Section Lecture video for Section 3.1 and Exercise Solutions Clip 21, 27, and 29.

Find the perimeter of the rectangle, using the appropriate formula.

4. **4.** ______________

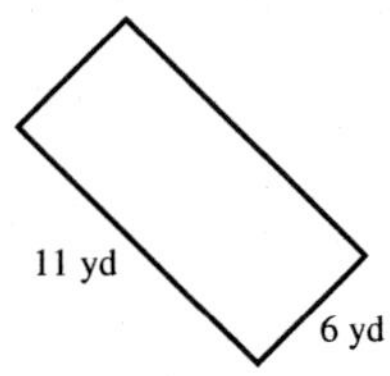

For each rectangle, you are given the perimeter and either the length or the width. Find the unknown measurement by using the appropriate formula.

5. The perimeter is 90 ft and the length is 32 ft. **5.** ______________

6. The width is 17 cm and the perimeter is 76 cm. **6.** ______________

Objective 3 Find the perimeter of parallelograms, triangles, and irregular shapes.

For extra help, see Examples 5–7 on page 168 of your text and Section Lecture video for Section 3.1 and Exercise Solutions Clip 35 and 43.

Find the perimeter of each shape.

7. Parallelogram **7.** ______________

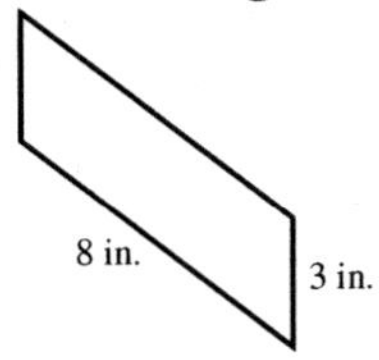

Name: Date:
Instructor: Section:

8.

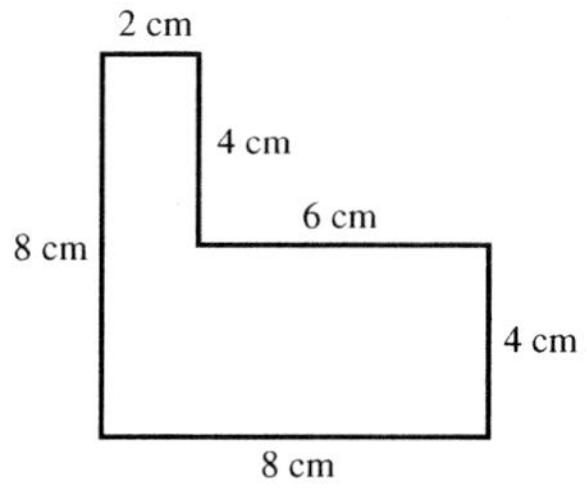

8. ____________________

9.

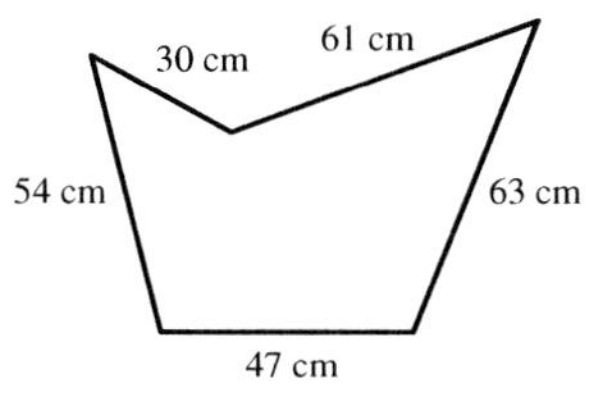

9. ____________________

Name: Date:
Instructor: Section:

Chapter 3 SOLVING APPLICATION PROBLEMS

3.2 Problem Solving: Area

Learning Objectives

1. Use the formula for area of a rectangle to find the area, the length, or the width.
2. Use the formula for area of a square to find the area or the length of one side.
3. Use the formula for area of a parallelogram to find the area, the base, or the height.
4. Solve application problems involving perimeter and area of rectangles, squares, or parallelograms.

Key Terms

Use the vocabulary terms listed below to complete each statement in exercises 1–4.

area **rectangle** **square** **parallelogram**

1. To find the area of a__________________, use the formula $A = s^2$.

2. To find the area of a __________________, use the formula $A = bh$.

3. To find the area of a __________________, use the formula $A = lw$.

4. The surface inside a two-dimensional (flat) shape is the __________________.

Guided Examples

Review these examples for Objective 1:

1. Find the area of each rectangle.

a. A rectangle measuring 9 m by 15 m.

The length of this rectangle is 15 m and the width is 9 m. Use the formula $A = lw$ to find the area.

$A = l \cdot w$

$A = 15 \text{ m} \cdot 9 \text{ m}$

$A = 135 \text{ m}^2$

The area of the rectangle is 135 m^2.

Now Try:

1. Find the area of each rectangle.

a. A rectangle measuring 11 m by 23 m.

____

Name: Date:
Instructor: Section:

b. A rectangle measuring 14 cm by 8 cm.

The length of this rectangle is 14 cm and the width is 8 cm. Then use the formula $A = lw$.

$$A = l \cdot w$$
$$A = 14 \text{ cm} \cdot 8 \text{ cm}$$
$$A = 112 \text{ cm}^2$$

The area of the rectangle is 112 cm^2.

b. A rectangle measuring 27 cm by 7 cm.

____

2. If the area of a rectangular rug is 24 yd^2, and the length is 6 yd, find the width.

Use the formula for area of a rectangle, $A = lw$. Replace A with 24 yd^2, and replace l with 6 yd.

$$A = l \cdot w$$
$$24 \text{ yd}^2 = 6 \text{ yd} \cdot w$$
$$\frac{24 \text{ yd} \cdot \cancel{\text{yd}}}{6 \cancel{\text{yd}}} = \frac{6 \text{ yd} \cdot w}{6 \text{ yd}}$$
$$4 \text{ yd} = w$$

The width of the rug is 4 yd.

Check To check the solution, use the area formula. Replace l with 6 yd and replace w with 4 yd.

$$A = l \cdot w$$
$$A = 6 \text{ yd} \cdot 4 \text{ yd}$$
$$A = 24 \text{ yd}^2$$

An area of 24 yd^2 matches the information in the original problem. So 4 yd is the correct width of the rug.

Now Try:

2. If the area of a rectangular flower garden is 35 ft^2, and the length is 7 ft, find the width.

____

Review this example for Objective 2:

3. Find the area of a square that is 5 ft on each side.

Use the formula for area of a square, $A = s^2$. Replace s with 5 ft.

$$A = s^2$$
$$A = s \cdot s$$
$$A = 5 \text{ ft} \cdot 5 \text{ ft}$$
$$A = 25 \text{ ft}^2$$

The area of the square is 25 ft^2.

Now Try:

3. Find the area of a square that is 13 in. on each side.

____

Name: Date:
Instructor: Section:

4. If the area of a square park is 81 mi^2, what is the length of one side of the park?

Use the formula for area of a square, $A = s^2$. The value of A is 81 mi^2.

$$A = s^2$$
$$81 \text{ mi}^2 = s^2$$
$$81 \text{ mi}^2 = s \cdot s$$
$$81 \text{ mi}^2 = 9 \text{ mi} \cdot 9 \text{ mi}$$

The value of s is 9 mi, so the length of one side of the park is 9 mi.

4. If the area of a square park is 25 mi^2, what is the length of one side of the park?

____

Review these examples for Objective 3:

5. Find the area of each parallelogram.

a. The base is 32 cm and the height is 15 cm.

Use the formula for the area of a parallelogram, $A = bh$. Replace b with 32 cm and replace h with 15 cm.

$$A = b \cdot h$$
$$A = 32 \text{ cm} \cdot 15 \text{ cm}$$
$$A = 480 \text{ cm}^2$$

The area of the parallelogram is 480 cm^2.

b. The base is 53 m and the height is 18 m.

Use the formula for the area of a parallelogram, $A = bh$. Replace b with 53 m and replace h with 18 m.

$$A = b \cdot h$$
$$A = 53 \text{ m} \cdot 18 \text{ m}$$
$$A = 954 \text{ m}^2$$

The area of the parallelogram is 954 m^2.

Now Try:

5. Find the area of each parallelogram.

a. The base is 19 cm and the height is 9 cm.

____

b. The base is 25 m and the height is 17 m.

____

Name: Date:
Instructor: Section:

6. The area of a parallelogram is 48 ft^2 and the base is 8 ft. Find the height.

Use the formula for the area of a parallelogram, $A = bh$. The value of A is 48 ft^2, and the value of b is 8 ft.

$$A = b \cdot h$$

$$48 \text{ ft}^2 = 8 \text{ ft} \cdot h$$

$$\frac{48 \text{ ft} \cdot \cancel{\text{ft}}}{8 \cancel{\text{ft}}} = \frac{8 \text{ ft} \cdot h}{8 \text{ ft}}$$

$$6 \text{ ft} = h$$

The height of the parallelogram is 6 ft.

Check To check the solution use the area formula.

$$A = b \cdot h$$

$$A = 8 \text{ ft} \cdot 6 \text{ ft}$$

$$A = 48 \text{ ft}^2$$

An area of 48 ft^2 matches the information in the original problem. So 6 ft is the correct height of the parallelogram.

6. The area of a parallelogram is 54 ft^2 and the base is 6 ft. Find the height.

____

Review this example for Objective 4:

7. Carla plans to adorn her afghan with a ribbon border. If the afghan measures 7 ft long and 5 ft wide, and the ribbon costs $3 per linear foot, how much is the total cost for this decoration?

The ribbon will go around the edges of the afghan, so you need to find the perimeter of the afghan. Replace l with 7 ft and replace w with 5 ft.

$$P = 2l + 2w$$

$$P = 2 \cdot 7 \text{ ft} + 2 \cdot 5 \text{ ft}$$

$$P = 14 \text{ ft} + 10 \text{ ft}$$

$$P = 24 \text{ ft}$$

The perimeter of the afghan is 24 ft, so Carla will need 24 ft of ribbon. The cost of the ribbon is $3 per linear foot, which means $3 for 1 foot. To find the cost for 24 ft, multiply $\$3 \cdot 24$ ft. The ribbon will cost $72.

Now Try:

7. A soybean field is 332 m wide and 512 m long. Find the perimeter and area of the field.

____

Name: Date:
Instructor: Section:

Objective 1 Use the formula for area of a rectangle to find the area, the length, or the width.

For extra help, see Examples 1–2 on pages 174–175 of your text and Section Lecture video for Section 3.2 and Exercise Solutions Clip 21.

Find the area of each rectangle using the appropriate formula.

1.

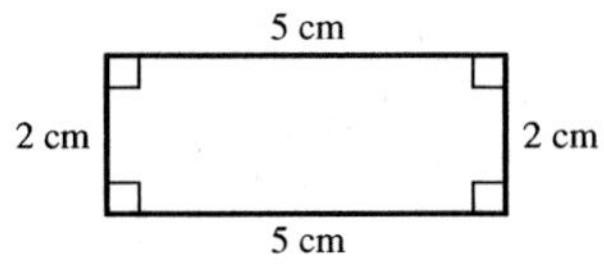

1. ______________

2.

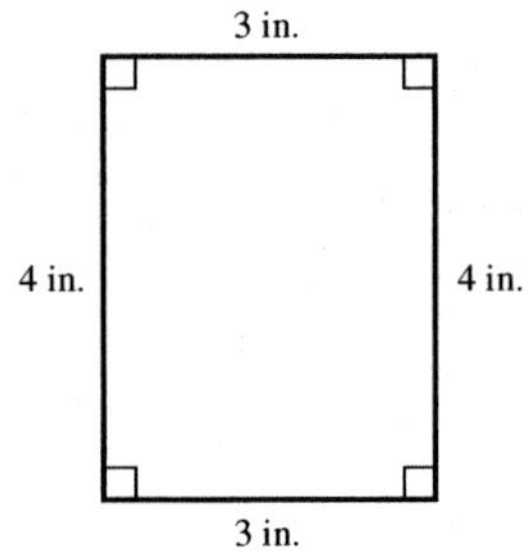

2. ______________

Use the area of the rectangle and either its length or width, and the appropriate formula, to find the other measurement.

3. The area of a food tray is 4104 cm^2 and the width is 54 cm. Find its length.

3. ______________

Objective 2 Use the formula for area of a square to find the area or the length of one side.

For extra help, see Examples 3–4 on pages 175–176 of your text and Section Lecture video for Section 3.2 and Exercise Solutions Clip 7.

Find the area of each square using the appropriate formula.

4.

4. ______________

Name: Date:
Instructor: Section:

5. 5. ____________

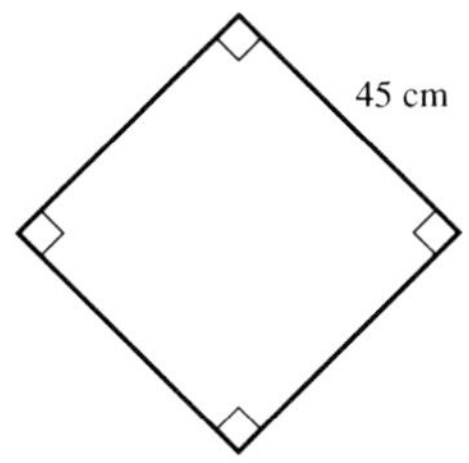

Given the area of the square, find the length of one side by inspection.

6. The area of a square pot holder is 49 in.^{2}. 6. ____________

Objective 3 Use the formula for area of a parallelogram to find the area, the base, or the height.

For extra help, see Examples 5–6 on pages 177–178 of your text and Section Lecture video for Section 3.2 and Exercise Solutions Clip 9, 13, and 27.

Find the area of each parallelogram using the appropriate formula.

7. 7. ____________

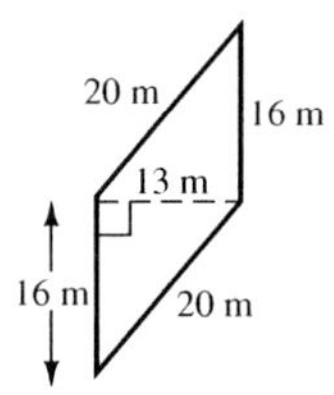

8. 8. ____________

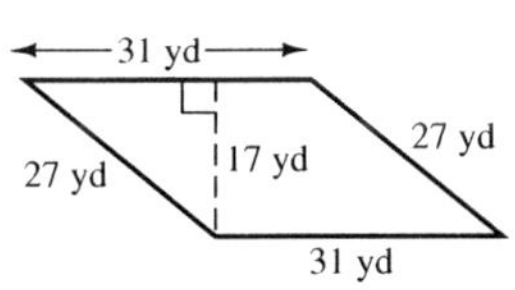

Use the area of the parallelogram and either its base or height, and the appropriate formula, to find the other measurement.

9. The area is 72 m^2, and the height is 12 m. Find the base. 9. ____________

Name: Date:
Instructor: Section:

Objective 4 Solve application problems involving perimeter and area of rectangles, squares, or parallelograms.

For extra help, see Example 7 on page 178 of your text and Section Lecture video for Section 3.2 and Exercise Solutions Clip 39.

Solve each application problem. You may need to find the perimeter, the area, or one of the side measurements

10. Chuck and Carla plan to carpet their 8 m by 4 m living room. If carpeting costs $9 per square meter, how much will carpeting this room cost?

10. ________________

11. Tabitha plans to sew a quilt measuring 88 ft^2 in area. If the quilt must be 11 ft long, how wide will the quilt be?

11. ________________

12. A rectangular dance floor measures 27 yd long and 15 yd wide. To cover the floor in wood paneling, it would cost $12 per square yard. How much is the total cost for covering this dance floor in wood paneling?

12. ________________

Name: Date:
Instructor: Section:

Chapter 3 SOLVING APPLICATION PROBLEMS

3.3 Solving Application Problems with One Unknown Quantity

Learning Objectives
1 Translate word phrases into algebraic expressions.
2 Translate sentences into equations.
3 Solve application problems with one unknown quantity.

Key Terms

Use the vocabulary terms listed below to complete each statement in exercises 1–4.

sum	**difference**	**product**	**quotient**
increased by	**less than**	**double**	**per**

1. ____________________ and ____________________ are words that mean addition.

2. ____________________ and ____________________ are words that mean multiplication.

3. ____________________ and ____________________ are words that mean division.

4. ____________________ and ____________________ are words that mean subtraction.

Guided Examples

Review these examples for Objective 1:

1. Write each phrase as an algebraic expression. Use x as the variable.

a. A number plus 15

Algebraic expression: $x + 15$ or $15 + x$

b. The sum of 18 and a number

Algebraic expression: $18 + x$ or $x + 18$

c. 110 more than a number

Algebraic expression: $x + 110$ or $110 + x$

d. –66 added to a number

Algebraic expression: $-66 + x$ or $x + (-66)$

Now Try:

1. Write each phrase as an algebraic expression. Use x as the variable.

a. A number plus 7

____

b. The sum of 43 and a number

____

c. 63 more than a number

____

d. –20 added to a number

____

Name: Date:
Instructor: Section:

e. A number increased by 39

Algebraic expression: $x + 39$ or $39 + x$

f. 17 less than a number

Algebraic expression: $x - 17$

g. A number subtracted from 73

Algebraic expression: $73 - x$

h. 73 subtracted from a number

Algebraic expression: $x - 73$

i. 34 fewer than a number

Algebraic expression: $x - 34$

j. A number decreased by 51

Algebraic expression: $x - 51$

k. 39 minus a number

Algebraic expression: $39 - x$

2. Write each phrase as an algebraic expression. Use x as the variable.

a. 112 times a number

Algebraic expression: $112x$

b. The product of 25 and a number

Algebraic expression: $25x$

c. Triple a number

Algebraic expression: $3x$

d. The quotient of –7 and a number

Algebraic expression: $\frac{-7}{x}$

e. A number divided by 9

Algebraic expression: $\frac{x}{9}$

e. A number increased by 75

____

f. 44 less than a number

____

g. A number subtracted from 89

____

h. 89 subtracted from a number

____

i. 16 fewer than a number

____

j. A number decreased by 36

____

k. 48 minus a number

____

2. Write each phrase as an algebraic expression. Use x as the variable.

a. 57 times a number

____

b. The product of 43 and a number

____

c. Double a number

____

d. The quotient of –13 and a number

____

e. A number divided by 19

____

Name: Date:
Instructor: Section:

f. 17 subtracted from 8 times a number

Algebraic expression: $8x - 17$

f. 81 subtracted from 6 times a number

____

Review this example for Objective 2:

3. If 6 times a number is added to 13, the result is 37. Find the number. Let x represent the number.

Let x represent the unknown number. Use the information in the problem to write an equation.

$$\underline{\text{6 times a number}}\ \underline{\text{added to}}\ \text{13 is 37.}$$

$$6x \quad + \quad 13 = 37$$

Next, solve the equation.

$$\begin{aligned} 6x + 13 &= \ 37 \\ \underline{\ -13} \quad & \ \underline{-13} \\ 6x + 0 &= \ 24 \\ \frac{6x}{6} &= \frac{24}{6} \\ x &= \ 4 \end{aligned}$$

The number is 4.

Check Go back to the words of the original problem.

$$\text{6 times}\ \underline{\text{a number}}\ \underline{\text{added to}}\ \text{13 is 37.}$$

$$6 \cdot 4 + 13 = 37$$

Since $6 \cdot 4 + 13 = 24 + 13 = 37$, then 4 is the correct solution.

Now Try:

3. If 7 times a number is added to 21, the result is 56. Write the equation and find the number.

____

Review these examples for Objective 3:

4. Shawn was paid a \$100 bonus check at work, which he promptly deposited into his checking account. After buying \$47 worth of groceries and concert tickets that cost a total of \$64, he found that he had a total of \$219 in his account. How much was in his account before he deposited the check?

Step 1 Read the problem. Find the balance of the bank account.

Step 2 Assign a variable. There is only one unknown quantity. Let a represent the account balance.

Now Try:

4. Betty left 9 pounds of bird seed in her garage one evening. During the night, mice ate some of the bird seed. Later, Betty bought a 30-pound bag of bird seed which she added to the bird seed left in the garage. At this point, she had 36 pounds of bird seed. How much bird seed did the mice eat?

____

Step 3 Write an equation.

Account balance	put in \$100	took out \$47	took out \$64	Ended up with \$219
a	$+100$	-47	-64	$=219$

Step 4 Solve.

$$
\begin{aligned}
a+100 - 47 - 64 &= 219 \\
a+100+(-47)+(-64) &= 219 \\
a+53+(-64) &= 219 \\
a - 11 &= 219 \\
+11 &\quad +11 \\
a + 0 &= 230 \\
a &= 230
\end{aligned}
$$

Step 5 State the answer. The original account balance was \$230.

Step 6 Check the solution. Go back to the original problem and insert the solution.
Start with \$230 in the account.
Put in \$100, so \$230 + \$100 = \$330.
Took out \$47, so \$330 – \$47 = \$283.
Took out \$64, so \$283 – \$64 = \$219.
Ended with \$219. This checks.

5. An office supply store bought nine boxes of black ink printer cartridges. One day the store sold 13 cartridges and the next day the store sold 22 cartridges. If nineteen cartridges were left on the shelf, how many cartridges were in each box?

Step 1 Read the problem. Find the number of cartridges in each box.

Step 2 Assign a variable. There is only one unknown quantity. Let *n* represent the number of cartridges in each box.

Step 3 Write an equation.

Number of boxes		cartridges in each box	took out 13	took out 22	Ended with 19
9	$\cdot$	n	-13	-22	$=19$

5. Before a camping weekend, Angela bought 4 packages of paper plates. On Saturday, one package was used up while on Sunday 30 paper plates were used. If two packages plus 25 plates remain, how many paper plates were there in each package?

____

Name: Date:
Instructor: Section:

Step 4 Solve.

$$\begin{aligned} 9n - 13 - 22 &= 19 \\ 9n + (-13) + (-22) &= 19 \\ 9n + (-35) &= 19 \\ +35 \quad & \quad +35 \\ 9n + 0 &= 54 \\ \frac{9n}{9} &= \frac{54}{9} \\ n &= 6 \end{aligned}$$

Step 5 State the answer. Each box contains 6 cartridges.

Step 6 Check the solution.
9 boxes each contains 6 cartridges, so $9 \cdot 6 = 54$.
13 were sold, so $54 - 13 = 41$.
Then 22 were sold, so $41 - 22 = 19$.
Ended with 19. This checks.

6. While grocery shopping, Teresa spent \$9 less than four times what Pepe spent. If Teresa spent \$23, how much did Pepe spend?

Step 1 Read the problem. Find the amount Pepe spent.

Step 2 Assign a variable. There is only one unknown quantity. Let x represent the amount Pepe spent.

Step 3 Write an equation.
\$9 less than four times what Pepe spent, is \$23

$$4x - 9 = 23$$

Step 4 Solve.

$$\begin{aligned} 4x - 9 &= 23 \\ +9 \quad & \quad +9 \\ 4x + 0 &= 32 \\ \frac{4x}{4} &= \frac{32}{4} \\ x &= 8 \end{aligned}$$

Step 5 State the answer. Pepe spent \$8.

Step 6 Check the solution. Go back to the original problem and insert the solution.

$$4 \cdot 8 - 9 \stackrel{?}{=} 32 - 9 = 23$$

This checks.

6. The price of a DVD is \$3.00 less than twice the cost of a book. If the DVD costs \$25.00, how much does the book cost?

____

Name: Date:
Instructor: Section:

Objective 1 Translate word phrases into algebraic expressions.

For extra help, see Examples 1–2 on page 185 of your text and Section Lecture video for Section 3.3 and Exercise Solutions Clip 5, 7, and 19.

Write an algebraic expression using x as the variable.

1. The product of –6 and a number **1.** ______________

2. The quotient of a number and 10 **2.** ______________

3. One more than three times a number **3.** ______________

Objective 2 Translate sentences into equations.

For extra help, see Example 3 on page 186 of your text and Section Lecture video for Section 3.3 and Exercise Solutions Clip 25 and 27.

Translate each sentence into an equation and solve it. Check your solution by going back to the words in the original problem.

4. If three times a number is decreased by eight, the result is 58. Find the number.

4.
Equation ____________

Solution ____________

5. The sum of three and seven times a number is 31. Find the number.

5.
Equation ____________

Solution ____________

6. If seven times a number is subtracted from twelve times the number, the result is –30. Find the number.

6.
Equation ____________

Solution ____________

Name: Date:
Instructor: Section:

Objective 3 Solve application problems with one unknown quantity.

For extra help, see Examples 4–6 on pages 187–189 of your text and Section Lecture video for Section 3.3 and Exercise Solutions Clip 35.

Solve each application problem. Use the six problem-solving steps listed in your text.

7. The number of bananas an adult gorilla ate is 15 less than three times the number a younger gorilla ate. If the adult gorilla ate 21 bananas, how many did the younger gorilla eat?

7. ________________

8. A local jeweler sold three gold watches for the same retail price and established a checking account with this money. The jeweler then bought a diamond for \$4000 and a pearl necklace for \$11,000 with money from this account. The jeweler now realized this new account was overdrawn for \$6000. For how much did she sell each of these gold watches?

8. ________________

9. There were 12 celery sticks in the vegetable crisper in Lance's refrigerator. After eating some of these celery sticks, Lance added 16 carrot sticks to the crisper. Later, Lance noticed there were a total of 21 celery and carrot sticks altogether. How many celery sticks did Lance eat?

9. ________________

Name: Date:
Instructor: Section:

Chapter 3 SOLVING APPLICATION PROBLEMS

3.4 Solving Application Problems with Two Unknown Quantities

Learning Objectives
1 Solve application problems with two unknown quantities.

Key Terms

Use the vocabulary terms listed below to complete each statement in exercises 1–4.

added to	**subtracted from**	**times**	**divided by**
minus	**more than**	**half**	**triple**

1. ____________________ and ____________________ are words that mean addition.

2. ____________________ and ____________________ are words that mean multiplication.

3. ____________________ and ____________________ are words that mean division.

4. ____________________ and ____________________ are words that mean subtraction.

Guided Examples

Review these examples for Objective 1:

1. Last year Lucas earned \$5000 less than his wife, Susanna. Together they earned \$75,320. How much did each of them earn?

   Step 1 Read the problem. Find the amount Lucas earned and the amount Susanna earned.

   Step 2 Assign a variable. There are two unknowns. Let s represent the amount Susanna earned. Let $s - 5000$ represent the amount Lucas earned.

   Step 3 Write an equation.

Susanna's earnings		Lucas' earning		Total earnings
s	$+$	$s - 5000$	$=$	$75{,}320$

Now Try:

1. Linda paid three times as much for her dog than for her cat. If she paid a total of \$220 for both, how much did she pay for each pet?

   ____

Step 4 Solve.

$$\begin{aligned} s+s \quad - \quad 5000 &= 75,320 \\ 2s+(-5000) &= 75,320 \\ +5000 \quad & \quad +5000 \\ \hline 2s \quad + \quad 0 &= 80,320 \\ \frac{2s}{2} &= \frac{80,320}{2} \\ s &= 40,160 \end{aligned}$$

Step 5 State the answer. Susanna earned \$40,160.
s – 5000 represents Lucas' earnings. Replace s with \$40,160.
\$40,160 – \$5000 = \$35,160, so Lucas earned \$35,160.

Step 6 Check the solution. The total earnings is \$40, 160 + \$35,160 = \$75,320. This checks.

2. A string is 89 cm long. Bob's cat, Reginald, bit the string into two pieces so that one piece is 17 cm longer than the other. Find the length of each piece.

Step 1 Read the problem. Find the length of a longer piece and a shorter piece.

Step 2 Assign a variable. There are two unknowns. Let x represent the length of the shorter piece. Let x + 17 represent the length of the longer piece.

Step 3 Write an equation.

shorter piece		longer piece		Total length
x	$+$	$x+17$	$=$	89

Step 4 Solve.

$$\begin{aligned} x+x \, + \, 17 &= 89 \\ 2x \quad +17 &= 89 \\ -17 \quad & \quad -17 \\ \hline 2x \, + \, 0 &= 72 \\ \frac{2x}{2} &= \frac{72}{2} \\ x &= 36 \end{aligned}$$

2. A telephone cable 98 meters in length is cut into two pieces. If one piece is 22 meters longer than the other, how long are the two pieces?

____

Step 5 State the answer. The shorter piece is 36 cm.
$x + 17$ represents the longer piece. Replace x with 36.
$36 + 17 = 53$, so the longer piece is 53 cm.

Step 6 Check the solution. The total length is 36 cm + 53 cm = 89 cm. This checks.

3. A site of a new building is rectangular in shape. The length is twice the width. If it will require 642 meters of fence to enclose the site, find the length and the width of the new building site.

Step 1 Read the problem. Find the length and the width.

Step 2 Assign a variable. There are two unknowns. Let x represent the width.
Let $2x$ represent the length.

Step 3 Write an equation. Use the formula for perimeter of a rectangle, $P = 2l + 2w$.
Replace P with 642, replace l with $2x$, and replace w with x.

$$P = 2\ l\ + 2w$$
$$642 = 2(2x) + 2x$$

Step 4 Solve.

$$642 = 2(2x) + 2x$$
$$642 = 4x + 2x$$
$$\frac{642}{6} = \frac{6x}{6}$$
$$107 = x$$

Step 5 State the answer. The width is 107 m.
$2x$ represents the length. Replace x with 107.
$2 \cdot 107 = 214$, so the length is 214 m.

Step 6 Check the solution.

$$P = 2 \cdot 214 + 2 \cdot 107$$
$$P = 428 + 214$$
$$P = 642$$

This matches the perimeter given in the original problem.

3. The length of a picture frame is 5 in. less than three times the width. If the perimeter of the frame is 86 in., find the length and the width of the picture frame.

____

Name: Date:
Instructor: Section:

Objective 1 Solve application problems with two unknown quantities.

For extra help, see Examples 1–3 on pages 194–197 of your text and Section Lecture video for Section 3.4 and Exercise Solutions Clip 5, 13, and 21.

Solve each application problem by using the six problem-solving steps.

1. The vote totals for the two candidates for county judge, Jones and Greer, differed by only 516 votes, with Jones receiving more votes. If there were 16,786 voters in the county, how many votes did each candidate receive?

1. Greer____________

Jones____________

2. After a 14-meter tall tree is chopped down, it is cut into four pieces. Three pieces are the same length, while the fourth piece is 2 m longer than each of the other three. Find the length of each piece.

2. ____________

Use the formula $\boldsymbol{P = 2l + 2w}$ *to solve. Make a sketch to help you solve each problem.*

3. A rectangular garden is three times as long as it is wide. The perimeter of the garden is 96 yd. Find the length and the width of the garden.

3. width____________

length ____________

Name: Date:
Instructor: Section:

Chapter 4 RATIONAL NUMBERS: POSITIVE AND NEGATIVE FRACTIONS

4.1 Introduction to Signed Fractions

Learning Objectives

1	Use a fraction to name part of a whole.
2	Identify numerators, denominators, proper fractions, and improper fractions.
3	Graph positive and negative fractions on a number line.
4	Find the absolute value of a fraction.
5	Write equivalent fractions.

Key Terms

Use the vocabulary terms listed below to complete each statement in exercises 1–6.

fraction	**numerator**	**denominator**
proper fraction	**improper fraction**	**equivalent fractions**

1. Two fractions are ______________________________ when they represent the same portion of a whole.

2. A number of the form $\frac{a}{b}$, where a and b are integers and $b \neq 0$ is called a ________________________.

3. A fraction whose numerator is larger than its denominator is called an ______________________.

4. In the fraction $\frac{2}{9}$, the 2 is the ___________________________.

5. A fraction whose denominator is larger than its numerator is called a ________________________.

6. The ___________________________ of a fraction shows the number of equal parts in a whole.

Name: Date:
Instructor: Section:

Guided Examples

Review these examples for Objective 1:

1. Use fractions to represent the shaded portion and the unshaded portion of the figure.

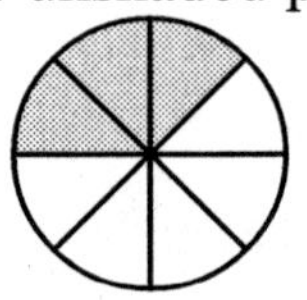

The figure has 8 equal parts. The 3 shaded parts are represented by the fraction $\frac{3}{8}$. The unshaded part is $\frac{5}{8}$ of the figure.

2. Use a fraction to represent the shaded parts.

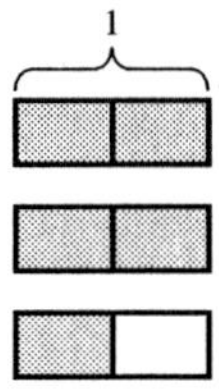

An area equal to 5 of the $\frac{1}{2}$ parts is shaded, so $\frac{5}{2}$ is shaded.

Now Try:

1. Use fractions to represent the shaded portion and the unshaded portion of the figure.

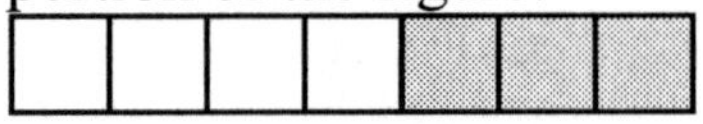

2. Use a fraction to represent the shaded parts.

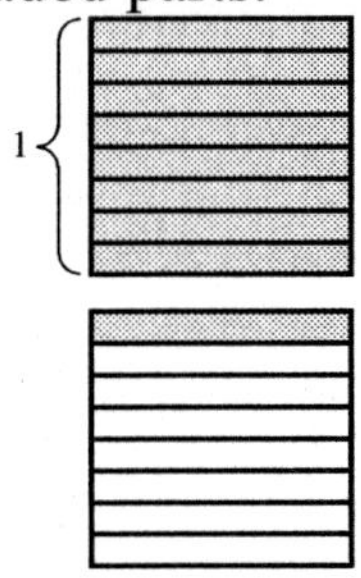

Review these examples for Objective 2:

3. Identify the numerator and denominator in each fraction. Then state the number of equal parts in the whole.

a. $\frac{9}{14}$

$\frac{9}{14}$ 9 ← Numerator, 14 ← Denominator

14 equal parts in the whole

b. $\frac{4}{15}$

$\frac{4}{15}$ 4 ← Numerator, 15 ← Denominator

15 equal parts in the whole

Now Try:

3. Identify the numerator and denominator in each fraction. Then state the number of equal parts in the whole.

a. $\frac{5}{17}$

b. $\frac{6}{23}$

Name: Date:
Instructor: Section:

4. $\frac{4}{7}, \frac{7}{10}, \frac{19}{6}, \frac{11}{11}, \frac{13}{27}, \frac{1}{8}, \frac{6}{5}$

a. Identify all the proper fractions in the list.

Proper fractions have a numerator that is less than the denominator.

$\frac{4}{7}, \frac{7}{10}, \frac{13}{27}, \frac{1}{8}$

b. Identify all the improper fractions in the list.

Improper fractions have a numerator that is equal or greater than the denominator.

$\frac{19}{6}, \frac{11}{11}, \frac{6}{5}$

4. $\frac{7}{8}, \frac{9}{11}, \frac{22}{7}, \frac{13}{13}, \frac{12}{29}, \frac{1}{5}, \frac{17}{4}$

a. Identify all the proper fractions in the list.

____

b. Identify all the improper fractions in the list.

____

Review this example for Objective 3:

5. Graph the fractions on the number line.

$-\frac{7}{9}, \frac{7}{9}$

For $-\frac{7}{9}$, the fraction is negative, so it is between 0 and –1. We divide that space into 9 equal parts. Then we start at 0 and count to the left 7 parts.

There is no sign in front of $\frac{7}{9}$, so it is positive.

Because $\frac{7}{9}$, is between 0 and 1, we divide that space into 9 equal parts. Then we start at 0 and count to the right 7 parts.

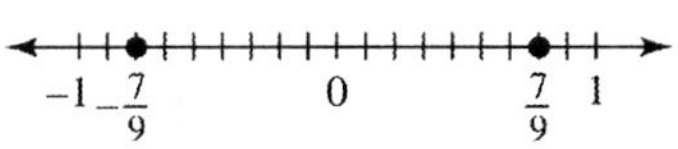

Now Try:

5. Graph the fractions on the number line.

$\frac{5}{6}, -\frac{5}{6}$

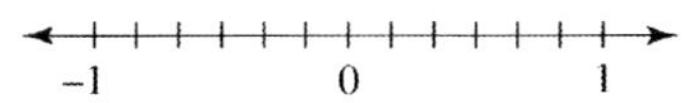

Review this example for Objective 4:

6. Find each absolute value: $\left|-\frac{3}{5}\right|$ and $\left|\frac{3}{5}\right|$.

The distance from 0 to $-\frac{3}{5}$ on the number line is $\frac{3}{5}$ space, so $\left|-\frac{3}{5}\right| = \frac{3}{5}$.

The distance from 0 to $\frac{3}{5}$ on the number line is also $\frac{3}{5}$ space, so $\left|\frac{3}{5}\right| = \frac{3}{5}$.

Now Try:

6. Find each absolute value:

$\left|\frac{9}{10}\right|$ and $\left|-\frac{9}{10}\right|$.

____

____

Name: Date:
Instructor: Section:

Review these examples for Objective 5:

7.

a. Write $-\frac{1}{3}$ as an equivalent fraction with a denominator of 18.

In other words, $-\frac{1}{3}=-\frac{?}{18}$.
The original denominator is 3. Multiplying 3 times 6 gives 18, the new denominator. To write an equivalent fraction, multiply both the numerator and denominator by 6.

$$-\frac{1}{3}=-\frac{1\cdot 6}{3\cdot 6}=-\frac{6}{18}$$

So, $-\frac{1}{3}$ is equivalent to $-\frac{6}{18}$.

b. Write $\frac{15}{18}$ as an equivalent fraction with a denominator of 6.

In other words, $\frac{15}{18}=\frac{?}{6}$.
The original denominator is 18. Dividing 18 by 3 gives 6, the new denominator. To write an equivalent fraction, divide both the numerator and denominator by 3.

$$\frac{15}{18}=\frac{15\div 3}{18\div 3}=\frac{5}{6}$$

So, $\frac{15}{18}$ is equivalent to $\frac{5}{6}$.

8. Simplify each fraction by dividing the numerator by the denominator.

a. $\frac{12}{12}$

Think of $\frac{12}{12}$ as $12\div 12$. The result is 1, so $\frac{12}{12}=1$.

b. $-\frac{20}{5}$

Think of $-\frac{20}{5}$ as $-20\div 5$. The result is –4, so $-\frac{20}{5}=-4$.

Now Try:

7.

a. Write $-\frac{4}{5}$ as an equivalent fraction with a denominator of 30.

____

b. Write $\frac{35}{42}$ as an equivalent fraction with a denominator of 6.

____

8. Simplify each fraction by dividing the numerator by the denominator.

a. $\frac{22}{22}$

____

b. $-\frac{42}{7}$

____

Name: Date:
Instructor: Section:

c. $\frac{7}{1}$

Think of $\frac{7}{1}$ as $7 \div 1$. The result is 7,

so $\frac{7}{1} = 7$.

c. $\frac{13}{1}$

____

Objective 1 Use a fraction to name part of a whole.

For extra help, see Examples 1–2 on pages 214–215 of your text and Section Lecture video for Section 4.1 and Exercise Solutions Clip 3 and 5.

Write the fractions that represent the shaded and unshaded portions of each figure.

1. 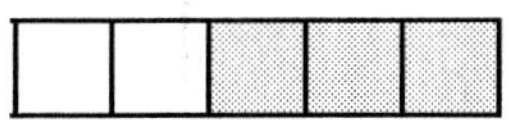

1. Shaded __________

Unshaded ________

2. 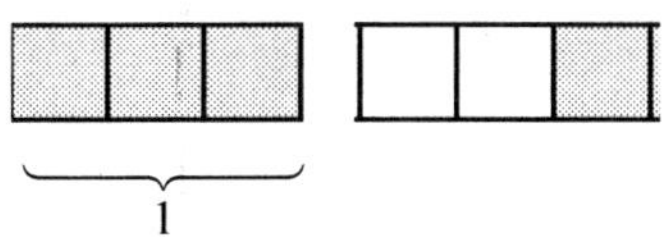

2. Shaded __________

Unshaded ________

3. 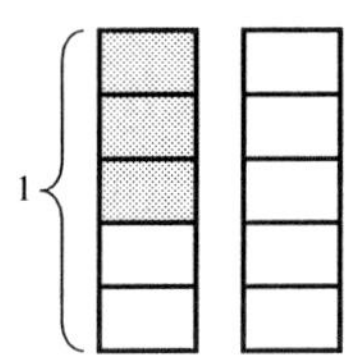

3. Shaded __________

Unshaded ________

Objective 2 Identify numerators, denominators, proper fractions, and improper fractions.

For extra help, see Examples 3–4 on pages 215–216 of your text and Section Lecture video for Section 4.1 and Exercise Solutions Clip 15 and 19.

Identify the numerator and denominator in each fraction.

4. $\frac{2}{9}$

4. numerator _______

denominator _____

5. $\frac{12}{5}$

5. numerator _______

denominator _____

Name: Date:
Instructor: Section:

List the proper and improper fractions in the group of numbers.

6. $\frac{2}{3}, \frac{5}{2}, \frac{4}{4}, \frac{6}{7}, \frac{8}{15}, \frac{20}{19}$

6. Proper __________

Improper ________

Objective 3 Graph positive and negative fractions on a number line.

For extra help, see Example 5 on page 217 of your text and Section Lecture video for Section 4.1 and Exercise Solutions Clip 21.

Graph each pair of fractions on a number line.

7. $\frac{2}{3}, -\frac{2}{3}$

7.

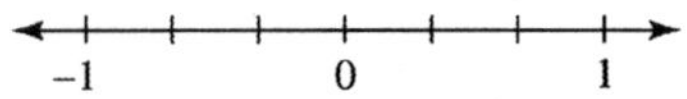

8. $-\frac{1}{4}, \frac{1}{4}$

8.

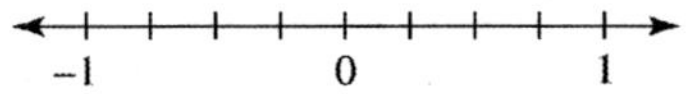

9. $-\frac{5}{8}, \frac{5}{8}$

9.

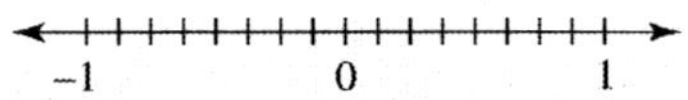

Objective 4 Find the absolute value of a fraction.

For extra help, see Example 6 on page 217 of your text and Section Lecture video for Section 4.1 and Exercise Solutions Clip 31 and 34.

Find each absolute value.

10. $\left|-\frac{15}{7}\right|$

10. ________________

11. $\left|\frac{8}{9}\right|$

11. ________________

12. $|0|$

12. ________________

Name: Date:
Instructor: Section:

Objective 5 Write equivalent fractions.

For extra help, see Examples 7–8 on pages 219–220 of your text and Section Lecture video for Section 4.1 and Exercise Solutions Clip 35 and 37.

Rewrite each fraction as an equivalent fraction with a denominator of 48.

13. $-\frac{2}{3}$ **13.** ____________

14. $\frac{3}{4}$ **14.** ____________

Rewrite the fraction as an equivalent fraction with a denominator of 5.

15. $-\frac{4}{20}$ **15.** ____________

Name: Date:
Instructor: Section:

Chapter 4 RATIONAL NUMBERS: POSITIVE AND NEGATIVE FRACTIONS

4.2 Writing Fractions in Lowest Terms

Learning Objectives	
1	Identify fractions written in lowest terms.
2	Write a fraction in lowest terms using common factors.
3	Write a number as a product of prime factors.
4	Write a fraction in lowest terms using prime factorization.
5	Write a fraction with variables in lowest terms.

Key Terms

Use the vocabulary terms listed below to complete each statement in exercises 1–4.

lowest terms **prime number** **composite number**

prime factorization

1. A ______________________ has at least one factor other than itself and 1.

2. In a ______________________________ every factor is a prime number.

3. The factors of a _____________________ are itself and 1.

4. A fraction is written in ____________________________ when its numerator and denominator have no common factor other than 1.

Guided Examples

Review these examples for Objective 1:

1. Are the following fractions in lowest terms? If not, find a common factor of the numerator and denominator (other than 1).

a. $\frac{3}{7}$

The numerator and denominator have no common factor other than 1, so the fraction is in lowest terms.

b. $\frac{28}{49}$

The numerator and denominator have a common factor of 7, so the fraction is not in lowest terms.

Now Try:

1. Are the following fractions in lowest terms? If not, find a common factor of the numerator and denominator (other than 1).

a. $\frac{8}{9}$

____

b. $\frac{27}{36}$

____

Name: Date:
Instructor: Section:

Review these examples for Objective 2:

2. Divide by a common factor to write each fraction in lowest terms.

a. $\frac{40}{48}$

The greatest common factor of 40 and 48 is 8. Divide both numerator and denominator by 8.

$$\frac{40}{48}=\frac{40\div 8}{48\div 8}=\frac{5}{6}$$

b. $\frac{35}{55}$

Divide both numerator and denominator by 5.

$$\frac{35}{55}=\frac{35\div 5}{55\div 5}=\frac{7}{11}$$

c. $-\frac{28}{49}$

Divide both numerator and denominator by 7. Keep the negative sign.

$$-\frac{28}{49}=-\frac{28\div 7}{49\div 7}=-\frac{4}{7}$$

d. $\frac{80}{96}$

Dividing by 4 does not give the greatest common factor of 80 and 96.

$$\frac{80}{96}=\frac{80\div 4}{96\div 4}=\frac{20}{24} \leftarrow \text{not lowest terms}$$

There is another common factor of 4.

$$\frac{20}{24}=\frac{20\div 4}{24\div 4}=\frac{5}{6} \leftarrow \text{lowest terms}$$

Divide by 16, the greatest common factor, in one step.

$$\frac{80}{96}=\frac{80\div 16}{96\div 16}=\frac{5}{6}$$

Now Try:

2. Divide by a common factor to write each fraction in lowest terms.

a. $\frac{20}{36}$ ____________

b. $\frac{35}{40}$ ____________

c. $-\frac{25}{65}$ ____________

d. $\frac{55}{60}$ ____________

Review these examples for Objective 3:

3. Label each number as prime or composite or neither.

1 3 4 6 17 23

First, 1 is neither prime nor composite.
Next, 3, 17, and 23 are prime.
The numbers 4 and 6 can be divided by 2, so 4 and 6 are composite.

Now Try:

3. Label each number as prime or composite or neither.

1 8 21 29 31

____

Name: Date:
Instructor: Section:

4. Find the prime factorization of each number.

a. 54

Start by dividing 54 by 2 and work your way up the chain of divisions.

$$\begin{array}{r} 1 \\ 3\overline{)3} \\ 3\overline{)9} \\ 3\overline{)27} \\ \text{Start here } 2\overline{)54} \end{array}$$

Because all the factors (divisors) are prime, the prime factorization of 54 is $54 = 2 \cdot 3 \cdot 3 \cdot 3$.

b. 84

Start by dividing 84 by 2.

$$\begin{array}{r} 1 \\ 7\overline{)7} \\ 3\overline{)21} \\ 2\overline{)42} \\ \text{Start here } 2\overline{)84} \end{array}$$

The prime factorization of 84 is $84 = 2 \cdot 2 \cdot 3 \cdot 7$.

5. Find the prime factorization of each number.

a. 180

Try to divide 180 by 2. The quotient is 90. Write the factors 2 and 90 under 180. Box the 2 (or circle it), because it is prime.

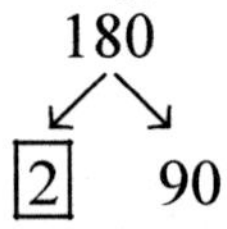

Try dividing 90 by 2. The quotient is 45. Write the factors 2 and 45 under the 90. Continue until you find the prime factorization.

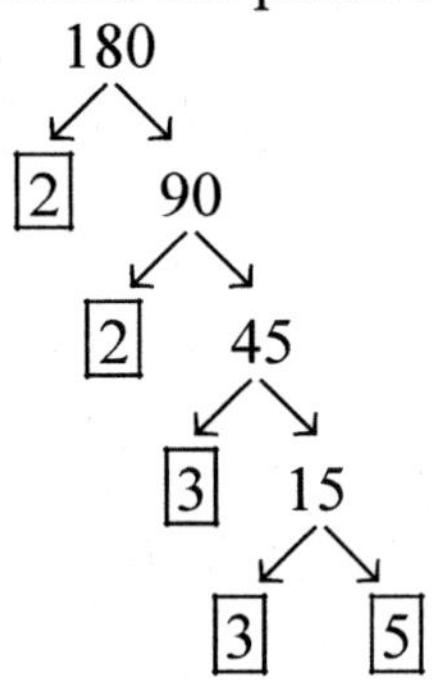

The prime factorization is $180 = 2 \cdot 2 \cdot 3 \cdot 3 \cdot 5$.

4. Find the prime factorization of each number.

a. 96

____

b. 56

____

5. Find the prime factorization of each number.

a. 152

____

b. 108

Divide 108 by 2, the first prime number.

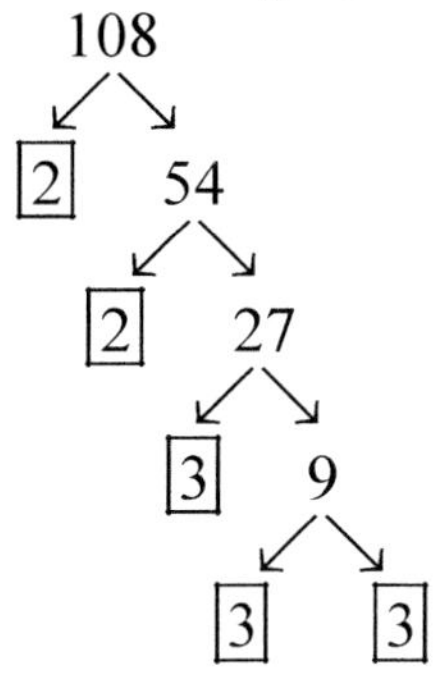

The prime factorization is $108 = 2 \cdot 2 \cdot 3 \cdot 3 \cdot 3$.

Now Try:

b. 168

____

Review these examples for Objective 4:

6. Write each fraction in lowest terms.

a. $\frac{28}{63}$

Write the prime factorizations of 28 and 63.

$$\frac{28}{63} = \frac{2 \cdot 2 \cdot 7}{3 \cdot 3 \cdot 7}$$

Divide the numerator and denominator by 7, the common factor.

$$\frac{28}{63} = \frac{2 \cdot 2 \cdot \overset{1}{\cancel{7}}}{3 \cdot 3 \cdot \underset{1}{\cancel{7}}} = \frac{4}{9}$$

b. $\frac{90}{126}$

Write the prime factorizations of 90 and 126. Then divide the numerator and denominator by the common factors.

$$\frac{90}{126} = \frac{\overset{1}{\cancel{2}} \cdot \overset{1}{\cancel{3}} \cdot \overset{1}{\cancel{3}} \cdot 5}{\underset{1}{\cancel{2}} \cdot \underset{1}{\cancel{3}} \cdot \underset{1}{\cancel{3}} \cdot 7} = \frac{5}{7}$$

c. $\frac{27}{180}$

$$\frac{27}{180} = \frac{3 \cdot \overset{1}{\cancel{3}} \cdot \overset{1}{\cancel{3}}}{2 \cdot 2 \cdot \underset{1}{\cancel{3}} \cdot \underset{1}{\cancel{3}} \cdot 5} = \frac{3}{20}$$

Now Try:

6. Write each fraction in lowest terms.

a. $\frac{75}{225}$

____

b. $\frac{42}{140}$

____

c. $\frac{36}{45}$

____

Name: Date:
Instructor: Section:

Review these examples for Objective 5:

7. Write each fraction in lowest terms.

a. $\frac{10}{15x}$

$$\frac{10}{15x} = \frac{2 \cdot \overset{1}{\cancel{5}}}{3 \cdot \underset{1}{\cancel{5}} \cdot x} = \frac{2}{3x}$$

b. $\frac{7xy}{28xy}$

$$\frac{7xy}{28xy} = \frac{\overset{1}{\cancel{7}} \cdot \overset{1}{\cancel{x}} \cdot \overset{1}{\cancel{y}}}{2 \cdot 2 \cdot \underset{1}{\cancel{7}} \cdot \underset{1}{\cancel{x}} \cdot \underset{1}{\cancel{y}}} = \frac{1}{4}$$

c. $\frac{6b^4}{12ab^2}$

$$\frac{6b^4}{12ab^2} = \frac{\overset{1}{\cancel{2}} \cdot \overset{1}{\cancel{3}} \cdot b \cdot b \cdot \overset{1}{\cancel{b}} \cdot \overset{1}{\cancel{b}}}{\underset{1}{\cancel{2}} \cdot 2 \cdot \underset{1}{\cancel{3}} \cdot a \cdot \underset{1}{\cancel{b}} \cdot \underset{1}{\cancel{b}}} = \frac{b^2}{2a}$$

d. $\frac{5a^2b}{x^2}$

$$\frac{5a^2b}{x^2} = \frac{5 \cdot a \cdot a \cdot b}{x \cdot x}$$

$\frac{5a^2b}{x^2}$ is already in lowest terms.

Now Try:

7. Write each fraction in lowest terms.

a. $\frac{25}{30x}$ ____________

b. $\frac{8ab}{24ab}$ ____________

c. $\frac{9b^5}{36ab^3}$ ____________

d. $\frac{9p^2q}{r^3}$ ____________

Objective 1 Identify fractions written in lowest terms.

For extra help, see Example 1 on page 225 of your text and Section Lecture video for Section 4.2.

Are the following fractions in lowest terms? If not, find a common factor of the numerator and denominator (other than 1).

1. $-\frac{7}{11}$ **1.** ____________

2. $\frac{20}{25}$ **2.** ____________

3. $\frac{12}{16}$ **3.** ____________

Name: Date:
Instructor: Section:

Objective 2 Write a fraction in lowest terms using common factors.

For extra help, see Example 2 on page 226 of your text and Section Lecture video for Section 4.2 and Exercise Solutions Clip 3.

Write in lowest terms.

4. $-\frac{24}{36}$ **4.** ________

5. $\frac{30}{42}$ **5.** ________

6. $-\frac{32}{72}$ **6.** ________

Objective 3 Write a number as a product of prime factors.

For extra help, see Examples 3–5 on pages 227–229 of your text and Section Lecture video for Section 4.2 and Exercise Solutions Clip 5 and 13.

Find the prime factorization of each number.

7. 105 **7.** ________

8. 168 **8.** ________

9. 320 **9.** ________

Objective 4 Write a fraction in lowest terms using prime factorization.

For extra help, see Example 6 on pages 230–231 of your text and Section Lecture video for Section 4.2.

Write each numerator and denominator as a product of prime factors. Then use the prime factorization to write the fraction in lowest terms.

10. $\frac{18}{99}$ **10.** ________

Name: Date:
Instructor: Section:

11. $\frac{63}{105}$ **11.** ______________

12. $\frac{36}{210}$ **12.** ______________

Objective 5 Write a fraction with variables in lowest terms.

For extra help, see Example 7 on page 232 of your text and Section Lecture video for Section 4.2 and Exercise Solutions Clip 49, 61, and 65.

Write each fraction in lowest terms.

13. $\frac{12r^2s}{4rs^3}$ **13.** ______________

14. $\frac{16b^2cd}{40b^2d}$ **14.** ______________

15. $\frac{8xy^2}{6x^2y^2}$ **15.** ______________

Name: Date:
Instructor: Section:

Chapter 4 RATIONAL NUMBERS: POSITIVE AND NEGATIVE FRACTIONS

4.3 Multiplying and Dividing Signed Fractions

Learning Objectives

1. Multiply signed fractions.
2. Multiply fractions that involve variables.
3. Divide signed fractions.
4. Divide fractions that involve variables.
5. Solve application problems involving multiplying and dividing fractions.

Key Terms

Use the vocabulary terms listed below to complete each statement in exercises 1–4.

reciprocals **indicator words** **of** **each**

1. Two numbers are ____________________ of each other if their product is 1.

2. The words "per" and "divided equally" are ____________________ for division.

3. When the word "_______________" follows a fraction, it means "multiply".

4. The word "__________" indicates division.

Guided Examples

Review these examples for Objective 1:

1. Find each product.

   a. $-\frac{7}{9} \cdot -\frac{5}{11}$

   Multiply the numerators and multiply the denominators.

   $-\frac{7}{9} \cdot -\frac{5}{11} = \frac{7 \cdot 5}{9 \cdot 11} = \frac{35}{99}$

   The answer is in lowest terms because 35 and 99 have no common factor other than 1.

   b. $\left(\frac{3}{5}\right)\left(-\frac{4}{7}\right)$

   $\left(\frac{3}{5}\right)\left(-\frac{4}{7}\right) = -\frac{3 \cdot 4}{5 \cdot 7} = -\frac{12}{35}$

Now Try:

1. Find each product.

   a. $-\frac{10}{11} \cdot -\frac{4}{13}$

   ____

   b. $\left(\frac{15}{22}\right)\left(-\frac{3}{4}\right)$

   ____

Name: Date:
Instructor: Section:

2. Multiply. Write the products in lowest terms.

a. $-\frac{9}{7}\left(\frac{7}{15}\right)$

Multiplying a negative number times a positive number gives a negative product.

$$-\frac{9}{7}\left(\frac{7}{15}\right)=-\frac{3\cdot 3\cdot 7}{7\cdot 3\cdot 5}=-\frac{\overset{1}{\cancel{3}}\cdot 3\cdot \overset{1}{\cancel{7}}}{\underset{1}{\cancel{7}}\cdot \underset{1}{\cancel{3}}\cdot 5}=-\frac{3}{5}$$

b. Find $\frac{3}{8}$ of $\frac{4}{9}$.

Recall that "of" indicates multiplication.

$$\frac{3}{8}\cdot\frac{4}{9}=\frac{3\cdot 2\cdot 2}{2\cdot 2\cdot 2\cdot 3\cdot 3}=\frac{\overset{1}{\cancel{3}}\cdot \overset{1}{\cancel{2}}\cdot \overset{1}{\cancel{2}}}{\underset{1}{\cancel{2}}\cdot \underset{1}{\cancel{2}}\cdot 2\cdot \underset{1}{\cancel{3}}\cdot 3}=\frac{1}{6}$$

3. Find $\frac{3}{7}$ of 14.

Write 14 in fraction form as $\frac{14}{1}$.

$$\frac{3}{7}\cdot\frac{14}{1}=\frac{3\cdot 2\cdot 7}{7\cdot 1}=-\frac{3\cdot 2\cdot \overset{1}{\cancel{7}}}{\underset{1}{\cancel{7}}\cdot 1}=\frac{6}{1}=6$$

2. Multiply. Write the products in lowest terms.

a. $-\frac{11}{7}\left(\frac{14}{33}\right)$ ____________

b. Find $\frac{2}{7}$ of $\frac{21}{40}$. ____________

3. Find $\frac{5}{9}$ of 27. ____________

Review these examples for Objective 2:

4. Find each product.

a. $\frac{6x}{7}\cdot\frac{5}{18x}$

$$\frac{6x}{7}\cdot\frac{5}{18x}=\frac{2\cdot 3\cdot x\cdot 5}{7\cdot 2\cdot 3\cdot 3\cdot x}=\frac{\overset{1}{\cancel{2}}\cdot \overset{1}{\cancel{3}}\cdot \overset{1}{\cancel{x}}\cdot 5}{7\cdot \underset{1}{\cancel{2}}\cdot \underset{1}{\cancel{3}}\cdot 3\cdot \underset{1}{\cancel{x}}}=\frac{5}{21}$$

b. $\left(\frac{9a}{5b}\right)\left(\frac{10b^2}{3a}\right)$

$$\left(\frac{9a}{5b}\right)\left(\frac{10b^2}{3a}\right)=\frac{3\cdot 3\cdot a\cdot 2\cdot 5\cdot b\cdot b}{5\cdot b\cdot 3\cdot a}$$

$$=\frac{\overset{1}{\cancel{3}}\cdot 3\cdot \overset{1}{\cancel{a}}\cdot 2\cdot \overset{1}{\cancel{5}}\cdot \overset{1}{\cancel{b}}\cdot b}{\underset{1}{\cancel{5}}\cdot \underset{1}{\cancel{b}}\cdot \underset{1}{\cancel{3}}\cdot \underset{1}{\cancel{a}}}$$

$$=6b$$

Now Try:

4. Find each product.

a. $\frac{12x}{5}\cdot\frac{7}{18x}$ ____________

b. $\left(\frac{7x}{9c}\right)\left(\frac{6c^2}{35x}\right)$ ____________

Name: Date:
Instructor: Section:

Review these examples for Objective 3:

5. Rewrite each division problem as a multiplication problem.

a. $\frac{5}{7} \div \frac{10}{3}$

$$\frac{5}{7} \div \frac{10}{3} = \frac{5}{7} \cdot \frac{3}{10} = \frac{\cancel{5}^{1} \cdot 3}{7 \cdot 2 \cdot \cancel{5}_{1}} = \frac{3}{14}$$

b. $6 \div \left(-\frac{1}{5}\right)$

$$6 \div \left(-\frac{1}{5}\right) = \frac{6}{1} \cdot \left(-\frac{5}{1}\right) = -\frac{6 \cdot 5}{1 \cdot 1} = -\frac{30}{1} = -30$$

c. $-\frac{7}{9} \div (-6)$

$$-\frac{7}{9} \div (-6) = -\frac{7}{9} \cdot \left(-\frac{1}{6}\right) = \frac{7 \cdot 1}{3 \cdot 3 \cdot 3 \cdot 2} = \frac{7}{54}$$

d. $\frac{5}{11} \div 0$

$\frac{5}{11} \div 0$ is undefined.

e. $0 \div \frac{7}{11}$

$$0 \div \frac{7}{11} = 0 \cdot \frac{11}{7} = 0$$

Now Try:

5. Rewrite each division problem as a multiplication problem.

a. $\frac{5}{9} \div \frac{20}{3}$ ____________

b. $10 \div \left(-\frac{1}{8}\right)$ ____________

c. $-\frac{6}{13} \div \left(-\frac{3}{26}\right)$ ____________

d. $\frac{13}{27} \div 0$ ____________

e. $0 \div \frac{2}{25}$ ____________

Review these examples for Objective 4:

6. Divide. (Assume that none of the variables represent zero.)

a. $\frac{x^3}{y^2} \div \frac{x^2}{4y}$

$$\frac{x^3}{y^2} \div \frac{x^2}{4y} = \frac{x^3}{y^2} \cdot \frac{4y}{x^2} = \frac{\cancel{x}^{1} \cdot \cancel{x}^{1} \cdot x \cdot 2 \cdot 2 \cdot \cancel{y}^{1}}{\cancel{y}_{1} \cdot y \cdot \cancel{x}_{1} \cdot \cancel{x}_{1}} = \frac{4x}{y}$$

b. $\frac{9b^2}{7} \div b^3$

$$\frac{9b^2}{7} \div b^3 = \frac{9b^2}{7} \cdot \frac{1}{b^3} = \frac{3 \cdot 3 \cdot \cancel{b}^{1} \cdot \cancel{b}^{1} \cdot 1}{7 \cdot \cancel{b}_{1} \cdot \cancel{b}_{1} \cdot b} = \frac{9}{7b}$$

Now Try:

6. Divide. (Assume that none of the variables represent zero.)

a. $\frac{a^2}{b^3} \div \frac{a}{5b^2}$ ____________

b. $\frac{10b}{9} \div b^3$ ____________

Name: Date:
Instructor: Section:

Review these examples for Objective 5:

7. Solve each application problem.

a. Signe gives $\frac{1}{12}$ of her income to charities. Last year she earned \$48,000. How much did she give to charities?

Because the word "of" follows the fraction, it indicates multiplication.

$$\frac{1}{12}\cdot 48,000=\frac{1}{12}\cdot\frac{48,000}{1}=\frac{1\cdot \cancel{12}\cdot 4000}{\underset{1}{\cancel{12}}\cdot 1}=4000$$

She gave \$4000 to charities.

b. How many times can a $1\frac{2}{3}$-quart spray bottle be filled before 20 quarts of water are used up?

$$20\div 1\frac{2}{3}=20\div\frac{5}{3}=\frac{20}{1}\cdot\frac{3}{5}=\frac{4\cdot \overset{1}{\cancel{5}}\cdot 3}{1\cdot \underset{1}{\cancel{5}}}=12$$

The spray bottle can be filled 12 times.

Now Try:

7. Solve each application problem.

a. A flower bed is $\frac{12}{11}$ m by $\frac{7}{8}$ m. Find its area.

____

b. How many $\frac{1}{3}$-lb servings will a 24-lb turkey provide?

____

Objective 1 Multiply signed fractions.

For extra help, see Examples 1–3 on pages 237–240 of your text and Section Lecture video for Section 4.3 and Exercise Solutions Clip 3, 5, and 7.

Multiply. Write the products in lowest terms.

1. $-\frac{10}{42}\cdot\frac{3}{5}$ **1.** ______________

2. $\frac{6}{18}\cdot\frac{9}{2}$ **2.** ______________

3. $\frac{5}{9}$ of 81 **3.** ______________

Name: Date:
Instructor: Section:

Objective 2 Multiply fractions that involve variables.

For extra help, see Example 4 on page 240 of your text and Section Lecture video for Section 4.3 and Exercise Solutions Clip 13.

Use prime factorization to find these products.

4. $\frac{3c}{5}\cdot\frac{c}{9}$ **4.** ________

5. $\left(\frac{m}{10}\right)\left(\frac{25}{m^2}\right)$ **5.** ________

6. $\frac{5a}{7b^2}\cdot\frac{14b}{10a^2}$ **6.** ________

Objective 3 Divide signed fractions.

For extra help, see Example 5 on pages 242–243 of your text and Section Lecture video for Section 4.3 and Exercise Solutions Clip 15, 19, and 31.

Divide. Write the quotients in lowest terms.

7. $\frac{7}{8}\div(-21)$ **7.** ________

8. $-\frac{5}{12}\div\frac{15}{8}$ **8.** ________

9. $-\frac{2}{3}\div\left(-\frac{7}{9}\right)$ **9.** ________

Name: Date:
Instructor: Section:

Objective 4 Divide fractions that involve variables.

For extra help, see Example 6 on page 243 of your text and Section Lecture video for Section 4.3 and Exercise Solutions Clip 23 and 25.

Divide. Write the quotients in lowest terms.

10. $\frac{3b}{5a} \div \frac{6}{7ab}$ **10.** ________________

11. $10x^2 \div \frac{5x}{3}$ **11.** ________________

12. $\frac{7c}{3d} \div 14c^2d$ **12.** ________________

Objective 5 Solve application problems involving multiplying and dividing fractions.

For extra help, see Example 7 on page 244 of your text and Section Lecture video for Section 4.3.

Solve each application problem.

13. Phillip saves $\frac{2}{9}$ of his income each month. How much did he save last month if he earned $2700? **13.** ________________

14. Rochelle sells hats at craft shows. She needs $\frac{2}{5}$ yd for each hat. How many hats can she make from 10 yards of fabric? **14.** ________________

Name: Date:
Instructor: Section:

Chapter 4 RATIONAL NUMBERS: POSITIVE AND NEGATIVE FRACTIONS

4.4 Adding and Subtracting Signed Fractions

Learning Objectives	
1	Add and subtract like fractions.
2	Find the lowest common denominator for unlike fractions.
3	Add and subtract unlike fractions.
4	Add and subtract unlike fractions that contain variables.

Key Terms

Use the vocabulary terms listed below to complete each statement in exercises 1–3.

like fractions **unlike fractions** **least common denominator**

1. Fractions with different denominators are called ____________________.

2. Fractions with the same denominator are called ________________________.

3. The _________________________________ of two whole numbers is the smallest whole number divisible by both of the numbers.

Guided Examples

Review these examples for Objective 1:

1. Find each sum or difference.

a. $\frac{1}{9}+\frac{5}{9}$

These are like fractions because they have a common denominator. Add the numerators and write the sum over the common denominator.

$$\frac{1}{9}+\frac{5}{9}=\frac{1+5}{9}=\frac{6}{9}$$

Now write $\frac{6}{9}$ in lowest terms.

$$\frac{6}{9}=\frac{2\cdot\overset{1}{\cancel{3}}}{3\cdot\underset{1}{\cancel{3}}}=\frac{2}{3}$$

b. $-\frac{4}{7}+\frac{6}{7}$

$$-\frac{4}{7}+\frac{6}{7}=\frac{-4+6}{7}=\frac{2}{7}$$

Now Try:

1. Find each sum or difference.

a. $\frac{1}{12}+\frac{7}{12}$

____

b. $-\frac{1}{4}+\frac{3}{4}$

____

Name: Date:
Instructor: Section:

c. $\frac{4}{15}-\frac{7}{15}$

Write the subtraction as adding the opposite.

$$\frac{4}{15}-\frac{7}{15}=\frac{4-7}{15}=\frac{4+(-7)}{15}=\frac{-3}{15}, \text{ or } -\frac{3}{15}$$

Now write $-\frac{3}{15}$ in lowest terms.

$$-\frac{3}{15}=-\frac{\overset{1}{\cancel{3}}}{\underset{1}{\cancel{3}}\cdot 5}=-\frac{1}{5}$$

d. $\frac{6}{x^3}-\frac{4}{x^3}$

$$\frac{6}{x^3}-\frac{4}{x^3}=\frac{6-4}{x^3}=\frac{2}{x^3}$$

c. $\frac{7}{20}-\frac{19}{20}$ __________

d. $\frac{9}{a^2}-\frac{5}{a^2}$ __________

Review these examples for Objective 2:

2.

a. Find the LCD for $\frac{2}{7}$ and $\frac{5}{21}$ by inspection.

Since 21 is divisible by 7, then 21 is the LCD.

b. Find the LCD for $\frac{7}{12}$ and $\frac{3}{8}$ by inspection.

Since 12 is not divisible by 8, use multiples of 12, that is 12, 24, and 36. Notice that 24 is divisible by both 8 and 12. The LCD is 24.

3.

a. What is the LCD for $\frac{11}{18}$ and $\frac{5}{24}$?

Write 18 and 24 as the product of prime factors.

$18 = 2\cdot 3\cdot 3$
$24 = 2\cdot 2\cdot 2\cdot 3$
$\text{LCD} = 2\cdot 2\cdot 2\cdot 3\cdot 3 = 72$

The LCD for $\frac{11}{18}$ and $\frac{5}{24}$ is 72.

b. What is the LCD for $\frac{9}{16}$ and $\frac{7}{36}$?

$16 = 2\cdot 2\cdot 2\cdot 2$
$36 = 2\cdot 2\cdot 3\cdot 3$
$\text{LCD} = 2\cdot 2\cdot 2\cdot 2\cdot 3\cdot 3 = 144$

The LCD for $\frac{9}{16}$ and $\frac{7}{36}$ is 144.

Now Try:

2.

a. Find the LCD for $\frac{2}{9}$ and $\frac{7}{27}$ by inspection. __________

b. Find the LCD for $\frac{3}{10}$ and $\frac{7}{15}$ by inspection. __________

3.

a. What is the LCD for $\frac{9}{20}$ and $\frac{8}{25}$? __________

b. What is the LCD for $\frac{7}{30}$ and $\frac{2}{45}$? __________

Name: Date:
Instructor: Section:

Review these examples for Objective 3:

4. Find each sum or difference.

a. $\frac{1}{2}+\frac{1}{6}$

Step 1 The larger denominator (6) is the LCD.

Step 2 $\frac{1}{2}=\frac{1\cdot 3}{2\cdot 3}=\frac{3}{6}$ and $\frac{1}{6}$ already has the LCD.

Step 3 Add the numerators.

$$\frac{1}{2}+\frac{1}{6}=\frac{3}{6}+\frac{1}{6}=\frac{3+1}{6}=\frac{4}{6}$$

Step 4 Write $\frac{4}{6}$ in lowest terms.

$$\frac{4}{6}=\frac{2\cdot \overset{1}{\cancel{2}}}{\underset{1}{\cancel{2}}\cdot 3}=\frac{2}{3}$$

b. $\frac{3}{8}-\frac{7}{12}$

Step 1 The LCD is 24.

Step 2 $\frac{3}{8}=\frac{3\cdot 3}{8\cdot 3}=\frac{9}{24}$ and $\frac{7}{12}=\frac{7\cdot 2}{12\cdot 2}=\frac{14}{24}$

Step 3 Subtract the numerators.

$$\frac{3}{8}-\frac{7}{12}=\frac{9}{24}-\frac{14}{24}=\frac{9-14}{24}=\frac{-5}{24}, \text{ or } -\frac{5}{24}$$

Step 4 $-\frac{5}{24}$ is in lowest terms.

c. $-\frac{7}{18}+\frac{5}{12}$

Step 1 Use prime factorization to find the LCD.

$18=2\cdot 3\cdot 3$

$12=2\cdot 2\cdot 3$

$\text{LCD}=2\cdot 2\cdot 3\cdot 3=36$

Step 2

$$-\frac{7}{18}=-\frac{7\cdot 2}{18\cdot 2}=-\frac{14}{36} \text{ and } \frac{5}{12}=\frac{5\cdot 3}{12\cdot 3}=\frac{15}{36}$$

Step 3 Add the numerators.

$$-\frac{7}{18}+\frac{5}{12}=-\frac{14}{36}+\frac{15}{36}=\frac{-14+15}{36}=\frac{1}{36}$$

Step 4 $\frac{1}{36}$ is in lowest terms.

Now Try:

4. Find each sum or difference.

a. $\frac{5}{9}+\frac{7}{18}$

____

b. $\frac{8}{15}-\frac{7}{10}$

____

c. $-\frac{5}{24}+\frac{7}{9}$

____

Name: Date:
Instructor: Section:

d. $5-\frac{3}{4}$

Step 1 The LCD for $\frac{5}{1}$ and $\frac{3}{4}$ is 4, the larger denominator.

Step 2 $\frac{5}{1}=\frac{5\cdot 4}{1\cdot 4}=\frac{20}{4}$ and $\frac{3}{4}$ already has the LCD.

Step 3 Subtract the numerators.

$$\frac{5}{1}-\frac{3}{4}=\frac{20}{4}-\frac{3}{4}=\frac{20-3}{4}=\frac{17}{4}$$

Step 4 $\frac{17}{4}$ is in lowest terms.

d. $9-\frac{2}{7}$ ____________

Review these examples for Objective 4:

5. Find each sum or difference.

a. $\frac{1}{3}+\frac{x}{2}$

Step 1 The LCD is 6.

Step 2 $\frac{1}{3}=\frac{1\cdot 2}{3\cdot 2}=\frac{2}{6}$ and $\frac{x}{2}=\frac{x\cdot 3}{2\cdot 3}=\frac{3x}{6}$

Step 3 $\frac{1}{3}+\frac{x}{2}=\frac{2}{6}+\frac{3x}{6}=\frac{2+3x}{6}$

Step 4 $\frac{2+3x}{6}$ is in lowest terms.

b. $\frac{5}{6}-\frac{7}{x}$

Step 1 The LCD is $6x$.

Step 2 $\frac{5}{6}=\frac{5\cdot x}{6\cdot x}=\frac{5x}{6x}$ and $\frac{7}{x}=\frac{7\cdot 6}{x\cdot 6}=\frac{42}{6x}$

Step 3 $\frac{5}{6}-\frac{7}{x}=\frac{5x}{6x}-\frac{42}{6x}=\frac{5x-42}{6x}$

Step 4 $\frac{5x-42}{6x}$ is in lowest terms.

Now Try:

5. Find each sum or difference.

a. $\frac{2}{3}+\frac{x}{5}$ ____________

b. $\frac{5}{8}-\frac{8}{x}$ ____________

Objective 1 Add and subtract like fractions.

For extra help, see Example 1 on page 250 of your text and Section Lecture video for Section 4.4 and Exercise Solutions Clip 13.

Write each sum or difference in lowest terms.

1. $-\frac{14}{15}+\frac{4}{15}$

1. ____________

Name: Date:
Instructor: Section:

2. $\frac{3}{4a}+\frac{1}{4a}$ **2.** ________________

3. $\frac{7}{y^2}-\frac{3}{y^2}$ **3.** ________________

Objective 2 Find the lowest common denominator for unlike fractions.

For extra help, see Examples 2–3 on page 252 of your text and Section Lecture video for Section 4.4.

Find the LCD for each pair of fractions.

4. $\frac{3}{7}$ and $\frac{3}{14}$ **4.** ________________

5. $\frac{8}{21}$ and $\frac{8}{9}$ **5.** ________________

6. $\frac{7}{12}$ and $\frac{3}{40}$ **6.** ________________

Objective 3 Add and subtract unlike fractions.

For extra help, see Example 4 on pages 253–254 of your text and Section Lecture video for Section 4.4 and Exercise Solutions Clip 11, 17, and 19.

Find each sum or difference. Write all answers in lowest terms.

7. $\frac{1}{6}+\frac{2}{15}$ **7.** ________________

8. $-\frac{1}{2}+\frac{7}{12}$ **8.** ________________

9. $\frac{33}{40}-\frac{7}{24}$ **9.** ________________

Name: Date:
Instructor: Section:

Objective 4 Add and subtract unlike fractions that contain variables.

For extra help, see Example 5 on pages 254–255 of your text and Section Lecture video for Section 4.4 and Exercise Solutions Clip 21 and 23.

Find each sum or difference. Write all answers in lowest terms.

10. $\frac{3}{n}+\frac{3}{5}$ **10.** ________________

11. $\frac{1}{6}+\frac{x}{3}$ **11.** ________________

12. $\frac{3}{a}-\frac{b}{4}$ **12.** ________________

Name: Date:
Instructor: Section:

Chapter 4 RATIONAL NUMBERS: POSITIVE AND NEGATIVE FRACTIONS

4.5 Problem Solving: Mixed Numbers and Estimating

Learning Objectives	
1	Identify mixed numbers and graph them on a number line.
2	Rewrite mixed numbers as improper fractions, or the reverse.
3	Estimate the answer and multiply or divide mixed numbers.
4	Estimate the answer and add or subtract mixed numbers.
5	Solve application problems containing mixed numbers.

Key Terms

Use the vocabulary terms listed below to complete each statement in exercises 1–2.

mixed number **improper fraction**

1. A(n) ____________________ includes a fraction and a whole number written together.

2. A mixed number can be rewritten as a(n) ____________________.

Guided Examples

Review this example for Objective 1:

1. Use a number line to show mixed numbers $\frac{5}{2}$ and $-\frac{5}{2}$.

$\frac{5}{2}$ is equivalent to $2\frac{1}{2}$.

$-\frac{5}{2}$ is equivalent to $-2\frac{1}{2}$.

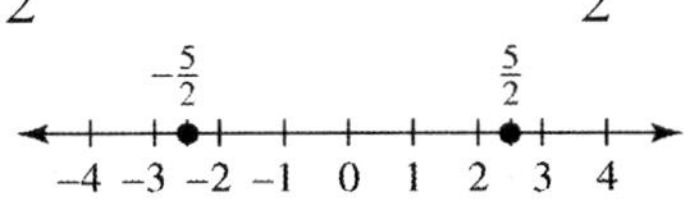

Now Try:

1. Use a number line to show mixed numbers $1\frac{1}{4}$ and $-1\frac{1}{4}$.

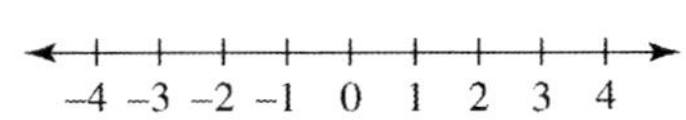

Review these examples for Objective 2:

2. Write $9\frac{3}{4}$ as an improper fraction.

Step 1 $9\frac{3}{4}$ $4 \cdot 9 = 36$ Then $36 + 3 = 39$

Step 2 $9\frac{3}{4} = \frac{39}{4}$

Now Try:

2. Write $5\frac{5}{6}$ as an improper fraction.

____

3. Write each improper fraction as an equivalent mixed number in simplest form.

a. $\frac{18}{7}$

Divide 18 by 7.

$$\begin{array}{r} 2 \\ 7\overline{)18} \\ \underline{14} \\ 4 \end{array}$$

The quotient 2 is the whole number part. The remainder 4 is the numerator of the fraction, and the denominator stays as 7.

$$\frac{18}{7} = 2\frac{4}{7}$$

b. $\frac{22}{6}$

Divide 22 by 6.

$$\begin{array}{r} 3 \\ 6\overline{)22} \\ \underline{18} \\ 4 \end{array} \quad \text{so } \frac{22}{6} = 3\frac{4}{6} = 3\frac{2}{3}$$

Or, write $\frac{22}{6}$ in lowest terms first.

$$\frac{22}{6} = \frac{\overset{1}{\cancel{2}} \cdot 11}{\underset{1}{\cancel{2}} \cdot 3} = \frac{11}{3} \quad \text{Then } \begin{array}{r} 3 \\ 3\overline{)11} \\ \underline{9} \\ 2 \end{array} \quad \text{so } \frac{11}{3} = 3\frac{2}{3}$$

3. Write each improper fraction as an equivalent mixed number in simplest form.

a. $\frac{27}{8}$ __________

b. $\frac{37}{9}$ __________

Review these examples for Objective 3:

4. Round each mixed number to the nearest whole number.

a. $1\frac{7}{12}$

$1\frac{7}{12}$ ← 7 is more than 6
← Half of 12 is 6

$1\frac{7}{12}$ rounds up to 2.

b. $8\frac{3}{7}$

$8\frac{3}{7}$ ← 3 is less than $3\frac{1}{2}$
← Half of 7 is $3\frac{1}{2}$

$8\frac{3}{7}$ rounds down to 8.

Now Try:

4. Round each mixed number to the nearest whole number.

a. $6\frac{11}{20}$ __________

b. $4\frac{13}{28}$ __________

Name: Date:
Instructor: Section:

5. First, round the numbers and estimate each answer. Then find the exact answer. Write exact answers in simplest form.

a. $2\frac{1}{3}\cdot 4\frac{5}{7}$

Estimate the answer by rounding the mixed numbers.

$2\frac{1}{3}$ rounds to 2 and $4\frac{5}{7}$ rounds to 5

$2\cdot 5 = 10 \leftarrow$ Estimated answer

To find the exact answer, first rewrite each mixed number as an improper fraction.

$2\frac{1}{3}=\frac{7}{3}$ and $4\frac{5}{7}=\frac{33}{7}$

Next, multiply.

$$2\frac{1}{3}\cdot 4\frac{5}{7}=\frac{7}{3}\cdot\frac{33}{7}=\frac{\overset{1}{\cancel{7}}\cdot\overset{1}{\cancel{3}}\cdot 11}{\underset{1}{\cancel{3}}\cdot\underset{1}{\cancel{7}}}=\frac{11}{1}=11$$

The estimate was 10, so an exact answer of 11 is reasonable.

b. $5\frac{2}{5}\cdot 1\frac{1}{9}$

First, round each mixed number and estimate the answer.

$5\frac{2}{5}$ rounds to 5 and $1\frac{1}{9}$ rounds to 1

$5\cdot 1 = 5 \leftarrow$ Estimated answer

Now find the exact answer.

$5\frac{2}{5}=\frac{27}{5}$ and $1\frac{1}{9}=\frac{10}{9}$

$$5\frac{2}{5}\cdot 1\frac{1}{9}=\frac{27}{5}\cdot\frac{10}{9}=\frac{\overset{1}{\cancel{9}}\cdot 3\cdot 2\cdot\overset{1}{\cancel{5}}}{\underset{1}{\cancel{5}}\cdot\underset{1}{\cancel{9}}}=\frac{6}{1}=6$$

The estimate was 5, so an exact answer of 6 is reasonable.

5. First, round the numbers and estimate each answer. Then find the exact answer. Write exact answers in simplest form.

a. $4\frac{9}{10}\cdot 3\frac{4}{7}$

Estimate ____________

Exact ____________

b. $6\frac{1}{8}\cdot 5\frac{5}{7}$

Estimate ____________

Exact ____________

Name: Date:
Instructor: Section:

6. First, round the numbers and estimate each answer. Then find the exact answer. Write exact answers in simplest form.

a. $3\frac{1}{5} \div 2\frac{2}{3}$

Estimate the answer by rounding the mixed numbers.

$3\frac{1}{5}$ rounds to 3 and $2\frac{2}{3}$ rounds to 3

$3 \div 3 = 1 \leftarrow$ Estimated answer

To find the exact answer, first rewrite each mixed number as an improper fraction.

$3\frac{1}{5} \div 2\frac{2}{3} = \frac{16}{5} \div \frac{8}{3}$

Now rewrite the problem as multiplying by the reciprocal of $\frac{8}{3}$.

$$\frac{16}{5} \div \frac{8}{3} = \frac{16}{5} \cdot \frac{3}{8} = \frac{\overset{1}{\cancel{8}} \cdot 2 \cdot 3}{5 \cdot \underset{1}{\cancel{8}}} = \frac{6}{5} = 1\frac{1}{5}$$

The estimate was 1, so an exact answer of $1\frac{1}{5}$ is reasonable.

b. $7\frac{5}{7} \div 9$

First, round the numbers and estimate the answer.

$7\frac{5}{7}$ rounds to 8

$8 \div 9 = \frac{8}{9} \leftarrow$ Estimated answer

Now find the exact answer.

$$7\frac{5}{7} \div 9 = \frac{54}{7} \div \frac{9}{1} = \frac{54}{7} \cdot \frac{1}{9} = \frac{\overset{1}{\cancel{9}} \cdot 6}{7 \cdot \underset{1}{\cancel{9}}} = \frac{6}{7}$$

The estimate was $\frac{8}{9}$, so an exact answer of $\frac{6}{7}$ is reasonable.

6. First, round the numbers and estimate each answer. Then find the exact answer. Write exact answers in simplest form.

a. $4\frac{4}{9} \div 6\frac{2}{3}$

Estimate ______________

Exact ______________

b. $6\frac{7}{8} \div 11$

Estimate ______________

Exact ______________

Name: Date:
Instructor: Section:

Review these examples for Objective 4:

7. First, estimate each answer. Then add or subtract to find the exact answer. Write exact answers in simplest form.

a. $3\frac{3}{10}+4\frac{4}{5}$

To estimate the answer, round each mixed number to the nearest whole number.

$3\frac{3}{10}+4\frac{4}{5}$

$3 \;+\; 5=8 \leftarrow$ Estimate

To find the exact answer, first rewrite each mixed number as an equivalent improper fraction.

$$3\frac{3}{10}+4\frac{4}{5}=\frac{33}{10}+\frac{24}{5}$$

The LCD for $\frac{33}{10}$ and $\frac{24}{5}$ is 10. Rewrite $\frac{24}{5}$ as an equivalent fraction with a denominator of 10.

$$\frac{33}{10}+\frac{24}{5}=\frac{33}{10}+\frac{48}{10}=\frac{33+48}{10}=\frac{81}{10}=8\frac{1}{10}$$

The estimate was 8, so an exact answer of $8\frac{1}{10}$ is reasonable.

b. $5\frac{1}{4}-3\frac{5}{6}$

Round each number and estimate the answer.

$5\frac{1}{4}-3\frac{5}{6}$

$5 \;-\; 4=1 \leftarrow$ Estimate

To find the exact answer, rewrite the mixed numbers as improper fractions and subtract.

$$5\frac{1}{4}-3\frac{5}{6}=\frac{21}{4}-\frac{23}{6}=\frac{63}{12}-\frac{46}{12}$$
$$=\frac{63-46}{12}=\frac{17}{12}=1\frac{5}{12}$$

The estimate was 1, so an exact answer of $1\frac{5}{12}$ is reasonable.

Now Try:

7. First, estimate each answer. Then add or subtract to find the exact answer. Write exact answers in simplest form.

a. $1\frac{2}{9}+5\frac{1}{3}$

Estimate ______________

Exact ______________

b. $12\frac{5}{12}-8\frac{5}{6}$

Estimate ______________

Exact ______________

c. $7-3\frac{2}{5}$

$7-3\frac{2}{5}$

$7-\ 3=4 \leftarrow$ Estimate

$7-3\frac{2}{5}=\frac{7}{1}-\frac{17}{5}=\frac{35}{5}-\frac{17}{5}$

$=\frac{35-17}{5}=\frac{18}{5}=3\frac{3}{5}$

The estimate was 4, so an exact answer of $3\frac{3}{5}$ is reasonable.

c. $9-4\frac{1}{7}$

Estimate ____________

Exact ____________

Review these examples for Objective 5:

8. First, estimate the answer to each application problem. Then find the exact answer.

a. Thomas worked $25\frac{1}{2}$ hours over the last four days. If he worked the same amount each day, how long was he at work each day?

First, round each mixed number to the nearest whole number.

$25\frac{1}{2}$ rounds to 26

Using the rounded numbers, we divide.

$26\div 4=6\frac{1}{2}\leftarrow$ Estimate

To find the exact answer, use the original mixed numbers and divide.

$25\frac{1}{2}\div 4=\frac{51}{2}\div\frac{4}{1}=\frac{51}{2}\cdot\frac{1}{4}=\frac{51\cdot 1}{2\cdot 4}=\frac{51}{8}=6\frac{3}{8}$

Thomas worked $6\frac{3}{8}$ hr each day. This is close to the estimate of $6\frac{1}{2}$ hr.

b. George's daughter grew $1\frac{1}{3}$ inches last year and $2\frac{1}{5}$ inches this year. How much has her height increased over the two years?

First, round each mixed number to the nearest whole number.

$1\frac{1}{3}$ rounds to 1 and $2\frac{1}{5}$ rounds to 2

Using the rounded numbers, we add.

$1+2=3\leftarrow$ Estimate

Now Try:

8. First, estimate the answer to each application problem. Then find the exact answer.

a. Katlyn used $2\frac{3}{8}$ packages of nuts in her salad recipe. Each package has $3\frac{1}{6}$ ounces of nuts. How many ounces of nuts did she use in the recipe?

Estimate ____________

Exact ____________

b. A plumber has three pieces of pipe measuring $2\frac{1}{5}$ ft, $3\frac{3}{4}$ ft, and $4\frac{1}{8}$ ft. What is the total length of pipe?

Estimate ____________

Exact ____________

Name: Date:
Instructor: Section:

To find the exact answer, use the original mixed numbers and add.

$$1\frac{1}{3}+2\frac{1}{5}=\frac{4}{3}+\frac{11}{5}=\frac{20}{15}+\frac{33}{15}=\frac{20+33}{15}$$

$$=\frac{53}{15}=3\frac{8}{15}$$

Her height increased $3\frac{8}{15}$ in. over the two years.

This result is close to the estimate of 3 in.

Objective 1 Identify mixed numbers and graph them on a number line.

For extra help, see Example 1 on page 262 of your text and Section Lecture video for Section 4.5 and Exercise Solutions Clip 1.

Graph the mixed numbers or improper fractions on a number line.

1. $-3\frac{1}{2}$ and $3\frac{1}{2}$

1.

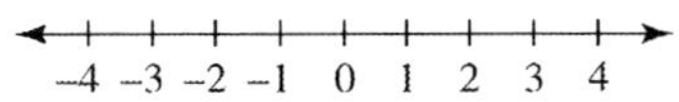

2. $\frac{10}{3}$ and $-\frac{10}{3}$

2.

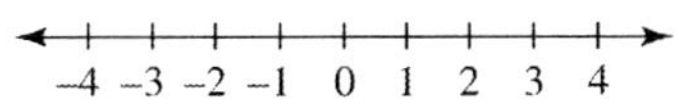

3. $-\frac{11}{4}$ and $\frac{11}{4}$

3.

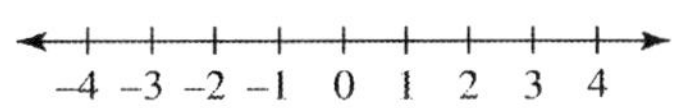

Objective 2 Rewrite mixed numbers as improper fractions, or the reverse.

For extra help, see Examples 2–3 on pages 263–265 of your text and Section Lecture video for Section 4.5 and Exercise Solutions Clip 23.

Write each mixed number as an improper fraction.

4. $-8\frac{2}{7}$ **4.** ________________

5. $-1\frac{7}{9}$ **5.** ________________

Write the improper fraction as a mixed number in simplest form.

6. $\frac{26}{3}$ **6.** ________________

Name: Date:
Instructor: Section:

Objective 3 Estimate the answer and multiply or divide mixed numbers.

For extra help, see Examples 4–6 on page 265–267 of your text and Section Lecture video for Section 4.5 and Exercise Solutions Clip 25 and 27.

First, round the mixed numbers to the nearest whole number and estimate each answer; then find the exact answer. Write exact answers in simplest form. Write each mixed number as an improper fraction.

7. $3\frac{2}{3} \cdot 1\frac{1}{7}$

7.
Estimate____________
Exact ____________

8. $3\frac{3}{4} \div 1\frac{3}{8}$

8.
Estimate____________
Exact ____________

9. $5\frac{1}{3} \div 2\frac{2}{5}$

9.
Estimate____________
Exact ____________

Objective 4 Estimate the answer and add or subtract mixed numbers.

For extra help, see Example 7 on pages 268–269 of your text and Section Lecture video for Section 4.5 and Exercise Solutions Clip 29 and 31.

First, round the mixed numbers to the nearest whole number and estimate each answer; then find the exact answer. Write exact answers in simplest form. Write each mixed number as an improper fraction.

10. $3\frac{5}{6} + 2\frac{3}{4}$

10.
Estimate____________
Exact ____________

11. $9-5\frac{4}{9}$

11.
Estimate_____________
Exact ______________

12. $9\frac{7}{18}-3\frac{5}{12}$

12.
Estimate_____________
Exact ______________

Objective 5 Solve application problems containing mixed numbers.

For extra help, see Example 8 on page 270 of your text and Section Lecture video for Section 4.5 and Exercise Solutions Clip 49.

First, estimate the answer to each application problem. Then find the exact answer.

13. A living room has dimensions $3\frac{3}{4}$ meters by $3\frac{1}{3}$ meters. What is the area of the room?

13.
Estimate_____________
Exact ______________

14. Suppose that a pair of pants requires $3\frac{1}{8}$ yd of material. How much material would be needed for 6 pairs of pants?

14.
Estimate_____________
Exact ______________

Name: Date:
Instructor: Section:

Chapter 4 RATIONAL NUMBERS: POSITIVE AND NEGATIVE FRACTIONS

4.6 Exponents, Order of Operations, and Complex Fractions

Learning Objectives	
1	Simplify fractions with exponents.
2	Use the order of operations with fractions.
3	Simplify complex fractions.

Key Terms

Use the vocabulary terms listed below to complete each statement in exercises 1–3.

complex fraction **exponent** **order of operations**

1. For problems or expressions with more than one operation, the ____________________________ tells what to do first, second, and so on, to obtain the correct answer.

2. An ______________________ tells how many times a number is used as a factor in repeated multiplication.

3. A ____________________________ is a fraction in which the numerator and/or denominator contain one or more fractions.

Guided Examples

Review these examples for Objective 1:

1. Simplify.

a. $\left(-\frac{1}{5}\right)^3$

$$\left(-\frac{1}{5}\right)^3=\left(-\frac{1}{5}\right)\left(-\frac{1}{5}\right)\left(-\frac{1}{5}\right)=\frac{1}{25}\left(-\frac{1}{5}\right)=-\frac{1}{125}$$

b. $\left(\frac{4}{5}\right)^3\left(\frac{5}{8}\right)^2$

$$\left(\frac{4}{5}\right)^3\left(\frac{5}{8}\right)^2=\left(\frac{4}{5}\right)\left(\frac{4}{5}\right)\left(\frac{4}{5}\right)\left(\frac{5}{8}\right)\left(\frac{5}{8}\right)$$

$$=\frac{\overset{1}{\cancel{2}}\cdot\overset{1}{\cancel{2}}\cdot\overset{1}{\cancel{4}}\cdot\overset{1}{\cancel{4}}\cdot\overset{1}{\cancel{5}}\cdot\overset{1}{\cancel{5}}}{\underset{1}{\cancel{5}}\cdot\underset{1}{\cancel{5}}\cdot 5\cdot\underset{1}{\cancel{2}}\cdot\underset{1}{\cancel{4}}\cdot\underset{1}{\cancel{2}}\cdot\underset{1}{\cancel{4}}}$$

$$=\frac{1}{5}$$

Now Try:

1. Simplify.

a. $\left(-\frac{1}{4}\right)^3$

____

b. $\left(\frac{5}{7}\right)^3\left(\frac{7}{10}\right)^2$

____

Name: Date:
Instructor: Section:

Review these examples for Objective 2:

2. Simplify.

a. $-\frac{1}{2}+\frac{2}{3}\left(\frac{4}{5}\right)$

$$-\frac{1}{2}+\frac{2}{3}\left(\frac{4}{5}\right)=-\frac{1}{2}+\frac{2\cdot 4}{3\cdot 5}=-\frac{1}{2}+\frac{8}{15}$$
$$=-\frac{15}{30}+\frac{16}{30}=\frac{-15+16}{30}$$
$$=\frac{1}{30}$$

b. $-4+\left(\frac{5}{6}-\frac{2}{3}\right)^2$

$$-4+\left(\frac{5}{6}-\frac{2}{3}\right)^2=-4+\left(\frac{5}{6}-\frac{4}{6}\right)^2$$
$$=-4+\left(\frac{5-4}{6}\right)^2=-4+\left(\frac{1}{6}\right)^2$$
$$=-4+\left(\frac{1}{36}\right)$$
$$=\frac{-144+1}{36}=-\frac{143}{36}$$

Now Try:

2. Simplify.

a. $-\frac{1}{6}+\frac{2}{3}\left(\frac{4}{5}\right)$ ____________

b. $-1+\left(\frac{4}{9}-\frac{2}{3}\right)^2$ ____________

Review these examples for Objective 3:

3. Simplify.

a. $\dfrac{-\frac{6}{7}}{-\frac{5}{14}}$

Rewrite the complex fraction using the ÷ symbol for division. Then follow the order of operations.

$$\frac{-\frac{6}{7}}{-\frac{5}{14}}=-\frac{6}{7}\div-\frac{5}{14}=-\frac{6}{7}\cdot-\frac{14}{5}$$
$$=\frac{6\cdot 2\cdot \overset{1}{\cancel{7}}}{\underset{1}{\cancel{7}}\cdot 5}=\frac{12}{5},\text{ or } 2\frac{2}{5}$$

b. $\dfrac{\left(\frac{5}{6}\right)^2}{10}$

Rewrite the complex fraction using the ÷

Now Try:

3. Simplify.

a. $\dfrac{-\frac{5}{9}}{-\frac{7}{12}}$ ____________

b. $\dfrac{\left(\frac{3}{5}\right)^2}{9}$ ____________

symbol for division. Then follow the order of operations.

$$\frac{\left(\frac{5}{6}\right)^2}{10}=\left(\frac{5}{6}\right)^2\div 10=\frac{25}{36}\div 10=\frac{25}{36}\cdot\frac{1}{10}$$

$$=\frac{\overset{1}{\cancel{5}}\cdot 5\cdot 1}{36\cdot 2\cdot \underset{1}{\cancel{5}}}=\frac{5}{72}$$

Objective 1 Simplify fractions with exponents.

For extra help, see Example 1 on page 277 of your text and Section Lecture video for Section 4.6 and Exercise Solutions Clip 7 and 13.

Simplify.

1. $3\left(-\frac{1}{3}\right)^3$ **1.** ______________

2. $\left(-\frac{2}{3}\right)^3\left(-\frac{3}{2}\right)^2$ **2.** ______________

3. $\left(-\frac{2}{3}\right)^2\left(\frac{1}{2}\right)^4$ **3.** ______________

Objective 2 Use the order of operations with fractions.

For extra help, see Example 2 on page 278 of your text and Section Lecture video for Section 4.6 and Exercise Solutions Clip 23 and 33.

Simplify.

4. $\frac{15}{16}-\frac{1}{8}-\left(\frac{3}{4}\right)^2$ **4.** ______________

5. $\left(\frac{3}{2}\right)^2-\frac{3}{10}\div\frac{1}{5}$ **5.** ______________

6. $\left(\frac{1}{3}\right)^2-\left(-\frac{1}{2}\right)^3$ **6.** ______________

Name: Date:
Instructor: Section:

Objective 3 Simplify complex fractions.

For extra help, see Example 3 on page 279 of your text and Section Lecture video for Section 4.6 and Exercise Solutions Clip 49.

Simplify.

7. $\dfrac{\frac{16}{35}}{-\frac{4}{70}}$

7. ________________

8. $\dfrac{-20}{\frac{2}{5}}$

8. ________________

9. $\dfrac{\left(\frac{2}{3}\right)^2}{\left(-\frac{4}{5}\right)^2}$

9. ________________

Name: Date:
Instructor: Section:

Chapter 4 RATIONAL NUMBERS: POSITIVE AND NEGATIVE FRACTIONS

4.7 Problem Solving: Equations Containing Fractions

Learning Objectives

1	Use the multiplication property of equality to solve equations containing fractions.
2	Use both the addition and the multiplication properties of equality to solve equations containing fractions.
3	Solve application problems using equations containing fractions.

Key Terms

Use the vocabulary terms listed below to complete each statement in exercises 1–2.

multiplication property of equality

division property of equality

1. The ______________________________ states that both sides of an equation may be divided by the same nonzero number and it will still be balanced.

2. The ______________________________ states that both sides of an equation may be multiplied by the same nonzero number and it will still be balanced.

Guided Examples

Review these examples for Objective 1:

1. Solve each equation and check each solution.

a. $\frac{1}{5}b = 12$

Multiply both sides by $\frac{5}{1}$.

$$\frac{1}{5}b = 12$$
$$\frac{5}{1}\left(\frac{1}{5}b\right) = \frac{5}{1}(12)$$
$$\left(\frac{5}{1}\cdot\frac{1}{5}\right)b = \frac{5}{1}\left(\frac{12}{1}\right)$$
$$1b = \frac{60}{1}$$
$$b = 60$$

Now Try:

1. Solve each equation and check each solution.

a. $\frac{1}{7}b = 3$

____

Name: Date:
Instructor: Section:

Check $\frac{1}{5}b=12$

$\frac{1}{5}(60)=12$

$\frac{1\cdot \cancel{5}^{1}\cdot 12}{\cancel{5}_{1}}=12$

$12=12$

The solution is 60.

b. $15=-\frac{3}{5}x$

Multiply both sides by $-\frac{5}{3}$.

$-\frac{5}{3}(15)=-\frac{5}{3}\left(-\frac{3}{5}x\right)$

$-25=x$

Check $15=-\frac{3}{5}x$

$15=-\frac{3}{5}(-25)$

$15=\frac{3\cdot \cancel{5}^{1}\cdot 5}{\cancel{5}_{1}}$

$15=15$

The solution is –25.

c. $-\frac{5}{6}n=-\frac{1}{4}$

$-\frac{6}{5}\left(-\frac{5}{6}n\right)=-\frac{6}{5}\left(-\frac{1}{4}\right)$

$n=\frac{2\cdot 3}{5\cdot 2\cdot 2}$

$n=\frac{3}{10}$

Check $-\frac{5}{6}n=-\frac{1}{4}$

$-\frac{5}{6}\left(\frac{3}{10}\right)=-\frac{1}{4}$

$-\frac{\cancel{5}^{1}\cdot \cancel{3}^{1}}{2\cdot \cancel{3}_{1}\cdot 2\cdot \cancel{5}_{1}}=-\frac{1}{4}$

$-\frac{1}{4}=-\frac{1}{4}$

The solution is $\frac{3}{10}$.

b. $25=-\frac{5}{6}x$

____

c. $-\frac{7}{9}n=-\frac{14}{45}$

____

Name: Date:
Instructor: Section:

Review these examples for Objective 2:

2. Solve each equation and check each solution.

a. $\frac{1}{5}c+2=5$

Add the opposite of 2, which is –2, to both sides.

$$\begin{aligned} \frac{1}{5}c+2 &= 5 \\ -2 &\quad -2 \\ \frac{1}{5}c+0 &= 3 \\ \frac{1}{5}c &= 3 \\ \frac{5}{1}\left(\frac{1}{5}c\right) &= \frac{5}{1}\left(\frac{3}{1}\right) \\ c &= 15 \end{aligned}$$

Check $\frac{1}{5}c+2=5$

$$\begin{aligned} \frac{1}{5}(15)+2 &= 5 \\ 3+2 &= 5 \\ 5 &= 5 \end{aligned}$$

The solution is 15.

b. $-5=\frac{7}{8}y+9$

To get the variable term by itself, add –9 to both sides.

$$\begin{aligned} -5 &= \frac{7}{8}y+9 \\ -9 &\qquad -9 \\ -14 &= \frac{7}{8}y+0 \\ \frac{8}{7}(-14) &= \frac{8}{7}\left(\frac{7}{8}y\right) \\ -16 &= y \end{aligned}$$

Check $-5=\frac{7}{8}y+9$

$$\begin{aligned} -5 &= \frac{7}{8}(-16)+9 \\ -5 &= -14+9 \\ -5 &= -5 \end{aligned}$$

The solution is –16.

Now Try:

2. Solve each equation and check each solution.

a. $\frac{1}{6}c+3=8$

____

b. $-9=\frac{5}{7}y+6$

____

Name: Date:
Instructor: Section:

Review this example for Objective 3:

3. An expression for the recommended weight of an adult is $\frac{11}{2}$(height in inches) – 220.

A man weighs 187 pounds. What is his height in inches?

Step 1 Read the problem. It is about weight and height.

Step 2 Assign a variable. Let h represent the height.

Step 3 Write an equation.

$$\frac{11}{2}(\text{height in inches}) - 220 \text{ is weight}$$

$$\frac{11}{2}h - 220 = 187$$

Step 4 Solve.

$$\frac{11}{2}h - 220 = 187$$

$$\underline{+220} \quad \underline{+220}$$

$$\frac{11}{2}h + 0 = 407$$

$$\frac{2}{11}\left(\frac{11}{2}h\right) = \frac{2}{11}(407)$$

$$h = 74$$

Step 5 State the answer. The man is 74 inches, or 6 ft 2 in.

Step 6 Check.
If the height is 74 inches, then find the weight.

$$\frac{11}{2}(74) - 220 = 407 - 220 = 187$$

The answer checks.

Now Try:

3. An expression for the recommended weight of an adult is $\frac{11}{2}$(height in inches) – 220.

A man weighs 220 pounds. What is his height in inches?

____

Objective 1 Use the multiplication property of equality to solve equations containing fractions.

For extra help, see Example 1 on pages 284–286 of your text and Section Lecture video for Section 4.7 and Exercise Solutions Clip 7.

Solve each equation and check each solution.

1. $-30 = \frac{5}{6}b$

1. ____________

Name: Date:
Instructor: Section:

2. $\frac{2}{9}n = 18$ **2.** ________________

3. $-\frac{8}{9}h = -\frac{1}{6}$ **3.** ________________

Objective 2 Use both the addition and the multiplication properties of equality to solve equations containing fractions.

For extra help, see Example 2 on pages 286–287 of your text and Section Lecture video for Section 4.7 and Exercise Solutions Clip 19 and 21.

Solve each equation and check each solution.

4. $\frac{1}{5}s - 15 = -10$ **4.** ________________

5. $-8 = \frac{5}{2}r + 2$ **5.** ________________

6. $-\frac{1}{6}n + 8 = -4$ **6.** ________________

Objective 3 Solve application problems using equations containing fractions.

For extra help, see Example 3 on page 288 of your text and Section Lecture video for Section 4.7.

In Exercises 7–9, find each person's age using the six problem-solving steps and this expression for approximate systolic blood pressure: $100 + \frac{age}{2}$. *Assume that all the people have normal blood pressure.*

7. A man has a systolic blood pressure of 110. How old is he? **7.** ________________

8. A woman has a systolic blood pressure of 126. How old is she? **8.** ________________

9. A man has a systolic blood pressure of 115. How old is he? **9.** ________________

Name: Date:
Instructor: Section:

Chapter 4 RATIONAL NUMBERS: POSITIVE AND NEGATIVE FRACTIONS

4.8 Geometry Applications: Area and Volume

Learning Objectives	
1	Find the area of triangle.
2	Find the volume of a rectangular solid.
3	Find the volume of a pyramid.

Key Terms

Use the vocabulary terms listed below to complete each statement in exercises 1–2.

volume **area**

1. ______________________________ is measured in square units.

2. ______________________________ is measured in cubic units.

Guided Examples

Review these examples for Objective 1:

1. Find the area of each triangle.

a.

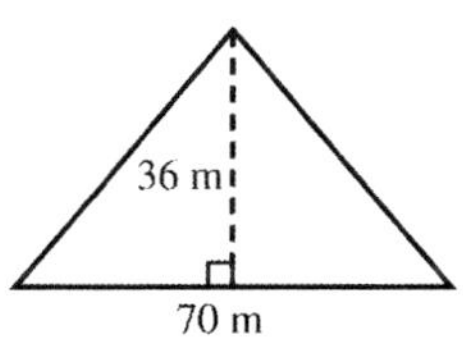

The base is 70 m and the height is 36 m.

$A = \frac{1}{2} \cdot b \cdot h$

$A = \frac{1}{\not{2}_1} \cdot \not{70}^{35} \text{ m} \cdot 36 \text{ m}$

$A = 1260 \text{ m}^2$

b.

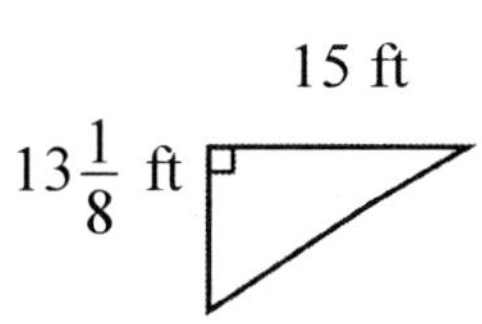

Two sides of the triangle are perpendicular to each other, so use those sides as the base and the height.

Now Try:

1. Find the area of each triangle.

a.

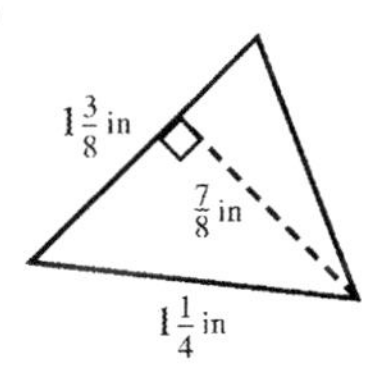

____

b.

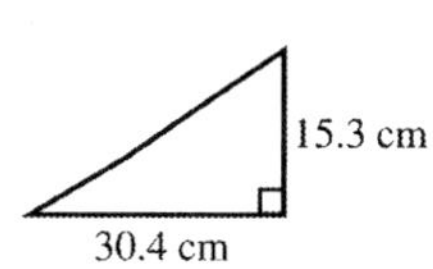

____

Name: Date:
Instructor: Section:

$$A = \frac{1}{2} \cdot 15 \text{ ft} \cdot 13\frac{1}{8} \text{ ft}$$

$$A = \frac{1}{2} \cdot \frac{15 \text{ ft}}{1} \cdot \frac{105 \text{ ft}}{8}$$

$$A = 98\frac{7}{16} \text{ ft}^2$$

2. Find the area of the shaded part of this figure.

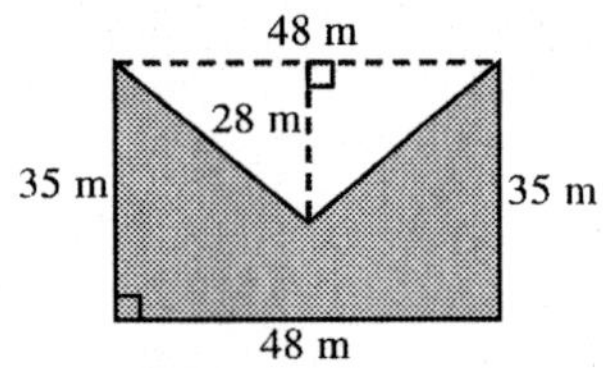

The entire figure is a rectangle. Find the area of the rectangle.

$$A = l \cdot w$$

$$A = 48 \text{ m} \cdot 35 \text{ m}$$

$$A = 1680 \text{ m}^2$$

The unshaded part is a triangle. Find the area of the triangle.

$$A = \frac{1}{2} \cdot b \cdot h$$

$$A = \frac{1}{\cancel{2}_1} \cdot \cancel{48}^{24} \text{ m} \cdot 28 \text{ m}$$

$$A = 672 \text{ m}^2$$

Subtract to find the area of the shaded part.

$$A = \underbrace{1680 \text{ m}^2}_{\text{Entire area}} - \underbrace{672 \text{ m}^2}_{\text{Unshaded part}} = \underbrace{1008 \text{ m}^2}_{\text{Shaded part}}$$

The area of the shaded part is 1008 m^2.

2. Find the area of the shaded part of this figure.

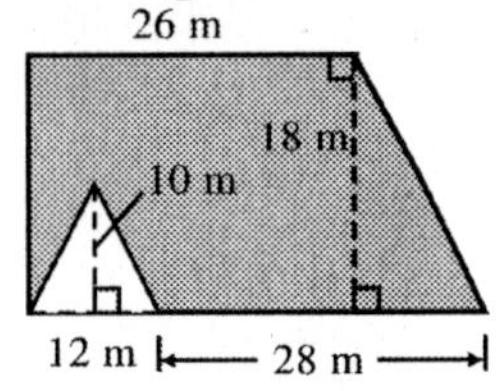

____

Review these examples for Objective 2:

3. Find the volume of each box.

a.

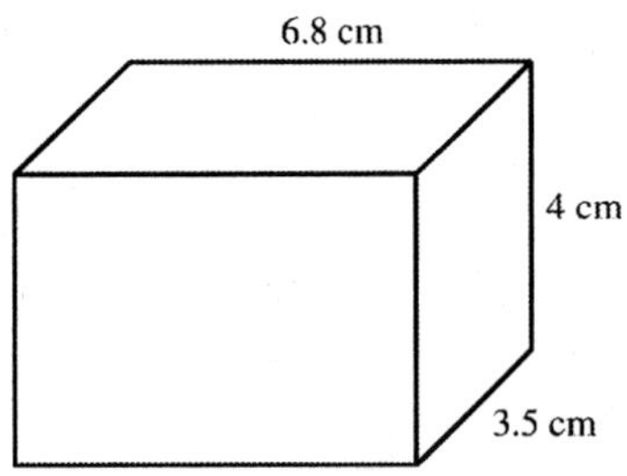

Use the formula $V = l \cdot w \cdot h$.

$$V = 6.8 \text{ cm} \cdot 3.5 \text{ cm} \cdot 4 \text{ cm}$$

$$V = 95.2 \text{ cm}^3$$

Now Try:

3. Find the volume of each box.

a.

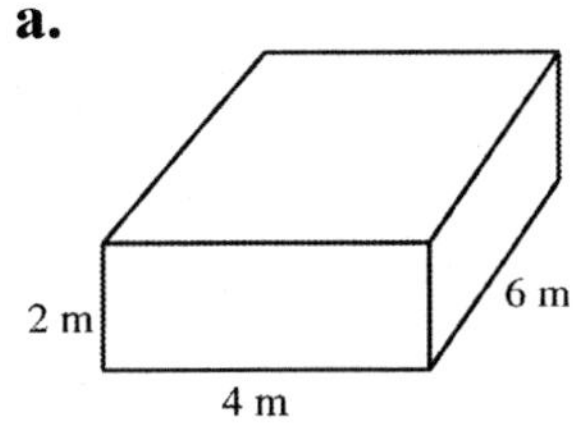

____

b.

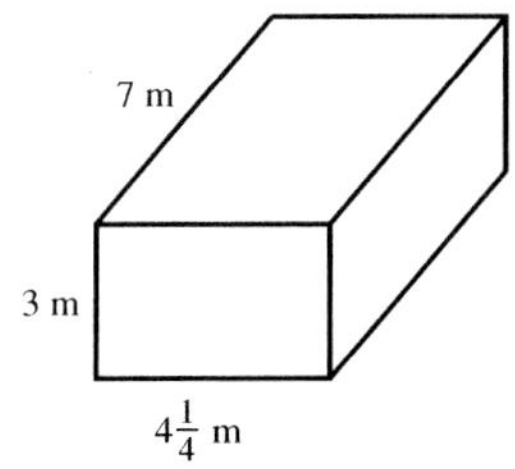

Use the formula $V = l \cdot w \cdot h$.

$$V = 7\text{ m} \cdot 4\frac{1}{4}\text{ m} \cdot 3\text{ m}$$

$$V = \frac{7\text{ m}}{1} \cdot \frac{17\text{ m}}{4} \cdot \frac{3\text{ m}}{1}$$

$$V = \frac{7\text{ m} \cdot 17\text{ m} \cdot 3\text{ m}}{4}$$

$$V = \frac{357}{4}\text{ m}^3\text{, or } 89\frac{1}{4}\text{ m}^3$$

b.

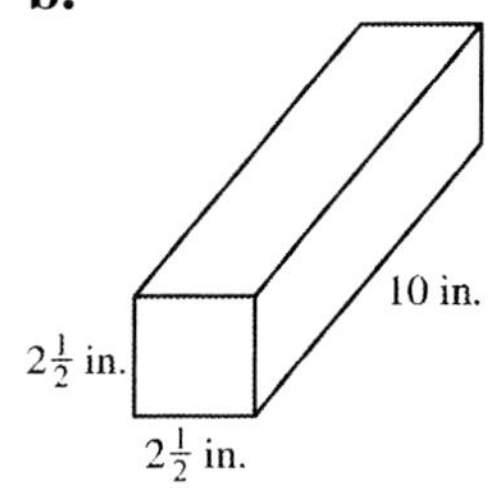

____

Review this example for Objective 3:

4. Find the volume of this pyramid with rectangular base. Round your answer to the nearest tenth.

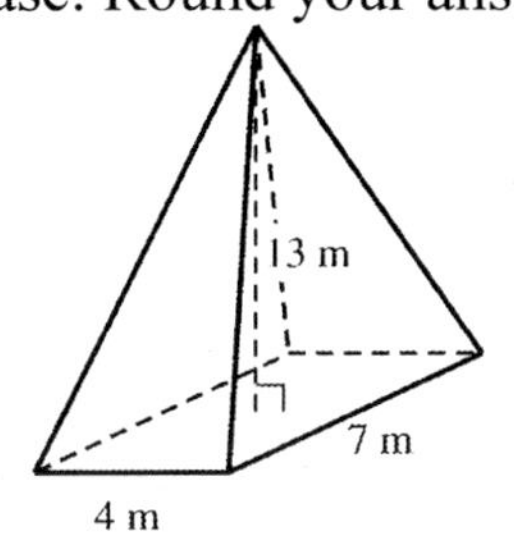

First find the value of B in the formula, which is the area of a rectangular base. Recall that the area of the rectangle is found by multiplying length times width.

$$B = 7\text{ m} \cdot 4\text{ m}$$

$$B = 28\text{ m}^2$$

Now find the volume.

$$V = \frac{B \cdot h}{3}$$

$$V \approx \frac{28\text{ m}^2 \cdot 13\text{ m}}{3}$$

$$V \approx 121.3\text{ m}^3$$

Now Try:

4. Find the volume of this pyramid with rectangular base. Round your answer to the nearest tenth.

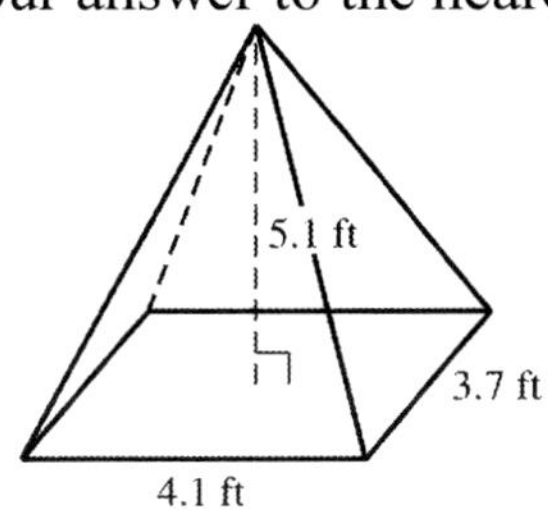

____

Name: Date:
Instructor: Section:

Objective 1 Find the area of triangle.

For extra help, see Examples 1–2 on pages 294–295 of your text and Section Lecture video for Section 4.8 and Exercise Solutions Clip 5.

Find the perimeter and area of each triangle.

1.

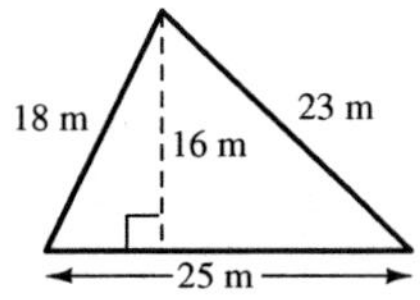

1. Perimeter________

Area ____________

2.

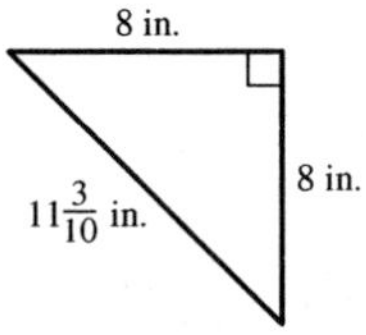

2. Perimeter________

Area ____________

Find the shaded area in the figure.

3.

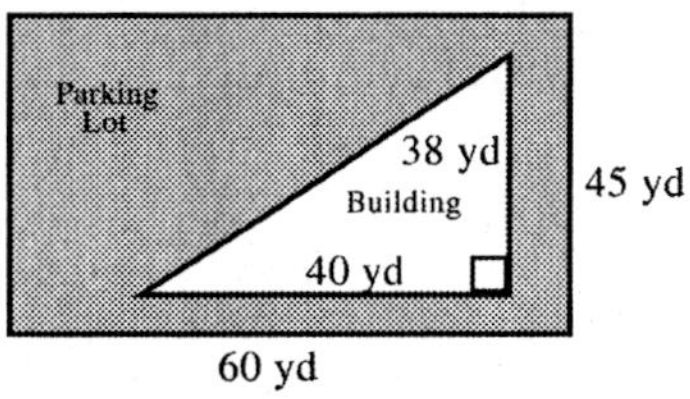

3. ________________

Objective 2 Find the volume of a rectangular solid.

For extra help, see Example 3 on page 296 of your text and Section Lecture video for Section 4.8 and Exercise Solutions Clip 17.

Find the volume of each figure.

4.

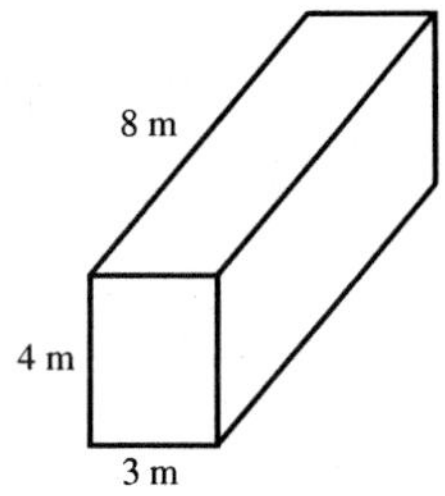

4. ________________

5.

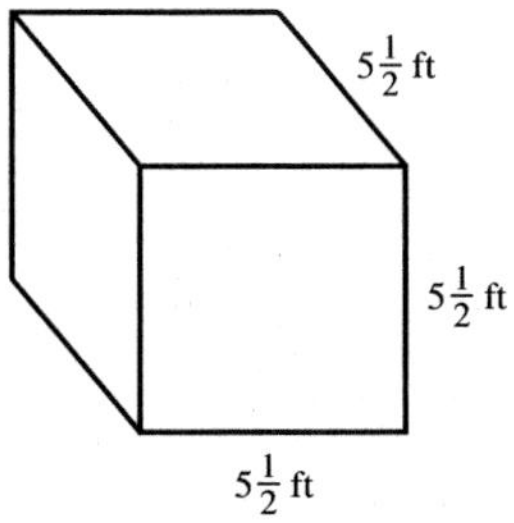

5. ________________

Name: Date:
Instructor: Section:

6.

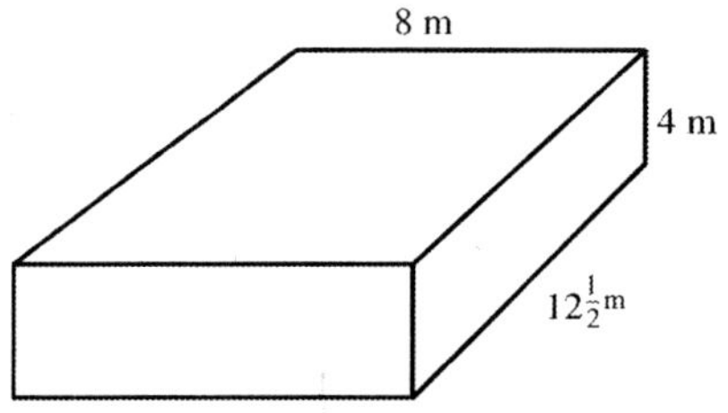

6. ______________

Objective 3 Find the volume of a pyramid.

For extra help, see Example 4 on page 297 of your text and Section Lecture video for Section 4.8 and Exercise Solutions Clip 21 and 27.

Find the volume of each figure.

7.

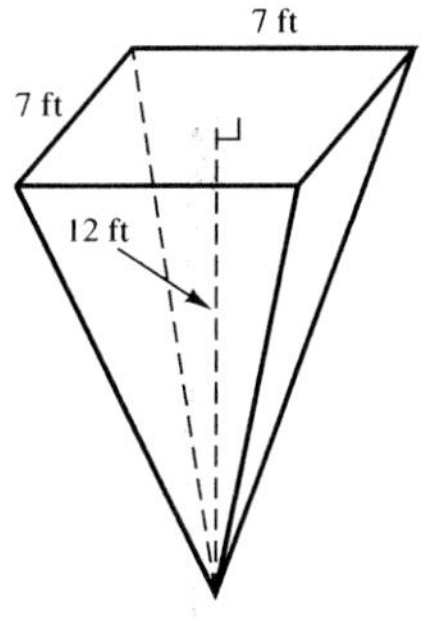

7. ______________

8.

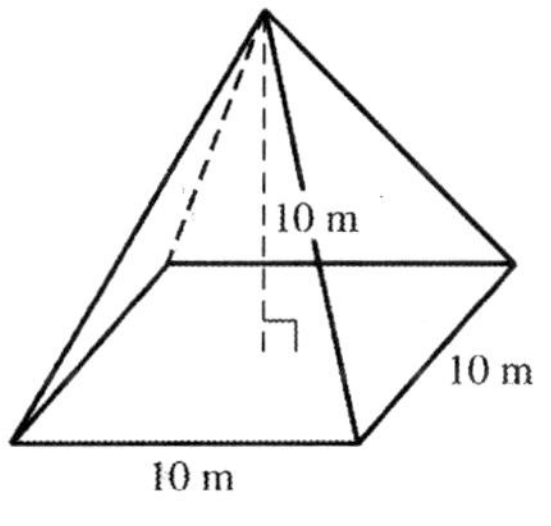

8. ______________

9.

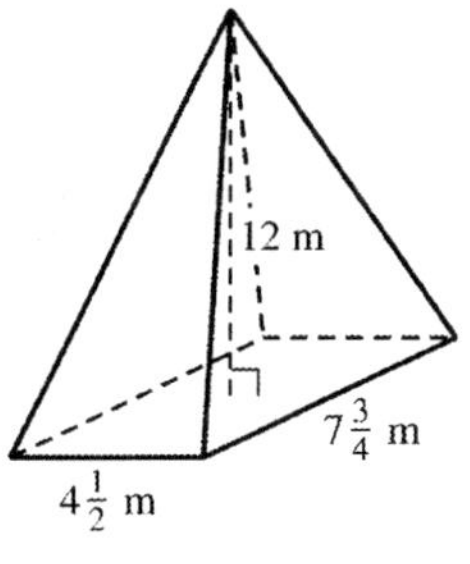

9. ______________

Name: Date:
Instructor: Section:

Chapter 5 RATIONAL NUMBERS: POSITIVE AND NEGATIVE DECIMALS

5.1 Reading and Writing Decimals

Learning Objectives

1. Write parts of a whole using decimals.
2. Identify the place value of a digit.
3. Read decimal numbers.
4. Write decimals as fractions or mixed numbers.

Key Terms

Use the vocabulary terms listed below to complete each statement in exercises 1−3.

decimals **decimal point** **place value**

1. We use ______________________ to show parts of a whole.
2. A ___________________________ is assigned to each place to the left or right of the decimal point.
3. The dot that separates the whole number part from the fractional part of a decimal number is called the ___________________________________.

Guided Examples

Review these examples for Objective 1:

1. Given a fraction and how it is read, write as a decimal.

a. $\frac{7}{10}$ seven tenths

Decimal: 0.7

b. $\frac{3}{100}$ three hundredths

Decimal: 0.03

c. $\frac{63}{100}$ sixty-three hundredths

Decimal: 0.63

Now Try:

1. Given a fraction and how it is read, write as a decimal.

a. $\frac{6}{10}$ six tenths

____

b. $\frac{5}{100}$ five hundredths

____

c. $\frac{37}{100}$ thirty-seven hundredths

____

d. $\frac{9}{1000}$ nine thousandths

Decimal: 0.009

e. $\frac{56}{1000}$ fifty-six thousandths

Decimal: 0.056

f. $\frac{943}{1000}$ nine hundred forty-three thousandths

Decimal: 0.943

d. $\frac{7}{1000}$ seven thousandths

____

e. $\frac{49}{1000}$ forty-nine thousandths

____

f. $\frac{518}{1000}$ five hundred eighteen thousandths

____

Review these examples for Objective 2:

2. Identify the place value of each digit.

a. 486.92

4 hundreds
8 tens
6 ones
.
9 tenths
2 hundredths

b. 0.00465

0 ones
.
0 tenths
0 hundredths
4 thousandths
6 ten-thousandths
5 hundred-thousandths

Now Try:

2. Identify the place value of each digit.

a. 862.93

____

b. 0.00769

____

Review these examples for Objective 3:

3. Tell how to read each decimal in words.

a. 0.7

Read it as: seven tenths

b. 0.62

Read it as: sixty-two hundredths

c. 0.09

Read it as: nine hundredths

Now Try:

3. Tell how to read each decimal in words.

a. 0.9

____

b. 0.53

____

c. 0.07

____

Name: Date:
Instructor: Section:

d. 0.352

Read it as: three hundred fifty-two thousandths

e. 0.0205

Read it as: two hundred five ten-thousandths

4. Read each decimal.

a. 17.8

seventeen and eight tenths

b. 543.71

five hundred forty-three and seventy-one hundredths

c. 0.089

eighty-nine thousandths

d. 12.4053

twelve and four thousand fifty-three ten-thousandths

d. 0.502 ____________

e. 0.0469 ____________

4. Read each decimal.

a. 4.7 ____________

b. 18.009 ____________

c. 0.0082 ____________

d. 57.906 ____________

Review these examples for Objective 4:

5. Write each decimal as a fraction or mixed number.

a. 0.16

The digits to the right of the decimal point, 16, are the numerator of the fraction. The denominator is 100 for hundredths because the rightmost digit is in the hundredths place.

$$0.16 = \frac{16}{100}$$

b. 0.739

The rightmost digit is in the thousandths place.

$$0.739 = \frac{739}{1000}$$

c. 5.0088

The whole number part stays the same. The rightmost digit is in the ten-thousandths place.

$$5.0088 = 5\frac{88}{10,000}$$

Now Try:

5. Write each decimal as a fraction or mixed number.

a. 0.27 ____________

b. 0.303 ____________

c. 3.2636 ____________

Name: Date:
Instructor: Section:

6. Write each decimal as a fraction or mixed number in lowest terms.

a. 0.6

$0.6 = \frac{6}{10}$ Write $\frac{6}{10}$ in lowest terms.

$\frac{6}{10} = \frac{6 \div 2}{10 \div 2} = \frac{3}{5}$

b. 0.88

$0.88 = \frac{88}{100} = \frac{88 \div 4}{100 \div 4} = \frac{22}{25}$

c. 17.216

$17.216 = 17\frac{216}{1000} = 17\frac{216 \div 8}{1000 \div 8} = 17\frac{27}{125}$

d. 53.6065

$53.6065 = 53\frac{6065}{10{,}000} = 53\frac{6065 \div 5}{10{,}000 \div 5} = 53\frac{1213}{2000}$

6. Write each decimal as a fraction or mixed number in lowest terms.

a. 0.2 ____________

b. 0.35 ____________

c. 6.04 ____________

d. 562.0404 ____________

Objective 1 Write parts of a whole using decimals.

For extra help, see Example 1 on page 317 of your text and Section Lecture video for Section 5.1.

Write the portion of each square that is shaded as a fraction, as a decimal, and in words.

1. 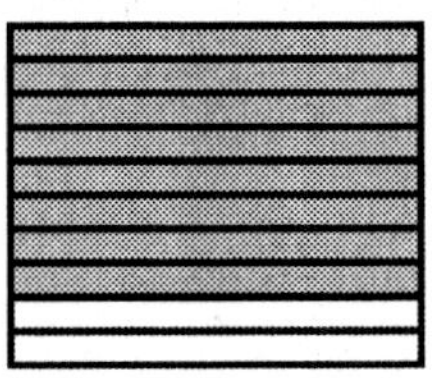

2. 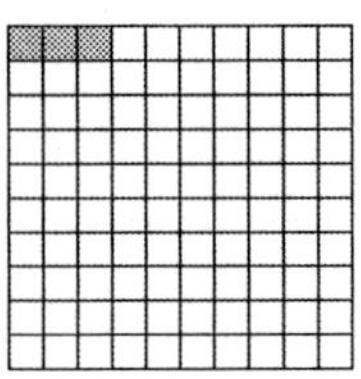

3. 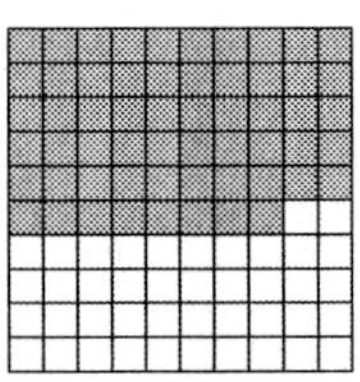

1. ____________
____
____

2. ____________
____
____

3. ____________
____
____

Name: Date:
Instructor: Section:

Objective 2 Identify the place value of a digit.

For extra help, see Example 2 on page 318 of your text and Section Lecture video for Section 5.1 and Exercise Solutions Clip 7 and 15.

Identify the digit that has the given place value.

4. 43.507 tenths **4.** ________________

hundredths ________________

5. 2.83714 thousandths **5.** ________________

ten-thousandths ________________

Identify the place value of each digit in these decimals.

6. 37.082 3 **6.** ________________

7 ________________

0 ________________

8 ________________

2 ________________

Objective 3 Read decimal numbers.

For extra help, see Examples 3–4 on pages 318–319 of your text and Section Lecture video for Section 5.1 and Exercise Solutions Clip 41 and 43.

Tell how to read each decimal in words.

7. 0.08 **7.** ________________

8. 10.835 **8.** ________________

9. 97.008 **9.** ________________

Objective 4 Write decimals as fractions or mixed numbers.

For extra help, see Examples 5–6 on page 320 of your text and Section Lecture video for Section 5.1 and Exercise Solutions Clip 21 and 29.

Write each decimal as a fraction or mixed number in lowest terms.

10. 0.001 **10.** ________________

11. 3.6 **11.** ________________

12. 0.95 **12.** ________________

Name: Date:
Instructor: Section:

Chapter 5 RATIONAL NUMBERS: POSITIVE AND NEGATIVE DECIMALS

5.2 Rounding Decimal Numbers

Learning Objectives

1 Learn the rules for rounding decimals.
2 Round decimals to any given place.
3 Round money amounts to the nearest cent or nearest dollar.

Key Terms

Use the vocabulary terms listed below to complete each statement in exercises 1–2.

rounding **decimal places**

1. ____________________ are the number of digits to the right of the decimal point.

2. When we "cut off" a number after a certain place value, we are ______________ that number.

Guided Examples

Review these examples for Objective 2:

1. Round 16.87453 to the nearest thousandth.

Step 1 Draw a "cut-off" line after the thousandths place.
16.874 | 53

Step 2 Look only at the first digit you are cutting off. Ignore the other digits you are cutting off.
16.874 | 53 (Ignore the 3)

Step 3 If the first digit you are cutting off is 5 or more, round up the part of the number you are keeping.

$$\begin{array}{r} 16.874 \mid 53 \\ +\ 0.001 \\ \hline 16.875 \end{array}$$

So, 16.87453 rounded to the nearest thousandth is 16.875. We can write $16.87453 \approx 16.875$.

2. Round to the place indicated.

a. 6.4387 to the nearest tenth

Step 1 Draw a cut-off line after the tenths place.
6.4 | 387

Step 2 Look only at the 3.
6.4 | 387 (Ignore the 8 and 7)

Now Try:

1. Round 43.80290 to the nearest thousandth.

____

2. Round to the place indicated.

a. 0.7976 to the nearest hundredth

____

Name: Date:
Instructor: Section:

Step 3 The first digit is 4 or less, so the part you are keeping stays the same.

$\underbrace{6.4}$ | 387

6.4

Rounding 6.4387 to the nearest tenth is 6.4. We can write $6.4387 \approx 6.4$.

b. 0.79846 to the nearest hundredth

Step 1 Draw a cut-off line after the hundredths place.

0.79 | 846

Step 2 Look only at the 8.

0.79 | 846

Step 3 The first digit is 5 or more, so round up by adding 1 hundredth to the part you are keeping.

$$\begin{array}{r} \overset{1}{0.79} \mid 846 \\ +\ 0.01 \\ \hline 0.80 \end{array}$$

0.79846 rounded to the nearest hundredth is 0.80. We can write $0.79846 \approx 0.80$.

c. 0.02709 to the nearest thousandth

0.027 | 09

The first digit cut is 4 or less, so the part you are keeping stays the same.

0.02709 rounded to the nearest thousandth is 0.027. We can write $0.02709 \approx 0.027$.

d. 64.983 to the nearest tenth

64.9 | 83

The first digit cut is 5 or more, so round up by adding 1 tenth to the part you are keeping.

$$\begin{array}{r} \overset{1}{64.9} \mid 83 \\ +\ 0.1 \\ \hline 65.0 \end{array}$$

64.983 rounded to the nearest tenth is 65.0. We can write $64.983 \approx 65.0$. You must write the 0 in the tenths place to show that the number was rounded to the nearest tenth.

b. 7.7804 to the nearest hundredth

____

c. 22.0397 to the nearest thousandth

____

d. 0.649 to the nearest tenth

____

Name: Date:
Instructor: Section:

Review these examples for Objective 3:

3. Round each money amount to the nearest cent.

a. \$6.5348

Is \$6.5348 closer to \$6.53 or to \$6.54?
The first digit cut is 4 or less, so the part you are keeping stays the same.

\$6.53 | 48

You pay \$6.53.

b. \$0.895

Is \$0.895 closer to \$0.89 or \$0.90?

\$0.89 | 5

The first digit cut is 5 or more, so round up.

$$\begin{array}{r} \overset{1}{\$0.89} \mid 5 \\ +\ \$0.01 \\ \hline \$0.90 \end{array}$$

You pay \$0.90.

4. Round to the nearest dollar.

a. \$56.76

Draw a cut-off line after the ones place.

\$56|.76

First digit cut is 5 or more, so round up by adding \$1.

$$\begin{array}{r} \$56 \mid .76 \\ +\quad 1 \\ \hline \$57 \end{array}$$

So, \$56.76 rounded to the nearest dollar is \$57.

b. \$697.41

Draw a cut-off line after the ones place.

\$697|.41

First digit cut is 4 or less, so the part you keep stays the same.
So, \$697.41 rounded to the nearest dollar is \$697.

c. \$599.66

Draw a cut-off line after the ones place.

\$599|.66

First digit cut is 5 or more, so round up by adding \$1.

Now Try:

3. Round each money amount to the nearest cent.

a. \$2.0849

____

b. \$425.0954

____

4. Round to the nearest dollar.

a. \$37.81

____

b. \$307.20

____

c. \$880.83

____

Name: Date:
Instructor: Section:

$599 | .66
+ 1
$600

So, $599.66 rounded to the nearest dollar is $600.

d. $5779.50

$5779 | .50

First digit cut is 5 or more, so round up by adding $1.

$5779 | .50
+ 1
$5780

$5779.50 rounded to the nearest dollar is $5780.

e. $0.53

$0 | .53

First digit cut is 5 or more, so round up.
$0.53 rounded to the nearest dollar is $1.

d. $6859.77

____

e. $0.61

____

Objective 1 Learn the rules for rounding decimals.

For extra help, see page 325 of your text and Section Lecture video for Section 5.2.

Select the phrase that makes the sentence correct.

1. When rounding a number to the nearest tenth, if the digit in the hundredths place is 5 or more, round the digit in the tenths place (up/down). **1.** ______________

2. When rounding a number to the nearest hundredth, look at the digit in the (tenth/thousandth) place. **2.** ______________

Objective 2 Round decimals to any given place.

For extra help, see Examples 1–2 on pages 325–327 of your text and Section Lecture video for Section 5.2 and Exercise Solutions Clip 7, 13, and 17.

Round each number to the place indicated.

3. 489.84 to the nearest tenth **3.** ______________

4. 54.4029 to the nearest hundredth **4.** ______________

5. 989.98982 to the nearest thousandth **5.** ______________

Name: Date:
Instructor: Section:

Objective 3 Round money amounts to the nearest cent or nearest dollar.

For extra help, see Examples 3–4 on pages 328–329 of your text and Section Lecture video for Section 5.2 and Exercise Solutions Clip 21, 23, 31, and 33.

Round to the nearest dollar.

6. \$28.39 **6.** ________________

7. \$11,839.73 **7.** ________________

Round to the nearest cent.

8. \$1028.6666 **8.** ________________

Name: Date:
Instructor: Section:

Chapter 5 RATIONAL NUMBERS: POSITIVE AND NEGATIVE DECIMALS

5.3 Adding and Subtracting Signed Decimal Numbers

Learning Objectives
1 Add and subtract positive decimals.
2 Add and subtract negative decimals.
3 Estimate the answer when adding or subtracting decimals.

Key Terms

Use the vocabulary terms listed below to complete each statement in exercises 1–2.

estimating **front end rounding**

1. With ______________________________, we round to the highest possible place.

2. Avoid common errors in working decimal problems by ____________________ the answer first.

Guided Examples

Review these examples for Objective 1:

1. Find each sum.

a. 29.73 and 56.84

Step 1 Write the numbers in columns with the decimal points lined up.

$$\begin{array}{r} 29.73 \\ +56.84 \\ \hline \end{array}$$

Step 2 Add as if these were whole numbers.
Step 3 Line up decimal point in answer under the decimal points in problem.

$$\begin{array}{r} {\scriptstyle 1\ 1} \\ 29.73 \\ +56.84 \\ \hline 86.57 \end{array}$$

b. 8.437 + 5.361 + 13.295

Write the numbers vertically with decimal points lined up. Then add.

$$\begin{array}{r} {\scriptstyle 1\ 1\ \ 1\ 1} \\ 8.437 \\ 5.361 \\ +13.295 \\ \hline 27.093 \end{array}$$

Now Try:

1. Find each sum.

a. 12.687 + 2.943

____

b. 0.428 + 16.005 + 5.276

____

Name: Date:
Instructor: Section:

2. Find each sum.

a. 6.7 + 0.41

There are two decimal places in 0.41, so write a 0 in the hundredths place in 6.7 so it has two decimal places also.

$$\begin{array}{r} 6.70 \\ +0.41 \\ \hline 7.11 \end{array}$$

b. 12 + 9.36 + 3.754

Write in zeros so that all the addends have three decimal places.

$$\begin{array}{r} 12.000 \\ 9.360 \\ +\ \ 3.754 \\ \hline 25.114 \end{array}$$

3. Find each difference.

a. 14.32 from 36.74

Step 1 Line up decimal points.

$$\begin{array}{r} 36.74 \\ -14.32 \\ \hline \end{array}$$

Step 2 Both numbers have two decimal places; no need to write in zeros.
Step 3 Line up decimal point in answer.

$$\begin{array}{r} 36.74 \\ -14.32 \\ \hline 22.42 \end{array}$$

b. 167.53 minus 69.85

Regrouping is needed here.

$$\begin{array}{r} {\scriptstyle 0\ 15\ 16\ \ 14\ 13} \\ \not{1}\,\not{6}\,\not{7}\,.\,\not{5}\,\not{3} \\ -\quad 6\ 9\ .\ 8\ 5 \\ \hline 9\ 7\ .\ 6\ 8 \end{array}$$

Line up decimal point in answer.

4. Find each difference.

a. 19.6 from 38.264

Line up decimal points and write in zeros so both numbers have three decimal places.

$$\begin{array}{r} 38.264 \\ -19.600 \\ \hline 18.664 \end{array}$$

2. Find each sum.

a. 7.53 + 29.314

____

b. 0.631 + 999.3 + 14

____

3. Find each difference.

a. 7.352 from 18.964

____

b. 50.43 minus 39.86

____

4. Find each difference.

a. 3.87 from 8.524

____

Name: Date:
Instructor: Section:

b. 28.8 – 19.963

Write in two zeros and subtract as usual.

$$\begin{array}{r} 28.800 \\ -\,19.963 \\ \hline 8.837 \end{array}$$

c. 16 less 7.54

Write a decimal point and two zeros after 16. Subtract as usual.

$$\begin{array}{r} 16.00 \\ -\quad 7.54 \\ \hline 8.46 \end{array}$$

b. 20 – 16.74

____

c. 1 less 0.499

____

Review these examples for Objective 2:

5. Find each sum.

a. $-5.8+(-21)$

Both addends are negative, so the sum will be negative. To begin, $|-5.8|$ is 5.8, and $|-21|$ is 21. Then add the absolute values.

$$\begin{array}{r} 5.8 \\ +\,21.0 \\ \hline 26.8 \end{array} \leftarrow \text{Write in a decimal point and one 0}$$

$-5.8+(-21)=-26.8$

b. $-4.36+0.973$

The addends have different signs. To begin, $|-4.36|$ is 4.36, and $|0.973|$ is 0.973. Then subtract the lesser absolute value from the greater.

$$\begin{array}{r} 4.360 \\ -\,0.973 \\ \hline 3.387 \end{array} \leftarrow \text{Write one 0}$$

$-4.36+0.973=-3.387$

6. Find each difference.

a. $5.4-13.89$

Rewrite subtractions as adding the opposite.

$5.4-13.89$

$5.4+(-13.89)$

–13.89 has the greater absolute value and is negative, so the answer will be negative.

$5.4+(-13.89)=-8.49$

$$\begin{array}{r} 13.89 \\ -\quad 5.40 \\ \hline 8.49 \end{array}$$

Now Try:

5. Find each sum.

a. $-6.4+(-32)$

____

b. $-13.57+2.984$

____

6. Find each difference.

a. $7.6-19.34$

____

Name: Date:
Instructor: Section:

b. $-2.74-(-5.9)$

Rewrite subtractions as adding the opposite.

$-2.74-(-5.9)$

$-2.74+5.9$

5.9 has the greater absolute value and is positive, so the answer will be positive.

$-2.74-(-5.9)=3.16$

$$\begin{array}{r} 5.90 \\ -\ 2.74 \\ \hline 3.16 \end{array}$$

c. $17.4-(1.58+0.63)$

Work the parentheses first.

$17.4-(1.58+0.63)$

$17.4- \quad (2.21)$

$17.4+ \ (-2.21)$

15.19

b. $-5.29-(-8.6)$

____

c. $21.8-(4.37+0.89)$

____

Review these examples for Objective 3:

7. Use front end rounding to round each number. Then add or subtract the rounded numbers to get an estimate answer. Finally, find the exact answer.

a. Find the sum of 295.8 and 7.894.

Estimate: *Exact:*

$$\begin{array}{rcr} 300 & \xleftarrow{\text{Rounds to}} & 295.800 \\ +\ 8 & \xleftarrow{\text{Rounds to}} & +\ 7.894 \\ \hline 308 & & 303.694 \end{array}$$

The estimate goes out to the hundreds place and so does the exact answer. Therefore, the decimal point is probably in the correct place in the exact answer.

b. Subtract \$16.95 from \$78.23.

Estimate: *Exact:*

$$\begin{array}{rcr} \$80 & \xleftarrow{\text{Rounds to}} & \$78.23 \\ -\ 20 & \xleftarrow{\text{Rounds to}} & -\ 16.95 \\ \hline \$60 & & \$61.28 \end{array}$$

Now Try:

7. Use front end rounding to round each number. Then add or subtract the rounded numbers to get an estimate answer. Finally, find the exact answer.

a. Find the sum of 5.74 and 8.107.

____

____

b. Subtract \$23.49 from \$65.04.

____

____

c. –2.753 – 8.4

Rewrite the subtraction as adding the opposite. Then use front end rounding to get an estimated answer.

$$-2.753-8.4$$
$$-2.753+(-8.4)$$
$$\text{Rounded} \rightarrow \quad -3 \quad + \quad (-8) = -11$$

To find the exact answer, add the absolute values.

$$\begin{array}{r} 2.753 \\ +\ 8.400 \\ \hline 11.153 \end{array}$$

The answer will be negative because both numbers are negative.

–2.753 – 8.4 = –11.153

The exact answer –11.153 is reasonable because it is close to the estimated answer of –11.

c. –1.497 – 9.6

____

____

Objective 1 Add and subtract positive decimals.

For extra help, see Examples 1–4 on pages 332–334 of your text and Section Lecture video for Section 5.3 and Exercise Solutions Clip 7, 11, and 23.

Find each sum or difference.

1. 43.96 + 48.53 **1.** ____________

2. 69.524 – 26.958 **2.** ____________

3. 45.83 + 20.923 + 5.7 **3.** ____________

Objective 2 Add and subtract negative decimals.

For extra help, see Examples 5–6 on pages 335–336 of your text and Section Lecture video for Section 5.3 and Exercise Solutions Clip 31 and 33.

Find each difference.

4. 18.1 – 84.6 **4.** ____________

Name: Date:
Instructor: Section:

5. $87.6 - (-90.4)$ **5.** ________________

6. $1.71 - 12.68$ **6.** ________________

Objective 3 Estimate the answer when adding or subtracting decimals.

For extra help, see Example 7 on pages 336–337 of your text and Section Lecture video for Section 5.3.

First, use front end rounding and estimate each answer. Then add or subtract to find the exact answer.

7.
$$\begin{array}{r} 593.8 \\ 27.93 \\ +\ 54.87 \\ \hline \end{array}$$

7.
Estimate______________

Exact ______________

8. $7.69 - 20.85$

8.
Estimate______________

Exact ______________

9. $-9.7 - 4.862$

9.
Estimate______________

Exact ______________

Name: Date:
Instructor: Section:

Chapter 5 RATIONAL NUMBERS: POSITIVE AND NEGATIVE DECIMALS

5.4 Multiplying Signed Decimal Numbers

Learning Objectives
1 Multiply positive and negative decimals.
2 Estimate the answer when multiplying decimals.

Key Terms

Use the vocabulary terms listed below to complete each statement in exercises 1–3.

decimal places **factor** **product**

1. Each number in a multiplication problem is called a ____________________.

2. When multiplying decimal numbers, first multiply the numbers, then find the total number of ______________ in both factors.

3. The answer to a multiplication problem is called the _________________.

Guided Examples

Review this example for Objective 1:

1. Find the product of 6.23 and –5.4.

Step 1 Multiply the numbers as if they were whole numbers.

$$\begin{array}{r} 6.\,2\,3 \\ \times \quad 5.\,4 \\ \hline 2\,4\,9\,2 \\ 3\,1\,1\,5 \\ \hline 3\,3\,6\,4\,2 \end{array}$$

Step 2 Count the total number of decimal places in both factors.

$$\begin{array}{rl} 6.\,2\,3 & \leftarrow 2 \text{ decimal places} \\ \times \quad 5.\,4 & \leftarrow 1 \text{ decimal place} \\ \hline 2\,4\,9\,2 & \;\; 3 \text{ total decimal places} \\ 3\,1\,1\,5 & \\ \hline 3\,3\,6\,4\,2 & \end{array}$$

Step 3 Count over 3 places in the product and write the decimal point. Count from right to left.

Now Try:

1. Find the product of 2.51 and –4.3.

____

$$\begin{array}{r} 6.23 \\ \times \quad 5.4 \\ \hline 2492 \\ 3115 \\ \hline 33.642 \end{array} \begin{array}{l} \leftarrow 2 \text{ decimal places} \\ \leftarrow 1 \text{ decimal place} \\ \hline 3 \text{ total decimal places} \end{array}$$

Step 4 The factors have different signs, so the product is negative. 6.23 times –5.4 is –33.642.

2. Find the product: $(-0.035)(-0.07)$.

Start by multiply, then count decimal places.

$$\begin{array}{r} 0.035 \\ \times \quad 0.07 \\ \hline 245 \end{array} \begin{array}{l} \leftarrow 3 \text{ decimal places} \\ \leftarrow 2 \text{ decimal places} \\ \hline 5 \text{ total decimal places} \end{array}$$

After multiplying, the answer has only three decimal places, but five are needed, so write two zeros on the left side of the answer. Then count over 5 places and write in the decimal point.

$$\begin{array}{r} 0.035 \\ \times \quad 0.07 \\ \hline .00245 \end{array} \begin{array}{l} \leftarrow 3 \text{ decimal places} \\ \leftarrow 2 \text{ decimal places} \\ \hline 5 \text{ total decimal places} \end{array}$$

The final product is 0.00245, which has five decimal places. The product is positive because the factors have the same sign.

2. Find the product $(-0.062)(-0.03)$.

____

Review this example for Objective 2:

3. First estimate the answer to $(74.56)(18.9)$ using front end rounding. Then find the exact answer.

Estimate:

$$\begin{array}{r} 70 \\ \times\ 20 \\ \hline 1400 \end{array} \begin{array}{l} \leftarrow \text{Rounds to} \\ \leftarrow \text{Rounds to} \end{array}$$

Exact:

$$\begin{array}{r} 74.56 \\ \times \quad 18.9 \\ \hline 67104 \\ 59648 \\ 7456 \\ \hline 1409.184 \end{array} \begin{array}{l} \leftarrow 2 \text{ decimal places} \\ \leftarrow 1 \text{ decimal place} \\ \hline 3 \text{ total decimal places} \end{array}$$

Both the estimate and the exact answer go out to the thousands place, so the decimal point in 1409.184 is probably in the correct place.

Now Try:

3. First estimate the answer to $(22.43)(5.03)$ using front end rounding. Then find the exact answer.

____

Name: Date:
Instructor: Section:

Objective 1 Multiply positive and negative decimals.

For extra help, see Examples 1–2 on pages 342–343 of your text and Section Lecture video for Section 5.4 and Exercise Solutions Clip 5, 7, and 19.

Find each product.

1. $\begin{array}{r} -19.3 \\ \times\ 4.7 \\ \hline \end{array}$ **1.** ________________

2. $\begin{array}{r} 0.682 \\ \times\ \ 3.9 \\ \hline \end{array}$ **2.** ________________

3. $(-0.074)(-0.05)$ **3.** ________________

Objective 2 Estimate the answer when multiplying decimals.

For extra help, see Example 3 on page 344 of your text and Section Lecture video for Section 5.4 and Exercise Solutions Clip 23.

First use front-end rounding and estimate the answer. Then multiply to find the exact answer.

4. $\begin{array}{r} 29.8 \\ \times\ 3.4 \\ \hline \end{array}$

4.
Estimate____________
Exact ____________

5. $\begin{array}{r} 32.53 \\ \times\ 23.26 \\ \hline \end{array}$

5.
Estimate____________
Exact ____________

6. $\begin{array}{r} 391.9 \\ \times\ \ 7.74 \\ \hline \end{array}$

6.
Estimate____________
Exact ____________

Name: Date:
Instructor: Section:

Chapter 5 RATIONAL NUMBERS: POSITIVE AND NEGATIVE DECIMALS

5.5 Dividing Signed Decimal Numbers

Learning Objectives	
1	Divide a decimal by an integer.
2	Divide a number by a decimal.
3	Estimate the answer when dividing decimals.
4	Use the order of operations with decimals.

Key Terms

Use the vocabulary terms listed below to complete each statement in exercises 1–4.

repeating decimal **quotient** **dividend** **divisor**

1. In a division problem, the number being divided is called the ________________.

2. The number $0.8\overline{3}$ is an example of a ____________________.

3. The answer to a division problem is called the ________________.

4. In the problem $6.39 \div 0.9$, 0.9 is called the ________________.

Guided Examples

Review these examples for Objective 1:

1. Find each quotient. Check the quotients by multiplying.

a. $24.48 \div (-4)$

First consider $24.48 \div 4$.
Rewrite the division problem.
Step 1 Write the decimal point in the quotient directly above the decimal point in the dividend.

$$4\overline{)24.48}$$

Step 2 Divide as if the numbers were whole numbers.

$$\begin{array}{r} 6.12 \\ 4\overline{)24.48} \end{array}$$

Check by multiplying the quotient times the divisor.

Now Try:

1. Find each quotient. Check the quotients by multiplying.

a. $9.891 \div (-7)$

____

Name: Date:
Instructor: Section:

$$\begin{array}{r} 6.12 \\ \times \quad 4 \\ \hline 24.48 \end{array}$$

Step 3 The quotient is –6.12 because the numbers have different signs.

$24.48 \div (-4) = -6.12$

b. $8\overline{)425.6}$

Write the decimal point in the quotient above the decimal point in the dividend. Then divide as if the number were whole numbers.

$$\begin{array}{r} 53.2 \\ 8\overline{)425.6} \\ \underline{40} \\ 25 \\ \underline{24} \\ 16 \\ \underline{16} \\ 0 \end{array} \qquad \text{Check: } \begin{array}{r} 53.2 \\ \times \quad 8 \\ \hline 425.6 \end{array}$$

The quotient is 53.2.

b. $5\overline{)75.15}$

____

2. Divide 1.35 by 4. Check the quotient by multiplying.

Divide.

$$\begin{array}{r} 0.33 \\ 4\overline{)1.35} \\ \underline{1\,2} \\ 15 \\ \underline{12} \\ 3 \end{array}$$

Write a 0 after the 5 in the dividend so you can continue dividing. Keep writing more zeros in the dividend, if needed.

$$\begin{array}{r} 0.3375 \\ 4\overline{)1.3500} \\ \underline{1\,2} \\ 15 \\ \underline{12} \\ 30 \\ \underline{28} \\ 20 \\ \underline{20} \\ 0 \end{array} \qquad \text{Check: } \begin{array}{r} 0.3375 \\ \times \quad 4 \\ \hline 1.3500 \end{array}$$

The quotient is 0.3375.

2. Divide 1008.9 by 50. Check the quotient by multiplying.

____

Name: Date:
Instructor: Section:

3. Divide 8.87 by 9. Round the quotient to the nearest thousandth.

Write extra zeros in the dividend so you can continue dividing.

$$\begin{array}{r} 0.9855 \\ 9\overline{)8.8700} \\ \underline{8\ 1} \\ 77 \\ \underline{72} \\ 50 \\ \underline{45} \\ 50 \\ \underline{45} \\ 5 \end{array}$$

Notice that the digit 5 in the answer is repeating. There are two ways to show that the answer is a repeating decimal that goes on forever.

$0.9855\ldots$ or $0.98\overline{5}$

To round to thousandths, divide out one more place, to ten-thousandths.

$8.87 \div 9 = 0.9855\ldots$ rounds to 0.986.

Check the answer by multiplying 0.986 by 9. Because 0.986 is a rounded answer, the check will not give exactly 8.87, but it should be very close.

$(0.986)(9) = 8.874$

3. Divide 302.24 by 18. Round the quotient to the nearest thousandth.

____

Review these examples for Objective 2:

4.

a. $\dfrac{41.2}{0.005}$

Move the decimal point in the divisor three places to the right so 0.005 becomes the whole number 5. Move the decimal point in the dividend the same number of places and write in two extra 0s.

$$\begin{array}{r} 8240. \\ 5\overline{)41200.} \end{array}$$

Now Try:

4.

a. $0.0024\overline{)48.984}$

____

Name: Date:
Instructor: Section:

b. Divide –7 by –2.4. Round to the nearest hundredth.

First consider $7 \div 2.4$. Move the decimal point in the divisor one place to the right so 2.4 becomes the whole number 24. The decimal point in the dividend starts on the right side of 7 and is also moved one place to the right. (Remember, in order to round to hundredths, divide out one more place, to thousandths.)

$$\begin{array}{r} 2.916 \\ 24\overline{)70.000} \\ \underline{48} \\ 220 \\ \underline{216} \\ 40 \\ \underline{24} \\ 160 \\ \underline{144} \\ 16 \end{array}$$

Round the quotient. The quotient is positive because both the divisor and the dividend have the same sign.

$-7 \div (-2.4) \approx 2.92$

b. Divide –8 by –4.5. Round to the nearest hundredth.

____

Review this example for Objective 3:

5. First use front end rounding to round each number and estimate the answer. Then divide to find the exact answer.

$767.38 \div 3.7$

Estimate:

$$\begin{array}{r} 200 \\ 4\overline{)800} \end{array}$$

Exact:

$$\begin{array}{r} 27.4 \\ 37\overline{)7673.8} \\ \underline{74} \\ 273 \\ \underline{259} \\ 148 \\ \underline{148} \\ 0 \end{array}$$

Notice that the estimate, which is in hundreds, is very different from the exact answer, which is in tens.

Find the error and rework.

The exact answer is 207.4, which fits the estimate of 200.

Now Try:

5. First use front end rounding to round each number and estimate the answer. Then divide to find the exact answer.

$185.22 \div 4.9$

____

Name: Date:
Instructor: Section:

Review these examples for Objective 4:

6. Use the order of operations to simplify each expression.

a. $4.5+(-8.2)^2+11.63$

Apply the exponent.
$4.5+67.24+11.63$
Add from left to right.
$71.74 + 11.63$
83.37

b. $1.73+(2.9-3.8)(6.5)$

Work inside the parentheses.
$1.73+(-0.9)(6.5)$
Multiply.
$1.73 + (-5.85)$
Add.
-4.12

c. $5.6^2-1.4\div 7(2.4)$

Apply the exponent.
$31.36-1.4\div 7(2.4)$
Multiply and divide from left to right.
$31.36-0.2(2.4)$
Multiply.
$31.36-0.48$
Subtract last.
30.88

Now Try:

6. Use the order of operations to simplify each expression.

a. $5.9-0.48+(-2.6)^2$

____

b. $4.06\div 1.4\times 7.8$

____

c. $0.07+0.3(6.99-8)$

____

Objective 1 Divide a decimal by an integer.

For extra help, see Examples 1–3 on pages 349–351 of your text and Section Lecture video for Section 5.5 and Exercise Solutions Clip 5.

Find each quotient. Round answers to the nearest thousandth, if necessary.

1. $5\overline{)34.8}$ **1.** ____________

2. $-11\overline{)46.98}$ **2.** ____________

3. $33\overline{)77.847}$ **3.** ____________

Name: Date:
Instructor: Section:

Objective 2 Divide a number by a decimal.

For extra help, see Example 4 on page 353 of your text and Section Lecture video for Section 5.5 and Exercise Solutions Clip 23.

Find each quotient. Round answers to the nearest thousandth, if necessary.

4. $0.9\overline{)3.4166}$ **4.** ____________

5. $3.4\overline{)436.05}$ **5.** ____________

6. $-0.07 \div (-0.00043)$ **6.** ____________

Objective 3 Estimate the answer when dividing decimals.

For extra help, see Example 5 on page 354 of your text and Section Lecture video for Section 5.5 and Exercise Solutions Clip 29.

Decide if each answer is ***reasonable*** *or* ***unreasonable*** *by rounding the numbers and estimating the answer.*

7. $126.2 \div 11.2 = 11.268$ **7.** ____________

8. $31.5 \div 8.4 = 37.5$ **8.** ____________

9. $8695.15 \div 98.762 = 88.0415$ **9.** ____________

Name: Date:
Instructor: Section:

Objective 4 Use the order of operations with decimals.

For extra help, see Example 6 on pages 354–355 of your text and Section Lecture video for Section 5.5 and Exercise Solutions Clip 57.

Use the order of operations to simplify each expression.

10. $(-3.1)^2 - 1.9 + 5.8$ **10.** ________________

11. $58.1 - (17.9 - 15.2) \times 1.8$ **11.** ________________

12. $9.1 - 0.07(2.1 \div 0.042)$ **12.** ________________

Name: Date:
Instructor: Section:

Chapter 5 RATIONAL NUMBERS: POSITIVE AND NEGATIVE DECIMALS

5.6 Fractions and Decimals

Learning Objectives
1. Write fractions as equivalent decimals.
2. Compare the size of fractions and decimals.

Key Terms

Use the vocabulary terms listed below to complete each statement in exercises 1–4.

numerator **denominator** **mixed number** **equivalent**

1. A fraction and a decimal that represent the same portion of a whole are ________________.

2. The ____________________ of a fraction is the dividend.

3. The __________________ of a fraction shows the number of equal parts in a whole.

4. A __________________ consists of a whole number part and a fractional or decimal part.

Guided Examples

Review these examples for Objective 1:

1. Write the fraction as a decimal.

   a. $\frac{7}{8}$

   $\frac{7}{8}$ means $7 \div 8$. Write it as $8\overline{)7}$. Write extra zeros in the dividend so you can continue dividing until the remainder is zero.

$$\begin{array}{r} 0.875 \\ 8\overline{)7.000} \\ \underline{6\,4} \\ 60 \\ \underline{56} \\ 40 \\ \underline{40} \\ 0 \end{array}$$

   Therefore, $\frac{7}{8} = 0.875$

Now Try:

1. Write the fraction as a decimal.

   a. $\frac{1}{16}$

   ____

Name: Date:
Instructor: Section:

b. $3\frac{3}{16}$

One method is to divide 3 by 16 to get 0.1875 for the fraction part. Then add the whole number part to 0.1875.

$$\frac{3}{16} \rightarrow \begin{array}{r} 0.1875 \\ 16\overline{)3.0000} \\ \underline{1\,6} \\ 140 \\ \underline{128} \\ 120 \\ \underline{112} \\ 80 \\ \underline{80} \\ 0 \end{array} \rightarrow \begin{array}{r} 3.0000 \\ +\ 0.1875 \\ \hline 3.1875 \end{array}$$

So $3\frac{3}{16} = 3.1875$.

A second method is to first write $3\frac{3}{16}$ as an improper fraction and then divide numerator by denominator.

$$3\frac{3}{16} = \frac{51}{16}$$

$$\frac{51}{16} \rightarrow 51 \div 16 \rightarrow 16\overline{)51} \rightarrow \begin{array}{r} 3.1875 \\ 16\overline{)51.0000} \\ \underline{48} \\ 3\,0 \\ \underline{1\,6} \\ 140 \\ \underline{128} \\ 120 \\ \underline{112} \\ 80 \\ \underline{80} \\ 0 \end{array}$$

So $3\frac{3}{16} = 3.1875$.

2. Write $\frac{5}{9}$ as a decimal and round to the nearest thousandth.

$\frac{5}{9}$ means $5 \div 9$. To round to thousandths, divide out one more place, to ten-thousandths.

b. $4\frac{3}{8}$

____

2. Write $\frac{5}{12}$ as a decimal and round to the nearest thousandth.

____

Name: Date:
Instructor: Section:

$$\frac{5}{9} \to 5 \div 9 \to 9\overline{)5} \to 9\overline{)5.0000}$$

$$\begin{array}{r} 0.5555 \\ 9\overline{)5.0000} \\ \underline{45} \\ 50 \\ \underline{45} \\ 50 \\ \underline{45} \\ 50 \\ \underline{45} \\ 5 \end{array}$$

Written as a repeating decimal, $\frac{5}{9} = 0.\overline{5}$.

Rounded to the nearest thousandth, $\frac{5}{9} = 0.556$.

Review these examples for Objective 2:

3. Write in <, >, or = in the blank between each pair of numbers.

a. 0.735 ____ 0.75

Because 0.735 is to the left of 0.75, use the < symbol.
0.735 is less than 0.75 can be written as
$0.735 < 0.75$

b. $\frac{3}{5}$ ____ 0.6

$\frac{3}{5}$ and 0.6 are the same point on the number line.
They are equivalent.
$\frac{3}{5} = 0.6$

c. $0.\overline{3}$ ____ 0.3

0.3 is to the right of $0.\overline{3}$ (which is actually 0.333...), so use the > symbol.
$0.\overline{3}$ is greater than 0.3 can be written as
$0.\overline{3} > 0.3$

d. $\frac{14}{18}$ ____ $0.\overline{7}$

Write $\frac{14}{18}$ in lowest terms as $\frac{7}{9}$.

We see that $\frac{7}{9} = 0.\overline{7}$.

Now Try:

3. Write in <, >, or = in the blank between each pair of numbers.

a. 0.85 ____ 0.7895

____

b. $\frac{5}{16}$ ____ 0.3125

____

c. $\frac{3}{7}$ ____ $0.\overline{4}$

____

d. $\frac{27}{33}$ ____ $0.\overline{81}$

____

Name: Date:
Instructor: Section:

4. Write each group of numbers in order, from least to greatest.

a. 0.58 0.575 0.5816

Write zeros to the right of 0.58 and 0.575, so they also have four decimal places. Then find the least and greatest number of ten-thousandths.

$0.58 = 0.5800 = 5800$ ten-thousandths middle

$0.575 = 0.5750 = 5750$ ten-thousandths least

$= 0.5816 = 5816$ ten-thousandths greatest

From least to greatest, the correct order is

0.575 0.58 0.5816

b. $4\frac{7}{8}$ 4.7 4.82

Write $4\frac{7}{8}$ as $\frac{39}{8}$ and divide to get the decimal form 4.875. Then because 4.875 has three decimal places, write zeros so all the numbers have three decimal places.

$4\frac{7}{8} = 4.875 = 4$ and 875 thousandths greatest

$4.7 = 4.700 = 4$ and 700 thousandths least

$4.82 = 4.820 = 4$ and 820 thousandths middle

From least to greatest, the correct order is

4.7 4.82 $4\frac{7}{8}$

4. Write each group of numbers in order, from least to greatest.

a. 0.3 0.307 0.3057

____

b. $\frac{2}{9}$, $\frac{3}{13}$, 0.23, $\frac{1}{5}$

____

Objective 1 Write fractions as equivalent decimals.

For extra help, see Examples 1–2 on pages 362–363 of your text and Section Lecture video for Section 5.6 and Exercise Solutions Clip 9, 17, 19, and 27.

Write each fraction or mixed number as a decimal. Round to the nearest thousandth, if necessary.

1. $\frac{1}{8}$

2. $4\frac{1}{9}$

3. $19\frac{17}{24}$

1. _______________

2. _______________

3. _______________

Name: Date:
Instructor: Section:

Objective 2 Compare the size of fractions and decimals.

For extra help, see Examples 3–4 on pages 364–365 of your text and Section Lecture video for Section 5.6 and Exercise Solutions Clip 57.

Write < or > to make a true statement.

4. $\frac{5}{6}$ ___ 0.83 **4.** ____________

Arrange in order from smallest to largest.

5. $\frac{3}{11}, \frac{1}{3}, 0.29$ **5.** ____________

6. $1.085, 1\frac{5}{11}, 1\frac{7}{20}$ **6.** ____________

Name: Date:
Instructor: Section:

Chapter 5 RATIONAL NUMBERS: POSITIVE AND NEGATIVE DECIMALS

5.7 Problem Solving with Statistics: Mean, Median, and Mode

Learning Objectives

1 Find the mean of a list of numbers.
2 Find a weighted mean.
3 Find the median.
4 Find the mode.

Key Terms

Use the vocabulary terms listed below to complete each statement in exercises 1–4.

mean **weighted mean** **median** **mode**

1. The ______________________________ is the value that occurs most often in a group of values.

2. A mean calculated so that each value is multiplied by its frequency is called a ______________________________.

3. The sum of all the values in a data set divided by the number of values in the data set is called the ______________________________.

4. The middle number in a group of values that are listed from smallest to largest is called the ______________________________.

Guided Examples

Review these examples for Objective 1:

1. Find the mean of 125, 234, 155, 275, and 141.

Use the formula for finding the mean. Add up all the values and then divide by the number of values.

$$\text{mean} = \frac{125+234+155+275+141}{5}$$

$$\text{mean} = \frac{930}{5}$$

$$\text{mean} = 186$$

The mean is 186.

Now Try:

1. Find the mean of 257, 261, 269, 274, 268, 280, and 295.

____

Name: Date:
Instructor: Section:

2. Find the mean of 31, 37, 44, 51, 52, 74, 69, and 83.

Find the mean (rounded to the nearest tenth).

$$\text{mean} = \frac{31+37+44+51+52+74+69+83}{8}$$

$$\text{mean} = \frac{441}{8}$$

$$\text{mean} \approx 55.1$$

The mean is 55.1.

2. Find the mean of 40.1, 32.8, 82.5, 51.2, 88.3, 31.7, 43.7, and 51.2.

____

Review these examples for Objective 2:

3. Use the following table to find the weighted mean.

Value	Frequency
13	4
12	2
19	5
15	3
21	1
27	5

To find the mean, multiply the value by the frequency. Then add the products. Next, add the numbers in the frequency column to find the total number of values.

Value	Frequency	Product
13	4	$(13 \cdot 4) = 52$
12	2	$(12 \cdot 2) = 24$
19	5	$(19 \cdot 5) = 95$
15	3	$(15 \cdot 3) = 45$
21	1	$(21 \cdot 1) = 21$
27	5	$(27 \cdot 5) = 135$
Totals	20	372

Finally, divide the totals. Round to the nearest tenth.

$$\text{mean} = \frac{372}{20} = 18.6$$

The mean is 18.6.

Now Try:

3. Use the following table to find the weighted mean.

Value	Frequency
35	1
36	2
39	5
40	4
42	3
43	5

____

Name: Date:
Instructor: Section:

4. Find the grade point average for this student. Assume A = 4, B = 3, C = 2, D = 1, F = 0.

Units	Grade
3	C
3	A
4	B
5	B
2	A

Multiply the credits and grades. Find the total.

	Units	Grade	Units · Grade
	3	C(= 2)	$3 \cdot 2 = 6$
	3	A(= 4)	$3 \cdot 4 = 12$
	4	B(= 3)	$4 \cdot 3 = 12$
	5	B(= 3)	$5 \cdot 3 = 15$
	2	A(= 4)	$2 \cdot 4 = 8$
Totals	17		53

It is common to round grade point averages to the nearest hundredth.

$$\text{GPA} = \frac{53}{17} \approx 3.12$$

4. Find the grade point average for this student. Assume A = 4, B = 3, C = 2, D = 1, F = 0.

Units	Grade
3	A
4	B
2	C
5	C
2	D

____

Review these examples for Objective 3:

5. Find the median for the list of values.
21, 32, 27, 23, 25, 29, 22

First arrange the numbers in numerical order from least to greatest.
21, 22, 23, 25, 27, 29, 32
Next, find the middle number in the list.

$$\underbrace{21,\ 22,\ 23,}_{\text{Three are below.}}\ \underset{\downarrow}{25,}\ \underbrace{27,\ 29,\ 32}_{\text{Three are above.}}$$

Middle number

The median value is 25.

6. Find the median for the list of values.
389, 464, 521, 610, 654, 672, 682, 712

First arrange the numbers in numerical order from least to greatest. Then find the middle two numbers.

$$389,\ 464,\ 521,\ \underbrace{610,\ 654,}_{\text{Middle two numbers}}\ 672,\ 682,\ 712$$

The median value is the mean of the two middle

Now Try:

5. Find the median for the list of values.
18, 12, 11, 19, 26

____

6. Find the median for the list of values.
0.02, 0.04, 0.12, 0.08

____

numbers.

$$\text{median} = \frac{610+654}{2} = \frac{1264}{2} = 632$$

The median value is 632.

Review these examples for Objective 4:

7. Find the mode for each list of numbers.

a. 37, 52, 41, 27, 41, 96

The number 41 occurs more often than any other number; therefore, 41 is the mode.

b. 964, 987, 973, 987, 921, 921, 975

Because both 987 and 921 occur twice, each is a mode.

c. \$10.71, \$11.67, \$12.39, \$21.54, \$13.98, \$14.66

No number occurs more than once. This list has no mode.

Now Try:

7. Find the mode for each list of numbers.

a. 5, 3, 6, 3, 7, 8, 5, 3, 4 ________

b. 16, 13, 21, 16, 18, 11, 13, 15, 14 ________

c. 1.5, 1.2, 1.3, 1.9, 1.8, 1.4, 1.7, 1.0 ________

Objective 1 Find the mean of a list of numbers.

For extra help, see Examples 1–2 on page 370 of your text and Section Lecture video for Section 5.7 and Exercise Solutions Clip 3.

Find the mean for each list of numbers. Round to the nearest tenth, if necessary.

1. 39, 50, 59, 61, 69, 73, 51, 80 **1.** ________

2. 62.7, 59.6, 71.2, 65.8, 63.1 **2.** ________

3. 216, 245, 268, 268, 280, 291, 304, 313 **3.** ________

Objective 2 Find a weighted mean.

For extra help, see Examples 3–4 on pages 371–372 of your text and Section Lecture video for Section 5.7 and Exercise Solutions Clip 7.

Find the weighted mean for each list of numbers. Round to the nearest tenth, if necessary.

4. **4.** ________

Value	*Frequency*
17	4
12	5
15	3
19	1

Name: Date:
Instructor: Section:

5.

Value	*Frequency*
1	2
2	3
4	5
5	7
6	4
7	2
8	1
9	1

5. ______________

Find the grade point average for this student. Assume A = 4, B = 3, C = 2, D = 1, F = 0.

6.

Units	*Grade*
5	B
4	C
3	B
2	C
2	C

6. ______________

Objective 3 Find the median.

For extra help, see Examples 5–6 on pages 372–373 of your text and Section Lecture video for Section 5.7 and Exercise Solutions Clip 15.

Find the median for each list of numbers.

7. 200, 215, 226, 238, 250, 283 **7.** ______________

8. 43, 69, 108, 32, 51, 49, 83, 57, 64 **8.** ______________

9. 200, 195, 302, 284, 256, 237, 239, 240 **9.** ______________

Objective 4 Find the mode.

For extra help, see Example 7 on page 373 of your text and Section Lecture video for Section 5.7 and Exercise Solutions Clip 25.

Find the mode for each list of numbers.

10. 4, 9, 3, 4, 7, 3, 2, 3, 9 **10.** ______________

11. 37, 24, 35, 35, 24, 38, 39, 28, 27, 39 **11.** ______________

12. 172.6, 199.7, 182.4, 167.1, 172.6, 183.4, 187.6 **12.** ______________

Name: Date:
Instructor: Section:

Chapter 5 RATIONAL NUMBERS: POSITIVE AND NEGATIVE DECIMALS

5.8 Geometry Applications: Pythagorean Theorem and Square Roots

Learning Objectives
1 Find square roots using the square root key on a calculator.
2 Find the unknown length in a right triangle.
3 Solve application problems involving right triangles.

Key Terms

Use the vocabulary terms listed below to complete each statement in exercises 1–3.

hypotenuse **legs** **right triangle**

1. A triangle with a 90° angle is called a ______________________________.

2. The side opposite the right angle in a right triangle is called the ____________________ of the triangle.

3. The two sides of the right angle in a right triangle are called the __________________ of the triangle.

Guided Examples

Review these examples for Objective 1:

1. Use a calculator to find each square root. Round answers to the nearest hundredth.

   a. $\sqrt{45}$

   Calculator shows 6.708203932; round to 6.71.

   b. $\sqrt{86}$

   Calculator shows 9.273618495; round to 9.27.

   c. $\sqrt{120}$

   Calculator shows 10.95445115; round to 10.95.

Now Try:

1. Use a calculator to find each square root. Round answers to the nearest hundredth.

   a. $\sqrt{20}$ ______________

   b. $\sqrt{92}$ ______________

   c. $\sqrt{134}$ ______________

Name: Date:
Instructor: Section:

Review these examples for Objective 2:

2. Find the unknown length in each right triangle. Round answers to the nearest tenth if necessary.

a.

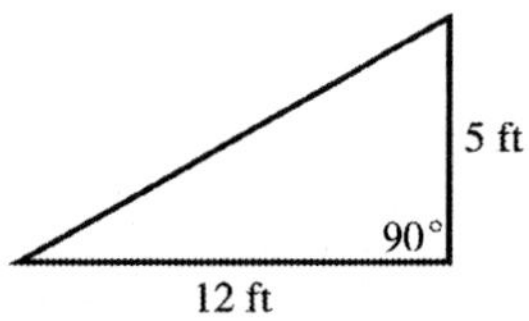

The unknown length is the side opposite the right angle, which is the hypotenuse. Use the formula for finding the hypotenuse.

$$\text{hypotenuse} = \sqrt{(\text{leg})^2 + (\text{leg})^2}$$

$$\begin{aligned}\text{hypotenuse} &= \sqrt{(5)^2 + (12)^2}\\ &= \sqrt{25 + 144}\\ &= \sqrt{169}\\ &= 13\end{aligned}$$

The hypotenuse is 13 ft long.

b.

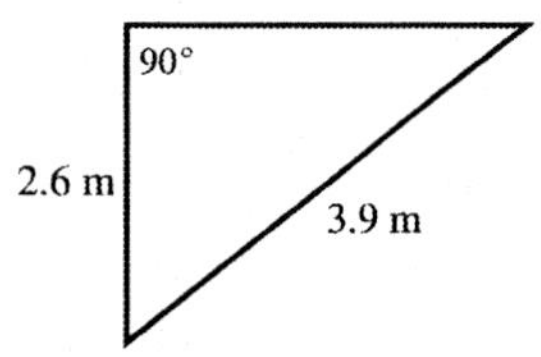

We do know the length of the hypotenuse (3.9 m), so it is the length of one of the legs that is unknown. Use the formula for finding the leg.

$$\text{leg} = \sqrt{(\text{hypotenuse})^2 - (\text{leg})^2}$$

$$\begin{aligned}\text{leg} &= \sqrt{(3.9)^2 - (2.6)^2}\\ &= \sqrt{15.21 - 6.76}\\ &= \sqrt{8.45}\\ &\approx 2.9\end{aligned}$$

The length of the leg is approximately 2.9 m.

Now Try:

2. Find the unknown length in each right triangle. Round answers to the nearest tenth if necessary.

a.

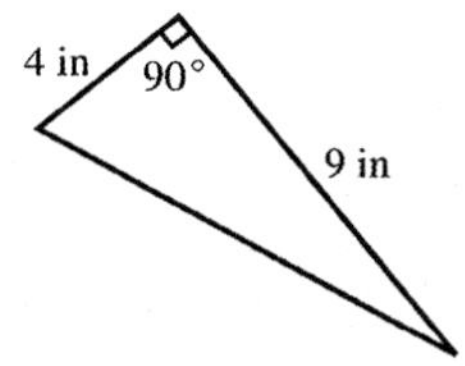

____

b.

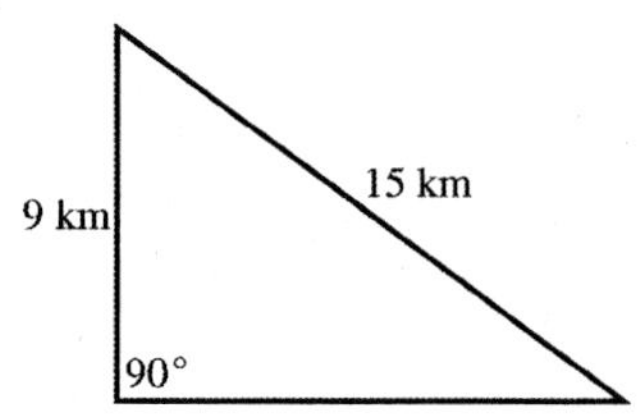

____

Name: Date:
Instructor: Section:

Review this example for Objective 3:

3. The base of a ladder is located 7 feet from a building. The ladder reaches 24 feet up the building. How long is the ladder? Round to the nearest tenth of a foot if necessary.

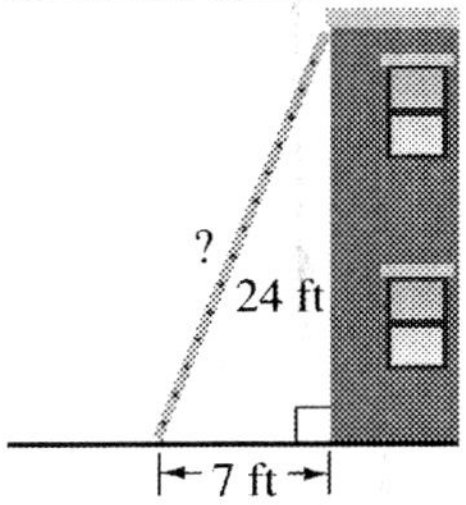

A right triangle is formed. The unknown side is the hypotenuse.

$$\text{hypotenuse} = \sqrt{(\text{leg})^2 + (\text{leg})^2}$$

$$\begin{aligned}\text{hypotenuse} &= \sqrt{(7)^2 + (24)^2}\\ &= \sqrt{49 + 576}\\ &= \sqrt{625}\\ &= 25\end{aligned}$$

The length of the ladder is 25 ft.

Now Try:

3. Find the unknown length in this roof plan. Round to the nearest tenth of a foot if necessary.

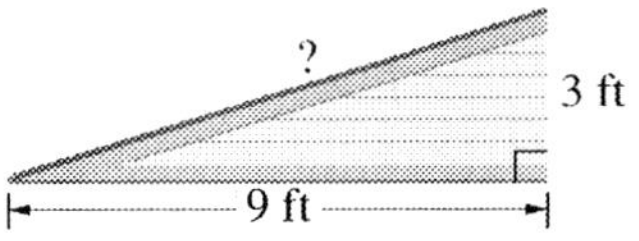

____

Objective 1 Find square roots using the square root key on a calculator.

For extra help, see Example 1 on page 378 of your text and Section Lecture video for Section 5.8 and Exercise Solutions Clip 9.

Find each square root. Use a calculator with a square root key. Round the answer to the nearest thousandth, if necessary.

1. $\sqrt{17}$ **1.** ______________

2. $\sqrt{75}$ **2.** ______________

3. $\sqrt{102}$ **3.** ______________

Name: Date:
Instructor: Section:

Objective 2 Find the unknown length in a right triangle.

For extra help, see Example 2 on page 379 of your text and Section Lecture video for Section 5.8 and Exercise Solutions Clip 19 and 23.

Find the unknown length in each right triangle. Use a calculator with a square root key. Round the answer to the nearest tenth, if necessary.

4. 4. ________________

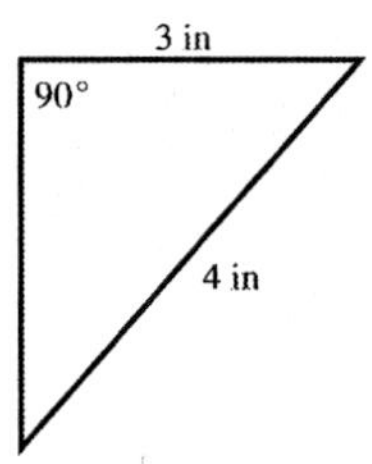

5. 5. ________________

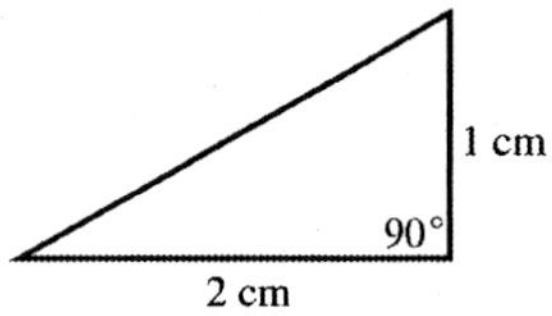

6. 6. ________________

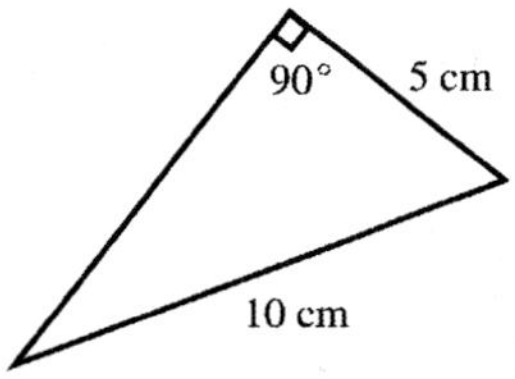

Objective 3 Solve application problems involving right triangles.

For extra help, see Example 3 on page 380 of your text and Section Lecture video for Section 5.8 and Exercise Solutions Clip 43.

Solve each application problem. Draw a diagram if one is not provided. Use a calculator with a square root key. Round the answer to the nearest tenth, if necessary.

7. Find the length of the loading ramp. 7. ________________

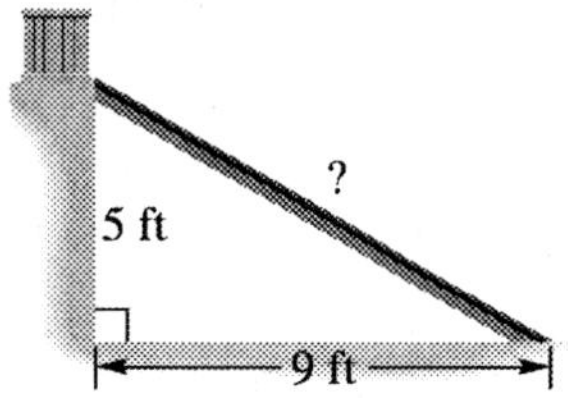

Name: Date:
Instructor: Section:

8. A kite is flying on 50 feet of string. If the horizontal distance of the kite from the person flying it is 40 feet, how far off the ground is the kite?

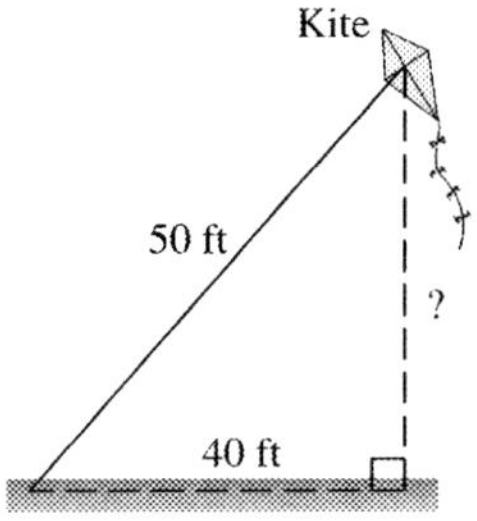

8. ________________

9. The base of a 17-ft ladder is located 15 ft from a building. How high up on the building will the ladder reach?

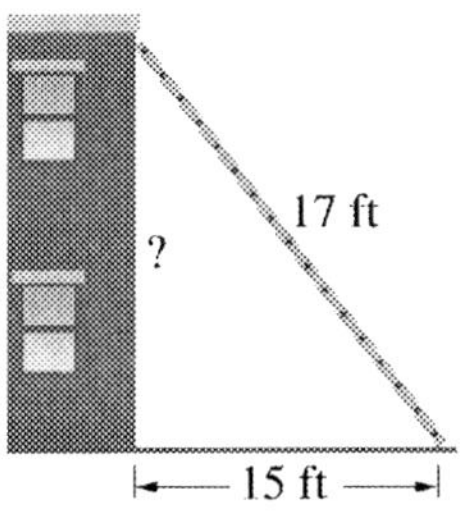

9. ________________

Name: Date:
Instructor: Section:

Chapter 5 RATIONAL NUMBERS: POSITIVE AND NEGATIVE DECIMALS

5.9 Problem Solving: Equations Containing Decimals

Learning Objectives

1 Solve equations containing decimals using the addition property of equality.
2 Solve equations containing decimals using the division property of equality.
3 Solve equations containing decimals using both properties of equality.
4 Solve application problems involving equations with decimals.

Key Terms

Use the vocabulary terms listed below to complete each statement in exercises 1–2.

addition property of equality

division property of equality

1. The ______________________________ states that both sides of an equation may be divided by the same nonzero number and it will still be balanced.

2. The ______________________________ states that the same number may be added to both sides of an equation and the equation will still be balanced.

Guided Examples

Review these examples for Objective 1:

1. Solve each equation and check each solution.

   a. $w+3.8=-0.7$

   Use the addition property to "get rid of" the 3.8 on the left side by adding its opposite, –3.8.

$$\begin{aligned} w+3.8 &= -0.7 \\ \underline{\quad -3.8} &\quad \underline{-3.8} \\ w+\ 0 &= -4.5 \\ w &= -4.5 \end{aligned}$$

   The solution is –4.5.

   Check. Go back to the original equation and replace w with –4.5.

$$\begin{aligned} w+3.8 &= -0.7 \\ -4.5+3.8 &= -0.7 \\ -0.7 &= -0.7 \end{aligned}$$

   So –4.5 is the correct solution.

Now Try:

1. Solve each equation and check each solution.

   a. $w+4.2=-9.7$

   ____

b. $8 = -5.6 + x$

To get x by itself on the right side of the equal sign, add 5.6 to both sides.

$$\begin{aligned} 8 &= -5.6 + x \\ +5.6 &\quad +5.6 \\ 13.6 &= 0 + x \\ 13.6 &= x \end{aligned}$$

The solution is 13.6.

Check. Go back to the original equation and replace x with 13.6.

$$\begin{aligned} 8 &= -5.6 + x \\ 8 &= -5.6 + 13.6 \\ 8 &= 8 \end{aligned}$$

So 13.6 is the correct solution.

b. $11 = -8.6 + x$

____

Review these examples for Objective 2:

2. Solve each equation and check each solution.

a. $6x = 25.5$

To undo the multiplication by 6, divide both sides by 6.

$$6x = 25.5$$

$$\frac{6 \cdot x}{6} = \frac{25.5}{6}$$

$$\frac{\overset{1}{\cancel{6}} \cdot x}{\underset{1}{\cancel{6}}} = 4.25$$

$$x = 4.25$$

The solution is 4.25.

Check. Go back to the original equation and replace x with 4.25.

$$\begin{aligned} 6x &= 25.5 \\ 6(4.25) &= 25.5 \\ 25.5 &= 25.5 \end{aligned}$$

So 4.25 is the correct solution.

b. $-10.5 = 1.4t$

Divide both sides by the coefficient of the variable term, 1.4.

Now Try:

2. Solve each equation and check each solution.

a. $8x = 24.4$

____

b. $-54.4 = 6.4t$

____

$$-10.5 = 1.4t$$

$$\frac{-10.5}{1.4} = \frac{\overset{1}{\cancel{1.4}}\,t}{\underset{1}{\cancel{1.4}}}$$

$$-7.5 = t$$

The solution is –7.5.

Check. Go back to the original equation and replace t with –7.5.

$$-10.5 = 1.4t$$

$$-10.5 = 1.4(-7.5)$$

$$-10.5 = -10.5$$

So –7.5 is the correct solution.

Review these examples for Objective 3:

3. Solve each equation and check each solution.

a. $3.5b + 0.45 = -6.55$

The first step is to get the variable term, $3.5b$, by itself on the left side of the equal sign.

$$\begin{aligned} 3.5b + 0.45 &= -6.55 \\ \underline{\quad -0.45} &\quad \underline{-0.45} \\ 3.5b + 0 &= -7.00 \\ 3.5b &= -7 \end{aligned}$$

The next step is to divide both sides by the coefficient of the variable term.

$$\frac{\overset{1}{\cancel{3.5}}\,b}{\underset{1}{\cancel{3.5}}} = \frac{-7}{3.5}$$

$$b = -2$$

The solution is –2.

Check. Go back to the original equation and replace b with –2.

$$3.5b + 0.45 = -6.55$$

$$3.5(-2) + 0.45 = -6.55$$

$$-7 \quad + 0.45 = -6.55$$

$$-6.55 = -6.55$$

So –2 is the correct solution.

b. $7x - 0.87 = 3x + 2.49$

Use the addition property to "get rid of" $3x$ on the right side by adding its opposite, $-3x$, to both sides.

Now Try:

3. Solve each equation and check each solution.

a. $4.2b - 0.79 = -7.51$

____

b. $9x - 1.46 = 3x + 1.72$

____

$$\begin{aligned} 7x - 0.87 &= 3x + 2.49 \\ \underline{-3x\quad} &\quad \underline{-3x\quad} \\ 4x - 0.87 &= 0 + 2.49 \\ 4x + (-0.87) &= 2.49 \\ \underline{+0.87} &\quad \underline{+0.87} \\ 4x + 0 &= 3.36 \\ \frac{\overset{1}{\cancel{4}}x}{\underset{1}{\cancel{4}}} &= \frac{3.36}{4} \\ x &= 0.84 \end{aligned}$$

The solution is 0.84.

Check. Go back to the original equation and replace x with 0.84.

$$\begin{aligned} 7x - 0.87 &= 3x + 2.49 \\ 7(0.84) - 0.87 &= 3(0.84) + 2.49 \\ 5.88 - 0.87 &= 2.52 + 2.49 \\ 5.01 &= 5.01 \end{aligned}$$

So 0.84 is the correct solution.

Review this example for Objective 4:

4. When using a particular calling card, the cost is \$1.15 per minute plus a \$0.75 service charge per call. If Julie was billed \$20.30 for one call, how long did the call last?

Step 1 Read the problem. It is about the cost of the call.

Step 2 Assign a variable. Let m be the number of minutes for the call.

Step 3 Write an equation.

Cost per minute		Number of minutes		Service charge		Total cost
1.15	$\cdot$	m	$+$	0.75	$=$	20.30

Step 4 Solve the equation.

$$\begin{aligned} 1.15m + 0.75 &= 20.30 \\ \underline{-0.75} &\quad \underline{-0.75} \\ 1.15m + 0 &= 19.55 \\ \frac{1.15m}{1.15} &= \frac{19.55}{1.15} \\ m &= 17 \end{aligned}$$

Now Try:

4. A telephone company charges \$4.25 plus an additional \$0.07 per minute for long distance calls. How many minutes did Sheila talk if her bill for a call was \$6.49?

____

Step 5 State the answer. The call was 17 min.

Step 6 Check the solution.
\$1.15 per minute times 17 minutes is \$19.55
\$19.55 plus \$0.75 service charge is \$20.30.
The answer checks.

Objective 1 Solve equations containing decimals using the addition property of equality.

For extra help, see Example 1 on page 385 of your text and Section Lecture video for Section 5.9.

Solve each equation and check each solution.

1. $-30.4 + n = -35$ **1.** ________________

2. $-7.1 = 0.32 + m$ **2.** ________________

3. $8.7 + h = 16.4$ **3.** ________________

Objective 2 Solve equations containing decimals using the division property of equality.

For extra help, see Example 2 on page 386 of your text and Section Lecture video for Section 5.9 and Exercise Solutions Clip 7 and 11.

Solve each equation and check each solution.

4. $-3y = -0.96$ **4.** ________________

5. $2.7r = 6.75$ **5.** ________________

6. $87.6 = -12r$ **6.** ________________

Objective 3 Solve equations containing decimals using both properties of equality.

For extra help, see Example 3 on page 387 of your text and Section Lecture video for Section 5.9 and Exercise Solutions Clip 19.

Solve each equation and check each solution.

7. $2.2 = 0.5y + 9.7$ **7.** ________________

Name: Date:
Instructor: Section:

8. $-11.2 - 0.7p = 0.9p + 5.6$ **8.** ________________

9. $0.6x + 4.98 = x - 6.78$ **9.** ________________

Objective 4 Solve application problems involving equations with decimals.

For extra help, see Example 4 on page 388 of your text and Section Lecture video for Section 5.9 and Exercise Solutions Clip 25.

Solve each application problem using the six problem-solving steps.

10. An air compressor can be rented for \$38.95 for the first three hours, and \$8.50 for each additional hour. William's rental charge was \$64.45. How many hours did he rent the compressor? **10.** ________________

11. Most adult medication doses are for a person weighing 150 pounds. For a 45-pound child, the adult dose should be multiplied by 0.3. If the child's dose of a decongestant is 9 milligrams, what is the adult dose? **11.** ________________

12. The bill for a three-night hotel stay was \$371.43. This included \$42.15 for room service and \$35.28 room tax. What was the room rate per night? **12.** ________________

Name: Date:
Instructor: Section:

Chapter 5 RATIONAL NUMBERS: POSITIVE AND NEGATIVE DECIMALS

5.10 Geometry Applications: Circles, Cylinders, and Surface Area

Learning Objectives	
1	Find the radius and diameter of a circle.
2	Find the circumference of a circle.
3	Find the area of a circle.
4	Find the volume of a cylinder.
5	Find the surface area of a rectangular solid.
6	Find the surface area of a cylinder.

Key Terms

Use the vocabulary terms listed below to complete each statement in exercises 1–6.

circle **radius** **diameter** **circumference** **π (pi)**

surface area

1. The __________________ is the distance from the center of a circle to any point on the circle.

2. The __________________ of a circle is the distance around the circle.

3. A figure whose points lie the same distance from a fixed center point is called a __________________.

4. The ratio of the circumference to the diameter of any circle equals ____________.

5. The area on the surface of a three-dimensional object is called its ____________.

6. The ______________________ of a circle is a segment connecting two points on a circle and passing through the center.

Guided Examples

Review these examples for Objective 1:

1. Find the unknown length of the diameter or radius of each circle.

a.

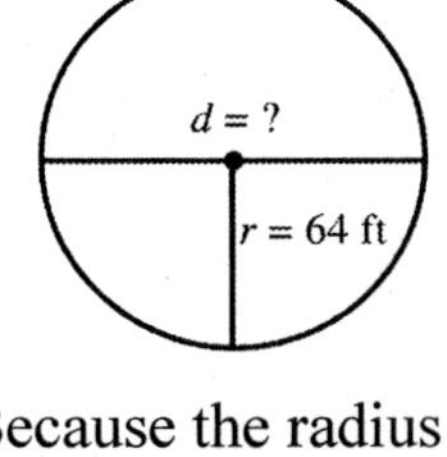

Because the radius is 64 ft, the diameter is twice

Now Try:

1. Find the unknown length of the diameter or radius of each circle.

a.

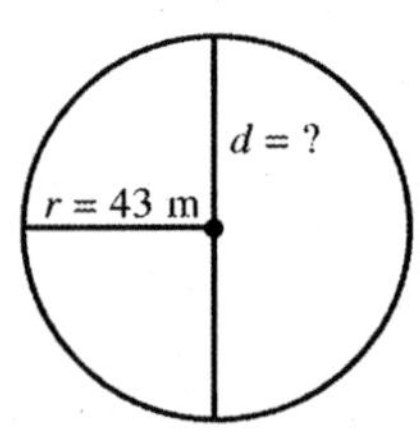

____

as long.

$d = 2 \cdot\ r$

$d = 2 \cdot 64 \text{ ft}$

$d = 128 \text{ ft}$

b.

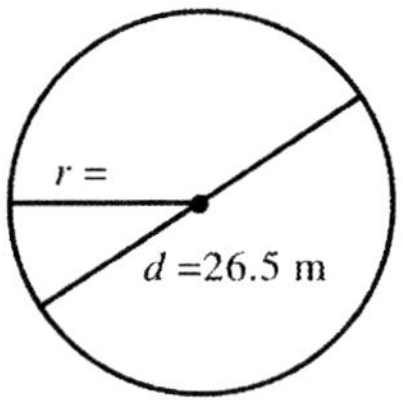

The radius is half the diameter.

$r = \dfrac{d}{2}$

$r = \dfrac{26.5 \text{ m}}{2}$

$r = 13.25 \text{ m}$

b.

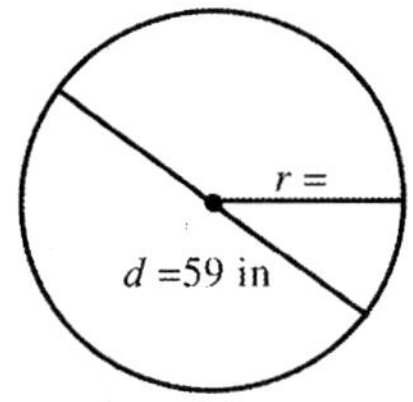

____

Review these examples for Objective 2:

2. Find the circumference of each circle. Use 3.14 as the approximate value for π. Round answers to the nearest tenth.

a.

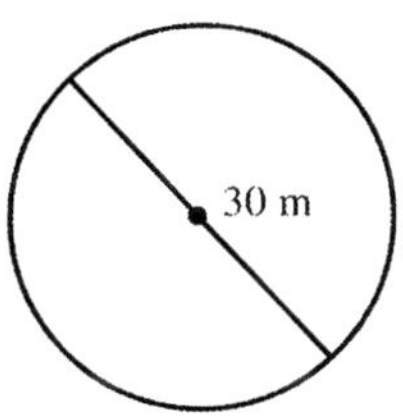

The diameter is 30 m, so use the formula with d in it.

$C = \ \pi \ \cdot \ d$

$C \approx 3.14 \cdot 30 \text{ m}$

$C \approx 94.2 \text{ m}$

b.

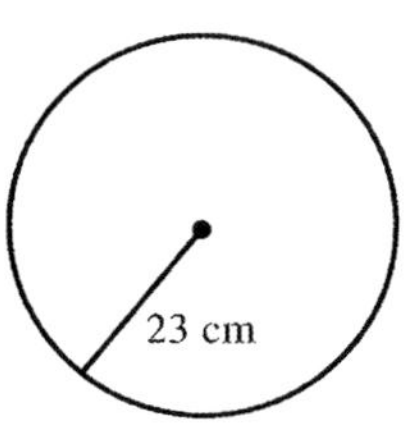

In this example, the radius is labeled, so it is easier to use the formula with r in it.

Now Try:

2. Find the circumference of each circle. Use 3.14 as the approximate value for π. Round answers to the nearest tenth.

a.

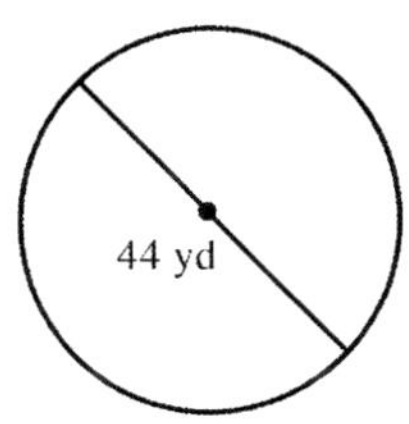

____

b.

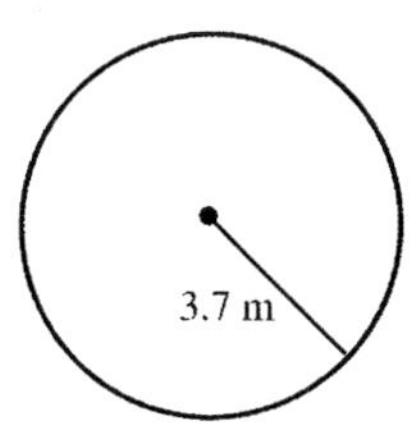

____

Name: Date:
Instructor: Section:

$C = 2 \cdot \pi \cdot r$

$C \approx 2 \cdot 3.14 \cdot 23$ cm

$C \approx 144.4$ cm Rounded

Review these examples for Objective 3:

3. Find the area of each circle. Use 3.14 as the approximate value for π. Round your answers to the nearest tenth.

a.

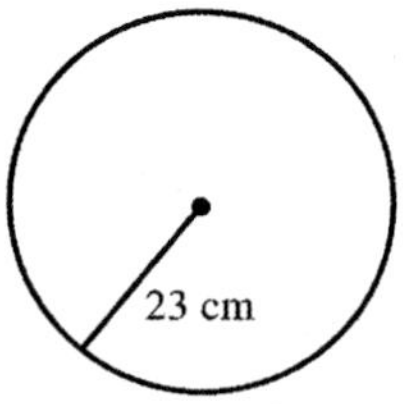

Use the formula $A = \pi \cdot r^2$, which means $\pi \cdot r \cdot r$.

$A = \pi \cdot r \cdot r$

$A \approx 3.14 \cdot 23 \text{ cm} \cdot 23 \text{ cm}$

$A \approx 1661.1 \text{ cm}^2$ Rounded

b.

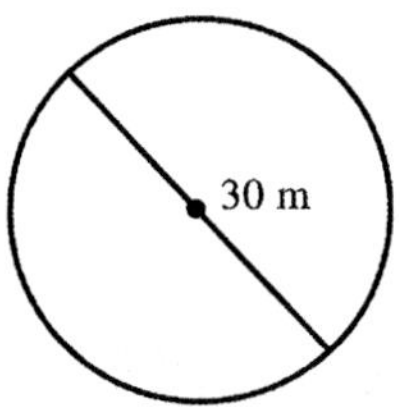

To use the area formula, you need to know the radius (r). In this circle, the diameter is 30 m. First find the radius.

$r = \frac{d}{2}$

$r = \frac{30 \text{ m}}{2} = 15 \text{ m}$

Now find the area.

$A = \pi \cdot r \cdot r$

$A \approx 3.14 \cdot 15 \text{ m} \cdot 15 \text{ m}$

$A \approx 706.5 \text{ m}^2$

Now Try:

3. Find the area of each circle. Use 3.14 as the approximate value for π. Round your answers to the nearest tenth.

a.

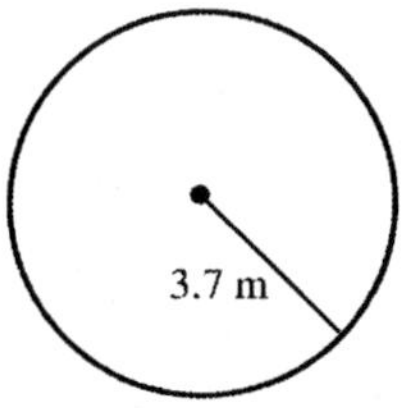

____

b.

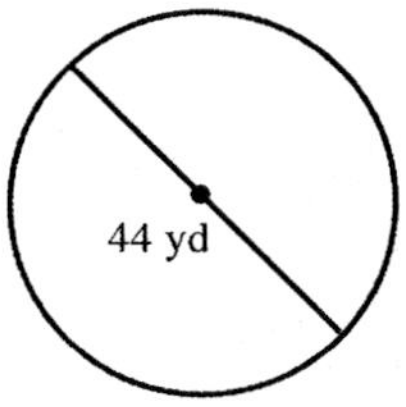

____

Name: Date:
Instructor: Section:

4. Find the area of a semicircle with radius 13 cm. Use 3.14 as the approximate value for π. Round your answers to the nearest tenth.

First, find the area of a whole circle with a radius of 13 cm.

$A = \pi \cdot r \cdot r$

$A \approx 3.14 \cdot 13 \text{ m} \cdot 13 \text{ m}$

$A \approx 530.66 \text{ m}^2$

Divide the area of the whole circle by 2 to find the area of the semicircle.

$\frac{530.66 \text{ m}^2}{2} = 265.33 \text{ m}^2$

The last step is rounding 265.33 to the nearest tenth.

Area of semicircle $\approx 265.3 \text{ m}^2$ Rounded

4. Find the area of a semicircle with radius 9 ft. Use 3.14 as the approximate value for π. Round your answers to the nearest tenth.

____

5. A circular piece of glass, for a customized table, is 3 ft in diameter. The cost of metal for the edge is \$3.50 per foot. What will it cost to add a metal edge to the glass? Use 3.14 for π.

Circumference $= \pi \cdot d$

$C \approx 3.14 \cdot 3 \text{ ft}$

$C \approx 9.42 \text{ ft}$

cost = cost per foot · circumference

$\text{cost} = \frac{\$3.50}{1 \cancel{\text{ft}}} \cdot \frac{9.42 \cancel{\text{ft}}}{1}$

cost = \$32.97

The cost of adding a metal edge to the glass is \$32.97.

5. A circular coaster is 5 inches in diameter. The cost of rubber for the edge is \$0.80 per inch. What will it cost to add rubber edge to the coaster? Use 3.14 for π.

____

6. Find the cost of covering the glass in Example 5 with a plastic cover. The material for the cover costs \$2.50 per square foot. Use 3.14 for π. Round your answer to the nearest cent.

First find the radius. $r = \frac{d}{2} = \frac{3 \text{ ft}}{2} = 1.5 \text{ ft}$

Then find the area. $A = \pi \cdot r^2$

$A \approx 3.14 \cdot 1.5 \text{ ft} \cdot 1.5 \text{ ft}$

$A \approx 7.065 \text{ ft}^2$

6. Find the cost of covering the coaster in Example 5 with a protective coating. The material for the cover costs \$0.10 per square inch. Use 3.14 for π. Round your answer to the nearest cent.

____

Name: Date:
Instructor: Section:

$$\text{cost} = \frac{\$2.50}{1\ \cancel{ft^2}} \cdot \frac{7.065\ \cancel{ft^2}}{1}$$

$$\text{cost} = \$17.66 \quad \text{Rounded}$$

The cost of the plastic cover is $17.66.

Review these examples for Objective 4:

7. Find the volume of each cylinder. Use 3.14 as the approximate value of π. Round your answers to the nearest tenth if necessary.

a.

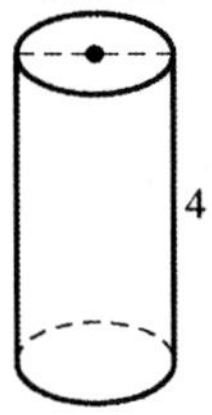

The diameter is 0.2 km, so the radius is $0.2 \text{ km} \div 2 = 0.1 \text{ km}$. The height is 4 km.

$$V = \pi \cdot r \cdot r \cdot h$$

$$V \approx 3.14 \cdot 0.1 \text{ km} \cdot 0.1 \text{ km} \cdot 4 \text{ km}$$

$$V \approx 0.1 \text{ km}^3$$

b.

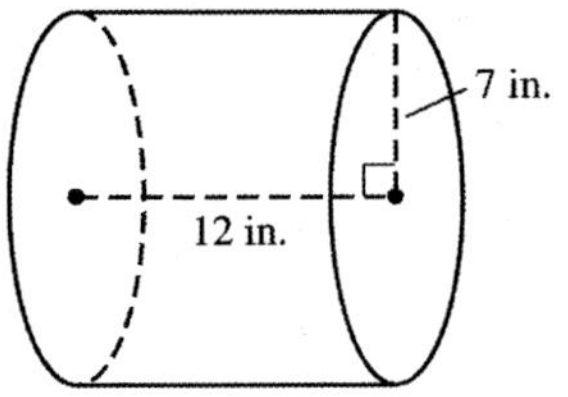

$$V = \pi \cdot r \cdot r \cdot h$$

$$V \approx 3.14 \cdot 7 \text{ in.} \cdot 7 \text{ in.} \cdot 12 \text{ in.}$$

$$V \approx 1846.3 \text{ in.}^3$$

Now Try:

7. Find the volume of each cylinder. Use 3.14 as the approximate value of π. Round your answers to the nearest tenth if necessary.

a. A cardboard mailing tube, diameter 5 centimeters and height 25 centimeters

____

b. An oil can, diameter 8 centimeters and height 13.5 centimeters

____

Review this example for Objective 5:

8. Find the volume and surface area of the figure.

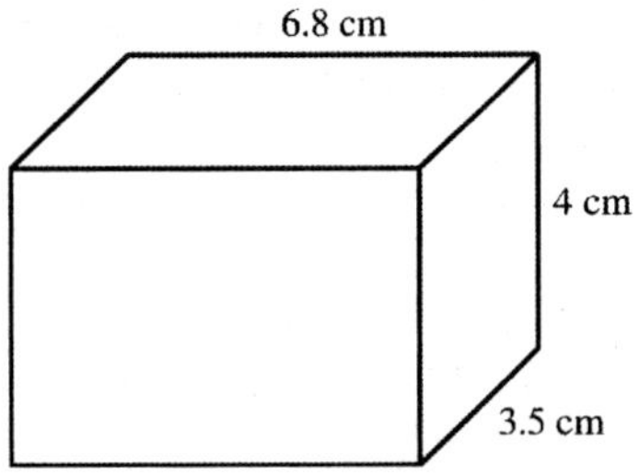

First find the volume.

Now Try:

8. Find the volume and surface area of the figure.

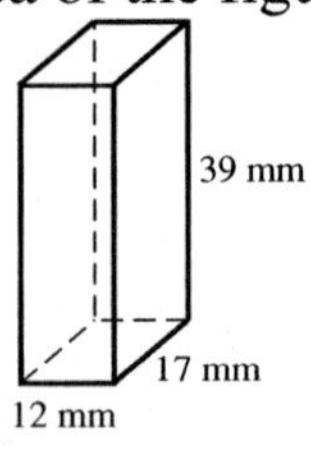

____

Name: Date:
Instructor: Section:

$V = lwh$

$V = 6.8 \text{ cm} \cdot 3.5 \text{ cm} \cdot 4 \text{ cm}$

$V = 95.2 \text{ cm}^3$

Now find the surface area.

$S = 2lw + 2lh + 2wh$

$S = (2 \cdot 6.8 \text{ cm} \cdot 3.5 \text{ cm}) + (2 \cdot 6.8 \text{ cm} \cdot 4 \text{ cm}) + (2 \cdot 3.5 \text{ cm} \cdot 4 \text{ cm})$

$S = 47.6 \text{ cm}^2 + 54.4 \text{ cm}^2 + 28 \text{ cm}^2$

$S = 130 \text{ cm}^2$

Review this example for Objective 6:

9. Find the volume and surface area of the figure. Use 3.14 as the approximate value of π. Round your answers to the nearest tenth if necessary.

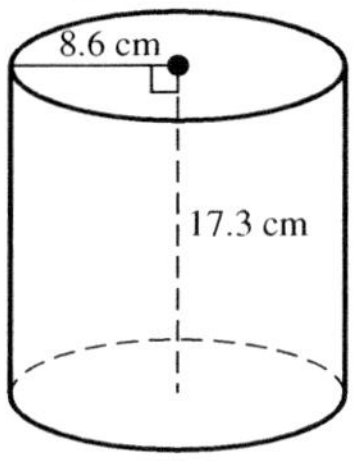

First find the volume.

$V = \pi r^2 h$

$V \approx 3.14 \cdot 8.6 \text{ cm} \cdot 8.6 \text{ cm} \cdot 17.3 \text{ cm}$

$V \approx 4017.7 \text{ cm}^3$

Now find the surface area.

$S = 2\pi rh + 2\pi r^2$

$S \approx (2 \cdot 3.14 \cdot 8.6 \text{ cm} \cdot 17.3 \text{ cm}) + (2 \cdot 3.14 \cdot 8.6 \text{ cm} \cdot 8.6 \text{ cm})$

$S \approx 934.3384 \text{ cm}^2 + 464.4688 \text{ cm}^2$

$S \approx 1398.8 \text{ cm}^2$

Now Try:

9. Find the volume and surface area of the figure. Use 3.14 as the approximate value of π. Round your answers to the nearest tenth if necessary.

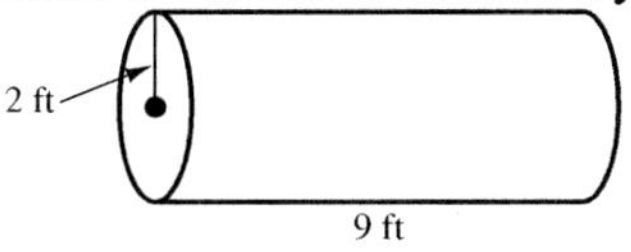

____

Objective 1 Find the radius and diameter of a circle.

For extra help, see Example 1 on pages 391–392 of your text and Section Lecture video for Section 5.10 and Exercise Solutions Clip 5.

Find the diameter or radius in each circle.

1. The diameter of a circle is 8 feet. Find its radius. **1.** ________________

2. The radius of a circle is 2.7 centimeters. Find its diameter. **2.** ________________

Name: Date:
Instructor: Section:

3. The diameter of a circle is $12\frac{1}{2}$ yards. Find its radius. **3.** ________________

Objective 2 Find the circumference of a circle.

For extra help, see Example 2 on page 393 of your text and Section Lecture video for Section 5.10 and Exercise Solutions Clip 13.

Find the circumference of each circle. Use 3.14 as an approximation for π. Round each answer to the nearest tenth.

4. A circle with a diameter of $4\frac{3}{4}$ inches **4.** ________________

5. A circle with a radius of 4.5 yards **5.** ________________

6. A circle with a radius of 8 cm **6.** ________________

Objective 3 Find the area of a circle.

For extra help, see Examples 3–6 on pages 394–396 of your text and Section Lecture video for Section 5.10 and Exercise Solutions Clip 13.

Find the area of each circle. Use 3.14 as an approximation for π. Round each answer to the nearest tenth.

7. A circle with diameter of $5\frac{1}{3}$ yards **7.** ________________

8. A circle with radius of 9.8 centimeters **8.** ________________

9. **9.** ________________

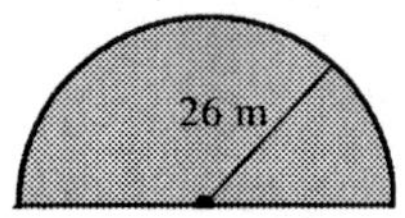

Name: Date:
Instructor: Section:

Objective 4 Find the volume of a cylinder.

For extra help, see Example 7 on page 396 of your text and Section Lecture video for Section 5.10 and Exercise Solutions Clip 35 and 39.

Find the volume of each figure. Use 3.14 as an approximation for π. Round answers to the nearest tenth, if necessary.

10.

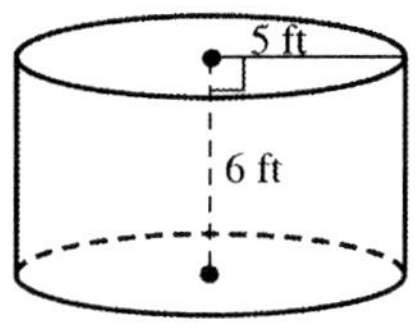

10. ________________

11.

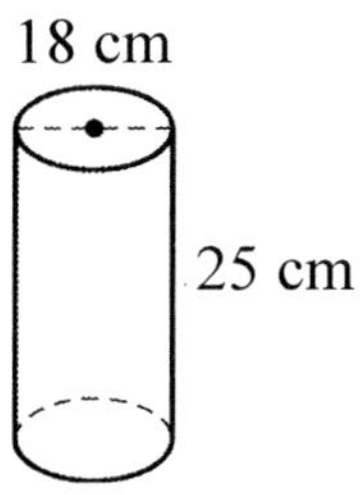

11. ________________

12. A soup can, diameter 6 inches and height 8 inches

12. ________________

Objective 5 Find the surface area of a rectangular solid.

For extra help, see Example 8 on page 397 of your text and Section Lecture video for Section 5.10 and Exercise Solutions Clip 47.

Find the surface area of each rectangular solid. Round your answers to the nearest tenth.

13.

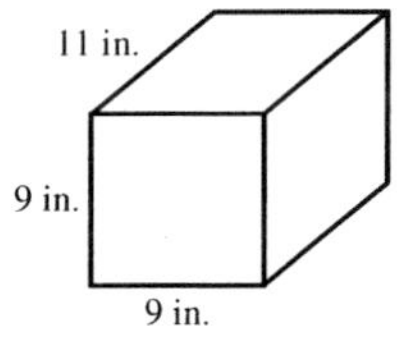

13. ________________

Name: Date:
Instructor: Section:

14.

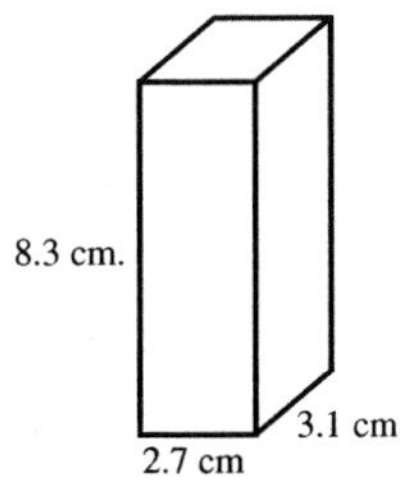

14. ________________

15.

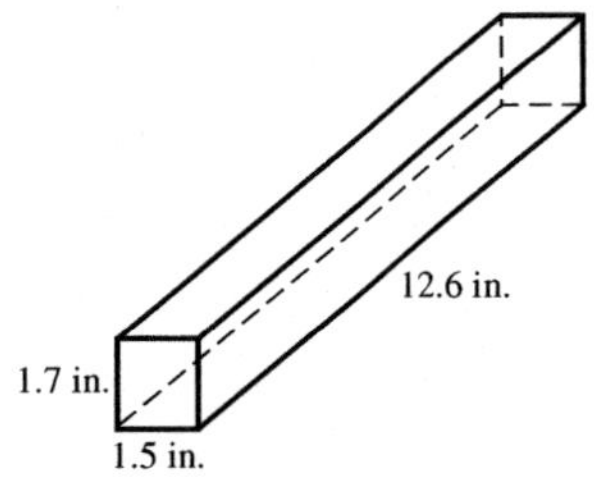

15. ________________

Objective 6 Find the surface area of a cylinder.

For extra help, see Example 9 on page 398 of your text and Section Lecture video for Section 5.10 and Exercise Solutions Clip 35 and 39.

Find the surface area of each cylinder. Use 3.14 as the approximate value for π. Round your answers to the nearest tenth.

16.

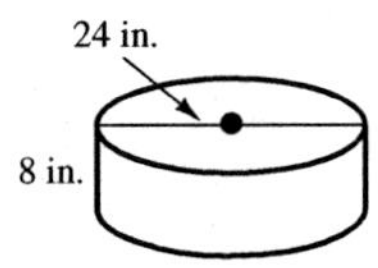

16. ________________

17.

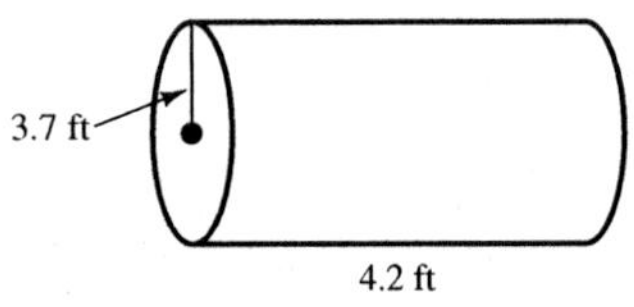

17. ________________

18.

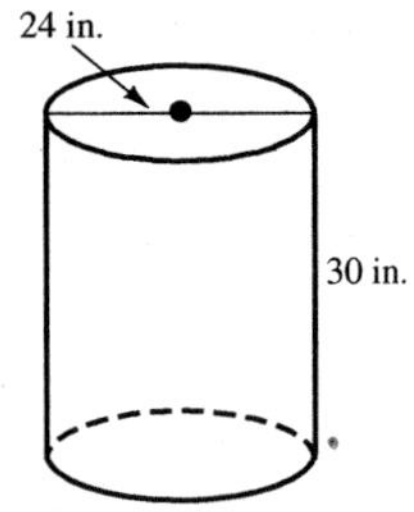

18. ________________

Name: Date:
Instructor: Section:

Chapter 6 RATIO, PROPORTION, AND LINE/ANGLE/TRIANGLE RELATIONSHIPS

6.1 Ratios

Learning Objectives

1 Write ratios as fractions.
2 Solve ratio problems involving decimals or mixed numbers.
3 Solve ratio problems after converting units.

Key Terms

Use the vocabulary terms listed below to complete each statement in exercises 1–2.

denominator **numerator** **ratio**

1. A ______________________ can be used to compare two measurements with the same type of units.

2. When writing the ratio to compare the width of a room to its height, the width goes in the ____________________ and the height goes in the ____________________.

Guided Examples

Review these examples for Objective 1:

1. A meeting room measures 22 feet long, 17 feet wide and 13 feet high. Write each ratio as a fraction, using the room measurements.

a. Ratio of length to width

The ratio of length to width is $\frac{22 \not{ft}}{17 \not{ft}} = \frac{22}{17}$

b. Ratio of width to height.

The ratio of width to height is $\frac{17 \not{ft}}{13 \not{ft}} = \frac{17}{13}$

Now Try:

1. Jacob spent \$5 for breakfast, \$8 for lunch, and \$17 for dinner. Write each ratio as a fraction.

a. Ratio of the amount spent on a dinner to the amount spent on lunch

____

b. Ratio of the amount spent on lunch to the amount spent on breakfast

____

Name: Date:
Instructor: Section:

2. Write each ratio in lowest terms.

a. 32 inches of snow to 8 inches of rain

The ratio is $\frac{32 \text{ inches}}{8 \text{ inches}}$. Divide out the common units. Then write this ratio in lowest terms by dividing the numerator and denominator by 8.

$$\frac{32 \cancel{\text{inches}}}{8 \cancel{\text{inches}}} = \frac{32}{8} = \frac{32 \div 8}{8 \div 8} = \frac{4}{1}$$

So, the ratio of 32 inches of snow to 8 inches of rain is 4 to 1, or $\frac{4}{1}$. For every 4 inches of snow there is 1 inch of rain.

b. 12 ounces of cleaner to 40 ounces of water

The ratio is $\frac{12 \text{ ounces}}{40 \text{ ounces}}$. Divide out the common units. Then write this ratio in lowest terms by dividing the numerator and denominator by 4.

$$\frac{12 \cancel{\text{ounces}}}{40 \cancel{\text{ounces}}} = \frac{12}{40} = \frac{12 \div 4}{40 \div 4} = \frac{3}{10}$$

So, the ratio of 12 ounces of cleaner to 40 ounces of water is $\frac{3}{10}$. For every 3 ounces of cleaner, there are 10 ounces of water.

c. 14 people at a large table to 8 people at a small table

The ratio is $\frac{14}{8} = \frac{14 \div 2}{8 \div 2} = \frac{7}{4}$

2. Write each ratio in lowest terms.

a. 16 seconds to 24 seconds

____

b. 150 grams to 75 grams

____

c. width of 14 m to length of 21 m

____

Review these examples for Objective 2:

3. The price of a newspaper increased from \$0.75 to \$1.00. Find the ratio of the increase in price to the original price.

new price − original price = increase

\$1.00 − \$0.75 = \$0.25

The words "the original price" are mentioned second, so the original price of \$0.75 is in the denominator.

The ratio of increase in price to original price is shown below.

$$\frac{0.25}{0.75}$$

Now rewrite the ratio as a ratio of whole

Now Try:

3. The original price of a burger is \$4.80 and the sale price is \$3.00. Find the ratio of the decrease in price to the original price.

____

numbers. Start by multiplying both the numerator and denominator by 100.

$$\frac{0.25}{0.75}=\frac{0.25\times 100}{0.75\times 100}=\frac{25}{75}=\frac{25\div 25}{75\div 25}=\frac{1}{3}$$

4. Write each ratio as a comparison of whole numbers in lowest terms.

a. 5 hours to $5\frac{1}{3}$ hours

Divide out the common units.

$$\frac{5\ \cancel{\text{hours}}}{5\frac{1}{3}\ \cancel{\text{hours}}}=\frac{5}{5\frac{1}{3}}$$

Next, write 5 as $\frac{5}{1}$ and $5\frac{1}{3}$ as the improper fraction $\frac{16}{3}$.

$$\frac{5}{5\frac{1}{3}}=\frac{\frac{5}{1}}{\frac{16}{3}}$$

Now, rewrite the problem in horizontal format, using the "$\div$" symbol for division. Finally, multiply by the reciprocal of the divisor.

$$\frac{\frac{5}{1}}{\frac{16}{3}}=\frac{5}{1}\div\frac{16}{3}=\frac{5}{1}\cdot\frac{3}{16}=\frac{15}{16}$$

The ratio, in lowest terms, is $\frac{15}{16}$.

b. $7\frac{1}{5}$ to $2\frac{1}{4}$

Write the ratio as $\frac{7\frac{1}{5}}{2\frac{1}{4}}$. Then write $7\frac{1}{5}$ and $2\frac{1}{4}$ as improper fractions.

$$7\frac{1}{5}=\frac{36}{5}\qquad\text{and}\qquad 2\frac{1}{4}=\frac{9}{4}$$

The ratio is shown here.

$$\frac{7\frac{1}{5}}{2\frac{1}{4}}=\frac{\frac{36}{5}}{\frac{9}{4}}$$

Rewrite as a division problem in horizontal

4. Write each ratio as a comparison of whole numbers in lowest terms.

a. $6\frac{1}{3}$ to 3

____

b. $3\frac{2}{3}$ to $2\frac{5}{6}$

____

format, using the "÷" symbol for division. Then multiply by the reciprocal of the divisor.

$$\frac{36}{5} \div \frac{9}{4} = \frac{\overset{4}{\cancel{36}}}{5} \cdot \frac{4}{\underset{1}{\cancel{9}}} = \frac{16}{5}$$

Review this example for Objective 3:

5. Write the ratio of length of a board 3 ft long to the length of another board that is 42 inches long.

First, express 3 ft in inches. Because 1 ft has 12 in., 3 ft is

$3 \cdot 12 \text{ in.} = 36 \text{ in.}$

The ratio of the lengths is

$$\frac{3 \text{ ft}}{42 \text{ in.}} = \frac{36 \cancel{\text{in.}}}{42 \cancel{\text{in.}}} = \frac{36}{42}$$

Write the ratio in lowest terms.

$$\frac{36}{42} = \frac{36 \div 6}{42 \div 6} = \frac{6}{7}$$

The shorter board is $\frac{6}{7}$ the length of the longer board.

Now Try:

5. Write the ratio of 20 days to 4 weeks.

____

Objective 1 Write ratios as fractions.

For extra help, see Examples 1–2 on pages 420–422 of your text and Section Lecture video for Section 6.1 and Exercise Solutions Clip 9.

Write each ratio as a fraction in lowest terms.

1. 125 cents to 95 cents — **1.** ____________

2. 80 miles to 30 miles — **2.** ____________

3. 5 men to 20 men — **3.** ____________

Objective 2 Solve ratio problems involving decimals or mixed numbers.

For extra help, see Examples 3–4 on pages 422–423 of your text and Section Lecture video for Section 6.1 and Exercise Solutions Clip 13 and 33.

Write each ratio as a fraction in lowest terms.

4. $4\frac{1}{8}$ to 3 — **4.** ____________

Name: Date:
Instructor: Section:

5. 11 to $2\frac{4}{9}$ **5.** ________________

Solve. Write each ratio as a fraction in lowest terms.

6. One car has a $15\frac{1}{2}$ gallon gas tank while another has a 22 gallon gas tank. Find the ratio of the amount the first tank holds to the amount the second tank holds. **6.** ________________

Objective 3 Solve ratio problems after converting units.

For extra help, see Example 5 on page 424 of your text and Section Lecture video for Section 6.1 and Exercise Solutions Clip 21.

Write each ratio as a fraction in lowest terms. Be sure to convert units as necessary.

7. 4 days to 2 weeks **7.** ________________

8. 6 yards to 10 feet **8.** ________________

9. 40 ounces to 3 pounds **9.** ________________

Name: Date:
Instructor: Section:

Chapter 6 RATIO, PROPORTION, AND LINE/ANGLE/TRIANGLE RELATIONSHIPS

6.2 Rates

Learning Objectives	
1	Write rates as fractions.
2	Find unit rates.
3	Find the best buy based on cost per unit.

Key Terms

Use the vocabulary terms listed below to complete each statement in exercises 1−3.

rate **unit rate** **cost per unit**

1. When the denominator of a rate is 1, it is called a ______________________.

2. The ______________________ is that rate that tells how much is paid for one item.

3. A ____________________ compares two measurements with different units.

Guided Examples

Review these examples for Objective 1:

1. Write each rate as a fraction in lowest terms.

a. 7 gallons for \$56

Write the units: gallons and dollars

$$\frac{7 \text{ gallons} \div 7}{56 \text{ dollars} \div 7} = \frac{1 \text{ gallon}}{8 \text{ dollars}}$$

b. 75 inches of growth in 15 weeks

$$\frac{75 \text{ inches} \div 15}{15 \text{ weeks} \div 15} = \frac{5 \text{ inches}}{1 \text{ week}}$$

c. 984 miles on 32 gallons of gas

$$\frac{984 \text{ miles} \div 8}{32 \text{ gallons} \div 8} = \frac{123 \text{ miles}}{4 \text{ gallons}}$$

Now Try:

1. Write each rate as a fraction in lowest terms.

a. \$7 for 35 pages

____

b. 300 strokes in 20 minutes

____

c. 396 strawberries for 24 cakes

____

Name: Date:
Instructor: Section:

Review these examples for Objective 2:

2. Find each unit rate.

a. 478.5 miles on 14.5 gallons of gas

Write the rate as a fraction.

$$\frac{478.5 \text{ miles}}{14.5 \text{ gallons}}$$

Divide 478.5 by 14.5 to find the unit rate.

$$145\overline{)4785}\quad 33$$

$$\frac{478.5 \text{ miles} \div 14.5}{14.5 \text{ gallons} \div 14.5} = \frac{33 \text{ miles}}{1 \text{ gallon}}$$

The unit rate is 33 miles per gallon or 33 miles/gallon.

b. 770 miles in 22 hours

$$\frac{770 \text{ miles}}{22 \text{ hours}} \quad \text{Divide: } 22\overline{)770}\quad 35$$

The unit rate is 35 miles/hour.

c. \$1240 in 8 days

$$\frac{1240 \text{ dollars}}{8 \text{ days}} \quad \text{Divide: } 8\overline{)1240}\quad 155$$

The unit rate is \$155/day.

Now Try:

2. Find each unit rate.

a. 294 miles on 10.5 gallons of gas

____

b. \$7.56 for 6 pounds of apples

____

c. \$580 in 4 days

____

Review these examples for Objective 3:

3. Determine the best price for peanut butter. For 18 ounces the price is \$2.89, for 28 ounces the price is \$3.99, and for 40 ounces, the price is \$6.18.

For 18 ounces, the cost per ounce is

$$\frac{\$2.89}{18 \text{ ounces}} \approx \$0.16$$

For 28 ounces, the cost per ounce is

$$\frac{\$3.99}{28 \text{ ounces}} \approx \$0.14$$

For 40 ounces, the cost per ounce is

$$\frac{\$6.18}{40 \text{ ounces}} \approx \$0.15$$

The lowest cost per ounce is \$0.14, so the 28-ounce container is the best buy.

Now Try:

3. Find the best buy:
2 pints for \$3.55,
3 pints for \$5.25, and
5 pints for \$8.50.

____

Name: Date:
Instructor: Section:

4. Solve each application problem.

a. Brand AA laundry detergent costs \$5.99 for 32 ounces. Brand ZZ laundry detergent costs \$13.29 for 100 ounces. Which choice is the best buy?

To find Brand AA's unit cost, divide \$5.99 by 32 ounces. Similarly, to find Brand ZZ's unit cost, divide \$13.29 by 100 ounces.

Brand AA $\frac{\$5.99}{32 \text{ ounces}} \approx 0.187$ per ounce

Brand ZZ $\frac{\$13.29}{100 \text{ ounces}} = 0.1329$ per ounce

Brand ZZ has the lower cost per ounce and is the better buy.

b. Special offers affect the best buy. Brand M of multivitamin costs \$3.99 for 200 + 100 tablets (a bonus of 100 tablets). Brand N of multivitamin costs \$5.47 for 200 tablets with a special Buy one Get one Free. Which choice is the best buy?

Brand M is \$3.99 for 300 tablets (200 + 100)
Brand N is \$5.47 for 400 tablets (200 + 200)

To find the best buy, divide to find the lowest cost per tablet.

Brand M $\frac{\$3.99}{300 \text{ tablets}} \approx \$0.0133$ per tablet

Brand N $\frac{\$5.47}{400 \text{ tablets}} \approx \$0.0137$ per tablet

Brand M has the lower cost per tablet and is the best buy.

4. Solve each application problem.

a. An eight-pack of AA-size batteries costs \$4.99. A twenty-pack of AA-size batteries costs \$12.99. Which battery pack is the best buy?

____

b. Which shampoo is the best buy? You have a coupon for 75¢ off Brand S and a coupon for \$1.25 off Brand T. Brand S is \$4.79 for 14 ounces. Brand T is \$5.69 for 18 ounces.

____

Objective 1 Write rates as fractions.

For extra help, see Example 1 on page 427 of your text and Section Lecture video for Section 6.2 and Exercise Solutions Clip 1, 3, and 5.

Write each rate as a fraction in lowest terms.

1. 119 pills for 17 patients **1.** __________

2. 28 dresses for 4 women **2.** __________

3. 256 pages for 8 chapters **3.** __________

Name: Date:
Instructor: Section:

Objective 2 Find unit rates.

For extra help, see Example 2 on page 428 of your text and Section Lecture video for Section 6.2 and Exercise Solutions Clip 9 and 11.

Find each unit rate.

4. $3500 in 20 days **4.** ________

5. $7875 for 35 pounds **5.** ________

6. 189.88 miles on 9.4 gallons **6.** ________

Objective 3 Find the best buy based on cost per unit.

For extra help, see Examples 3–4 on pages 428–430 of your text and Section Lecture video for Section 6.2 and Exercise Solutions Clip 21 and 23.

Find the best buy (based on cost per unit) for each item.

7. Peanut butter: 18 ounces for $1.77; 24 ounces for $2.08 **7.** ________

8. Batteries: 4 for $2.79; 10 for $4.19 **8.** ________

9. Soup: 3 cans for $1.75; 5 cans for $2.75; 8 cans for $4.55 **9.** ________

Name: Date:
Instructor: Section:

Chapter 6 RATIO, PROPORTION, AND LINE/ANGLE/TRIANGLE RELATIONSHIPS

6.3 Proportions

Learning Objectives

1	Write proportions.
2	Determine whether proportions are true or false.
3	Find the unknown number in a proportion.

Key Terms

Use the vocabulary terms listed below to complete each statement in exercises 1−3.

cross products **proportion** **ratio**

1. A ______________________________ shows that two ratios or rates are equivalent.

2. To see whether a proportion is true, determine if the ____________________ are equal.

3. A _____________________ is a comparison of two quantities with the same units.

Guided Examples

Review these examples for Objective 1:

1. Write each proportion.

a. 7 m is to 13 m as 28 m is to 52 m

$$\frac{7\ \cancel{m}}{13\ \cancel{m}} = \frac{28\ \cancel{m}}{52\ \cancel{m}} \quad \text{so} \quad \frac{7}{13} = \frac{28}{52}$$

b. \$14 is to 8 gallons as \$7 is to 4 gallons

$$\frac{\$14}{8 \text{ gallons}} = \frac{\$7}{4 \text{ gallons}}$$

Now Try:

1. Write each proportion.

a. 24 ft is to 17 ft as 72 ft is to 51 ft

____

b. \$10 is to 7 cans as \$60 is to 42 cans

____

Name: Date:
Instructor: Section:

Review these examples for Objective 2:

2. Determine whether each proportion is true or false by writing both ratios in lowest terms.

a. $\frac{7}{11} = \frac{16}{24}$

Write each ratio in lowest terms.

$\frac{7}{11} \leftarrow$ Already in lowest terms $\quad \frac{16 \div 8}{24 \div 8} = \frac{2}{3} \leftarrow$ Lowest terms

Because $\frac{7}{11}$ is not equivalent to $\frac{2}{3}$, the proportion is false.

b. $\frac{9}{15} = \frac{21}{35}$

Write each ratio in lowest terms.

$\frac{9 \div 3}{15 \div 3} = \frac{3}{5} \quad \frac{21 \div 7}{35 \div 7} = \frac{3}{5}$

Both ratios are equivalent to $\frac{3}{5}$, so the proportion is true.

3. Use cross products to see whether each proportion is true or false.

a. $\frac{5}{8} = \frac{30}{48}$

Multiply along one diagonal, then multiply along the other diagonal.

$\frac{5}{8} = \frac{30}{48}$ $\quad \nearrow 8 \cdot 30 = 240$ $\quad \searrow 5 \cdot 48 = 240$

The cross products are equal, so the proportion is true.

b. $\frac{3\frac{1}{5}}{4\frac{2}{3}} = \frac{7}{10}$

$\frac{3\frac{1}{5}}{4\frac{2}{3}} = \frac{7}{10}$

$\nearrow 4\frac{2}{3} \cdot 7 = \frac{14}{3} \cdot \frac{7}{1} = \frac{98}{3} = 32\frac{2}{3}$

$\searrow 3\frac{1}{5} \cdot 10 = \frac{16}{\cancel{5}_1} \cdot \frac{\cancel{10}^2}{1} = \frac{32}{1} = 32$

Now Try:

2. Determine whether each proportion is true or false by writing both ratios in lowest terms.

a. $\frac{36}{28} = \frac{24}{18}$

____

b. $\frac{4}{12} = \frac{9}{27}$

____

3. Use cross products to see whether each proportion is true or false.

a. $\frac{6}{17} = \frac{18}{51}$

____

b. $\frac{3.2}{5} = \frac{7}{10}$

____

The cross products are unequal, so the proportion is false.

Review these examples for Objective 3:

4. Find the unknown number in each proportion. Round answers to the nearest hundredth when necessary.

a. $\frac{14}{x}=\frac{21}{18}$

Recall that ratios can be rewritten in lowest terms. Write $\frac{21}{18}$ in lowest terms as $\frac{7}{6}$, which gives the proportion $\frac{14}{x}=\frac{7}{6}$.

Step 1 Find the cross product.

$$\frac{14}{x}=\frac{7}{6} \quad \nearrow x\cdot 7 \quad \searrow 14\cdot 6$$

Step 2 Show that cross products are equal.

$$x\cdot 7=14\cdot 6$$

$$x\cdot 7=84$$

Step 3 Divide both sides by 7.

$$\frac{x\cdot \overset{1}{\cancel{7}}}{\underset{1}{\cancel{7}}}=\frac{84}{7}$$

$$x=12$$

Step 4 Check in original proportion.

$$\frac{14}{12}=\frac{21}{18} \quad \nearrow 12\cdot 21=252 \quad \searrow 14\cdot 18=252$$

The cross products are equal, so 12 is the correct solution.

b. $\frac{9}{13}=\frac{21}{x}$

Step 1 Find the cross product.

$$\frac{9}{13}=\frac{21}{x} \quad \nearrow 13\cdot 21=273 \quad \searrow 9\cdot x$$

Step 2 Show that cross products are equal.

$$9\cdot x=273$$

Step 3 Divide both sides by 9.

Now Try:

4. Find the unknown number in each proportion. Round answers to the nearest hundredth when necessary.

a. $\frac{24}{x}=\frac{9}{12}$

____

b. $\frac{2}{3}=\frac{x}{16}$

____

Name: Date:
Instructor: Section:

$$\frac{\overset{1}{\cancel{9}} \cdot x}{\underset{1}{\cancel{9}}} = \frac{273}{9}$$

$x = 30.33$ rounded to the nearest hundredth

Step 4 Check in original proportion.

$$\frac{9}{13} = \frac{21}{30.33} \quad \begin{matrix} \nearrow 13 \cdot 21 = 273 \\ \searrow 9 \cdot 30.33 = 272.97 \end{matrix}$$

The cross products are slightly different because of the rounded value of x. However, they are close enough to see that the problem was done correctly and that 30.33 is the approximate solution.

5. Find the unknown number in each proportion.

a. $\frac{3\frac{1}{4}}{8} = \frac{x}{12}$

$$\frac{3\frac{1}{4}}{8} = \frac{x}{12} \quad \begin{matrix} \nearrow 8 \cdot x \\ \searrow 3\frac{1}{4} \cdot 12 \end{matrix}$$

Change $3\frac{1}{4}$ to an improper fraction and write in lowest terms.

$$3\frac{1}{4} \cdot 12 = \frac{13}{4} \cdot \frac{12}{1} = \frac{13}{\underset{1}{\cancel{4}}} \cdot \frac{\overset{3}{\cancel{12}}}{1} = \frac{39}{1} = 39$$

Show that the cross products are equal.

$8 \cdot x = 39$

Divide both sides by 8.

$$\frac{\overset{1}{\cancel{8}} \cdot x}{\underset{1}{\cancel{8}}} = \frac{39}{8}$$

Write the solution as a mixed number in lowest terms.

$$x = \frac{39}{8} = 4\frac{7}{8}$$

The unknown number is $4\frac{7}{8}$.

Check in original proportion.

5. Find the unknown number in each proportion.

a. $\frac{4\frac{1}{3}}{5} = \frac{x}{3}$

____

Name: Date:
Instructor: Section:

$$\frac{3\frac{1}{4}}{8} = \frac{4\frac{7}{8}}{12} \quad \begin{array}{l} \nearrow 8 \cdot 4\frac{7}{8} = \frac{\overset{1}{\cancel{8}}}{1} \cdot \frac{39}{\underset{1}{\cancel{8}}} = 39 \\ \searrow 3\frac{1}{4} \cdot 12 = \frac{13}{\underset{1}{\cancel{4}}} \cdot \frac{\overset{3}{\cancel{12}}}{1} = 39 \end{array}$$

The cross products are equal, so $4\frac{7}{8}$ is the correct solution.

b. $\frac{3.5}{1.4} = \frac{4}{x}$

Show that the products are equal.

$(3.5)(x) = (1.4)(4)$

$(3.5)(x) = 5.6$

Divide both sides by 3.5.

$$\frac{\overset{1}{\cancel{(3.5)}} \cdot (x)}{\underset{1}{\cancel{3.5}}} = \frac{5.6}{3.5}$$

$$x = \frac{5.6}{3.5}$$

Complete the division.

$x = 1.6$ $\quad 35\overline{)56.0}$ (quotient 1.6)

So, the unknown number is 1.6. Write the solution in the original proportion and check it by finding the cross products.

$$\frac{3.5}{1.4} = \frac{4}{1.6} \quad \begin{array}{l} \nearrow 1.4 \cdot 4 = 5.6 \\ \searrow 3.5 \cdot 1.6 = 5.6 \end{array}$$

The cross products are equal, so 1.6 is the correct solution.

b. $\frac{x}{8} = \frac{1.2}{1.5}$

____

Objective 1 Write proportions.

For extra help, see Example 1 on page 435 of your text and Section Lecture video for Section 6.3 and Exercise Solutions Clip 1, 3, and 5.

Write each proportion.

1. 50 is to 8 as 75 is to 12. **1.** _______________

Name: Date:
Instructor: Section:

2. 36 is to 45 as 8 is to 10. **2.** ________________

3. 3 is to 33 as 12 is to 132. **3.** ________________

Objective 2 Determine whether proportions are true or false.

For extra help, see Examples 2–3 on pages 435–437 of your text and Section Lecture video for Section 6.3 and Exercise Solutions Clip 9, 11, 15, and 25.

Determine whether each proportion is true or false by writing the ratios in lowest terms. Show the simplified ratios and then write **true** *or* **false**.

4. $\frac{48}{36}=\frac{3}{4}$ **4.** ________________

5. $\frac{30}{25}=\frac{6}{5}$ **5.** ________________

Use cross products to determine whether the proportion is true or false. Show the cross products and then write **true** *or* **false**.

6. $\frac{4\frac{3}{5}}{9}=\frac{18\frac{2}{5}}{36}$ **6.** ________________

Objective 3 Find the unknown number in a proportion.

For extra help, see Examples 4–5 on pages 438–441 of your text and Section Lecture video for Section 6.3 and Exercise Solutions Clip 31, 33, and 39.

Find the unknown number in the proportion.

7. $\frac{9}{7}=\frac{x}{28}$ **7.** ________________

Find the unknown number in each proportion.

8. $\frac{2}{3\frac{1}{4}}=\frac{8}{x}$ **8.** ________________

9. $\frac{3}{x}=\frac{0.8}{5.6}$ **9.** ________________

Name: Date:
Instructor: Section:

Chapter 6 RATIO, PROPORTION, AND LINE/ANGLE/TRIANGLE RELATIONSHIPS

6.4 Problem Solving with Proportions

Learning Objectives
1 Use proportions to solve application problems.

Key Terms

Use the vocabulary terms listed below to complete each statement in exercises 1–2.

rate **ratio**

1. A statement that compares a number of inches to a number of inches is a ________________.

2. A statement that compares a number of gallons to a number of miles is a ____________________.

Guided Examples

Review these examples for Objective 1:

1. Alexis drove 343 miles on 7.5 gallons of gas. How far can she travel on a full tank of 12 gallons of gas?

Step 1 Read the problem. The problem asks for the number of miles the car can travel on 12 gallons of gas.

Step 2 Work out a plan. Decide what is being compared and write a proportion using the two rates.

$$\frac{343 \text{ miles}}{7.5 \text{ gallons}} = \frac{x \text{ miles}}{12 \text{ gallons}}$$

Step 3 Estimate a reasonable answer. To estimate the answer, notice that 12 is a little more than 1.5 times 7.5. So use $343 \cdot 1.5 = 514.5$ miles, as an estimate.

Step 4 Solve the problem. Ignore the units while solving for x.

$$\frac{343 \text{ miles}}{7.5 \text{ gallons}} = \frac{x \text{ miles}}{12 \text{ gallons}}$$
$$(7.5)(x) = (343)(12)$$
$$(7.5)(x) = 4116$$

Now Try:

1. Aiden spends 23 hours painting 4 apartments. How long will it take him to paint the other 16 apartments?

____

Name: Date:
Instructor: Section:

$$\frac{\overset{1}{\cancel{(7.5)}}(x)}{\underset{1}{\cancel{7.5}}} = \frac{4116}{7.5}$$

$x = 548.8$ Round to 549.

Step 5 State the answer. Rounded to the nearest mile, the car can travel about 549 miles on a full tank of gas.

Step 6 Check your work. The answer 549 miles, is a little more than the estimate of 514.5 miles, so it is reasonable.

2. There are 24 women in a college class of 39. At that rate, how many of the college's 10,400 students are women?

Step 1 Read the problem. The problem asks how many of the 10,400 students are women.

Step 2 Work out a plan. Decide what is being compared and write a proportion using the two rates.

$$\frac{24 \text{ women}}{39 \text{ students}} = \frac{x \text{ women}}{10,400 \text{ students}}$$

Step 3 Estimate a reasonable answer. To estimate the answer, notice that 24 is a little more than half 39. Half of 10,400 is $10,400 \div 2 = 5200$, so our estimate is more than 5200 students.

Step 4 Solve the problem. Ignore the units while solving for x.

$$\frac{24 \text{ women}}{39 \text{ students}} = \frac{x \text{ women}}{10,400 \text{ students}}$$

$$39 \cdot x = 24 \cdot 10,400$$

$$39 \cdot x = 249,600$$

$$\frac{\overset{1}{\cancel{39}} \cdot x}{\underset{1}{\cancel{39}}} = \frac{249,600}{39}$$

$x = 6400$ No need to round.

Step 5 State the answer. There are 6400 woman at the college.

Step 6 Check your work. The answer 6400 women, is a little more than the estimate of 5200 women, so it is reasonable.

2. A survey showed that 4 out of 5 smokers have tried to quit smoking. At this rate, how many people in a group of 540 have tired to quit smoking?

____

Name: Date:
Instructor: Section:

Objective 1 Use proportions to solve application problems.

For extra help, see Examples 1–2 on pages 448–450 of your text and Section Lecture video for Section 6.4 and Exercise Solutions Clip 5, 17, and 23.

Set up and solve a proportion for each problem.

1. If 22 hats cost \$198, find the cost of 12 hats. **1.** ________________

2. If 150 square yards of carpet cost \$3142.50, find the cost of 210 square yards of the carpet. **2.** ________________

3. A biologist tags 50 deer and releases them in a wildlife preserve area. Over the course of a two-week period, she observes 80 deer, of which 12 are tagged. What is the estimate for the population of deer in this particular area? **3.** ________________

Name: Date:
Instructor: Section:

Chapter 6 RATIO, PROPORTION, AND LINE/ANGLE/TRIANGLE RELATIONSHIPS

6.5 Geometry: Lines and Angles

Learning Objectives

1 Identify and name lines, line segments, and rays.
2 Identify parallel and intersecting lines.
3 Identify and name angles.
4 Classify angles as right, acute, straight, or obtuse.
5 Identify perpendicular lines.
6 Identify complementary angles and supplementary angles and find the measure of a complement or supplement of a given angle.
7 Identify congruent angles and vertical angles and use this knowledge to find the measures of angles.
8 Identify corresponding angles and alternate interior angles and use this knowledge to find the measures of angles.

Key Terms

Use the vocabulary terms listed below to complete each statement in exercises 1–19.

point	**line**	**line segment**	**ray**
parallel lines	**intersecting lines**	**angle**	
degrees	**straight angle**	**right angle**	
acute angle	**obtuse angle**	**perpendicular lines**	
complementary angles	**supplementary angles**		
congruent angles	**vertical angles**		
corresponding angles	**alternate interior angles**		

1. A ____________________________ is a part of a line that has one endpoint and which extends infinitely in one direction.

2. Two lines that intersect to form a right angle are ____________________________.

3. An angle whose measure is between 90° and 180° is an ________________________.

4. A ________________________ is a location in space.

5. Two rays with a common endpoint form an __________________________.

6. A set of points that form a straight path that extends infinitely in both directions is called a ____________________________.

7. An angle that measures less than 90° is called an __________________________.

Name: Date:
Instructor: Section:

8. Angles are measured using ______________________.

9. Two lines in the same plane that never intersect are ______________________.

10. A part of a line with two endpoints is a ______________________.

11. Two lines that cross at one point are ______________________.

12. An angle whose measure is exactly 90° is a ______________________.

13. The nonadjacent angles formed by two intersecting lines are called ______________________.

14. Angles whose measures are equal are called ______________________.

15. Two angles whose measures sum to 180° are ______________________.

16. Two angles whose measures sum to 90° are ______________________.

17. When two parallel lines are cut by a transversal, the angles between the parallel lines on opposite sides of the transversal are called ______________________.

18. When two parallel lines are cut by a transversal, the angles in the same relative position with regard to the parallel lines and the transversal are called ______________________.

19. An angle whose measure is exactly 180° is a ______________________.

Guided Examples

Review these examples for Objective 1:

1. Identify each figure below as a line, line segment, or ray, and name it using the appropriate symbol.

a.

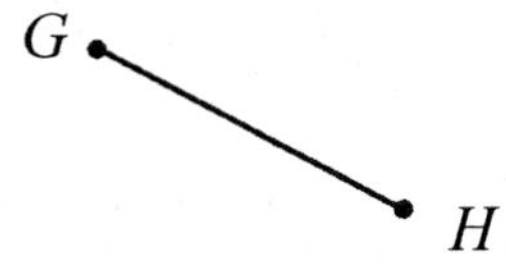

This figure has two endpoints, so it is a line segment named $\overline{GH}$ or $\overline{HG}$.

Now Try:

1. Identify each figure below as a line, line segment, or ray, and name it using the appropriate symbol.

a.

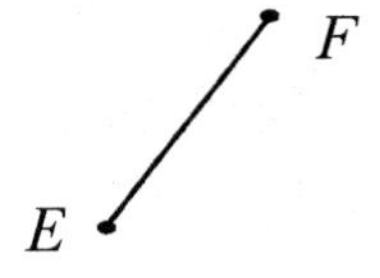

____

Name: Date:
Instructor: Section:

b.

This figure starts at point B and goes on forever in one direction, so it is a ray named $\overrightarrow{BA}$.

c.

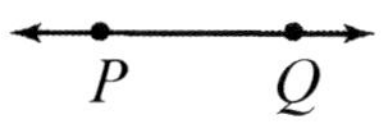

This figure goes on forever in both directions, so it is a line named $\overleftrightarrow{PQ}$ or $\overleftrightarrow{QP}$.

b.

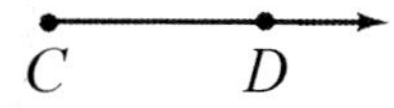

____

c.

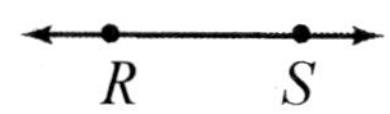

____

Review these examples for Objective 2:

2. Label each pair of the lines as appearing to be parallel or as intersecting.

a.

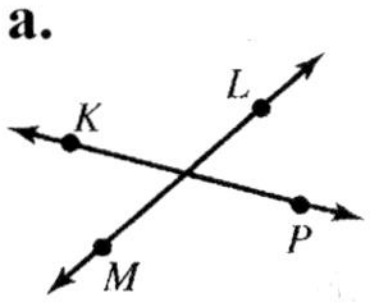

The lines in this figure cross, so they are intersecting lines.

b.

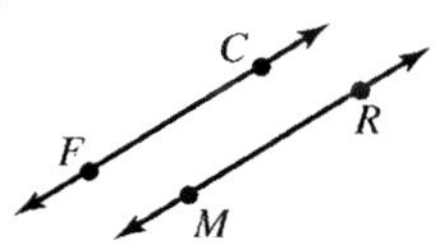

The lines in this figure do not intersect; they appear to be parallel.

Now Try:

2. Label each pair of the lines as appearing to be parallel or as intersecting.

a.

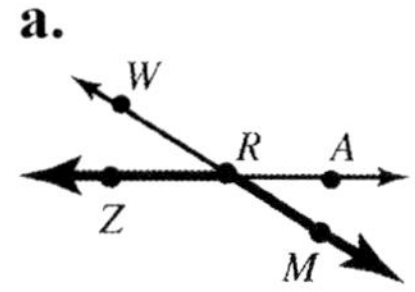

____

b.

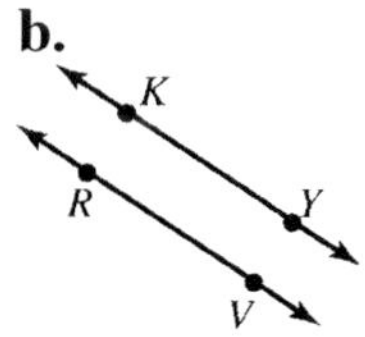

____

Review this example for Objective 3:

3. Name the highlighted angle in two different ways.

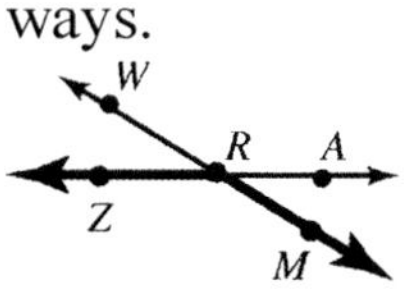

The angle can be named $\angle ZRM$ or $\angle MRZ$. It cannot be named $\angle R$, using the vertex alone, because four different angles have R as their vertex.

Now Try:

3. Name the highlighted angle in two different ways.

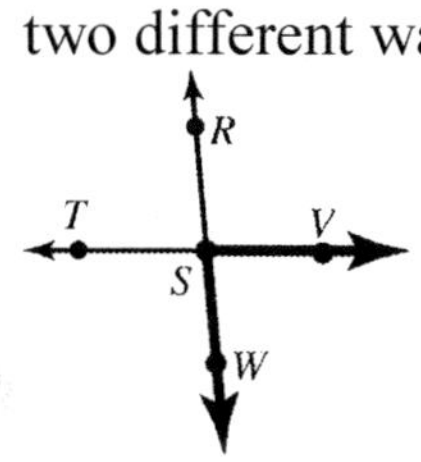

____

Name: Date:
Instructor: Section:

Review these examples for Objective 4:

4. Label each angle as acute, right, obtuse, or straight.

a.

This figure shows a straight angle (exactly 180°).

b.

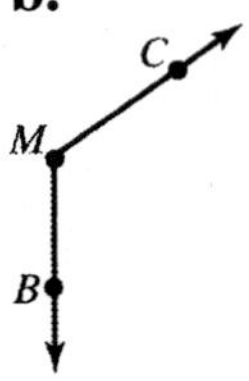

This figure shows an obtuse angle (more than 90° but less than 180°).

c.

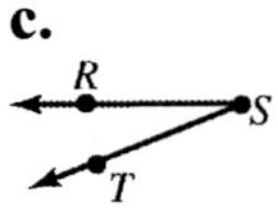

This figure shows an acute angle (less than 90°).

d.

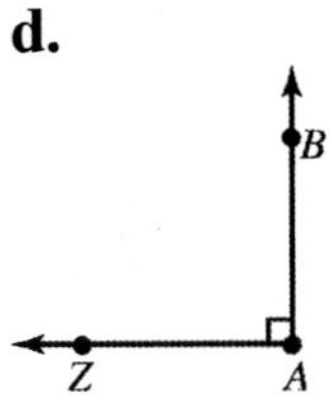

This figure shows a right angle (exactly 90° and identified by a small square at the vertex).

Now Try:

4. Label each angle as acute, right, obtuse, or straight.

a.

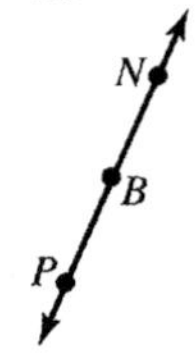

b.

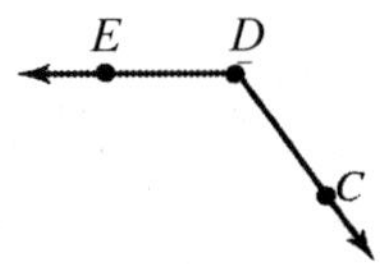

c.

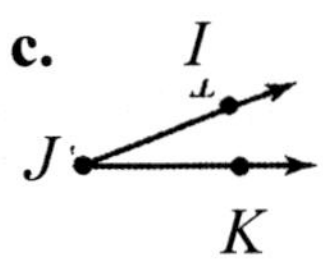

d.

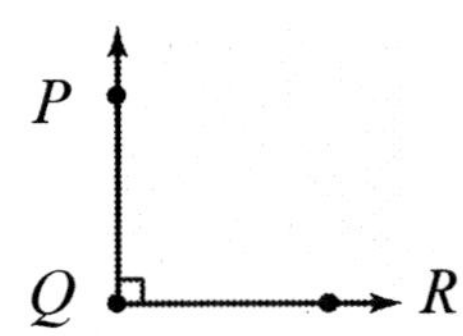

Review these examples for Objective 5:

5. Which pairs of lines are perpendicular?

a.

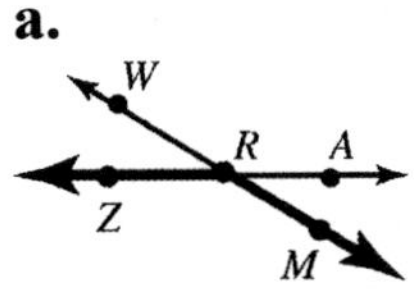

The lines in this figure are intersecting lines, but they are not perpendicular because they do not form a right angle.

Now Try:

5. Which pairs of lines are perpendicular?

a.

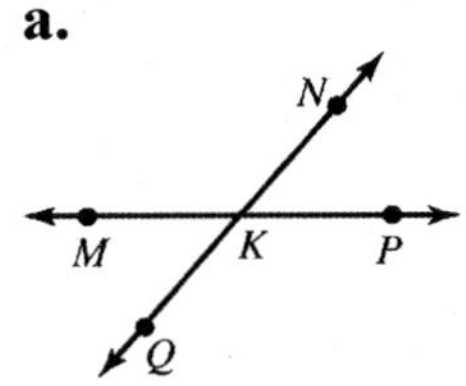

Name: Date:
Instructor: Section:

b.

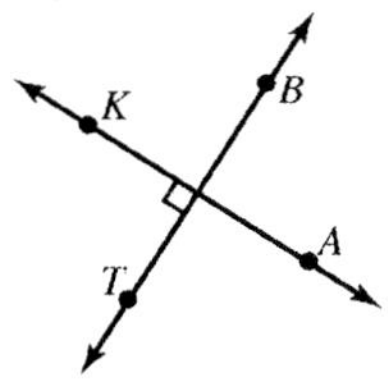

The lines in this figure are perpendicular to each other because they intersect at right angles.

b.

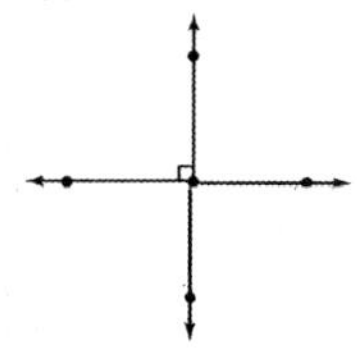

____

Review these examples for Objective 6:

6. Identify each pair of complementary angles.

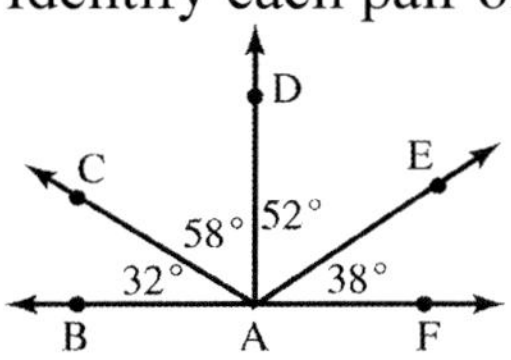

$\angle BAC\,(32°)$ and $\angle CAD\,(58°)$ are complementary angles because
$32° + 58° = 90°$

$\angle DAE\,(52°)$ and $\angle EAF\,(38°)$ are complementary angles because
$52° + 38° = 90°$

7. Find the complement of each angle.

a. 66°

Find the complement of 66° by subtracting.
$90° - 66° = 24°$

b. 4°

Find the complement of 4° by subtracting.
$90° - 4° = 86°$

8. In the figures below, $\angle ABC$ measures 150°, $\angle WXY$ measures 30°, $\angle MKQ$ measures 30°, and $\angle QKP$ measures 150°. Identify each pair of supplementary angles.

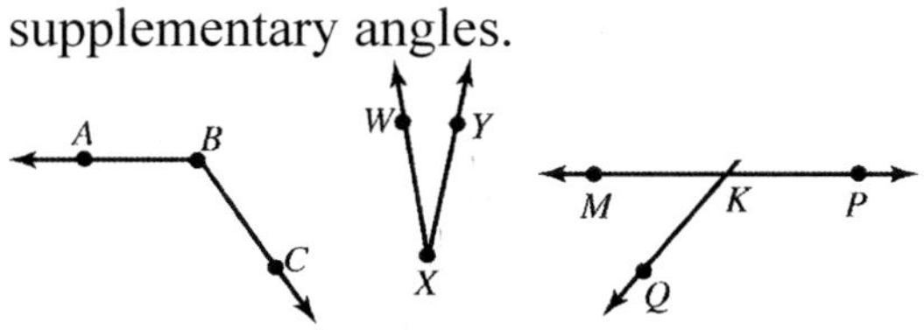

$\angle ABC$ and $\angle WXY$, because $150° + 30° = 180°$
$\angle MKQ$ and $\angle QKP$, because $30° + 150° = 180°$
$\angle ABC$ and $\angle MKQ$, because $150° + 30° = 180°$
$\angle WXY$ and $\angle QKP$, because $30° + 150° = 180°$

Now Try:

6. Identify each pair of complementary angles.

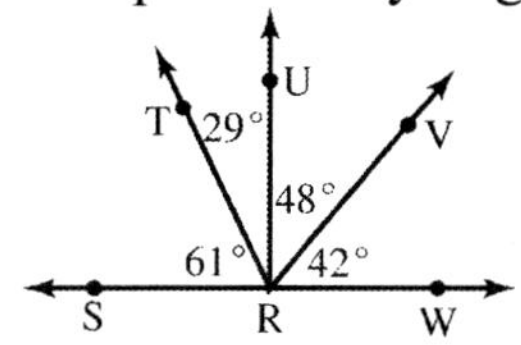

____

7. Find the complement of each angle.

a. 72°

____

b. 12°

____

8. In the figures below, $\angle OQP$ measures 35°, $\angle PQN$ measures 145°, $\angle RST$ measures 35°, and $\angle BMC$ measures 145°. Identify each pair of supplementary angles.

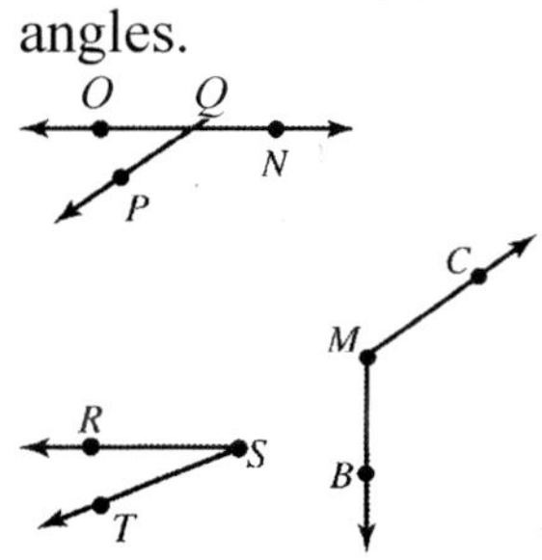

____

Name: Date:
Instructor: Section:

9. Find the supplement of each angle.

a. 38°

Find the supplement of 38° by subtracting.
$180° - 38° = 142°$

b. 121°

Find the supplement of 121° by subtracting.
$180° - 121° = 59°$

9. Find the supplement of each angle.

a. 82° ________

b. 168° ________

Review this example for Objective 7:

10. Identify the angles that are congruent.

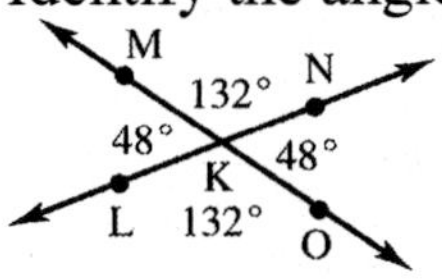

Congruent angles measure the same number of degrees.
$\angle LKM \cong \angle OKN$ and $\angle LKO \cong \angle MKN$.

11. Identify the vertical angles in this figure.

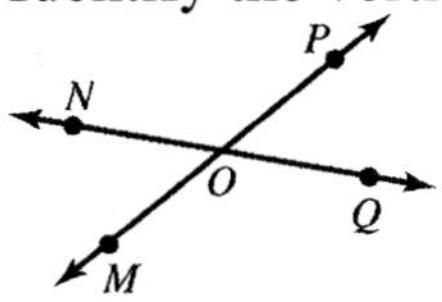

$\angle POQ$ and $\angle NOM$ are vertical angles because they do not share a common side and they are formed by two intersecting lines $(\overleftrightarrow{NQ}$ and $\overleftrightarrow{MP})$.

$\angle NOP$ and $\angle MOQ$ are also vertical angles.

Now Try:

10. Identify the angles that are congruent.

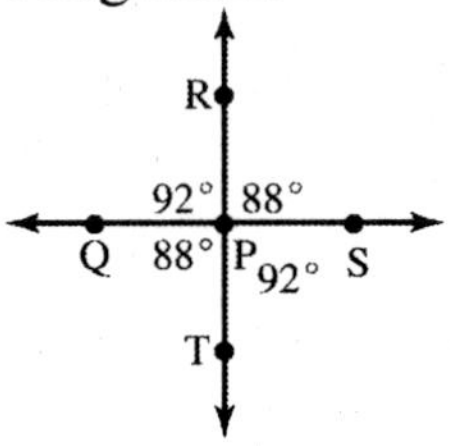

____

11. Identify the vertical angles in this figure.

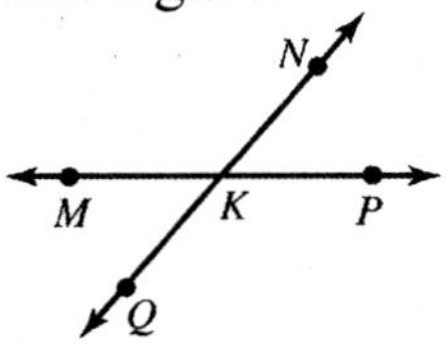

____

Name: Date:
Instructor: Section:

12. In the figure below, $\angle APB$ measures 93° and $\angle BPC$ measures 37°. Find the measures of the indicated angles.

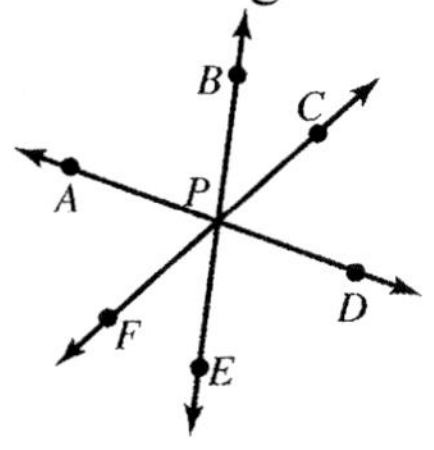

a. $\angle EPF$

$\angle EPF$ and $\angle BPC$ are vertical angles, so they are congruent. This means they measure the same number of degrees.
The measure of $\angle BPC$ is 37°, so the measure of $\angle EPF$ is 37° also.

b. $\angle DPE$

$\angle DPE$ and $\angle APB$ are vertical angles, so they are congruent.
The measure of $\angle APB$ is 93°, so the measure of $\angle DPE$ is 93° also.

c. $\angle CPD$

Look at $\angle CPD$, $\angle BPC$, and $\angle APB$. Notice that $\overrightarrow{PA}$ and $\overrightarrow{PD}$ go in opposite directions. Therefore, $\angle APD$ is a straight angle and measures 180°. To find the measure of $\angle CPD$, subtract the sum of the other two angles from 180°.

$$180° - (93° + 37°) = 180° - (130°) = 50°$$

The measure of $\angle CPD$ is 50°.

d. $\angle FPA$

$\angle FPA$ and $\angle CPD$ are vertical angles, so they are congruent. We know from part (c) that the measure of $\angle CPD$ is 50°, so the measure of $\angle FPA$ is 50° also.

12. In the figure below, $\angle AGF$ measures 33° and $\angle BGC$ measures 105°. Find the measures of the indicated angles.

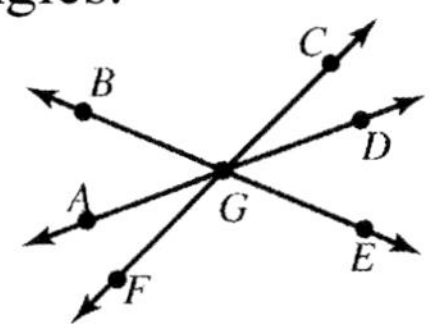

a. $\angle CGD$

____

b. $\angle EGF$

____

c. $\angle DGE$

____

d. $\angle BGA$

____

Name: Date:
Instructor: Section:

Review these examples for Objective 8:

13. In each figure, line *m* is parallel to line *n*.

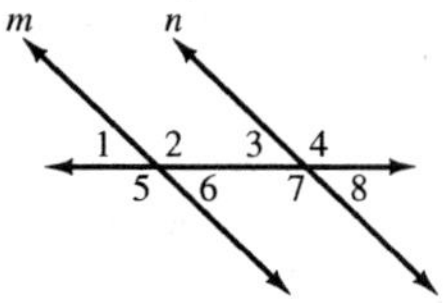

a. Identify all pairs of corresponding angles.

There are four pair of corresponding angles:

$\angle 1$ and $\angle 3$ $\quad$ $\angle 2$ and $\angle 4$

$\angle 5$ and $\angle 7$ $\quad$ $\angle 6$ and $\angle 8$

b. Identify all pairs of alternate interior angles.

There are two pair of alternate interior angles:

$\angle 2$ and $\angle 7$

$\angle 6$ and $\angle 3$

14. In the figure below, line *m* is parallel to line *n* and the measure of $\angle 3$ is 44°. Find the measure of the other angles.

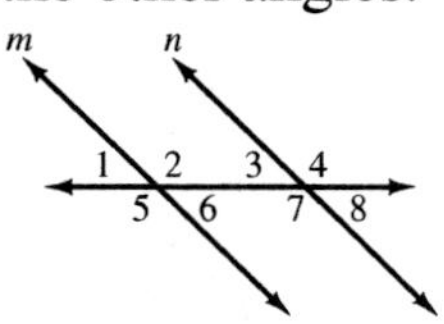

We know that $\angle 3$ is 44°.

$\angle 3 \cong \angle 1$ (corresponding angles), so the measure of $\angle 1$ is also 44°.

$\angle 3 \cong \angle 6$ (alternate interior angles), so the measure of $\angle 6$ is also 44°.

$\angle 6 \cong \angle 8$ (corresponding angles), so the measure of $\angle 8$ is also 44°.

Notice that the exterior sides of $\angle 3$ and $\angle 4$ form a straight line. Therefore, $\angle 3$ and $\angle 4$ are supplementary angles. If $\angle 3$ is 44°, then $\angle 4$ must be $180° - 44° = 136°$.

$\angle 4 \cong \angle 2$ (corresponding angles), so the measure of $\angle 4$ is also 136°.

$\angle 7 \cong \angle 2$ (alternate interior angles), so the measure of $\angle 2$ is also 136°.

$\angle 7 \cong \angle 5$ (corresponding angles), so the measure of $\angle 5$ is also 136°.

Now Try:

13. In each figure, line *m* is parallel to line *n*. Identify all pairs of corresponding angles and all pairs of alternate interior angles.

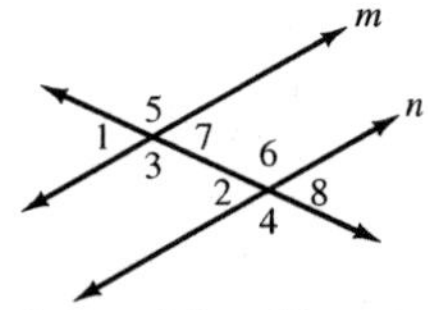

a. Identify all pairs of corresponding angles.

____

____

b. Identify all pairs of alternate interior angles.

____

____

14. In the figure below, line *m* is parallel to line *n* and the measure of $\angle 6$ is 125°. Find the measure of the other angles.

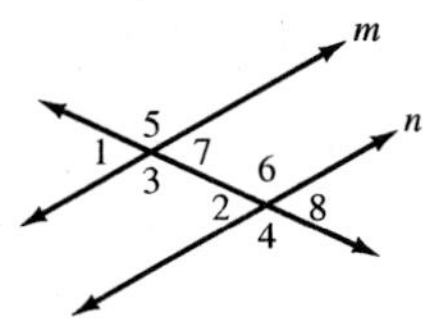

____

____

____

____

Name: Date:
Instructor: Section:

Objective 1 Identify and name lines, line segments, and rays.

For extra help, see Example 1 on page 455 of your text and Section Lecture video for Section 6.5.

Identify each figure as a line, line segment, or ray, and name it.

1.

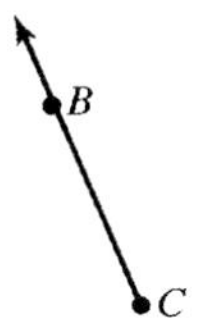

1. ____________________

2.

2. ____________________

3.

3. ____________________

Objective 2 Identify parallel and intersecting lines.

For extra help, see Example 2 on page 456 of your text and Section Lecture video for Section 6.5.

Label each pair of lines as appearing to be **parallel** *or* **intersecting**.

4.

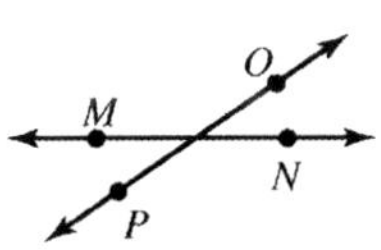

4. ____________________

5.

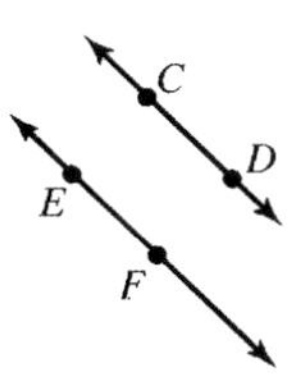

5. ____________________

6.

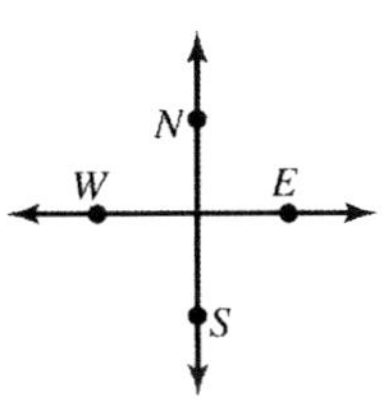

6. ____________________

Name: Date:
Instructor: Section:

Objective 3 Identify and name angles.

For extra help, see Example 3 on page 457 of your text and Section Lecture video for Section 6.5 and Exercise Solutions Clip 15.

Name each angle drawn with darker rays by using the three-letter form of identification.

7.

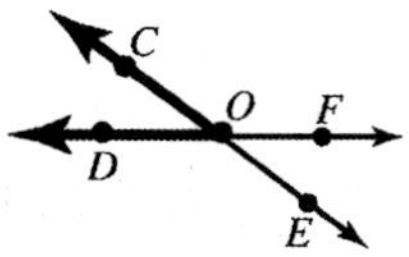

7. ________________

8.

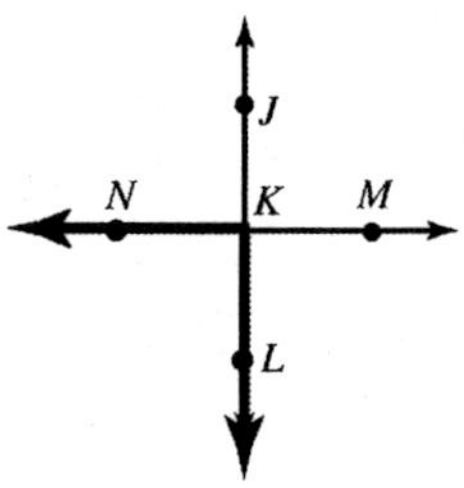

8. ________________

9.

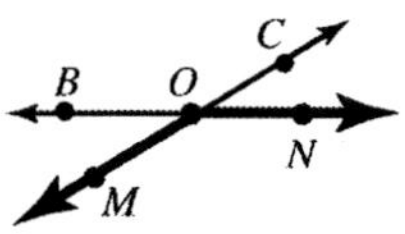

9. ________________

Objective 4 Classify angles as right, acute, straight, or obtuse.

For extra help, see Example 4 on page 458 of your text and Section Lecture video for Section 6.5.

Label each angle as **acute**, **right**, **obtuse**, *or* **straight**.

10.

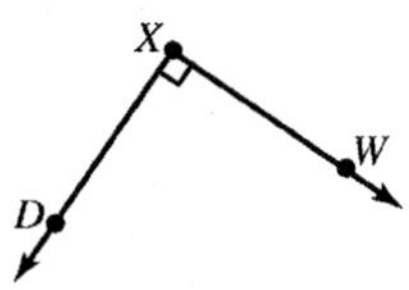

10. ________________

11.

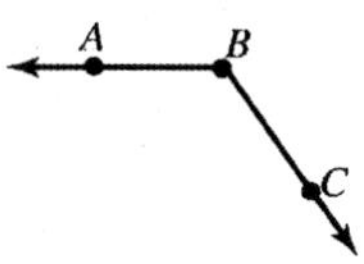

11. ________________

12.

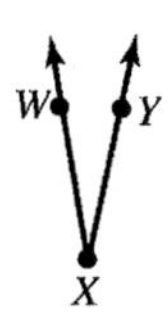

12. ________________

Name: Date:
Instructor: Section:

Objective 5 Identify perpendicular lines.

For extra help, see Example 5 on page 459 of your text and Section Lecture video for Section 6.5.

Label each pair of lines as appearing to be **parallel**, **perpendicular**, *or* **intersecting**.

13.

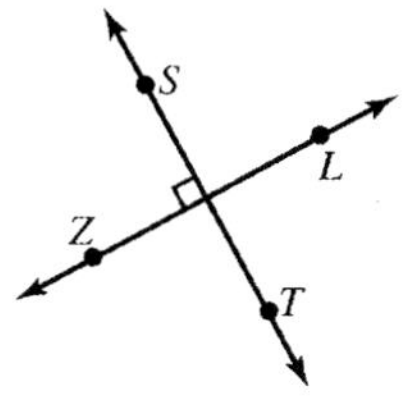

13. ________________

14.

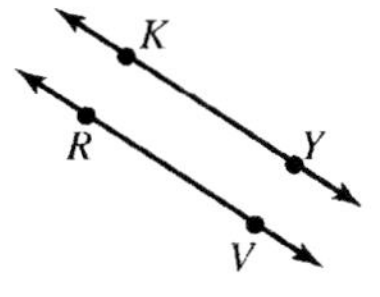

14. ________________

15.

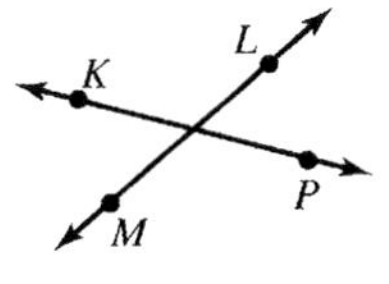

15. ________________

Objective 6 Identify complementary angles and supplementary angles and find the *measure* of a complement or supplement of a given angle.

For extra help, see Examples 6–9 on pages 459–460 of your text and Section Lecture video for Section 6.5 and Exercise Solutions Clip 25 and 27.

Find the complement of the angle.

16. 43°

16. ________________

Find the supplement of the angle.

17. 16°

17. ________________

Identify each pair of supplementary angles.

18.

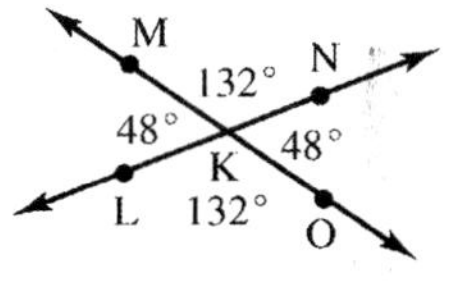

18. ________________

Name: Date:
Instructor: Section:

Objective 7 Identify congruent angles and vertical angles and use this knowledge to find the measures of angles.

For extra help, see Examples 10–12 on pages 461–462 of your text and Section Lecture video for Section 6.5 and Exercise Solutions Clip 37.

In the figure below, identify the angles that are congruent.

19. 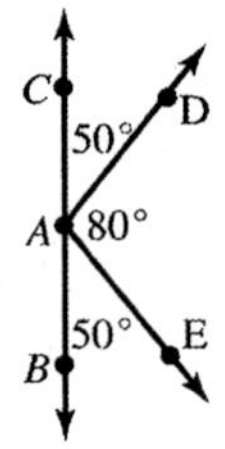

19. ________________

In the figure below, identify all the vertical angles.

20. 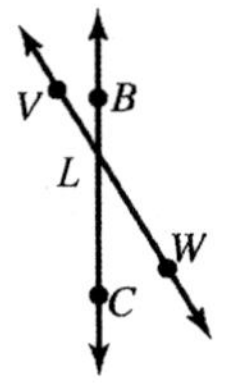

20. ________________

In the figure below, $\angle ABE$ measures 73° and $\angle FEB$ measures 107°. Find the measures of the indicated angles.

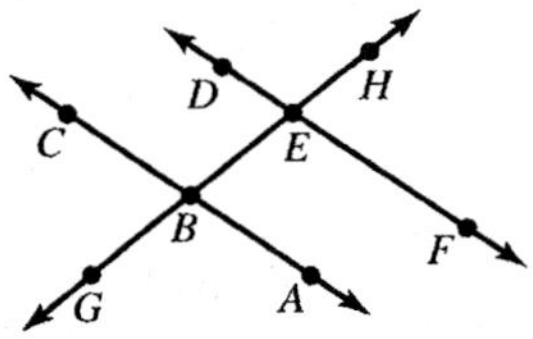

21. $\angle CBG$ and $\angle DEH$

21. ________________

Name: Date:
Instructor: Section:

Objective 8 Identify corresponding angles and alternate interior angles and use this knowledge to find the measures of angles.

For extra help, see Examples 13–14 on pages 463–464 of your text and Section Lecture video for Section 6.5 and Exercise Solutions Clip 47 and 53.

In each figure, line m is parallel to line n. List the corresponding angles and the alternate interior angles. Then find the measure of each angle.

22. ∠4 measures 100°.

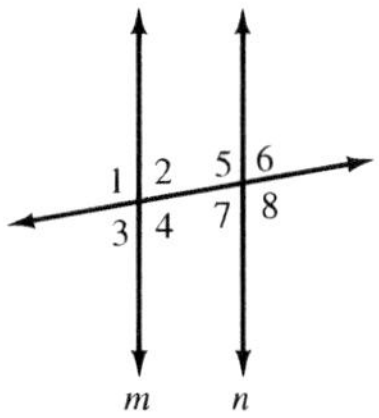

22.
Corresponding angles:

____

Alternate interior angles:

____

Angle measures:

____

23. ∠7 measures 37°.

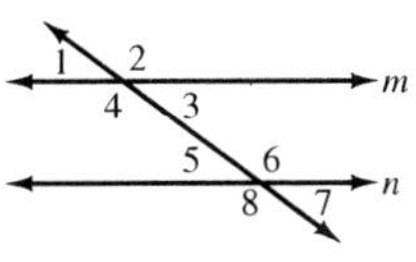

23.
Corresponding angles:

____

Alternate interior angles:

____

Angle measures:

____

Name: Date:
Instructor: Section:

Chapter 6 RATIO, PROPORTION, AND LINE/ANGLE/TRIANGLE RELATIONSHIPS

6.6 Geometry Applications: Congruent and Similar Triangles

Learning Objectives

1	Identify corresponding parts of congruent triangles.
2	Prove that triangles are congruent using SAS, SSS, or ASA.
3	Identify corresponding parts of similar triangles.
4	Find the unknown lengths of sides in similar triangles.
5	Solve application problems involving similar triangles.

Key Terms

Use the vocabulary terms listed below to complete each statement in exercises 1−4.

congruent figures **similar figures**

congruent triangles **similar triangles**

1. Triangles with the same shape and size are ________________________.

2. ________________________________ are identical both in shape and in size.

3. Triangles with the same shape but not necessarily the same size are ____________________________.

4. ____________________________ have the same shape but are different sizes.

Guided Examples

Review this example for Objective 1:

1. Identify corresponding angles and sides in these congruent triangles.

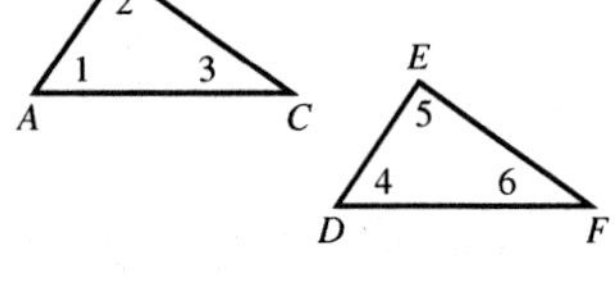

The corresponding parts are:

$\angle 1$ and $\angle 4$ $\overline{AB}$ and $\overline{DE}$

$\angle 2$ and $\angle 5$ $\overline{BC}$ and $\overline{EF}$

$\angle 3$ and $\angle 6$ $\overline{AC}$ and $\overline{DF}$

Now Try:

1. Identify corresponding angles and sides in these congruent triangles.

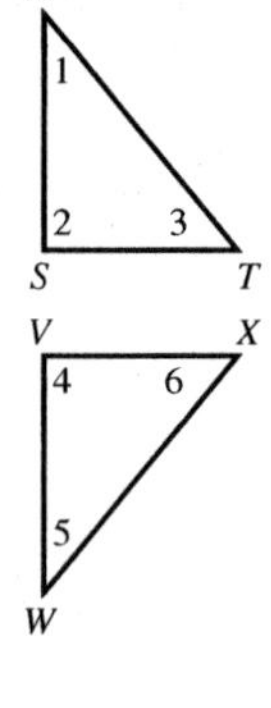

____

Name: Date:
Instructor: Section:

Review these examples for Objective 2:

2. Explain which method can be used to prove that each pair of triangles is congruent. Choose from ASA, SSS, and SAS.

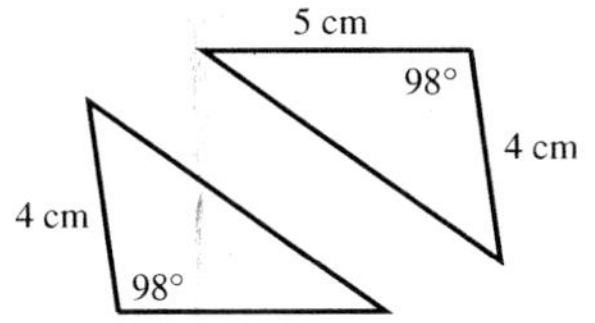

a.

On both triangles, two corresponding sides and the angle between them measure the same, so Side-Angle-Side (SAS) method can be used to prove that the triangles are congruent.

b.

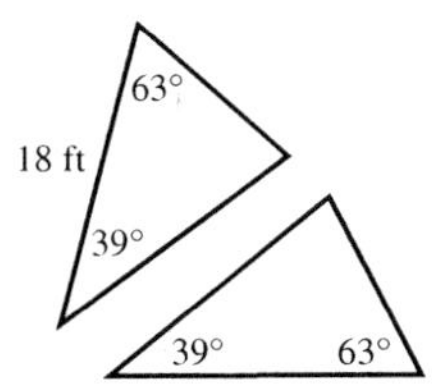

On both triangles, two corresponding angles and the side between them measure the same, so the Angle-Side-Angle (ASA) method can be used to prove that the triangles are congruent.

c.

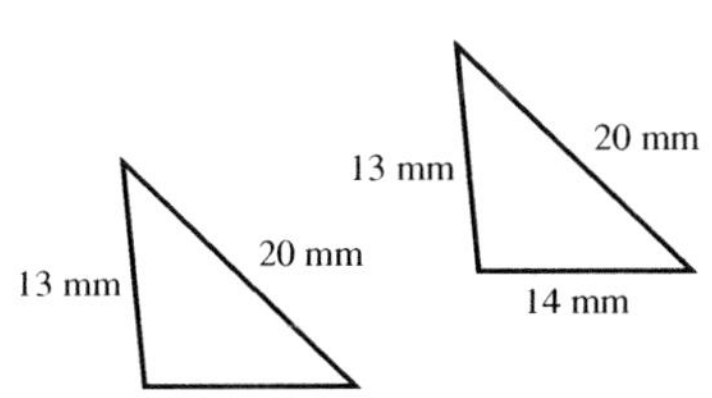

Each pair of corresponding sides has the same length, so the Side-Side-Side (SSS) method can be used to prove that the triangles are congruent.

Now Try:

2. Explain which method can be used to prove that each pair of triangles is congruent. Choose from ASA, SSS, and SAS.

a.

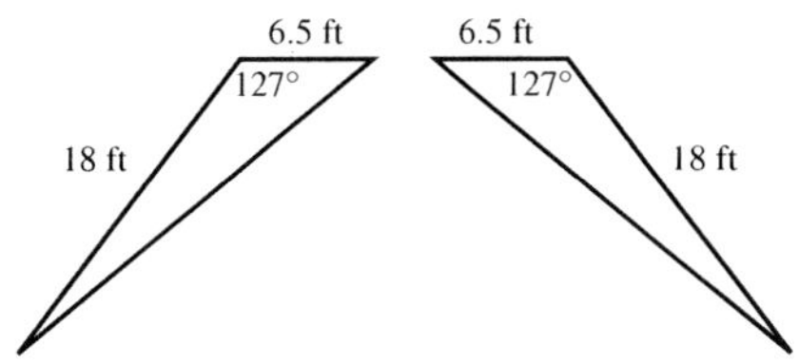

____

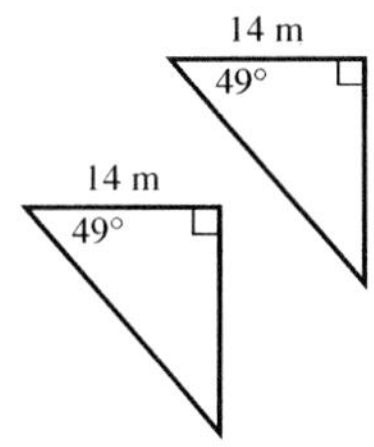

b.

____

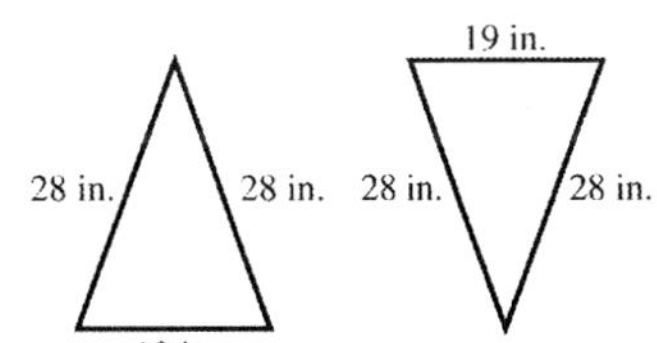

c.

____

Name: Date:
Instructor: Section:

Review this example for Objective 3:

3. Identify corresponding angles and sides in these similar triangles.

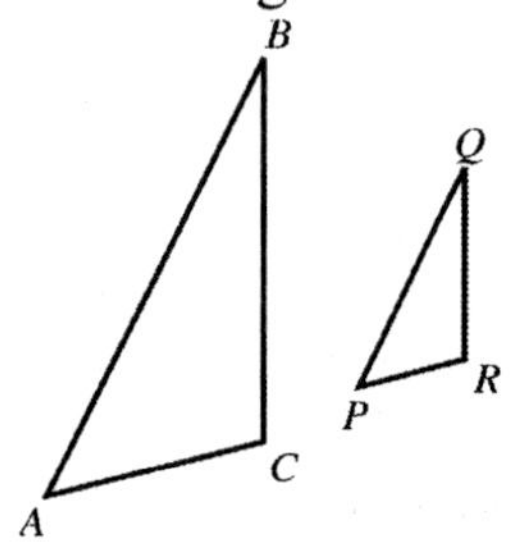

$\overline{AB}$ and $\overline{PQ}$; $\overline{AC}$ and $\overline{PR}$; $\overline{BC}$ and $\overline{QR}$;
$\angle A$ and $\angle P$; $\angle B$ and $\angle Q$; $\angle C$ and $\angle R$

Now Try:

3. Identify corresponding angles and sides in these similar triangles.

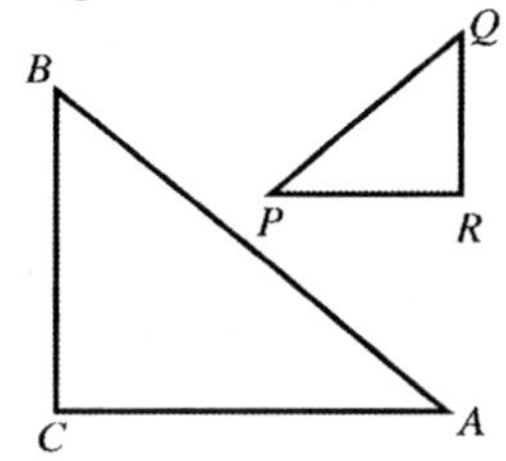

____

Review these examples for Objective 4:

4. Find the length of b in the smaller triangle. Assume the triangles are similar.

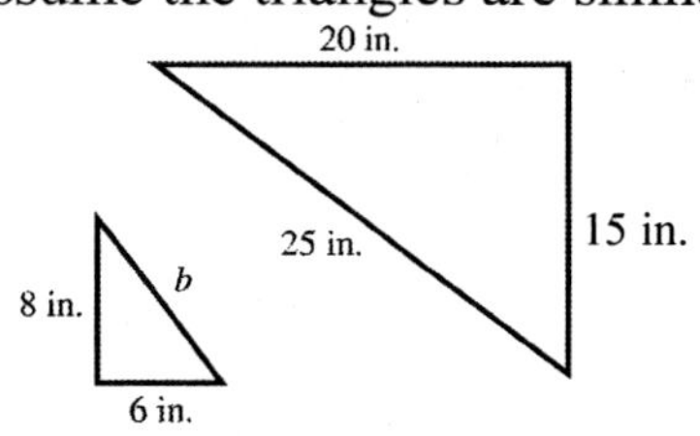

The length you want to find in the smaller triangle is side b, and it corresponds to 25 in. in the larger triangle. The smaller triangle is turned "upside down" compared to the larger triangle, so be careful when identifying corresponding sides. Then notice that 6 in. in the smaller triangle corresponds to 15 in. in the larger triangle, and you know both of their lengths. Because the ratios of the lengths of corresponding sides are equal, you can set up a proportion.

$$\frac{b}{25} = \frac{6}{15}$$

$$\frac{b}{25} = \frac{2}{5} \quad \text{Write } \frac{6}{15} \text{ in lowest terms as } \frac{2}{5}.$$

Find the cross products.

$$b \cdot 5 = 25 \cdot 2$$

$$\frac{b \cdot \overset{1}{\cancel{5}}}{\underset{1}{\cancel{5}}} = \frac{50}{5}$$

$$b = 10$$

Side b has length of 10 in.

Now Try:

4. Find the length of x in the smaller triangle. Assume the triangles are similar.

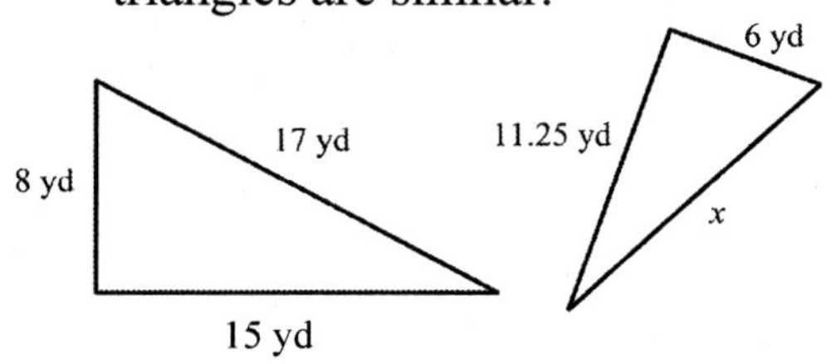

____

Name: Date:
Instructor: Section:

5. Find the perimeter of the smaller triangle. Assume the triangles are similar.

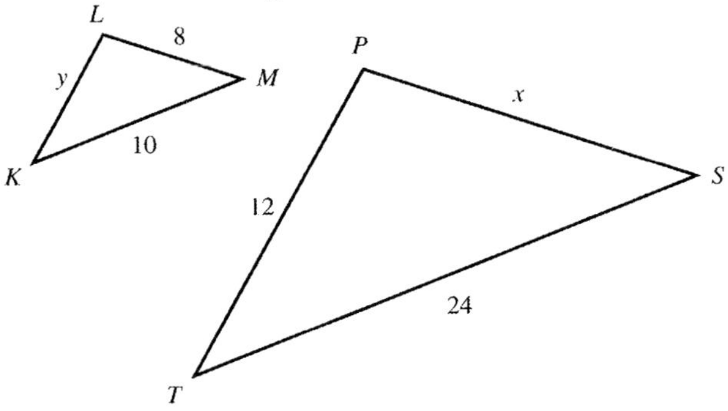

First, find y, the length of $\overline{KL}$, then add the lengths of all three sides to find the perimeter. The ratios of the lengths of corresponding sides are equal, so you can set up a proportion.

$$\begin{matrix} KL \rightarrow \\ KM \rightarrow \end{matrix} \frac{y}{10} = \frac{12}{24} \begin{matrix} \leftarrow PT \\ \leftarrow TS \end{matrix}$$

$$\frac{y}{10} = \frac{1}{2} \quad \text{Write } \frac{12}{24} \text{ in lowest terms as } \frac{1}{2}.$$

Find the cross products.

$$\frac{y}{10} = \frac{1}{2}$$

$$y \cdot 2 = 10 \cdot 1$$

$$\frac{y \cdot \overset{1}{\cancel{2}}}{\underset{1}{\cancel{2}}} = \frac{10}{2}$$

$$y = 5$$

$\overline{KL}$ has a length of 5.

Now add the lengths of all three sides to find the perimeter of the smaller triangle.

Perimeter $= 8 + 10 + 5 = 23$

5. Find the perimeter of the smaller triangle. Assume the triangles are similar.

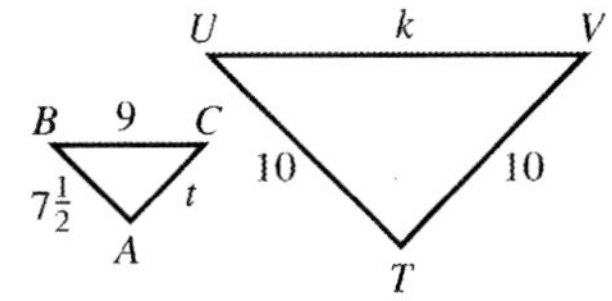

____

Name: Date:
Instructor: Section:

Review this example for Objective 5:

6. The height of the house shown here can be found by using similar triangles and proportion. Find the height of the house by writing a proportion and solving it.

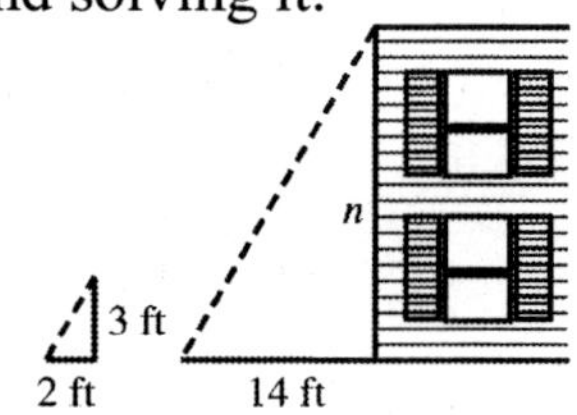

The triangles shown are similar, so write a proportion to find *n*.

$$\frac{n}{3} = \frac{14}{2}$$

height of larger triangle → n; length of larger triangle ← 14
height of smaller triangle → 3; length of smaller triangle ← 2

Find the cross products and show that they are equal.

$$n \cdot 2 = 3 \cdot 14$$
$$n \cdot 2 = 42$$
$$\frac{n \cdot \overset{1}{\cancel{2}}}{\underset{1}{\cancel{2}}} = \frac{42}{2}$$
$$n = 21$$

The height of the house is 21 ft.

Now Try:

6. A flagpole casts a shadow 77 feet long at the same time that a pole 15 feet tall casts a shadow 55 ft long. Find the height of the flagpole.

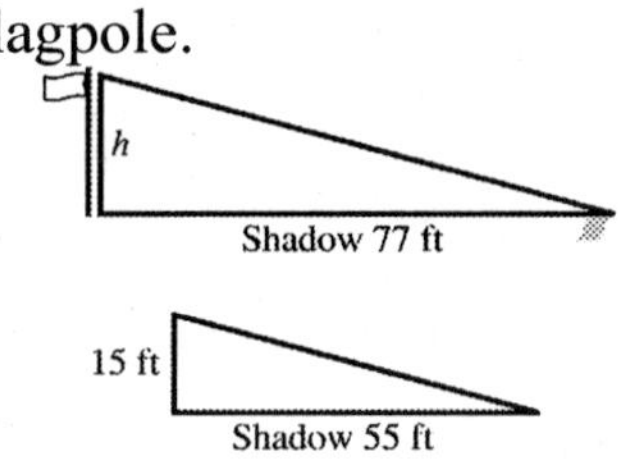

____

Objective 1 Identify corresponding parts of congruent triangles.

For extra help, see Example 1 on pages 469–470 of your text and Section Lecture video for Section 6.6.

Each pair of triangles is congruent. List the corresponding angles and the corresponding sides.

1.

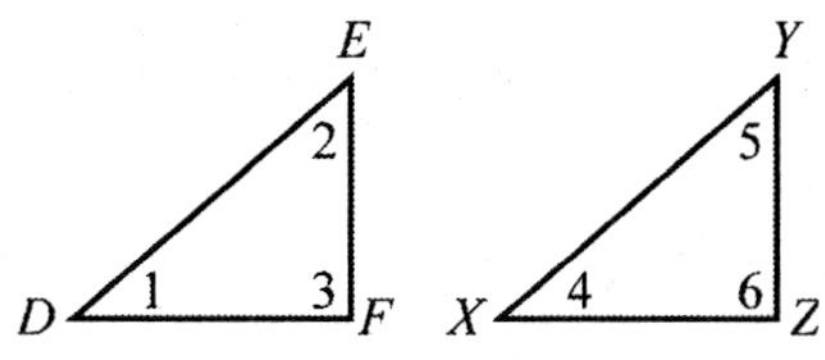

1. ________________

2.

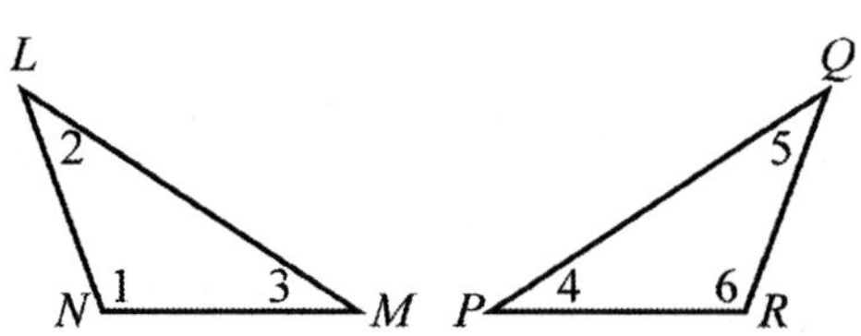

2. ________________

Name: Date:
Instructor: Section:

Objective 2 Prove that triangles are congruent using SAS, SSS, and ASA.

For extra help, see Example 2 on page 471 of your text and Section Lecture video for Section 6.6.

Determine which of these methods can be used to prove that each pair of triangles is congruent: Angle-Side-Angle (ASA), Side-Side-Side (SSS), or Side-Angle-Side (SAS).

3.

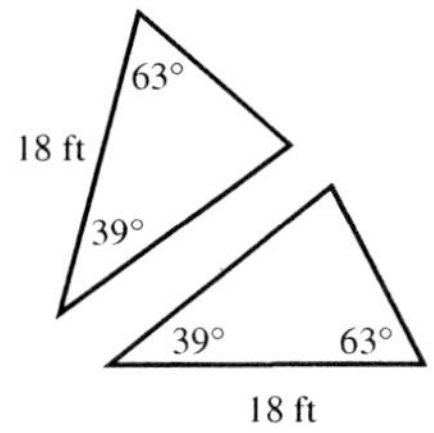

3. ________________

4.

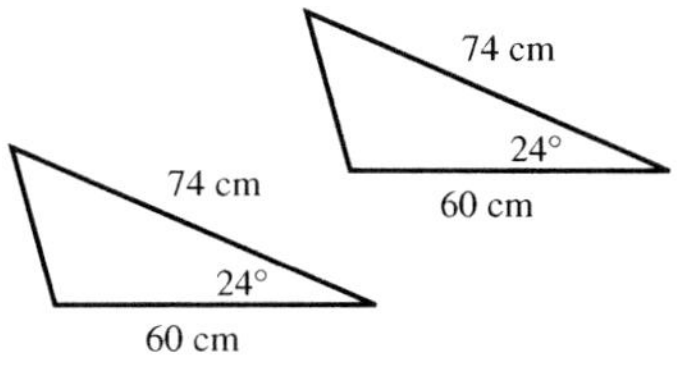

4. ________________

Objective 3 Identify the corresponding parts of similar triangles.

For extra help, see page 471 of your text and Section Lecture video for Section 6.6.

Name the corresponding angles and the corresponding sides in each pair of similar triangles.

5.

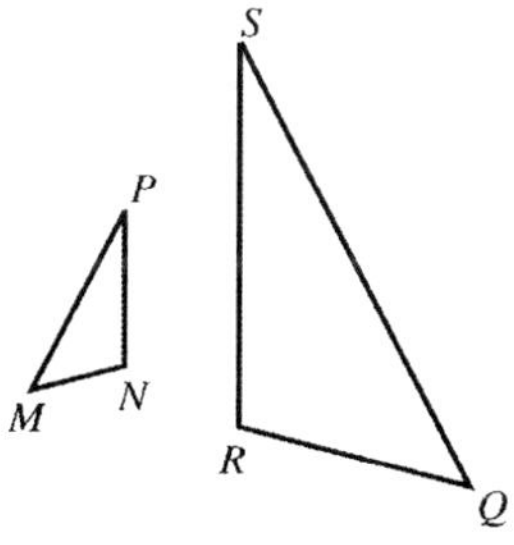

5. ________________

6.

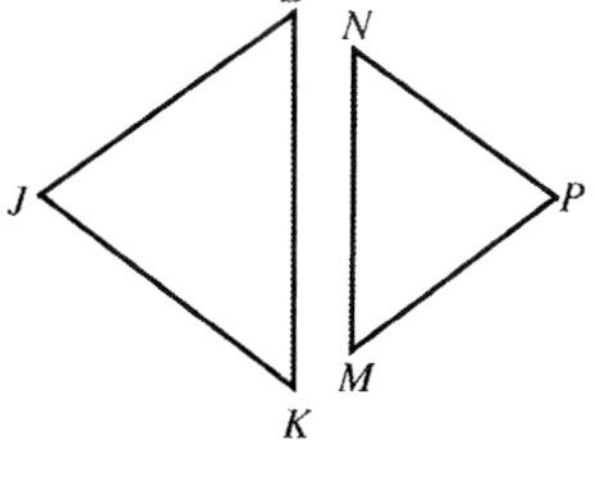

6. ________________

Name: Date:
Instructor: Section:

7. 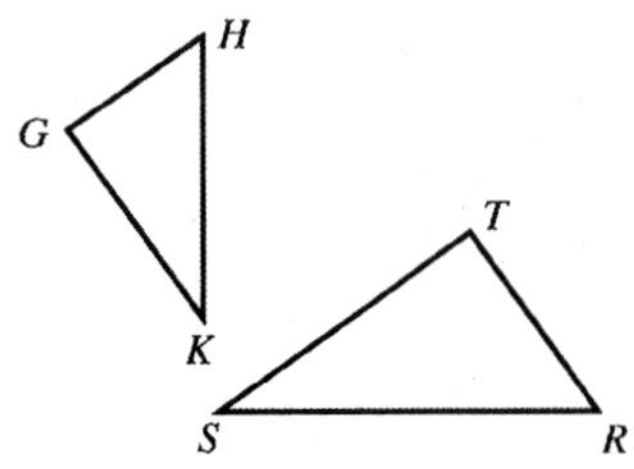

7. ______________

Objective 4 Find the unknown lengths of sides in similar triangles.

For extra help, see Examples 3–4 on pages 472–473 of your text and Section Lecture video for Section 6.6 and Exercise Solutions Clip 21 and 23.

Find the unknown lengths in each pair of similar triangles.

8. 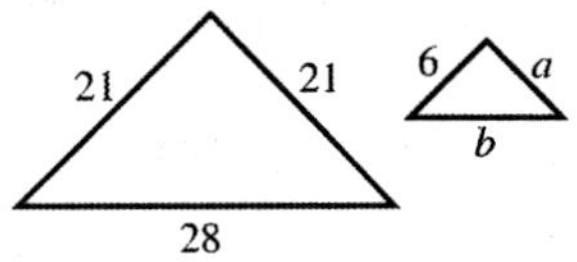

8. ______________

9. 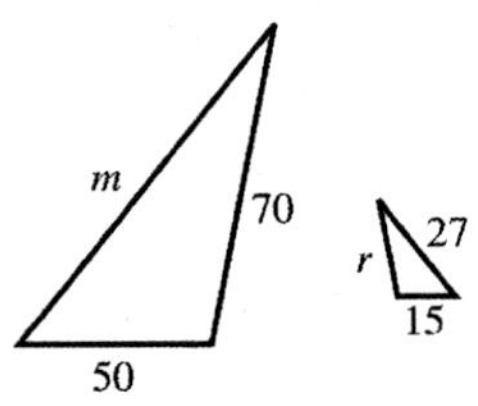

9. ______________

Find the perimeter of each triangle. Assume the triangles are similar.

10. 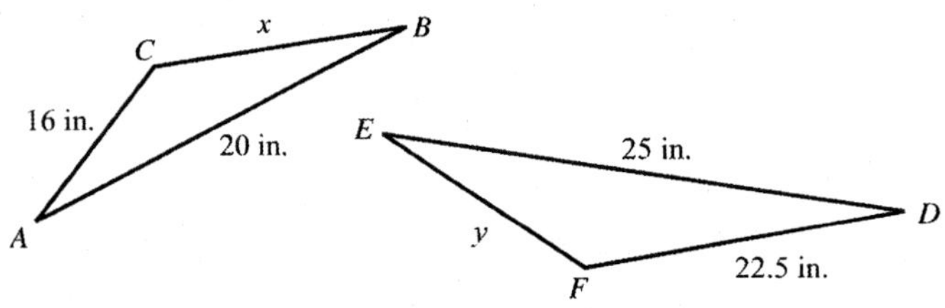

10. *ABC* ______________

EDF ______________

Name: Date:
Instructor: Section:

Objective 5 Solve application problems involving similar triangles.

For extra help, see Example 5 on page 474 of your text and Section Lecture video for Section 6.6 and Exercise Solutions Clip 29.

Solve each application problem.

11. A sailor on the USS Ramapo saw one of the highest waves ever recorded. He used the height of the ship's mast, the length of the deck and similar triangles to find the height of the wave. Using the information in the figure, write a proportion and then find the height of the wave.

11. ________________

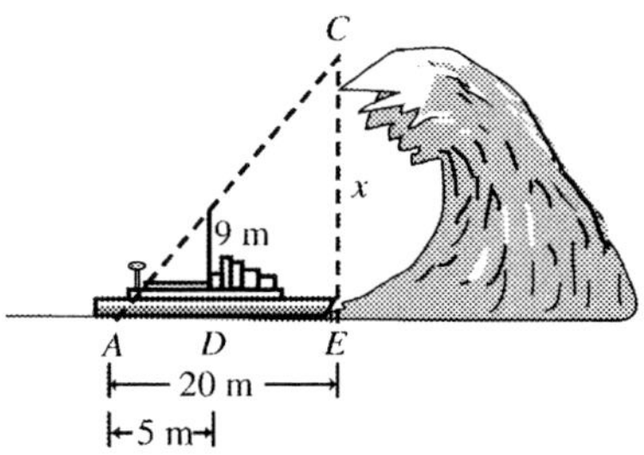

12. A fire lookout tower provides an excellent view of the surrounding countryside. The height of the tower can be found by lining up the top of the tower with the top of a 3-meter stick. Use similar triangles to find the height of the tower.

12. ________________

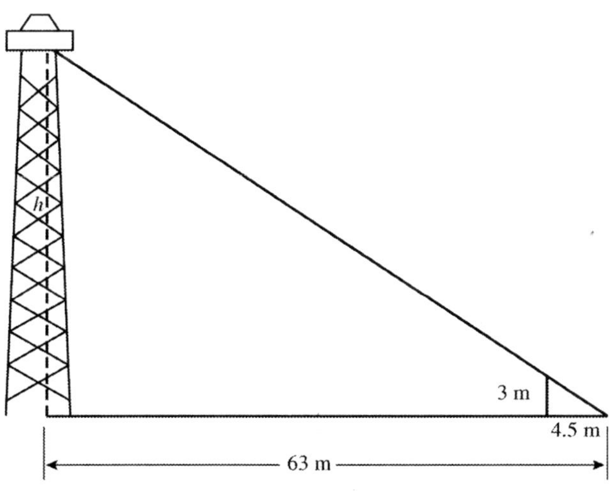

Name: Date:
Instructor: Section:

13. A 30 m ladder touches the side of a building at a height of 25 m. At what height would a 12-m ladder touch the building if it makes the same angle with the ground?

13. ________________

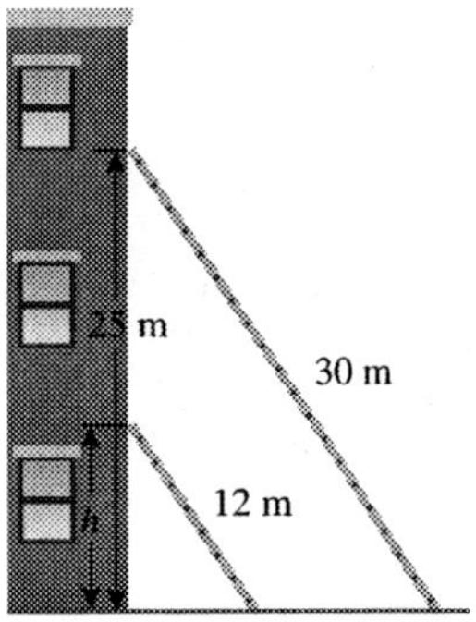

Name: Date:
Instructor: Section:

Chapter 7 PERCENT

7.1 The Basics of Percent

Learning Objectives

1	Learn the meaning of percent.
2	Write percents as decimals.
3	Write decimals as percents.
4	Write percents as fractions.
5	Write fractions as percents.
6	Use 100% and 50%.

Key Terms

Use the vocabulary terms listed below to complete each statement in exercises 1–3.

percent **ratio** **decimals**

1. To compare two quantities that have the same type of units, use a ____________.

2. ____________________ means per one hundred.

3. ____________________ represent parts of a whole.

Guided Examples

Review these examples for Objective 1:

1. Write a percent to describe each situation.

a. You leave a $25 tip when the restaurant bill was $100. What percent is the tip?

The tip is $25 per $100, or $\frac{25}{100}$. The percent is 25%.

b. You pay $8 tax on a $100 catering order. What is the tax rate?

The tax is $8 per $100, or $\frac{8}{100}$. The percent is 8%.

c. You earn 91 points on a 100-point math test. What percent is your score?

Then your score is 91 per 100, or $\frac{91}{100}$, or 91%.

Now Try:

1. Write a percent to describe each situation.

a. You give $10 to a charity per $100. What is the percent given to charity?

____

b. The tax on a hotel room of $100 is $18. What is the tax rate?

____

c. You earn 85 points on a 100-point math quiz. What percent is your score?

____

Name: Date:
Instructor: Section:

Review these examples for Objective 2:

2. Write each percent as a decimal.

a. 39%

$39\% = 39 \div 100 = 0.39$

b. 7%

$7\% = 7 \div 100 = 0.07$

c. 89.9%

$89.9\% = 89.9 \div 100 = 0.899$

d. 400%

$400\% = 400 \div 100 = 4.00$

e. 169%

$169\% = 169 \div 100 = 1.69$

3. Write each percent as a decimal by dropping the percent symbol and moving the decimal point two places to the left.

a. 56%

Drop the percent sign and move the decimal point two places to the left.

$56\% = 56.\% = 0.56$

b. 120%

$120\% = 120.\% = 1.20$ or 1.2

c. 1.3%

0 is attached so the decimal point can be moved two places to the left.

$1.3\% = 0.013$

d. 0.2%

Two zeros are attached so the decimal point can be moved two places to the left.

$0.2\% = 0.002$

Now Try:

2. Write each percent as a decimal.

a. 23% __________

b. 8% __________

c. 15.3% __________

d. 500% __________

e. 302% __________

3. Write each percent as a decimal by dropping the percent symbol and moving the decimal point two places to the left.

a. 48% __________

b. 240% __________

c. 2.5% __________

d. 0.8% __________

Name: Date:
Instructor: Section:

Review these examples for Objective 3:

4. Write each decimal as a percent by moving the decimal point two places to the right.

a. 0.31

Decimal point is moved two places to the right and percent symbol is attached.
$0.31 = 31\%$

b. 1.6

0 is attached so the decimal point can be moved two places to the right.
$1.6 = 1.60 = 160\%$

c. 0.904

$0.904 = 90.4\%$

Now Try:

4. Write each decimal as a percent by moving the decimal point two places to the right.

a. 0.43 __________

b. 2.3 __________

c. 0.751 __________

Review these examples for Objective 4:

5. Write each percent as a fraction or mixed number in lowest terms.

a. 12%

Recall writing 12% as a decimal.
$12\% = 12 \div 100 = 0.12$
Because 0.12 means 12 hundredths,

$$0.12 = \frac{12}{100} = \frac{12 \div 4}{100 \div 4} = \frac{3}{25}$$

b. 86%

Write 86% as $\frac{86}{100}$.

Write $\frac{86}{100}$ in lowest terms.

$$\frac{86}{100} = \frac{86 \div 2}{100 \div 2} = \frac{43}{50}$$

c. 350%

$$350\% = \frac{350}{100} = \frac{350 \div 50}{100 \div 50} = \frac{7}{2} = 3\frac{1}{2}$$

Now Try:

5. Write each percent as a fraction or mixed number in lowest terms.

a. 30% __________

b. 78% __________

c. 175% __________

Name: Date:
Instructor: Section:

6. Write each percent as a fraction in lowest terms.

a. 62.5%

Write 62.5 over 100.

$$62.5\% = \frac{62.5}{100}$$

To get a whole number in the numerator, multiply the numerator and denominator by 10.

$$\frac{62.5}{100} = \frac{62.5(10)}{100(10)} = \frac{625}{1000}$$

Now write the fraction in lowest terms.

$$\frac{625}{1000} = \frac{625 \div 125}{1000 \div 125} = \frac{5}{8}$$

b. $16\frac{2}{3}\%$

Write $16\frac{2}{3}\%$ over 100.

$$16\frac{2}{3}\% = \frac{16\frac{2}{3}}{100}$$

When there is a mixed number in the numerator, rewrite the mixed number as an improper fraction.

$$\frac{16\frac{2}{3}}{100} = \frac{\frac{50}{3}}{100}$$

Next, rewrite the division problem in a horizontal form. Finally, multiply by the reciprocal of the divisor.

$$\frac{\frac{50}{3}}{100} = \frac{50}{3} \div 100 = \frac{50}{3} \div \frac{100}{1} = \frac{50}{3} \cdot \frac{1}{100} = \frac{1}{6}$$

6. Write each percent as a fraction in lowest terms.

a. 43.6%

____

b. $22\frac{2}{9}\%$

____

Review these examples for Objective 5:

7. Write each fraction as a percent. Round to the nearest tenth as necessary.

a. $\frac{3}{5}$

$$\frac{3}{5} = \left(\frac{3}{5}\right)(100\%) = \left(\frac{3}{5}\right)\left(\frac{100}{1}\%\right) = \left(\frac{3}{5}\right)\left(\frac{20 \cdot 5}{1}\%\right)$$

$$= \frac{60}{1}\% = 60\%$$

Now Try:

7. Write each fraction as a percent. Round to the nearest tenth as necessary.

a. $\frac{3}{20}$

____

Name: Date:
Instructor: Section:

b. $\frac{5}{16}$

$$\frac{5}{16}=\left(\frac{5}{16}\right)(100\%)=\left(\frac{5}{16}\right)\left(\frac{100}{1}\%\right)$$
$$=\left(\frac{5}{4\cdot 4}\right)\left(\frac{4\cdot 25}{1}\%\right)$$
$$=\frac{125}{4}\%=31\frac{1}{4}\%$$

c. $\frac{1}{15}$

$$\frac{1}{15}=\left(\frac{1}{15}\right)(100\%)=\left(\frac{1}{15}\right)\left(\frac{100}{1}\%\right)$$
$$=\left(\frac{1}{3\cdot 5}\right)\left(\frac{5\cdot 20}{1}\%\right)$$
$$=\frac{20}{3}\%=6\frac{2}{3}\%$$

$6\frac{2}{3}\%$ rounds to 6.7%.

b. $\frac{15}{16}$

____

c. $\frac{1}{18}$

____

Review these examples for Objective 6:

8. Fill in the blanks.

a. 100% of $69 is ______ .

100% is all of the money. So 100% of $69 is $69.

b. 100% of 25 miles is ______ .

100% is all of the miles. So 100% of 25 miles is 25 miles.

9. Fill in the blanks.

a. 50% of $68 is _____ .

50% is half of the money. So 50% of $68 is $34.

b. 50% of 13 hours is _____ .

50% is half of the hours. So 50% of 13 hours is $6\frac{1}{2}$ hours.

Now Try:

8. Fill in the blanks.

a. 100% of $6.15 is _____ .

____

b. 100% of $23\frac{1}{2}$ days is _____.

____

9. Fill in the blanks.

a. 50% of $7.30 is _____ .

____

b. 50% of 19 pages is _____ .

____

Name: Date:
Instructor: Section:

Objective 1 Learn the meaning of percent.

For extra help, see Example 1 on page 496 of your text and Section Lecture video for Section 7.1

Write as a percent.

1. 68 people out of 100 drive small cars. **1.** __________

2. The tax is \$8 per \$100. **2.** __________

3. The cost for labor was \$45 for every \$100 spent to manufacture an item. **3.** __________

Objective 2 Write percents as decimals.

For extra help, see Examples 2–3 on pages 497–498 of your text and Section Lecture video for Section 7.1 and Exercise Solutions Clip 5, 11, and 17.

Write each percent as a decimal.

4. 42% **4.** __________

5. 310% **5.** __________

6. 18.9% **6.** __________

Objective 3 Write decimals as percents.

For extra help, see Example 4 on pages 498–499 of your text and Section Lecture video for Section 7.1 and Exercise Solutions Clip 19, 21, 27, and 31.

Write each decimal as a percent.

7. 0.2 **7.** __________

8. 0.564 **8.** __________

9. 4.93 **9.** __________

Objective 4 Write percents as fractions.

For extra help, see Examples 5–6 on pages 499–501 of your text and Section Lecture video for Section 7.1 and Exercise Solutions Clip 39 and 47.

Write each percent as a fraction or mixed number in lowest terms.

10. 140% **10.** __________

11. $18\frac{1}{3}\%$ **11.** ________________

12. 55.6% **12.** ________________

Objective 5 Write fractions as percents.

For extra help, see Example 7 on pages 501–503 of your text and Section Lecture video for Section 7.1.

Write each fraction or mixed number as a percent. If you're using a calculator, first work each one by hand. Then use your calculator and round to the nearest tenth of a percent, if necessary.

13. $\frac{47}{50}$ **13.** ________________

14. $\frac{11}{40}$ **14.** ________________

15. $\frac{64}{75}$ **15.** ________________

Objective 6 Use 100% and 50%.

For extra help, see Examples 8–9 on page 503 of your text and Section Lecture video for Section 7.1 and Exercise Solutions Clip 103.

Fill in the blanks. Remember that 100% is all of something and 50% is half of it.

16. 50% of 260 miles is _____ . **16.** ________________

17. 100% of $520 is _____ . **17.** ________________

18. 50% of 40 acres is _____ . **18.** ________________

Name: Date:
Instructor: Section:

Chapter 7 PERCENT

7.2 The Percent Proportion

Learning Objectives
1 Identify the percent, whole, and part.
2 Solve percent problems using the percent proportion.

Key Terms

Use the vocabulary terms listed below to complete each statement in exercises 1–3.

percent proportion **whole** **part**

1. The ______________________________ in a percent problem is the entire quantity.

2. The ____________________ in a percent problem is the portion being compared with the whole.

3. Part is to whole as percent is to 100 is called the ________________________.

Guided Examples

Review these examples for Objective 1:

1. Find the percent in the following.

   a. 43% of 1100 people exercise daily.

   The percent is 43.

   b. $255 is 29 percent of what number?

   The percent is 29, because 29 appears with the word percent.

   c. What percent of 6500 tons is 275 tons?

   The word percent has no number with it, so the percent is the unknown part of the problem.

Now Try:

1. Find the percent in the following.

   a. 950 tons of trash is 59% of what number of tons?

   ____

   b. Find the amount of sales tax by multiplying $650 and 7.75 percent.

   ____

   c. If 2650 children attend preschool, what percent of 3280 children attend preschool?

   ____

Name: Date:
Instructor: Section:

2. Identify the whole in the following.

a. 43% of 1100 people exercise daily.

The whole is 1100.

b. \$255 is 29 percent of what number?

The whole is the unknown part of the problem.

c. What percent of 6500 tons is 275 tons?

The whole is 6500.

3. Identify the part in the following.

a. 43% of 1100 people exercise daily.

The part is the unknown part of the problem.

b. \$255 is 29 percent of what number?

The part is \$255.

c. What percent of 6500 tons is 275 tons?

The part is 275.

2. Identify the whole in the following.

a. Find the amount of sales tax by multiplying \$650 and 7.75 percent.

____

b. 950 tons of trash is 59% of what number of tons?

____

c. If 2650 children attend preschool, what percent of 3280 children attend preschool?

____

3. Identify the part in the following.

a. Find the amount of sales tax by multiplying \$650 and 7.75 percent.

____

b. 950 tons of trash is 59% of what number of tons?

____

c. If 2650 children attend preschool, what percent of 3280 children attend preschool?

____

Review these examples for Objective 2:

4. Find 24% of \$650.

Here the percent is 24 and the whole is \$650. Now find the part. Let x represent the unknown part.

$$\frac{x}{650}=\frac{24}{100} \text{ or } \frac{x}{650}=\frac{6}{25}$$

Find the cross products in the proportion and show that they are equal.

Now Try:

4. Find 38% of 6500.

____

Name: Date:
Instructor: Section:

$$x \cdot 25 = 650 \cdot 6$$
$$x \cdot 25 = 3900$$
$$\frac{x \cdot 25}{25} = \frac{3900}{25}$$
$$x = 156$$

24% of $650 is $156.

5. Use the percent proportion to answer this question.
9 pounds is what percent of 450 pounds?

The percent proportion is $\frac{\text{percent}}{100} = \frac{\text{part}}{\text{whole}}$. Set up the proportion using p as the variable representing the unknown percent. Then find the cross products.

$$\frac{p}{100} = \frac{9}{450}$$
$$p \cdot 450 = 900$$
$$\frac{p \cdot \cancel{450}}{\cancel{450}} = \frac{900}{450}$$
$$p = 2$$

The percent is 2%. So 9 pounds is 2% of 450 pounds.

6. Use the percent proportion to answer this question.
150 students is 75% of how many students?

The percent proportion is $\frac{\text{percent}}{100} = \frac{\text{part}}{\text{whole}}$.
Here the percent is 75%, the part is 150, and the whole is unknown.

$$\frac{75}{100} = \frac{150}{n}$$
$$75 \cdot n = 15{,}000$$
$$\frac{\cancel{75} \cdot n}{\cancel{75}} = \frac{15{,}000}{75}$$
$$n = 200$$

The whole is 200 students. So 150 students is 75% of 200 students.

5. Use the percent proportion to answer this question. $28 is what percent of $700?

____

6. Use the percent proportion to answer this question. 45 cars is 30% of how many cars?

____

Name: Date:
Instructor: Section:

7. Use the percent proportion to answer each question.

a. How many cats is 275% of 20 cats?

Here, the percent is 275%, the part is unknown, and the whole is 20. Use the percent proportion.

$$\frac{275}{100} = \frac{n}{20}$$
$$100 \cdot n = 5500$$
$$\frac{\cancel{100} \cdot n}{\cancel{100}} = \frac{5500}{100}$$
$$n = 55$$

The part is 55 cats.

b. What percent of \$40 is \$46?

Here, the percent is unknown, the part is \$46, and the whole is \$40. Use the percent proportion.

$$\frac{p}{100} = \frac{46}{40}$$
$$p \cdot 40 = 4600$$
$$\frac{p \cdot \cancel{40}}{\cancel{40}} = \frac{4600}{40}$$
$$p = 115$$

The percent is 115%.

7. Use the percent proportion to answer each question.

a. 66 days is what percent of 30 days?

____

b. 185% of what amount is \$173?

____

Objective 1 Identify the percent, whole, and part.

For extra help, see Examples 1–3 on pages 511–512 of your text and Section Lecture video for Section 7.2 and Exercise Solutions Clip 5.

In each of the following, identify the percent, the whole, and the part. Do not try to solve for any unknowns.

1. \$50 is 250% of what number?

1. percent __________

whole __________

part __________

2. 336 students is what percent of 840 students?

2. percent __________

whole __________

part __________

Name: Date:
Instructor: Section:

3. The state sales tax is $7\frac{3}{4}$ percent of the \$895 price.

3. percent __________

whole __________

part __________

Objective 2 Solve percent problems using the percent proportion.

For extra help, see Examples 4–7 on pages 512–515 of your text and Section Lecture video for Section 7.2 and Exercise Solutions Clip 7, 9, and 21.

Use the percent proportion to answer these questions. If necessary, round money answers to the nearest cent and percent answers to the nearest tenth of a percent.

4. 165% of what number of feet is 12.7 feet? Round the answer to the nearest tenth.

4. __________

5. What percent of 78 bicyclists is 110 bicyclists?

5. __________

6. What number is 43% of 2200?

6. __________

Name: Date:
Instructor: Section:

Chapter 7 PERCENT

7.3 The Percent Equation

Learning Objectives

1 Estimate answers to percent problems involving 25%.
2 Find 10% and 1% of a number by moving the decimal point.
3 Solve basic percent problems using the percent equation.

Key Terms

Use the vocabulary terms listed below to complete each statement in exercises 1–2.

percent equation **percent**

1. A number written with a ______________ sign means "divided by 100".

2. The ______________________ is $\text{part} = \text{percent} \cdot \text{whole}$.

Guided Examples

Review these examples for Objective 1:

1. Estimate the answer to each question.

a. What is 25% of \$423?

Use front end rounding to round \$423 to \$400. Then divide \$400 by 4. The estimate is \$100.

b. Find 25% of 82.7 miles.

Use front end rounding to round 82.7 miles to 80 miles. Then divide 80 by 4. The estimate is 20 miles.

c. 25% of 89 days is how long?

Round 89 days to 90 days using front end rounding. Then divide 90 by 4 to get 22.5 days.

Now Try:

1. Estimate the answer to each question.

a. Find 25% of \$205.

____

b. What is 25% of 59 hours?

____

c. 25% of 97 pounds is how many pounds?

____

Review these examples for Objective 2:

2. Find the exact answer to each question by moving the decimal point.

a. What is 10% of \$674?

To find 10% of \$674, divide \$674 by 10. Do the division by moving the decimal point one place to the left.

10% of \$674 = \$67.40.

Now Try:

2. Find the exact answer to each question by moving the decimal point.

a. What is 10% of \$932?

____

Name: Date:
Instructor: Section:

b. Find 10% of 37.6 miles.

To find 10% of 37.6 miles, divide 37.6 by 10. Move the decimal point one place to the left.
10% of 37.6 miles = 3.76 miles

b. Find 10% of 693 miles.

____

3. Find the exact answer to each question by moving the decimal point.

a. What is 1% of $947?

To find 1% of $947, divide $947 by 100. Do the division by moving the decimal point two places to the left.
1% of $947 = $9.47

b. Find 1% of 68.9 miles.

To find 1% of 68.9 miles, divide 68.9 by 100. Move the decimal point two places to the left.
1% of 68.9 = 0.689 mile

3. Find the exact answer to each question by moving the decimal point.
a. What is 1% of $549?

____

b. Find 1% of 30.7 miles.

____

Review these examples for Objective 3:

4. Write and solve a percent equation to answer each question.

a. 35% of $245 is how much money?

Translate the sentence into an equation, where *of* indicates multiplication and *is* translates to the equal sign. The percent must be written in decimal form.

$$\underbrace{35\%}\ \text{of}\ \$245\ \text{is}\ \underbrace{\text{how much money?}}$$
$$\downarrow\quad\downarrow\quad\downarrow\quad\downarrow\quad\downarrow$$
$$0.35 \cdot 245 = n$$

To solve, simplify the left side.
$$(0.35)(245) = n$$
$$85.75 = n$$
So, 35% of $245 is $85.75.

b. How many books is 310% of 70 books?

Translate the sentence into an equation. Write the percent in decimal form.

$$\underbrace{\text{How many books}}\ \text{is}\ 310\%\ \text{of}\ \underbrace{70\ \text{books?}}$$
$$\downarrow\quad\downarrow\quad\downarrow\quad\downarrow\quad\downarrow$$
$$n = 3.10 \cdot 70$$

This time the two sides of the percent equation are reversed, so

Now Try:

4. Write and solve a percent equation to answer each question.
a. 89% of $310 is how much money?

____

b. How many miles is 180% of 60 miles?

____

$$\text{part} = \text{percent} \cdot \text{whole}$$

To solve the equation, simplify the right side, multiplying 3.10 by 70.

$$n = (3.10)(70)$$

$$n = 217$$

So 217 books is 310% of 70 books.

5. Write and solve a percent equation to answer each question.

a. 16 pounds is what percent of 200 pounds?

Translate the sentence into an equation. This time the percent is unknown.

$$\underbrace{\text{16 pounds}}_{\downarrow \atop 16} \quad \underset{\downarrow \atop =}{\text{is}} \quad \underbrace{\text{what percent}}_{\downarrow \atop p} \quad \underset{\downarrow \atop \cdot}{\text{of}} \quad \underbrace{\text{200 pounds?}}_{\downarrow \atop 200}$$

To solve the equation, divide both sides by 200.

$$\frac{16}{200} = \frac{p \cdot \cancel{200}}{\cancel{200}}$$

$$0.08 = p$$

$$0.08 = 8\%$$

So 16 pounds is 8% of 200 pounds.

b. What percent of \$80 is \$132?

Translate the sentence into an equation and solve it.

$$\underbrace{\text{What percent}}_{\downarrow \atop p} \quad \underset{\downarrow \atop \cdot}{\text{of}} \quad \underset{\downarrow \atop 80}{\$80} \quad \underset{\downarrow \atop =}{\text{is}} \quad \underset{\downarrow \atop 132}{\$132}$$

To solve, divide both sides by 80.

$$\frac{p \cdot \cancel{80}}{\cancel{80}} = \frac{132}{80}$$

$$p = 1.65$$

$$1.65 = 165\%$$

So 165% of \$80 is \$132.

5. Write and solve a percent equation to answer each question.

a. 66 hours is what percent of 30 hours?

____

b. What percent of 100 is 33?

____

Name: Date:
Instructor: Section:

6. Write and solve a percent equation to answer each question.

a. 126 credits is 70% of how many credits?

Translate the sentence into an equation. Write the percent in decimal form.

126 credits	is	70%	of	how many credits?
↓	↓	↓	↓	↓
126	=	0.70	·	n

Recall that 0.70 is equivalent to 0.7, so use 0.7 in the equation. To solve, divide both sides by 0.7.

$$\frac{126}{0.7} = \frac{\cancel{0.7} \cdot n}{\cancel{0.7}}$$

$$180 = n$$

So, 126 credits is 70% of 180 credits.

b. 475% of what amount is $190?

Translate the sentence into an equation. Write the percent in decimal form.

475%	of	what amount	is	$190?
↓	↓	↓	↓	↓
4.75	·	n	=	190

To solve, divide both sides by 4.75.

$$\frac{\cancel{4.75} \cdot n}{\cancel{4.75}} = \frac{190}{4.75}$$

$$n = 40$$

So 475% of $40 is $190.

6. Write and solve a percent equation to answer each question.

a. 48 tons is 16% of how many tons?

____

b. 350% of what amount is 273?

____

Objective 1 Estimate answers to percent problems involving 25%.

For extra help, see Example 1 on page 518 of your text and Section Lecture video for Section 7.3.

Estimate the answer to each question.

1. Find 25% of 97 minutes. **1.** ______________

2. 25% of 43 inches is how many inches? **2.** ______________

3. What is 25% of $7823? **3.** ______________

Name: Date:
Instructor: Section:

Objective 2 Find 10% and 1% of a number by moving the decimal point.

For extra help, see Examples 2–3 on pages 518–519 of your text and Section Lecture video for Section 7.3 and Exercise Solutions Clip 9.

Find the exact answer to each question by moving the decimal point.

4. 1% of 400 homes is _____ . **4.** ________________

5. 10% of 4920 televisions is _____ . **5.** ________________

6. 1% of $98 is _____ . **6.** ________________

Objective 3 Solve basic percent problems using the percent equation.

For extra help, see Examples 4–6 on pages 520–523 of your text and Section Lecture video for Section 7.3 and Exercise Solutions Clip 21 and 29.

Write and solve an equation to answer each question.

7. 195 calls is what percent of 260 calls? **7.** ________________

8. 3% of $720 is how much? **8.** ________________

9. 24 magazines is 6% of what number of magazines? **9.** ________________

Name: Date:
Instructor: Section:

Chapter 7 PERCENT

7.4 Problem Solving with Percent

Learning Objectives
1. Solve percent application problems.
2. Solve problems involving percent of increase or decrease.

Key Terms

Use the vocabulary terms listed below to complete each statement in exercises 1–3.

percent of increase or decrease **percent proportion**

percent equation

1. The statement part = percent · whole is called the ____________________.

2. In a ____________________ problem, the increase or decrease is expressed as a percent of the original amount.

3. The statement $\frac{\text{part}}{\text{whole}} = \frac{\text{percent}}{100}$ is called the ____________________.

Guided Examples

Review this example for Objective 1:

1. 20% of the students at Lakeville Community College are married. How many of the 3260 students enrolled this year are married?

Step 1 Read the problem. Find the number of married students.

Step 2 Assign a variable. Let n be the number of married students.

Step 3 Write an equation. Use the percent equation.

percent · whole = part

$$\underbrace{20\%}\ \text{of}\ 3260\ \text{is}\ \underbrace{\text{how many students?}}$$
$$\downarrow \quad \downarrow \quad \downarrow \quad \downarrow \quad \downarrow$$
$$0.20 \cdot 3260 = n$$

Step 4 Solve the equation. Simplify the left side, multiplying 0.20 and 3260.

$$(0.20)(3260) = n$$
$$652 = n$$

Now Try:

1. There were 75 points on the first statistics test. Steven's score was 88% correct. How many points did Steven earn?

____

Name: Date:
Instructor: Section:

Step 5 State the answer. The number of married students is 652.

Step 6 Check the solution. 10% of 3260 is 326. 20% is two times 326, or 652.

2. Louise Trent earns \$450 per week and has 14% of this amount withheld for taxes. Find the amount withheld.

Step 1 Read the problem. Find the amount withheld for taxes.

Step 2 Assign a variable. Let n be the amount withheld.

Step 3 Write an equation. Use the percent equation.

percent · whole = part

$$\underbrace{14\%}_{\downarrow\ 0.14} \quad \underset{\downarrow\ \cdot}{\text{of}} \quad \underset{\downarrow\ 450}{450} \quad \underset{\downarrow\ =}{\text{is}} \quad \underbrace{\text{how much?}}_{\downarrow\ n}$$

Step 4 Solve the equation. Simplify the left side, multiplying 0.14 and 450.

$$(0.14)(450) = n$$
$$63 = n$$

Step 5 State the answer. The amount withheld is \$63.

Step 6 Check the solution. 10% of \$450 is \$45 and 5% of \$450 is \$22.50. So then \$45 + \$22.50 = \$67.50, so a little less than \$67.50 should be withheld.

2. A survey at an intersection found that, of 2450 drivers, 12% were talking on a cell phone. How many drivers in the survey were talking on the phone?

____

3. Lin needs 60 credits to graduate from her community college. So far she has earned 51 credits. What percent of the required credits does she have?

Step 1 Read the problem. Determine the credits earned.

Unknown: percent earned

Known: earned 51 out of 60 credits.

Step 2 Assign a variable. Let p be the unknown percent.

Step 3 Write an equation.

3. Jane eats 1500 calories a day. If she eats 350 calories for breakfast, what percent of her daily calories is her breakfast?

____

percent	·	whole	=	part
↓		↓		↓
p	·	60	=	51

Step 4 Solve the equation.

$$p \cdot 60 = 51$$

$$\frac{p \cdot \cancel{60}}{\cancel{60}} = \frac{51}{60}$$

$$p = 0.85$$

$$0.85 = 85\%$$

Step 5 State the answer. Lin earned 85% of credits needed.

Step 6 Check the solution. Lin earned most of the credits, so the percent should be close to 100%.

4. Janelle had budgeted \$250 for new school clothes but ended up spending \$390. The amount she spent was what percent of her budget?

Step 1 Read the problem. It is about comparing her budget to the amount spent.
Unknown: The percent of her budget
Known: \$250 budgeted, \$390 spent.

Step 2 Assign a variable. Let p be the unknown percent.

Step 3 Write an equation.

Amount spent	is	what percent	of	amount budgeted?
↓	↓	↓	↓	↓
390	=	p	·	250

Step 4 Solve the equation.

$$390 = p \cdot 250$$

$$\frac{390}{250} = \frac{p \cdot \cancel{250}}{\cancel{250}}$$

$$1.56 = p$$

$$1.56 = 156\%$$

Step 5 State the answer. Janelle spent 156% of her budget.

Step 6 Check the solution. 100% is \$250, and 50% is \$125. So \$250 + \$125 = \$350, or 150%, which is close to 156%.

4. Total Fitness Club predicted that 240 new members would join after Christmas. It actually had 396 new members join. The actual number joining is what percent of the predicted number?

____

5. David has 4.5% of his earnings deposited into a money market. If this amounts to \$146.25 per month, find his monthly earnings.

Step 1 Read the problem. It is about earnings.
Unknown: monthly salary
Known: \$146.25 is 4.5% of monthly earnings.

Step 2 Assign a variable. Let n = monthly salary.

Step 3 Write an equation.

$$\underbrace{4.5\%}\ \text{of}\ \underbrace{\text{how much}}\ \text{is}\ \$146.25$$

$$\downarrow\quad\downarrow\quad\downarrow\quad\downarrow\quad\downarrow$$

$$0.045 \cdot n = 146.25$$

Step 4 Solve the equation.

$$0.045 \cdot n = 146.25$$

$$\frac{\cancel{0.045} \cdot n}{\cancel{0.045}} = \frac{146.25}{0.045}$$

$$n = 3250$$

Step 5 State the answer. David's monthly earnings is \$3250.

Step 6 Check the solution. Round 4.5% to 5%. 10% of \$3250 is \$325, then 5% is half of \$325 or \$162.50, which is close to the number given.

5. A multivitamin contains 10 micrograms of vitamin K. If this is 13% of the recommended daily dosage, what is the recommended daily dosage of vitamin K? Round the answer to the nearest whole number.

____

Review this example for Objective 2:

6. Over the last three years, Calvin's salary has increased from \$2700 per month to \$3200. What is the percent increase?

Step 1 Read the problem. It is about a salary increase.
Unknown: percent of increase
Known: original salary was \$2700;
new salary is \$3200.

Step 2 Assign a variable. Let p be the percent of increase.

Step 3 Write an equation. First, subtract \$3200 - \$2700 to find the amount of increase.

$$\$3200 - \$2700 = \$500$$

$$\underbrace{\text{percent}}\ \text{of}\ \underbrace{\text{original wage}}\ \text{is}\ \underbrace{\text{amount of increase}}$$

$$\downarrow\quad\downarrow\quad\downarrow\quad\downarrow\quad\downarrow$$

$$p \cdot 2700 = 500$$

Now Try:

6. A business increased the number of phone lines from 4 to 9. What is the percent increase?

____

Step 4 Solve the equation.

$$p \cdot 2700 = 500$$

$$\frac{p \cdot \cancel{2700}}{\cancel{2700}} = \frac{500}{2700}$$

$$p \approx 0.185$$

$$0.185 = 18.5\%$$

Step 5 State the answer. Calvin's salary increased by 18.5%.

Step 6 Check the solution. Round 18.5% to 20%. Then 10% of \$2700 is \$270, and 20% is two times 10%, or \$540, which is close to the number given.

7. During the holiday season, average daily attendance at a health club fell from 495 members to 230 members. What was the percent decrease?

Step 1 Read the problem. It is about decrease in attendance.

Unknown: percent of decrease
Known: original attendance: 495
new attendance: 230.

Step 2 Assign a variable. Let p be the percent of decrease.

Step 3 Write an equation. First subtract 230 from 495 to find the amount of decrease.

$$495 - 230 = 265$$

$$\underbrace{\text{percent}}_{\downarrow\ p} \quad \underset{\cdot}{\overset{\text{of}}{\downarrow}} \quad \underbrace{\text{original attendance}}_{\downarrow\ 495} \quad \underset{=}{\overset{\text{is}}{\downarrow}} \quad \underbrace{\text{amount of decrease}}_{\downarrow\ 265}$$

Step 4 Solve the equation.

$$p \cdot 495 = 265$$

$$\frac{p \cdot \cancel{495}}{\cancel{495}} = \frac{265}{495}$$

$$p \approx 0.535$$

$$0.535 = 53.5\%$$

Step 5 State the answer. The attendance decreased by 53.5%.

Step 6 Check the solution. 50% is half of 495, or 247.5, so 53.5% is a reasonable solution.

7. The earnings per share of Amy's Cosmetic Company decreased from \$1.20 to \$0.86 in the last year. Find the percent of decrease.

____

Name: Date:
Instructor: Section:

Objective 1 Solve percent application problems.

For extra help, see Examples 1–5 on pages 530–533 of your text and Section Lecture video for Section 7.4 and Exercise Solutions Clip 11 and 13.

Use the six problem-solving steps to answer each question. Round percent answers to the nearest tenth of a percent.

1. Members who are between 25 and 45 years of age make up 92% of the total membership of an organization. If there are 850 total members in the organization, find the number of members in the 25 to 45 age group.

1. ________________

2. Payroll deductions are 35% of Jason's gross pay. If his deductions total $350, what is his gross pay?

2. ________________

3. Vera's Antique Shoppe says that of its 5100 items in stock, 4233 are just plain junk, while the rest are antiques. What percent of the number of items in stock is antiques?

3. ________________

Name: Date:
Instructor: Section:

Objective 2 Solve problems involving percent of increase or decrease.

For extra help, see Examples 6–7 on pages 534–535 of your text and Section Lecture video for Section 7.4 and Exercise Solutions Clip 27.

Use the six problem-solving steps to find the percent increase or decrease. Round your answers to the nearest tenth of a percent.

4. Students at Withrow's College were charged \$1560 for tuition this semester. If the tuition was \$1480 last semester, find the percent of increase.

4. ________________

5. Tomika's part-time work schedule has been reduced to 20 hours per week. She had been working 28 hours per week. What is the percent decrease?

5. ________________

6. During a sale, the price of a futon was cut from \$1250 to \$999. Find the percent of decrease in price.

6. ________________

Name: Date:
Instructor: Section:

Chapter 7 PERCENT

7.5 Consumer Applications: Sales Tax, Tips, Discounts, and Simple Interest

Learning Objectives	
1	Find sales tax and total cost.
2	Estimate and calculate restaurant tips.
3	Find the discount and sale price.
4	Calculate simple interest and the total amount due on a loan.

Key Terms

Use the vocabulary terms listed below to complete each statement in exercises 1–8.

sales tax	**tax rate**	**discount**
interest	**principal**	**interest rate**
simple interest	**interest formula**	

1. The charge for money borrowed or loaned, expressed as a percent, is called ________________.

2. A fee paid for borrowing or lending money is called ____________________.

3. The formula $I = p \cdot r \cdot t$ is the ____________________________.

4. Use the formula $I = p \cdot r \cdot t$ to compute the amount of ____________________ due on a loan.

5. The amount of money borrowed or loaned is called the ____________________.

6. The percent of the total sales charged as tax is called the ___________________.

7. The percent of the original price that is deducted from the original price is called the ______________________.

8. The percent used when calculating the amount of tax is called the ______________________.

Guided Examples

Review these examples for Objective 1:

1. A television sells for \$750 plus 8% sales tax. Find the price of the TV including sales tax.

   Step 1 Read the problem. It asks for the tax and the price of the TV.

   Step 2 Assign a variable. Let n be the amount of

Now Try:

1. A cell phone is \$89. The sales tax is 7%. What is the tax and total?

   ____

tax.

Step 3 Write an equation. Use the sales tax equation.

$$\underbrace{\text{tax rate}}_{0.08} \cdot \underbrace{\text{cost of item}}_{750} = \underbrace{\text{sales tax}}_{n}$$

Step 4 Solve.

$$(0.08)(750) = n$$
$$60 = n$$

Add the sales tax to the cost of the TV to find the total cost.

$$\underbrace{\text{cost of item}}_{\$750} + \underbrace{\text{sales tax}}_{\$60} = \underbrace{\text{total cost}}_{\$810}$$

Step 5 State the answer. The tax is \$60 and the total cost of the TV, including tax, is \$810.

Step 6 Check. Use estimation to check that the amount of sales tax is reasonable. Round 8% to 10%. Then 10% of \$750 is \$75. It is reasonable.

2. A gold bracelet costs \$1300 not including a sales tax of \$71.50. Find the sales tax rate.

Step 1 Read the problem. It asks for the tax rate.

Step 2 Assign a variable. Let p be the tax rate (the percent).

Step 3 Write an equation.

$$\underbrace{\text{tax rate}}_{p} \cdot \underbrace{\text{cost of item}}_{1300} = \underbrace{\text{sales tax}}_{71.50}$$

Step 4 Solve.

$$p \cdot 1300 = 71.50$$
$$\frac{p \cdot \cancel{1300}}{\cancel{1300}} = \frac{71.50}{1300}$$
$$p = 0.055$$
$$0.055 = 5.5\%$$

Step 5 State the answer. The sales tax rate is 5.5%.

Step 6 Check. Use estimation to check that the answer is reasonable. 1% of \$1300 is \$13. Round 5.5% to 6%. Then 6% is six times \$13 or \$78. The solution is reasonable.

2. The rate for a hotel room is \$124. The tax is \$14.88. Find the tax rate.

____

Name: Date:
Instructor: Section:

Review these examples for Objective 2:

3. First estimate each tip. Then calculate the exact answer.

a. Find a 15% tip for a restaurant bill of \$87.22.

Estimate: Round \$87.22 to \$90. Then 10% of \$90 is \$9, and 5% is half of \$9, or \$4.50. So the estimate is \$9 + \$4.50 = \$13.50.
Exact: Use the percent equation. Write 15% as a decimal. The bill for food and beverages is the whole and the tip is the part.

$$\begin{array}{ccccc} \text{percent} & \cdot & \text{whole} & = & \text{part} \\ \downarrow & & \downarrow & & \downarrow \\ 15\% & \cdot & \$87.22 & = & n \end{array}$$

$$(0.15)(87.22) = n$$

$$13.083 = n$$

Round \$13.083 to \$13.08 (nearest cent), which is close to the estimate of \$13.50.

b. Find a 20% tip for a restaurant bill of \$87.22.

Estimate: Round \$87.22 to \$90. Then 10% of \$90 is \$9, and 20% is two times \$9 is \$18.
Exact: Use the percent equation. Write 20% as a decimal. The bill for food and beverages is the whole and the tip is the part.

$$\begin{array}{ccccc} \text{percent} & \cdot & \text{whole} & = & \text{part} \\ \downarrow & & \downarrow & & \downarrow \\ 20\% & \cdot & \$87.22 & = & n \end{array}$$

$$(0.20)(87.22) = n$$

$$17.444 = n$$

Round \$17.444 to \$17.44 (nearest cent), which is close to the estimate of \$18.

Now Try:

3. First estimate each tip. Then calculate the exact answer.

a. Find a 15% tip for a restaurant bill of \$23.75.

____

b. Find a 20% tip for a restaurant bill of \$23.75.

____

Review this example for Objective 3:

4. Mike Lee can purchase a new car at 8% below window sticker price. Find the sale price on a car with a window sticker price of \$17,600.

Step 1 Read the problem. The problem asks for the price of the car after a discount of 8%.

Step 2 Work out a plan. The problem is solved in two steps. First, find the amount of the discount, that is, the amount that will be "taken off" (subtracted), by multiplying the original price (\$17,600) by the rate of the discount (8%).

Now Try:

4. A hard-cover book with an original price of \$24.95 is on sale at 60% off. Find the sale price of the book.

____

The second step is to subtract the amount of discount from the original price. This gives the sale price, which is what Mike will actually pay for the car.

Step 3 Estimate a reasonable answer. Round the original price from \$17,600 to \$20,000, and the rate of discount from 8% to 10%. Since 10% is equivalent to $\frac{1}{10}$, the estimated discount is $\$20,000 \div 10 = \$2000$, so the estimated sale price is \$20,000 – \$2000 = \$18,000.

Step 4 Solve the problem. First find the exact amount of the discount.

amount of discount = rate of discount · original price

$$a = (8\%)(\$17,600)$$
$$a = (0.08)(\$17,600)$$
$$a = \$1408$$

Now, find the sale price of the car by subtracting the amount of the discount (\$1408) from the original price.

sale price = original price — amount of discount

$$= \$17,600 - \$1408$$
$$= \$16,192$$

Step 5 State the answer. The sale price of the car is \$16,192.

Step 6 Check. The exact answer, \$16,192, is close to the estimate of \$18,000.

Review these examples for Objective 4:

5. Find the simple interest and total amount due on \$4000 at 4% for 1 year.

The amount borrowed, or principal (p), is \$4000. The interest rate (r), is 4%, which is 0.04 as a decimal, and the time of the loan (t) is 1 year. Use the formula.

$$I = p \cdot r \cdot t$$
$$I = (4000)(0.04)(1)$$
$$I = \$160$$

The interest is \$160.

Now add the principal and the interest to find the

Now Try:

5. Find the interest and total amount due on \$80 at 5% for 1 year.

____

total amount due.

amount due = principal + interest

$= \$4000 + \$160$

$= \$4160$

The total amount due is \$4160.

6. Find the simple interest and total amount due on \$5600 at 5% for two and a half years.

The principal (p) is \$5600. The rate (r) is 5%, or 0.05 as a decimal, and the time (t) is $2\frac{1}{2}$ or 2.5 years. Use the formula.

$I = p \cdot r \cdot t$

$I = (5600)(0.05)(2.5)$

$I = \$700$

The interest is \$700.

Now add the principal and the interest to find the total amount due.

amount due = principal + interest

$= \$5600 + \$700$

$= \$6300$

The total amount due is \$6300.

6. Find the simple interest and total amount due on \$620 at 16% for one and a quarter years.

____

7. Find the simple interest and total amount due on \$720 at $3\frac{1}{2}\%$ for 8 months.

The principal is \$720. The rate is $3\frac{1}{2}\%$ or 0.035 as a decimal, and the time is $\frac{8}{12}$ of a year. Use the formula $I = p \cdot r \cdot t$.

$I = (720)(0.035)\left(\frac{8}{12}\right)$

$= 25.2\left(\frac{2}{3}\right)$

$= 16.80$

The interest is \$16.80.

The total amount due is \$720 + \$16.80 = \$736.80.

7. Find the simple interest and total amount due on \$840 at $8\frac{1}{2}\%$ for 5 months.

____

Name: Date:
Instructor: Section:

Objective 1 Find sales tax and total cost.

For extra help, see Examples 1–2 on pages 540–541 of your text and Section Lecture video for Section 7.5 and Exercise Solutions Clip 7.

Find the amount of sales tax and the total cost. Round answers to the nearest cent, if necessary.

	Amount of sale	Tax Rate
1.	$350	6.5%

1. Tax____________

Total____________

Find the sales tax rate. Round answers to the hundredth, if necessary.

	Amount of sale	Amount of Tax
2.	$450	$36
3.	$215	$10.75

2. ________________

3. ________________

Objective 2 Estimate and calculate restaurant tips.

For extra help, see Example 3 on pages 541–542 of your text and Section Lecture video for Section 7.5 and Exercise Solutions Clip 49.

For each restaurant bill, estimate a 15% tip and a 20% tip. Then find the exact amounts for a 15% tip and a 20% tip. Round exact amounts to the nearest cent if necessary.

4. $43.16

4. 15% estimate _____

15% exact________

20% estimate _____

20% exact________

5. $63.85

5. 15% estimate _____

15% exact________

20% estimate _____

20% exact________

Name: Date:
Instructor: Section:

6. $72.81

6. 15% estimate _____

15% exact________

20% estimate _____

20% exact________

Objective 3 Find the discount and sale price.

For extra help, see Example 4 on page 543 of your text and Section Lecture video for Section 7.5.

Find the amount of discount and the amount paid after the discount. Round money answers to the nearest cent, if necessary.

	Original price	Rate of Discount
7.	$200	15%

7. Discount________

Amount paid ________

Solve each application problem. Round money answers to the nearest cent, if necessary.

8. A "Super 35% Off Sale" begins today. What is the price of a hair dryer normally priced at $15?

8. ________________

9. Geishe's Shoes sells shoes at 33% off the regular price. Find the price of a pair of shoes normally priced at $54, after the discount is given.

9. ________________

Objective 4 Calculate simple interest and the total amount due on a loan.

For extra help, see Examples 5–7 on pages 544–545 of your text and Section Lecture video for Section 7.5 and Exercise Solutions Clip 35 and 47.

Find the simple interest and the total amount due. Round to the nearest cent, if necessary.

	Principal	Rate	Time in Years
10.	$5280	9%	1
11.	$780	10%	$2\frac{1}{2}$
12.	$14,400	7%	7 months

10. ________________

____

11. ________________

____

12. ________________

____

Name: Date:
Instructor: Section:

Chapter 8 MEASUREMENT

8.1 Problem Solving with U.S. Measurement Units

Learning Objectives

1. Learn the basic U.S. measurement units.
2. Convert among U.S. measurement units using multiplication or division.
3. Convert among measurement units using unit fractions.
4. Solve application problems using U.S. measurement units.

Key Terms

Use the vocabulary terms listed below to complete each statement in exercises 1–3.

U.S. measurement units **unit fractions** **metric system**

1. The ____________________ is based on units of ten.
2. ____________________ are used to convert among different measurements.
3. ____________________ include inches, feet, quarts, and pounds.

Guided Examples

Review these examples for Objective 1:

1. Memorize the U.S. measurement conversions from the text. Then fill in the blanks.

   a. 2 c = __________ pt

   Answer: 1 pt

   b. 1 mi = __________ ft

   Answer: 5280 ft

Now Try:

1. Memorize the U.S. measurement conversions from the text. Then fill in the blanks.

   a. 16 oz = __________ lb

   ____

   b. 1 gal = __________ qt

   ____

Review these examples for Objective 2:

2. Convert each measurement.

   a. 16 yd to feet

   You are converting from a larger unit to a smaller unit, so multiply.
   Because 1 yd = 3 ft, multiply by 3.
   $16 \text{ yd} = 16 \cdot 3 = 48 \text{ ft}$

Now Try:

2. Convert each measurement.

   a. $17\frac{1}{2}$ ft to inches

   ____

b. 6 T to pounds

You are converting from a larger unit to a smaller unit, so multiply.
Because 1 T = 2000 lb, multiply by 2000.

$$6\text{ T} = 6 \cdot 2000 = 12{,}000\text{ lb}$$

c. 8 pt to quarts

You are converting from a smaller unit to a larger unit, so divide.
Because 2 pt = 1 qt, divide by 2.

$$8\text{ pt} = \frac{8}{2} = 4\text{ qt}$$

d. 3960 ft to miles

You are converting from a smaller unit to a larger unit, so divide.
Because 5280 ft = 1 mi, divide by 5280.

$$3960\text{ ft} = \frac{3960}{5280} = \frac{3}{4}\text{ mi}$$

b. $3\frac{1}{2}$ lb to ounces

____

c. 75 sec to minutes

____

d. 380 min to hours

____

Review these examples for Objective 3:
3.

a. Convert 64 oz to pounds.

Use a unit fraction with pounds (the unit for your answer) in the numerator, and ounces (the unit being changed) in the denominator. Because 1 lb = 16 oz, the necessary unit fraction is

$$\frac{1\text{ lb}}{16\text{ oz}} \begin{array}{l} \leftarrow \text{Unit for your answer is pounds.} \\ \leftarrow \text{Unit being changed is ounces.} \end{array}$$

Next, multiply 64 oz times this unit fraction.
Write 64 oz as the fraction $\frac{64\text{ oz}}{1}$ and divide out common units and factors wherever possible.

$$\frac{64\text{ oz}}{1} \cdot \frac{1\text{ lb}}{16\text{ oz}} = \frac{\overset{4}{\cancel{64}}\ \cancel{\text{oz}}}{1} \cdot \frac{1\text{ lb}}{\underset{1}{\cancel{16}}\ \cancel{\text{oz}}} = \frac{4 \cdot 1\text{ lb}}{1} = 4\text{ lb}$$

b. Convert 8 lb to ounces.

Select the correct unit fraction to change 8 lb to ounces.

$$\frac{16\text{ oz}}{1\text{ lb}} \begin{array}{l} \leftarrow \text{Unit for your answer is ounces.} \\ \leftarrow \text{Unit being changed is pounds.} \end{array}$$

Multiply 8 lb times the unit fraction.

$$\frac{8\text{ lb}}{1} \cdot \frac{16\text{ oz}}{1\text{ lb}} = \frac{8\ \cancel{\text{lb}}}{1} \cdot \frac{16\text{ oz}}{1\ \cancel{\text{lb}}} = \frac{8 \cdot 16\text{ oz}}{1} = 128\text{ oz}$$

Now Try:
3.

a. Convert 12 yd to inches.

____

b. Convert 4 in. to feet.

____

Name: Date:
Instructor: Section:

4. Convert using unit fractions.

a. Convert 6 qt to gallons.

First select the correct unit fraction.

$\frac{1\text{ gal}}{4\text{ qt}}$ ← Unit for your answer is gallons.
← Unit being changed is quarts.

Next multiply.

$$\frac{6\text{ qt}}{1}\cdot\frac{1\text{ gal}}{4\text{ qt}}=\frac{\overset{3}{\cancel{6}}\ \cancel{\text{qt}}}{1}\cdot\frac{1\text{ gal}}{\underset{2}{\cancel{4}}\ \cancel{\text{qt}}}=\frac{3}{2}\text{ gal}=1\frac{1}{2}\text{ gal}$$

b. Convert $6\frac{1}{2}$ lb to ounces.

Write $6\frac{1}{2}$ lb as an improper fraction.

$$\frac{6\frac{1}{2}\ \cancel{\text{lb}}}{1}\cdot\frac{16\text{ oz}}{1\ \cancel{\text{lb}}}=\frac{13}{2}\cdot\frac{16}{1}\text{ oz}$$
$$=\frac{13}{\underset{1}{\cancel{2}}}\cdot\frac{\overset{8}{\cancel{16}}}{1}\text{ oz}$$
$$=104\text{ oz}$$

c. Convert 340 hr to days.

$$\frac{\overset{85}{\cancel{340}}\ \cancel{\text{hr}}}{1}\cdot\frac{1\text{ day}}{\underset{6}{\cancel{24}}\ \cancel{\text{hr}}}=\frac{85}{6}\text{ day}=14\frac{1}{6}\text{ day}$$

5. Sometimes you may need to use two or three unit fractions to complete a conversion.

a. Convert 110 pt to gallons.

Use the unit fraction $\frac{1\text{ qt}}{2\text{ pt}}$ to change pints to quarts and the unit fraction $\frac{1\text{ gal}}{4\text{ qt}}$ to change quarts to gallons. Notice how all the units divide out except gallons, which is the unit you want in the answer.

$$\frac{110\ \cancel{\text{pt}}}{1}\cdot\frac{1\ \cancel{\text{qt}}}{2\ \cancel{\text{pt}}}\cdot\frac{1\text{gal}}{4\ \cancel{\text{qt}}}=\frac{110}{8}\text{gal}=\frac{110\div2}{8\div2}\text{gal}=13\frac{3}{4}\text{gal}$$

13.75 gal is equivalent to $13\frac{3}{4}$ gal.

4. Convert using unit fractions.

a. Convert 4 mi to feet.

____

b. Convert 3000 lb to tons.

____

c. Convert $3\frac{1}{4}$ gal to quarts.

____

5. Sometimes you may need to use two or three unit fractions to complete a conversion.

a. Convert 12 cups to gallons.

____

Name: Date:
Instructor: Section:

b. Convert 3 miles to inches.

Use three unit fractions. The first one changes from miles to yards, the next one changes yards to feet, and the last one changes feet to inches. All the units divide out except inches, which is the unit you want in your answer.

$$\frac{3\,\cancel{\text{mi}}}{1} \cdot \frac{1760\,\cancel{\text{yd}}}{1\,\cancel{\text{mi}}} \cdot \frac{3\,\cancel{\text{ft}}}{1\,\cancel{\text{yd}}} \cdot \frac{12\text{ in.}}{1\,\cancel{\text{ft}}} = 190{,}080 \text{ inches}$$

b. Convert 5 days to minutes.

____

Review these examples for Objective 4:

6. Answer each application problem.

a. Lee paid \$4.65 for 14 oz of honey baked ham. What is the price per pound, to the nearest cent?

Step 1 Read the problem. The problem asks for the price per pound of the ham.

Step 2 Work out a plan. The weight of the ham is given in ounces, but the answer must be cost per pound. Convert ounces to pounds. The word per indicates division. You need to divide the cost by the number of pounds.

Step 3 Estimate a reasonable answer. To estimate, round \$4.65 to \$5, and round 14 ounces to 16 ounces. Then there are 16 oz in a pound, so there is about 1 pound. Finally, $\$5 \div 1 = \$5$ per pound is our estimate.

Step 4 Solve the problem. Use a unit fraction to convert 14 oz to pounds.

$$\frac{\overset{7}{\cancel{14}}\,\cancel{\text{oz}}}{1} \cdot \frac{1\text{ lb}}{\underset{8}{\cancel{16}}\,\cancel{\text{oz}}} = \frac{7}{8}\text{ lb} = 0.875 \text{ lb}$$

Then divide to find the cost per pound.

$$\frac{\$4.65}{0.875} \approx \$5.31$$

Step 5 State the answer. The ham is \$5.31 per pound (nearest cent).

Step 6 Check your work. The exact answer, \$5.31 is close to the estimate of \$5.

Now Try:

6. Answer each application problem.

a. Clarissa paid \$1.79 for 4.5 oz of nuts. What is the cost per pound, to the nearest cent?

____

Name: Date:
Instructor: Section:

b. At a preschool, each of 20 children drinks about $\frac{3}{4}$ c of juice with their snack each day. The school is open 5 days a week. How many quarts of juice are needed for 1 week of snacks?

Step 1 Read the problem. The problem asks for the number of quarts of juice needed for 1 week.

Step 2 Work out a plan. Multiply to find the number of cups of juice for one week. Then convert cups to quarts.

Step 3 Estimate a reasonable answer. To estimate, round $\frac{3}{4}$ cups to 1 cup. Then 1 cup times 20 children times 5 days is 100 cups. There are 4 cups in a quart. So, $100 \text{ cups} \div 4 = 25 \text{ quarts}$ is our estimate.

Step 4 Solve the problem. First multiply. Then use unit fractions to convert.

$$\frac{3}{4} \cdot 20 \cdot 5 = \frac{3}{\cancel{4}_1} \cdot \frac{\cancel{20}^{5}}{1} \cdot \frac{5}{1} = 75 \text{ cups}$$

$$\frac{75 \cancel{\text{cups}}}{1} \cdot \frac{1 \cancel{\text{pt}}}{2 \cancel{\text{cups}}} \cdot \frac{1 \text{ qt}}{2 \cancel{\text{pt}}} = \frac{75}{4} \text{ qt} = 18\frac{3}{4} \text{ qt}$$

Step 5 State the answer. The preschool needs $18\frac{3}{4}$ qt of juice for 1 week.

Step 6 Check your work. The exact answer of $18\frac{3}{4}$ qt is close to our estimate of 25 qt.

b. Sweet Suzie's Shop makes 49,000 lb of fudge each week. How many tons of fudge are produced each day of the 7-day week?

____

Objective 1 Learn the basic U.S. measurement units.

For extra help, see Example 1 on page 566 of your text and Section Lecture video for Section 8.1.

Fill in the blanks.

1. 1 T = _____ lb

1. _______________

2. _____ qt = 1 gal

2. _______________

Name: Date:
Instructor: Section:

3. 1 c = _____ fl oz **3.** ________________

Objective 2 Convert among U.S. measurement units using multiplication or division.

For extra help, see Example 2 on page 567 of your text and Section Lecture video for Section 8.1 and Exercise Solutions Clip 11.

Convert each measurement using multiplication or division.

4. 12 ft to yards **4.** ________________

5. 40 pt to gallons **5.** ________________

6. 30 in to yards **6.** ________________

Objective 3 Convert among measurement units using unit fractions.

For extra help, see Examples 3–5 on pages 568–570 of your text and Section Lecture video for Section 8.1 and Exercise Solutions Clip 39 and 41.

Convert each measurement using unit fractions.

7. 28 pt to gallons **7.** ________________

8. 38 c to pints **8.** ________________

9. 60 oz to pounds **9.** ________________

Name: Date:
Instructor: Section:

Objective 4 Solve application problems using U.S. measurement units.

For extra help, see Example 6 on pages 570–571 of your text and Section Lecture video for Section 8.1 and Exercise Solutions Clip 57.

Solve each application problem using the six problem-solving steps.

10. Tony paid \$2.84 for 6.5 oz of fudge. What is the cost per pound, to the nearest cent? **10.** ____________

11. At an office, each of 15 workers drinks about $1\frac{3}{4}$ c of coffee with each day. The office is open 4 days a week. How many quarts of coffee are needed for a week? **11.** ____________

Name: Date:
Instructor: Section:

Chapter 8 MEASUREMENT

8.2 The Metric System—Length

Learning Objectives

1	Learn the basic metric units of length.
2	Use unit fractions to convert among metric units.
3	Move the decimal point to convert among metric units.

Key Terms

Use the vocabulary terms listed below to complete each statement in exercises 1–3.

meter **prefix** **metric conversion line**

1. Attaching a ______________________ such as "kilo-" or "milli" to the words "meter", "liter", or "gram" gives the names of larger or smaller units.

2. A line showing the various metric measurement prefixes and their size relationship to each other is called a ______________________.

3. The basic unit of length in the metric system is the ____________________.

Guided Examples

Review these examples for Objective 1:

1. Write the most reasonable metric unit in each blank. Choose from km, m, cm, and mm.

   a. The man's foot is 28 _____ .

   28 cm because cm are used instead of inches. 28 cm is about 11 inches.

   b. The living room is 4 _____ .

   4 m because m are used instead of feet. 4 m is about 13 feet.

   c. Sam rode his bike 8 _____ on the trail.

   8 km because km are used instead of miles. 8 km is about 5 miles.

Now Try:

1. Write the most reasonable metric unit in each blank. Choose from km, m, cm, and mm.

   a. A strand of hair is 1 ____ wide.

   ____

   b. The truck sped down the highway at 120 _____ per hour.

   ____

   c. The width of a piece of paper is 21.6 _____ .

   ____

Name: Date:
Instructor: Section:

Review these examples for Objective 2:

2. Convert each measurement using unit fractions.

a. 8 m to cm

Put the unit for the answer (cm) in the numerator of the unit fraction; put the unit you want to change (m) in the denominator.

$$\frac{100\text{ cm}}{1\text{ m}} \begin{matrix}\leftarrow \text{Unit for answer}\\ \leftarrow \text{Unit being changed}\end{matrix}$$

Multiply. Divide out common units where possible.

$$8\text{m}\cdot\frac{100\text{cm}}{1\text{m}}=\frac{8\cancel{\text{m}}}{1}\cdot\frac{100\text{cm}}{1\cancel{\text{m}}}=\frac{8\cdot 100\text{cm}}{1}=800\text{cm}$$

8 m = 800 cm

b. 13.5 mm to cm

Multiply by a unit fraction that allows you to divide out millimeters.

$$\frac{13.5\ \cancel{\text{mm}}}{1}\cdot\frac{1\text{ cm}}{10\ \cancel{\text{mm}}}=\frac{13.5\text{ cm}}{10}=1.35\text{ mm}$$

13.5 mm = 1.35 cm

There are 10 mm in a cm, so 13.5 mm will be a smaller part of a cm.

Now Try:

2. Convert each measurement using unit fractions.

a. 5400 m to km

____

b. 7.6 cm to mm

____

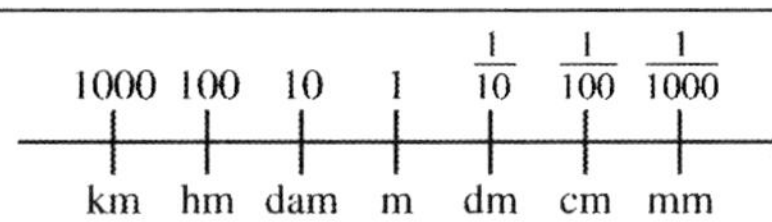

Review these examples for Objective 3:

3. Use the metric conversion line to make the following conversions.

a. 62.892 km to m

Find km on the metric conversion line. To get m, you move three places to the right. So move the decimal point in 62.892 three places to the right.

62.892 km = 62,892 m

b. 47.6 cm to m

Find cm on the conversion line. To get m, move two places to the left. So move the decimal point two places to the left.

47.6 cm = 0.476 m

c. 52.8 mm to cm

From mm to cm is one place to the left.

52.8 mm = 5.28 cm

Now Try:

3. Use the metric conversion line to make the following conversions.

a. 67.5 cm to mm

____

b. 986.5 m to km

____

c. 4.31 m to cm

____

Name: Date:
Instructor: Section:

4. Convert using the metric conversion line.

a. 1.79 m to mm

Moving from m to mm is going three places to the right. In order to move the decimal point in 1.79 three places to the right, you must add a 0 as a place holder.

1.790

1.79 m = 1790 mm

b. 50 cm to m

From cm to m is two places to the left. The decimal point in 50 starts at the far right side because 50 is a whole number. Then move it two places to the left.

50 cm = 0.50 m and 0.50 m is equivalent to 0.5 m.

c. 19 m to km

From m to km is three places to the left. The decimal point in 19 starts at the far right side. In order to move it three places to the left, you must write in one zero as a placeholder.

19 m = 0.019 km

4. Convert using the metric conversion line.

a. 23 m to mm

____

b. 4.2 cm to m

____

c. 950 m to km

____

Objective 1 Learn the basic metric units of length.

For extra help, see Example 1 on page 577 of your text and Section Lecture video for Section 8.2 and Exercise Solutions Clip 13, 15, 17, and 19.

Choose the most reasonable metric unit. Choose from **km**, **m**, **cm**, *or* **mm**.

1. the width of a twin bed **1.** ______________

2. the thickness of a dime **2.** ______________

3. the distance driven in 2 hours **3.** ______________

Name: Date:
Instructor: Section:

Objective 2 Use unit fractions to convert among metric units.

For extra help, see Example 2 on page 578 of your text and Section Lecture video for Section 8.2.

Convert each measurement using unit fractions.

4. 25.87 m to centimeters **4.** ________________

5. 450 m to kilometers **5.** ________________

6. 140 millimeters to meters **6.** ________________

Objective 3 Move the decimal point to convert among metric units.

For extra help, see Examples 3–4 on pages 579–580 of your text and Section Lecture video for Section 8.2 and Exercise Solutions Clip 29, 31, and 35.

Convert each measurement using the metric conversion line.

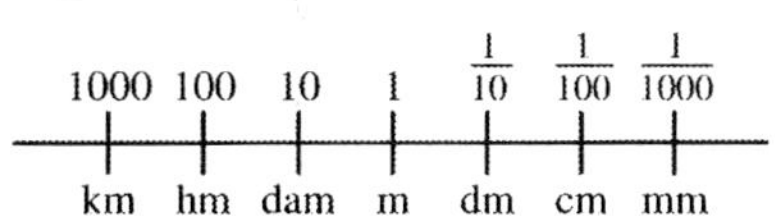

7. 1.94 cm to millimeters **7.** ________________

8. 10.35 km to meters **8.** ________________

9. 3.5 cm to kilometers **9.** ________________

Name: Date:
Instructor: Section:

Chapter 8 MEASUREMENT

8.3 The Metric System—Capacity and Weight (Mass)

Learning Objectives	
1	Learn the basic metric units of capacity.
2	Convert among metric capacity units.
3	Learn the basic metric units of weight (mass).
4	Convert among metric weight (mass) units.
5	Distinguish among basic metric units of length, capacity, and weight (mass).

Key Terms

Use the vocabulary terms listed below to complete each statement in exercises 1–2.

liter **gram**

1. The basic unit of weight (mass) in the metric system is the ________________.

2. The basic unit of capacity in the metric system is the ____________________.

Guided Examples

Review these examples for Objective 1:

1. Write the most reasonable metric unit in each blank. Choose from L and mL.

a. The bottle of sunscreen held 118 _____ .

118 mL because 118 L would be about 118 qts, which is too much.

b. The soda bottle has 2 _____ .

2 L because 2 mL would be less than a teaspoon.

Now Try:

1. Write the most reasonable metric unit in each blank. Choose from L and mL.

a. Ty added 5 _____ to the tea.

____

b. The water tank holds 100 _____ .

____

Review these examples for Objective 2:

2. Convert using the metric conversion line or unit fractions.

a. 5.4 L to mL

Using the metric conversion line:
From L to mL is three places to the right.
5.4 L = 5400 mL

Using unit fractions:
Multiply by a unit fraction that allows you to divide out liters.

$$\frac{5.4 \not{L}}{1}\cdot\frac{1000 \text{ mL}}{1 \not{L}} = 5400 \text{ mL}$$

Now Try:

2. Convert using the metric conversion line or unit fractions.

a. 973 mL to L

____

b. 92 mL to L

Using the metric conversion line:
From mL to L is three places to the left.
92 mL = 0.092 L

Using unit fractions:
Multiply by a unit fraction that allows you to divide out mL.

$$\frac{9.2\ \cancel{\text{mL}}}{1} \cdot \frac{1\ \text{L}}{1000\ \cancel{\text{mL}}} = 0.092\ \text{L}$$

b. 3.85 L to mL

____

Review these examples for Objective 3:

3. Write the most reasonable metric unit in each blank. Choose from kg, g, and mg.

a. The watermelon weighed 18 _____ .

18 kg because kilograms are used instead of pounds.
18 kg is about 40 pounds.

b. The vitamin tablet was 400 _____ .

400 mg because 400 g would be more than two hamburgers, which is too much.

c. A birthday card weighs 8 _____ .

8 grams because 8 kg would be too heavy.

Now Try:

3. Write the most reasonable metric unit in each blank. Choose from kg, g, and mg.

a. A fashion model weighs 48 _____ .

____

b. Jacob's football weighs 590 _____ .

____

c. The pencil lead weighs 3 _____ .

____

Review these examples for Objective 4:

4. Convert using the metric conversion line or unit fractions.

a. 26 mg to g

Using the metric conversion line:
From mg to g is three places to the left.
26 mg = 0.026 g

Using unit fractions:
Multiply by a unit fraction that allows you to divide out mg.

$$\frac{26\ \cancel{\text{mg}}}{1} \cdot \frac{1\ \text{g}}{1000\ \cancel{\text{mg}}} = 0.026\ \text{g}$$

Now Try:

4. Convert using the metric conversion line or unit fractions.

a. 3.72 g to mg

____

b. 97.34 kg to g

Using the metric conversion line:
From kg to g is three places to the right.
97.34 kg = 97,340 g

Using unit fractions:
Multiply by a unit fraction that allows you to divide out kg.

$$\frac{97.34\ \cancel{kg}}{1}\cdot\frac{1000\ g}{1\ \cancel{kg}}=97{,}340\ g$$

b. 84 g to kg

____

Review these examples for Objective 5:

5. First decide which type of unit is needed: length, capacity, or weight. Then write the most appropriate metric unit in the blank. Choose from km, m, cm, mm, L, mL, kg, g, and mg.

a. The tree is 9 _____ high.

Use length units because of the word high.
The tree is 9 m high.

b. A ten-carat diamond weighs 2 _____ .

Use weight units because of the word weigh.
The ten-carat diamond weighs 2 g.

c. The milk container has 3.78 _____ .

Use capacity units because milk is a liquid.
The milk container has 3.78 L.

Now Try:

5. First decide which type of unit is needed: length, capacity, or weight. Then write the most appropriate metric unit in the blank. Choose from km, m, cm, mm, L, mL, kg, g, and mg.

a. The moisturizer jar has 125 _____ .

____

b. This is a 1.19 _____ box of breakfast cereal.

____

c. The boy walked 400 _____ to the bus stop.

____

Objective 1 Learn the basic metric units of capacity.

For extra help, see Example 1 on page 584 of your text and Section Lecture video for Section 8.3 and Exercise Solutions Clip 11.

Choose the most reasonable metric unit. Choose from **L**, *or* **ml**.

1. the amount of soda in a can **1.** ____________

2. the amount of orange juice in a large bottle **2.** ____________

3. the amount of cough syrup in one dose **3.** ____________

Name: Date:
Instructor: Section:

Objective 2 Convert among metric capacity units.

For extra help, see Example 2 on pages 584–585 of your text and Section Lecture video for Section 8.3.

Convert each measurement. Use unit fractions or the metric conversion line.

4. 2.5 L to milliliters **4.** ________________

5. 836 kL to liters **5.** ________________

6. 7863 mL to liters **6.** ________________

Objective 3 Learn the basic metric units of weight (mass).

For extra help, see Example 3 on page 586 of your text and Section Lecture video for Section 8.3 and Exercise Solutions Clip 7 and 13.

Choose the most reasonable metric unit. Choose from **kg**, **g**, *or* **mg**.

7. the weight (mass) of a vitamin pill **7.** ________________

8. the weight (mass) of a car **8.** ________________

9. the weight (mass) of an egg **9.** ________________

Objective 4 Convert among metric weight (mass) units.

For extra help, see Example 4 on page 587 of your text and Section Lecture video for Section 8.3 and Exercise Solutions Clip 39, 41, and 43.

Convert each measurement. Use unit fractions or the metric conversion line.

10. 27,000 g to kilograms **10.** ________________

11. 0.76 kg to grams **11.** ________________

Name: Date:
Instructor: Section:

12. 4.7 g to milligrams **12.** ________________

Objective 5 Distinguish among basic metric units of length, capacity, and weight (mass).

For extra help, see Example 5 on page 588 of your text and Section Lecture video for Section 8.3 and Exercise Solutions Clip 57 and 61.

Choose the most reasonable metric unit. Choose from ***km, m, cm, mm, mL, L, kg, g,*** *or* ***mg****.*

13. Buy a 5 _____ bottle of water. **13.** ________________

14. The piece of wood weighs 5 ___________. **14.** ________________

15. A paperclip is 3 __________ long. **15.** ________________

Name: Date:
Instructor: Section:

Chapter 8 MEASUREMENT

8.4 Problem Solving with Metric Measurement

Learning Objectives
1 Solve application problems involving metric measurements.

Key Terms

Use the vocabulary terms listed below to complete each statement in exercises 1−3.

meter **liter** **gram**

1. A ______________________ is the weight of 1 mL of water.

2. A ________________________ is a little longer than a yard.

3. A ________________________ is a little more than one quart.

Guided Examples

Review these examples for Objective 1:

1. Colby cheese is on sale at $9.89 per kilogram. Nelson bought 750 g of the cheese. How much did he pay, to the nearest cent?

   Step 1 Read the problem. The problem asks for the cost of 750 g of cheese.

   Step 2 Work out a plan. The price is $9.89 per kilogram, but the amount Nelson bought is given in grams. Convert grams to kilograms. Then multiply the weight by the cost per kilogram.

   Step 3 Estimate a reasonable answer. Round the cost of 1 kg from $9.89 to $10. There are 1000 g in a kilogram, so 750 g is about $\frac{3}{4}$ of a kilogram. Nelson buys about $\frac{3}{4}$ of a kilogram, so $\frac{3}{4}$ of $10 = $7.50 is our estimate.

   Step 4 Solve the problem. Use a unit fraction to convert 750 g to kilograms.

   $$\frac{750\ \cancel{g}}{1} \cdot \frac{1\text{ kg}}{1000\ \cancel{g}} = 0.75\text{ kg}$$

   Now multiply 0.75 kg times the cost per kilogram.

Now Try:

1. A ball of yarn weighs 140 g. Ellie knitted a sweater for her boyfriend that used 11 balls of yarn. How many kilograms did the finished sweater weigh?

   ____

$$\frac{0.75\ \cancel{kg}}{1} \cdot \frac{\$9.89}{1\ \cancel{kg}} = \$7.4175 \approx \$7.42$$

Step 5 State the answer. Nelson paid $7.42, rounded to the nearest cent.

Step 6 Check your work. The exact answer of $7.42 is close to our estimate of $7.50.

2. A 70-L drum is filled with oil which is to be packaged into 140-mL bottles. How many bottles can be filled.?

Step 1 Read the problem. The problem asks for the number of filled bottles.

Step 2 Work out a plan. The given amount is in liters, but the capacity of the bottles is in mL. Convert liters to milliliters, then divide by 140 mL (the capacity of the bottles).

Step 3 Estimate a reasonable answer. To estimate round 140 mL to 100 mL. Then 70 L = 70,000 mL, and $70,000 \div 100 = 700$ bottles.

Step 4 Solve the problem. On the metric conversion line, moving from L to mL is three places to the right, so move the decimal point in 70 L three places to the right. Then divide by 140.

70 L = 70,000 mL

$$\frac{70,000\ \text{mL}}{140\ \text{mL}} = 500\ \text{bottles}$$

Step 5 State the answer. 500 bottles will be filled.

Step 6 Check your work. The exact answer of 500 bottles is close to our estimate of 700 bottles.

2. If 1.8 kg of candy is to be divided equally among 9 children, how many grams will each child receive?

____

3. Lucy purchased 1 m 60 cm of fabric at $6.25 per meter for a jacket and 1m 80 cm of fabric at $4.75 per m for a dress. How much did she spend in total?

Step 1 Read the problem. Lucy purchased two quantities of two different kinds of fabric. Find the total she spent on fabric.

Step 2 Work out a plan. The lengths involve two

3. A fish tank can hold up to 75.6 L of water. If there are 70,000 mL of water in the tank, how many more milliliters of water can the tank hold?

____

units, m and cm. Rewrite both lengths in meters, determine the price, and find the total.

Step 3 Estimate a reasonable answer. To estimate, 1 m 60 cm can be rounded to 2 m. Round 1 m 80 cm to 2 m. Also, \$6.25 can be rounded to \$6 and \$4.75 can be rounded to \$5. Then, $2\times\$6+2\times\$5=\$12+\$10=\$22$, our estimate.

Step 4 Solve the problem. Rewrite the lengths in meters.

$$\begin{array}{rr} 1\text{ m}\rightarrow & 1.0\text{ m} \\ \text{plus } 60\text{ cm}\rightarrow & \underline{+\ 0.6\text{ m}} \\ & 1.6\text{ m} \end{array} \qquad \begin{array}{rr} 1\text{ m}\rightarrow & 1.0\text{ m} \\ \text{plus } 80\text{ cm}\rightarrow & \underline{+\ 0.8\text{ m}} \\ & 1.8\text{ m} \end{array}$$

Jacket: $1.6\times\$6.25=\$10$
Dress: $1.8\times\$4.75=\underline{\$\ 8.55}$
$\$18.55$ total

Step 5 State the answer. The total spent is \$18.55.

Step 6 Check your work. The exact answer of \$18.55 is close to our estimate of \$22.

Objective 1 Solve application problems involving metric measurements.

For extra help, see Examples 1–3 on pages 595–596 of your text and Section Lecture video for Section 8.4 and Exercise Solutions Clip 3, 9, 11, and 13.

Solve each application problem. Round money answers to the nearest cent.

1. Metal chain costs \$5.26 per meter. Find the cost of 2 m 47 cm of the chain. Round your answer to the nearest cent.

1. ________________

2. Henry bowls with a 7-kg bowling ball, while Denise uses a ball that weighs 5 kg 750g. How much heavier is Henry's bowling ball than Denise's?

2. ________________

3. The label on a bottle of pills says that there are 3.5 mg of the medication in 5 pills. If a patient needs to take 8.4 mg of the medication, how many pills does he need to take?

3. ________________

Name: Date:
Instructor: Section:

Chapter 8 MEASUREMENT

8.5 Metric–U.S. Measurement Conversions and Temperature

Learning Objectives	
1	Use unit fractions to convert between metric and U.S. measurement units.
2	Learn common temperatures on the Celsius scale.
3	Use formulas to convert between Celsius and Fahrenheit temperatures.

Key Terms

Use the vocabulary terms listed below to complete each statement in exercises 1–2.

Celsius **Fahrenheit**

1. The ____________________ scale is used to measure temperature in the metric system.

2. The ____________________ scale is used to measure temperature in the U.S. customary system.

Guided Examples

Review these example for Objective 1:

1. Convert from 16 in. to centimeters.

We're changing from a U.S length unit to a metric length unit. In the "U.S. to Metric Units" side of the table, you see that 1 inch $\approx$ 2.54 centimeters. Two unit fractions can be written using that information.

$$\frac{1 \text{ in.}}{2.54 \text{ cm}} \text{ or } \frac{2.54 \text{ cm}}{1 \text{ in.}}$$

Multiply by the unit fraction that allows you to divide out inches.

$$16 \text{ in.} \cdot \frac{2.54 \text{ cm}}{1 \text{ in.}} = \frac{16 \cancel{\text{in.}}}{1} \cdot \frac{2.54 \text{ cm}}{1 \cancel{\text{in.}}} = 40.64 \text{ cm}$$

16 in. $\approx$ 40.64 cm

Now Try:

1. Convert from 12.9 m to feet.

____

Name: Date:
Instructor: Section:

2. Convert using unit fractions. Round your answers to the nearest tenth.

a. 22.5 pounds to kilograms

Look in the "U.S. to Metric Units" side of the table to see that 1 pound ≈ 0.45 kilograms. Use this information to write a unit fraction that allows you to divide out pounds.

$$\frac{22.5 \not{lb}}{1} \cdot \frac{0.45 \text{ kg}}{1 \not{lb}} = 10.125 \text{ kg}$$

22.5 lb $\approx$ 10.1 kg

b. 46.8 L to quarts

Look in the "Metric to U.S. Units" side of the table to see that 1 L $\approx$ 1.06 quarts. Write a unit fraction that allows you to divide out liters.

$$\frac{46.8 \not{L}}{1} \cdot \frac{1.06 \text{ qt}}{1 \not{L}} = 49.608 \text{ qt}$$

46.8 L $\approx$ 49.6 qt

2. Convert using unit fractions. Round your answers to the nearest tenth.

a. 9.68 kg to pounds

____

b. 20 gallons to liters

____

Review these examples for Objective 2:

3. Choose the metric temperature that is most reasonable for each situation.

a. hot water in a bathtub
27°C 40°C 100°C

27°C is too cold and 100°C is too hot.
40°C is reasonable.

b. fall day
13°C 50°C 65°C

50°C and 65°C are too hot.
13°C is reasonable.

Now Try:

3. Choose the metric temperature that is most reasonable for each situation.

a. hot cocoa
10°C 35°C 65°C

____

b. ice cream
–10°C 4°C 30°C

____

Name: Date:
Instructor: Section:

Review these examples for Objective 3:

4. Convert 77°F to Celsius.

Use the formula and follow the order of operations.

$$C = \frac{5(F-32)}{9}$$
$$= \frac{5(77-32)}{9}$$
$$= \frac{5(45)}{9}$$
$$= \frac{5(\overset{5}{\cancel{45}})}{\underset{1}{\cancel{9}}}$$
$$= 25$$

Thus, 77°F = 25°C.

5. Convert 30°C to Fahrenheit.

Use the formula and follow the order of operations.

$$F = \frac{9 \cdot C}{5} + 32$$
$$= \frac{9 \cdot 30}{5} + 32$$
$$= \frac{9 \cdot \overset{6}{\cancel{30}}}{\underset{1}{\cancel{5}}} + 32$$
$$= 54 + 32$$
$$= 86$$

Thus, 30°C = 86°F.

Now Try:

4. Convert 122°F to Celsius. __________

5. Convert 150°C to Fahrenheit. __________

Objective 1 Use unit fractions to convert between metric and U.S. measurement units.

For extra help, see Examples 1–2 on pages 599–600 of your text and Section Lecture video for Section 8.5 and Exercise Solutions Clip 7, 9, and 13.

Use the table in your textbook and unit fractions to make the following conversions. Round answers to the nearest tenth.

1. 291 mi to kilometers **1.** __________

2. 7 L to gallons **2.** __________

Name: Date:
Instructor: Section:

3. 26 oz to grams **3.** ________________

Objective 2 Learn common temperatures on the Celsius scale.

For extra help, see Example 3 on page 601 of your text and Section Lecture video for Section 8.5.

Choose the most reasonable temperature for each situation.

4. hot coffee **4.** ________________
35°C 60°C 100°C

5. Normal body temperature **5.** ________________
37°C 37°F

6. Oven temperature **6.** ________________
300°C 300°F

Objective 3 Use formulas to convert between Celsius and Fahrenheit temperatures.

For extra help, see Examples 4–5 on page 602 of your text and Section Lecture video for Section 8.5 and Exercise Solutions Clip 31.

Use the conversions formulas and the order of operations to convert Fahrenheit temperatures to Celsius and Celsius temperatures to Fahrenheit. Round your answers to the nearest degree, if necessary.

7. 62°F **7.** ________________

8. 10°C **8.** ________________

Solve the application problem. Round to the nearest degree, if necessary.

9. A recipe for roast beef calls for an oven temperature of 400°F. What is the temperature in degrees Celsius? **9.** ________________

Name: Date:
Instructor: Section:

Chapter 9 GRAPHS AND GRAPHING

9.1 Problem Solving with Tables and Pictographs

Learning Objectives
1 Read and interpret data presented in a table.
2 Read and interpret data from a pictograph.

Key Terms

Use the vocabulary terms listed below to complete each statement in exercises 1–2.

table **pictograph**

1. A graph that uses pictures or symbols to display information is called a ____________________.

2. A display of facts in rows and columns is called a ____________________.

Guided Examples

The table below shows information about the performance of the eight U.S. airlines during the year 2011 (January–December).

PERFORMANCE DATA FOR SELECTED U.S. AIRLINES
January–December 2011

Airline	On-Time Performance	Luggage Handling*
American	78%	3.6
Atlantic Southeast	75%	5.5
Continental	77%	3.4
Delta	82%	2.7
Hawaiian	93%	2.6
JetBlue	73%	2.2
Southwest	81%	3.7
United	80%	3.7

*Luggage problems per 1000 passengers
Source: Department of Transportation Air Travel Consumer Report

Review these examples for Objective 1:

1. Use the table above to answer these questions.

a. What percent of United's flights were on time?

Look across the row labeled United. 80% of its flights were on time.

Now Try:

1. Use the table above to answer these questions.

a. What percent of Hawaiian's flights were on time?

____

Name: Date:
Instructor: Section:

b. Which airline had the best luggage handling record?

Look down the column headed Luggage Handling. To find the best record, look for the lowest number, which is 2.2. Then look to the left to find the airline, which is JetBlue.

b. What airline(s) had the second worst luggage handling record?

____

c. What was the average percent of on-time flights for the three airlines with the worst performance, to the nearest whole percent?

Look down the column headed On-Time Performance to find the three lowest numbers: 73%, 75%, and 77%. To find the average, add the values and divide by 3.

$$\frac{73+75+77}{3}=\frac{225}{3}=75$$

The average on-time performance for the three worst airlines was 75%.

c. What was the average on-time performance for all eight airlines, to the nearest percent?

____

The table below shows the maximum cab fares in five different cities in 2011. The "flag drop" change is made when the driver starts the meter. "Wait time" is the charge for having to wait in the middle of a ride.

MAXIMUM TAXICAB FARES ALLOWED IN SELECTED CITIES IN 2011

City	Flag Drop	Price per Mile	Wait Time (per Hour)
Chicago	$2.25	$1.80	$20
Denver	$2.50	$2.25	$24
Miami	$2.50	$2.40	$24
New York	$2.50	$2	$24
San Francisco	$3.50	$2.75	$27

Source: taxifarefinder.com

2. Use the table above to answer these questions.

a. What is the maximum fare for a 15-mile ride in Denver that includes having the cab wait 10 minutes?

The price per mile in Denver is $2.25, so the cost for 15 miles is 15($2.25) = $33.75. Then add the flag drop charge of $2.50. Finally, figure out the cost of the wait time. One way is to set up a proportion.

2. Use the table above to answer these questions.

a. What is the maximum fare for a 25-mile ride in Miami that includes having the cab wait 5 minutes?

____

$$\begin{array}{r} \text{Cost} \rightarrow \\ \text{Wait time} \rightarrow \end{array} \frac{\$24}{60 \text{ min}} = \frac{\$x}{10 \text{ min}} \begin{array}{l} \leftarrow \text{Cost} \\ \leftarrow \text{Wait time} \end{array}$$

$$60 \cdot x = 24 \cdot 10$$

$$\frac{60x}{60} = \frac{240}{60}$$

$$x = \$4$$

Total fare = $33.75 + $2.50 + $4 = $40.25

b. It is customary to give the cab driver a tip. Find the total cost of the cab ride in part (a) if the passenger added at 10% tip, rounded to the nearest quarter (nearest $0.25).

Use the percent equation to find the exact tip.

percent · whole = part

$10\%(\$40.25) = n$

$0.10(\$40.25) = n$

$\$4.025 = n$ Round to $4 (nearest $0.25)

The total cost of the cab ride is $40.25 + $4 tip = $44.25.

b. It is customary to give the cab driver a tip. Find the total cost of the cab ride in part (a) if the passenger added at 20% tip, rounded to the nearest quarter (nearest $0.25).

____

This pictograph shows the approximate number of passenger arrivals and departures at selected U.S. airports in 2006.

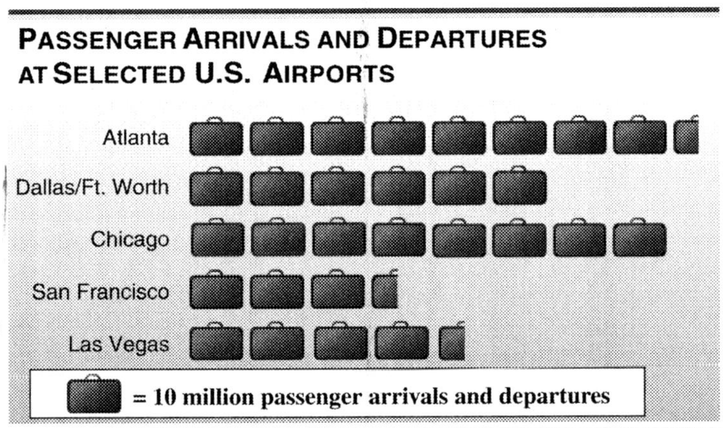

Source: Airports Council International—North America.

Review these examples for Objective 2:

3. Use the pictograph to answer these questions.

a. Approximately how many passenger arrivals and departures took place at the San Francisco airport?

The passenger arrivals and departures for San Francisco shows 3 whole symbols ($3 \cdot 10$ million = 30 million) plus half of a symbol $\left(\frac{1}{2}\right.$ of 10 million is 5 million) for a total of 35 million.

Now Try:

3. Use the pictograph to answer these questions.

a. Approximately how many passenger arrivals and departures took place at the Las Vegas airport?

____

Name: Date:
Instructor: Section:

b. What is the difference in the number of arrivals and departures at Dallas/Ft. Worth airport and Atlanta airport?

Dallas/Fort Worth shows 6 symbols and Atlanta shows $8\frac{1}{2}$ symbols. So Atlanta has $2\frac{1}{2}$ more symbols than Dallas/Fort Worth, and $2\frac{1}{2} \cdot 10$ million $= 25$ million. Thus, Atlanta has 25 million more passenger arrivals and departures than Dallas/Fort Worth.

b. What is the difference in the number of arrivals and departures at Chicago airport and San Francisco airport?

____

Objective 1 Read and interpret data presented in a table.

For extra help, see Examples 1–2 on pages 620–621 of your text and Section Lecture video for Section 9.1 and Exercise Solutions Clip 13 and 19.

Use the table below for exercises 1–3. Use proportions to help solve exercise 3.

Weight of Exerciser	**123 lbs**	**130 lbs**	**143 lbs**
	Calories burned in 30 minutes		
Cycling	168	177	195
Running	324	342	375
Jumping Rope	273	288	315
Walking	162	171	189

Source: Fitness magazine

1. How many calories are burned by a 130-pound adult in 30 minutes of cycling? **1.** _______________

2. Which activity will burn at least 300 calories when performed by a 123-pound adult? **2.** _______________

3. How many calories will a 123-pound adult burn in 60 minutes of walking and 15 minutes of running? **3.** _______________

Name: Date:
Instructor: Section:

Objective 2 Read and interpret data from a pictograph.

For extra help, see Example 3 on page 622 of your text and Section Lecture video for Section 9.1 and Exercise Solutions Clip 27.

The pictograph below shows the population of five cities in 2010. Use the pictograph to answer exercises 4–6.

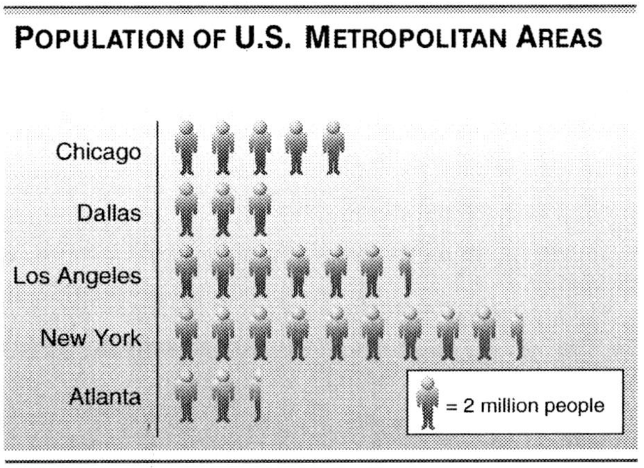

Source: U.S. Bureau of the Census, U.S. Department of Commerce.

4. What is the approximate population of Dallas? **4.** ________________

5. Approximately how much greater is the population of Los Angeles than the population of Chicago? **5.** ________________

6. What is the approximate total population of all five cities? **6.** ________________

Name: Date:
Instructor: Section:

Chapter 9 GRAPHS AND GRAPHING

9.2 Reading and Constructing Circle Graphs

Learning Objectives
1. Read a circle graph.
2. Use a circle graph.
3. Use a protractor to draw a circle graph.

Key Terms

Use the vocabulary terms listed below to complete each statement in exercises 1–2.

circle graph **protractor**

1. A ______________________________ shows how a total amount is divided into parts or sectors.

2. A ______________________________ is a device used to measure the number of degrees in angles or parts of a circle.

Guided Examples

The circle graph shows the number of students enrolled in certain majors at a college. The entire circle represents 11,600 students.

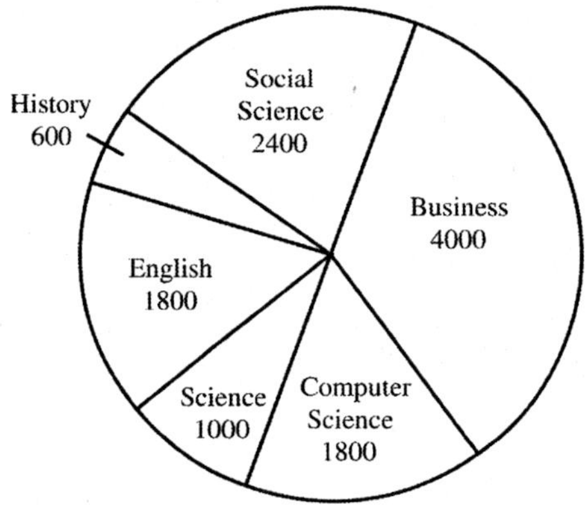

Review these examples for Objective 2:

1. Find the ratio of social science majors to the total number of students. Write the ratio as a fraction in lowest terms.

The circle graph shows 2400 social science majors of the 11,600 total students. The ratio of social science students to the total number of students is shown below.

$$\frac{2400 \text{ students (social science)}}{11{,}600 \text{ students (total)}}$$

$$= \frac{2400 \text{ \sout{students}}}{11{,}600 \text{ \sout{students}}} = \frac{2400 \div 400}{11{,}600 \div 400} = \frac{6}{29}$$

Now Try:

1. Find the ratio of history majors to the total number of students. Write the ratio as a fraction in lowest terms.

____

Name: Date:
Instructor: Section:

2. Use the circle graph to find the ratio of history majors to business majors. Write the ratio as a fraction in lowest terms.

The circle graph shows 600 history majors and 4000 business majors. The ratio of history majors to business majors is shown below.

$$\frac{600 \text{ students (history)}}{4000 \text{ students (business)}}$$

$$= \frac{600 \cancel{\text{students}}}{4000 \cancel{\text{students}}} = \frac{600 \div 200}{4000 \div 200} = \frac{3}{20}$$

2. Use the circle graph to find the ratio of science majors to computer science majors. Write the ratio as a fraction in lowest terms.

____

The circle graph shows the expenses involved in keeping a sales force on the road. Each expense item is expressed as a percent of the total sales force cost of $950,000.

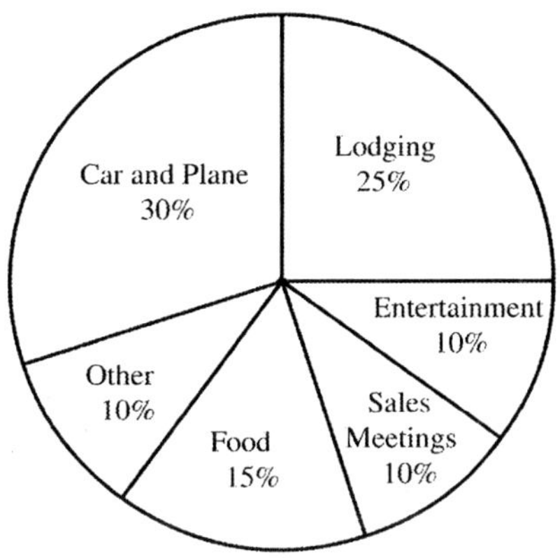

3. Use the circle graph above on expenses to find the amount spent on lodging.

Recall the percent equation.

percent · whole = part

The total expenses is $950,000, so the whole is $950,000. The percent is 25%, or as a decimal, 0.25. Find the part.

$$\text{percent} \cdot \text{whole} = \text{part}$$
$$(0.25)(950{,}000) = n$$
$$237{,}500 = n$$

The amount spent on lodging was $237,500.

3. Use the circle graph above on expenses to find the amount spent on food.

____

Name: Date:
Instructor: Section:

Review this example for Objective 3:

4. A family recorded its expenses for a year, with the following results: 40% for Housing, 20% for Food, 14% for Automobile, 8% for Clothing, 6% for Medical, 8% for Savings, and 4% for Other.

a. Find the number of degrees in a circle graph for each type of expenses.

Recall that a complete circle has 360°. Because "Housing" makes up 40% of the expenses, the number of degrees needed for the "Housing" sector of the circle graph is 40% of 360°.

Housing
$(360°)(40\%)=(360°)(0.40)=144°$

Food
$(360°)(20\%)=(360°)(0.20)=72°$

Automobile
$(360°)(14\%)=(360°)(0.14)=50.4°$

Clothing
$(360°)(8\%)=(360°)(0.08)=28.8°$

Medical
$(360°)(6\%)=(360°)(0.06)=21.6°$

Savings
$(360°)(8\%)=(360°)(0.08)=28.8°$

Other
$(360°)(4\%)=(360°)(0.04)=14.4°$

b. Draw a circle graph showing this information.

Use a protractor to make the circle graph.

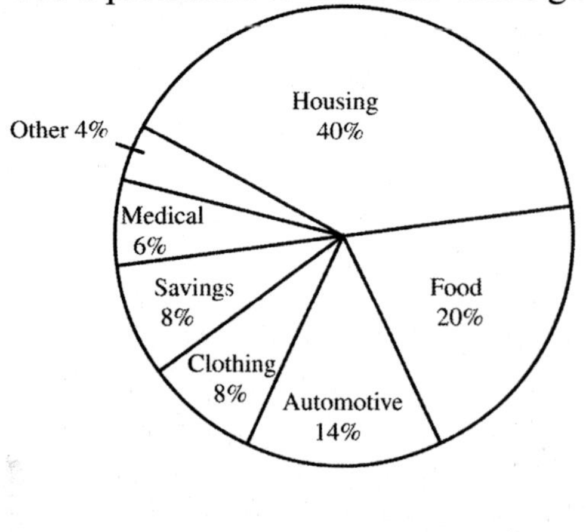

Now Try:

4. A book publisher had 30% of its sales in mysteries, 15% in biographies, 10% in cookbooks, 25% in romance novels, 15% in science, and the rest in business books.

a. Find the number of degrees in a circle graph for each type of book.

mysteries ________

biographies ______

cookbooks _______

romance _________

science __________

business _________

b. Draw a circle graph showing this information.

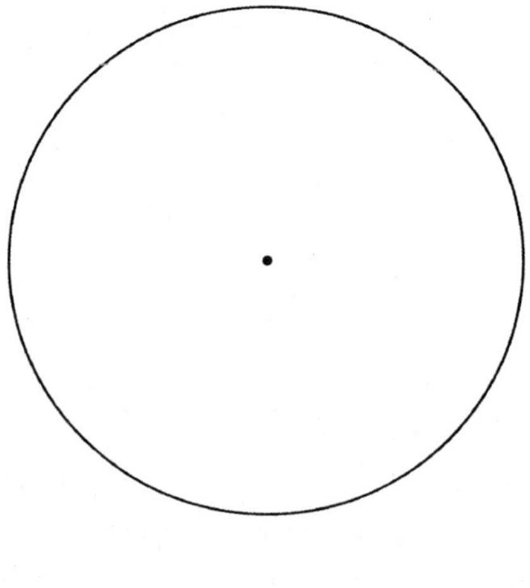

Name: Date:
Instructor: Section:

Objective 1 Read a circle graph.

For extra help, see page 627 of your text and Section Lecture video for Section 9.2.

The circle graph shows the cost of remodeling a kitchen. Use the graph to answer exercises 1–3.

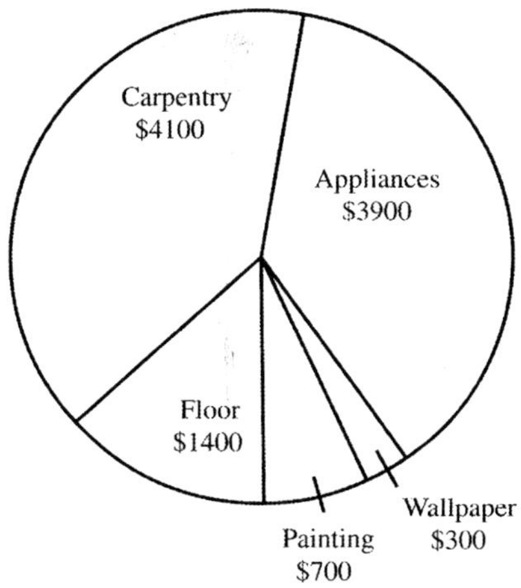

1. Find the total cost of remodeling the kitchen. **1.** ____________

2. What is the largest single expense in remodeling the kitchen? **2.** ____________

3. How much less does the wallpaper cost than painting? **3.** ____________

Objective 2 Use a circle graph.

For extra help, see Examples 1–3 on pages 627–628 of your text and Section Lecture video for Section 9.2 and Exercise Solutions Clip 21, 22, and 23.

The circle graph shows the number of students enrolled in certain majors at a college. Use the graph to answer exercises 4–5. The entire circle represents 11,600 students.

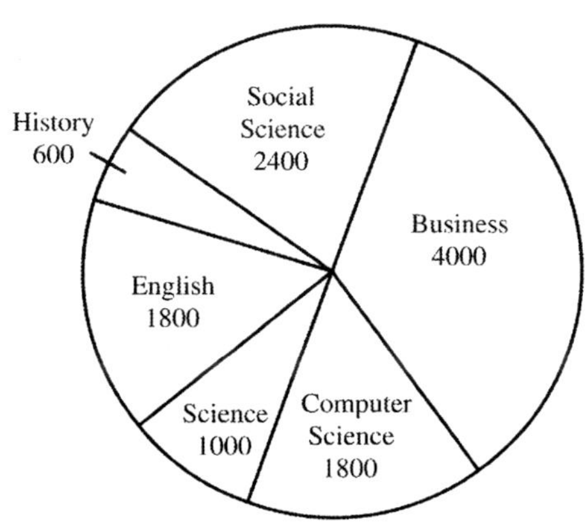

Name: Date:
Instructor: Section:

4. Find the ratio of the number of business majors to the total number of students. **4.** ________________

5. Find the ratio of the number of science majors to the number of English majors. **5.** ________________

The circle graph shows the expenses involved in keeping a sales force on the road. Each expense item is expressed as a percent of the total sales force cost of $950,000. Find the number of dollars of expense for the category.

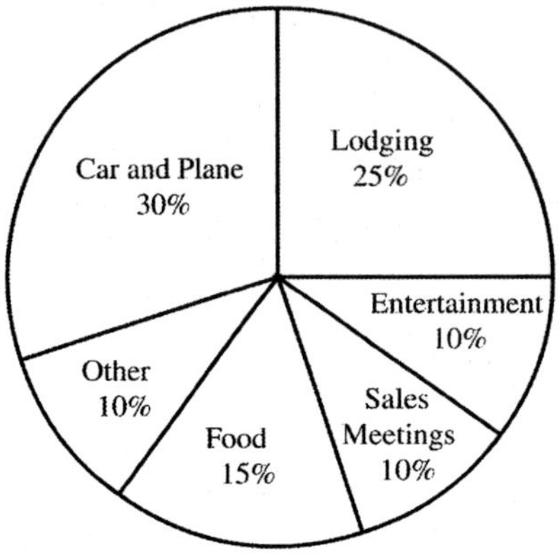

6. Car and plane **6.** ________________

Objective 3 Use a protractor to draw a circle graph.

For extra help, see Example 4 on pages 629–630 of your text and Section Lecture video for Section 9.2 and Exercise Solutions Clip 31.

Use the given information to draw a circle graph.

Jensen Manufacturing Company has its annual sales divided into five categories as follows. The total sales for a year is $400,000.

Item	*Annual Sales*
Parts	$20,000
Hand tools	80,000
Bench tools	100,000
Brass fittings	140,000
Cabinet hardware	60,000

Name: Date:
Instructor: Section:

7. Find the percent of the total sales for each item.

7. parts ____________

hand tools ________

bench tools _______

brass fittings ______

hardware _________

8. Find the number of degrees in a circle graph for each item.

8. parts ____________

hand tools ________

bench tools _______

brass fittings ______

hardware _________

9. Make a circle graph showing this information.

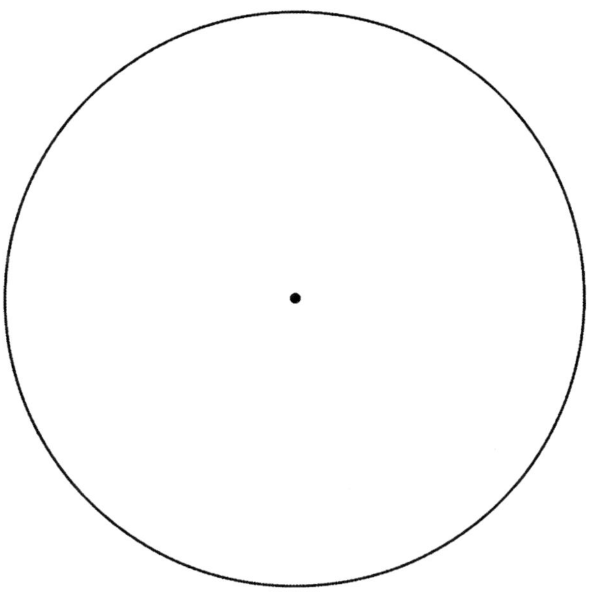

Name: Date:
Instructor: Section:

Chapter 9 GRAPHS AND GRAPHING

9.3 Bar Graphs and Line Graphs

Learning Objectives
1. Read and understand a bar graph.
2. Read and understand a double-bar graph.
3. Read and understand a line graph.
4. Read and understand a comparison line graph.

Key Terms

Use the vocabulary terms listed below to complete each statement in exercises 1−4.

bar graph **double-bar graph** **line graph** **comparison line graph**

1. A ______________________________ uses dots connected by a line to show trends.

2. A ________________________________ compares two sets of data by showing two sets of bars.

3. A ________________________________ uses bars of various heights or lengths to show quantity or frequency.

4. A ________________________________ shows how two sets of data relate to each other by showing a line graph for each item.

Guided Examples

The bar graph shows the enrollment for the fall semester at a small college from 2002 to 2006.

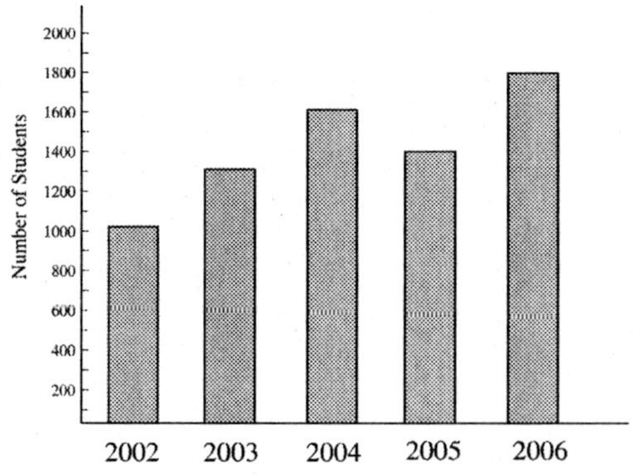

Review this example for Objective 1:

1. What was the fall enrollment for 2003?

The bar for 2003 rises to 1300. So the enrollment in 2003 was 1300 students.

Now Try:

1. What was the fall enrollment for 2006?

____

Name: Date:
Instructor: Section:

The double-bar graph shows the enrollment by gender in each class at a small college.

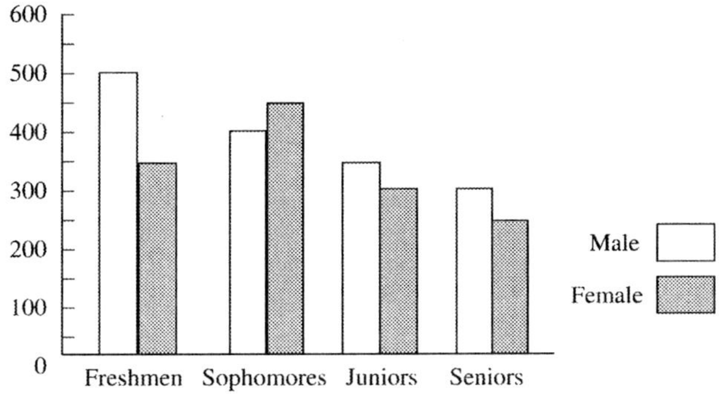

Review these examples for Objective 2:

2. Use the double-bar graph to find the following.

a. The number of male sophomores enrolled

There are two bars for the sophomore class. The color code to the right tells you that white bars represent male enrollments. So the white bar on the left for Sophomores represents the number of male sophomores. It rises to 400.
So the sophomore enrollment of males is 400.

b. The number of female juniors enrolled

The gray column for juniors rises to 300.
So the junior enrollment of females is 300.

Now Try:

2. Use the double-bar above to find the following.

a. The number of female seniors enrolled

____

b. The number of male freshmen enrolled

____

The line graph gives the value of one share of stock of Microchip Computer Corporation on the first trading day of the month for six consecutive months.

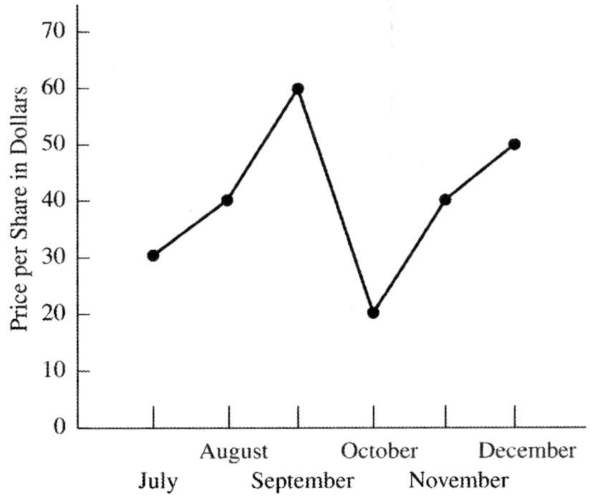

Name: Date:
Instructor: Section:

Review these examples for Objective 3:

3. Use the line graph on the previous page to find the following.

a. In which month was the value of the stock the lowest?

The lowest point on the graph is the dot directly over October, so the lowest value of the stock occurred in October.

b. Find the value of one share of stock on the first trading day in August.

Use a ruler or straightedge to line up the August dot with the numbers along the left edge of the graph. The August dot is directly across from 40. So in August, the price per share was $40.

Now Try:

3. Use the line graph on the previous page to find the following.

a. Find the value of the stock in September.

____

b. In which month was the value of the stock $50?

____

The comparison line graph shows annual sales for two different stores from 2002 to 2006.

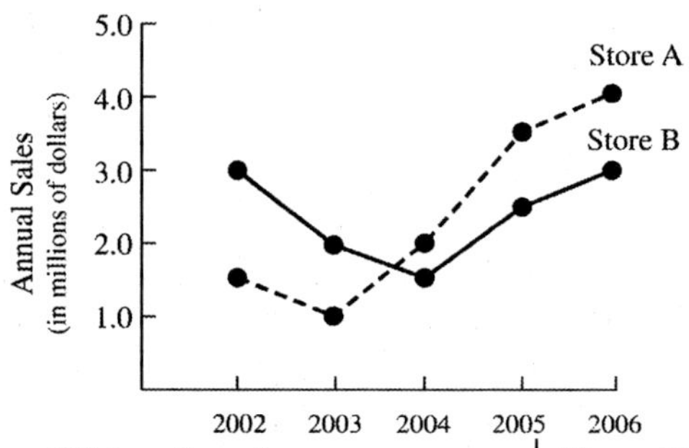

Review these examples for Objective 4:

4. Use the comparison line graph to find the following.

a. The number of annual sales for store A in 2004

Find the dot on the dotted line above 2004. It rises to $2.0. Multiply 2.0 by $1,000,000 because the label on the left side of the graph says in millions of dollars.
So, in 2004 the annual sales for store A was $2,000,000.

b. The number of annual sales for store B in 2006

The solid line on the graph shows 3.0 times $1,000,000 or $3,000,000 was the annual sales for store B in 2006.

Now Try:

4. Use the comparison line graph to find the following.

a. The number of annual sales for store B in 2004

____

b. The number of annual sales for store A in 2006

____

Name: Date:
Instructor: Section:

c. Find the amount of decrease and the percent of decrease in Store B's annual sales from 2003 to 2004. Round to the nearest whole percent, if necessary.

Use subtraction to find the amount of decrease.

$$\begin{array}{ccccc} \text{Store B} & - & \text{Store B} & = & \text{amount} \\ \text{sales in 2003} & & \text{sales in 2004} & & \text{of decrease} \\ 2{,}000{,}000 & - & 1{,}500{,}000 & = & 500{,}000 \end{array}$$

Now use the percent equation to find the percent of decrease.

percent of original sales = amount of decrease

$$p \cdot 2{,}000{,}000 = 500{,}000$$

$$\frac{p \cdot 2{,}000{,}000}{2{,}000{,}000} = \frac{500{,}000}{2{,}000{,}000}$$

$$p = 0.25$$

$$p = 0.25 = 25\%$$

The amount of decrease is $500,000.
The percent of decrease is 25%.

c. The amount of decrease and the percent of decrease in Store B's annual sales from 2002 to 2003. Round to the nearest whole percent, if necessary.

____

____

Objective 1 Read and understand a bar graph.

For extra help, see Example 1 on page 636 of your text and Section Lecture video for Section 9.3.

The bar graph shows the enrollment for the fall semester at a small college from 2002 to 2006. Use this graph for problems 1–3.

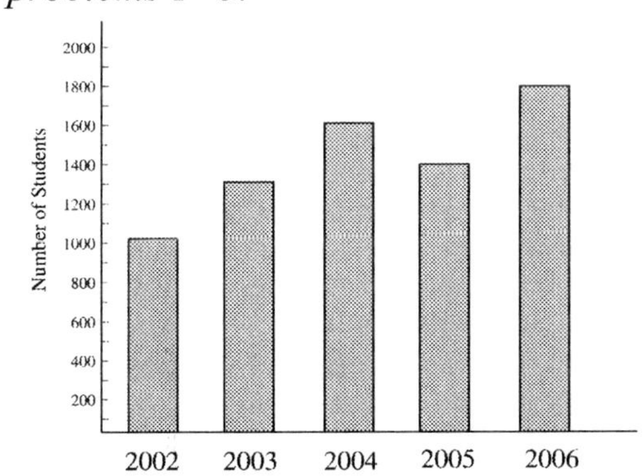

1. What was the fall enrollment for 2004? **1.** ______________

2. What year had the greatest enrollment? **2.** ______________

3. By how many students did the enrollment increase from 2005 to 2006? **3.** ______________

Name: Date:
Instructor: Section:

Objective 2 Read and understand a double-bar graph.

For extra help, see Example 2 on page 637 of your text and Section Lecture video for Section 9.3.

The double-bar graph shows the enrollment by gender in each class at a small college. Use the double-bar graph for Problems 4–6.

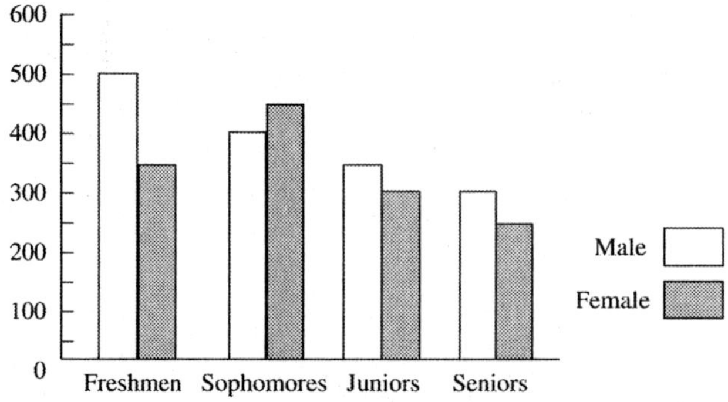

4. How many female freshmen are enrolled? **4.** ________________

5. Which class has a greater female enrollment than male enrollment? **5.** ________________

6. Find the total number of juniors enrolled. **6.** ________________

Objective 3 Read and understand a line graph.

For extra help, see Example 3 on page 637 of your text and Section Lecture video for Section 9.3 and Exercise Solutions Clip 25 and 27.

The line graph gives the value of one share of stock of Microchip Computer Corporation on the first trading day of the month for six consecutive months. Use the line graph for Problems 7–9.

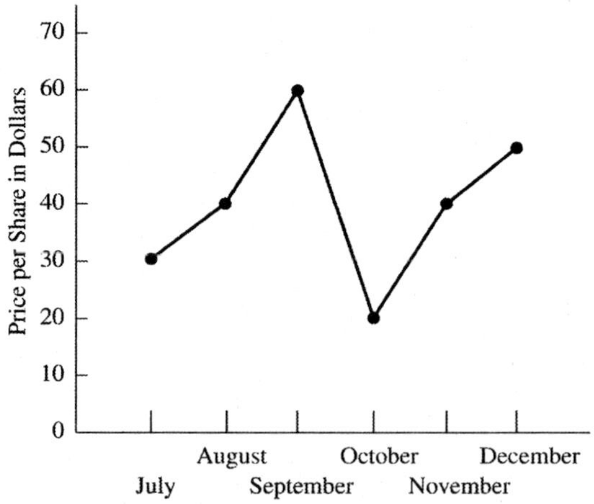

7. In which month was the value of the stock highest? **7.** ________________

Name: Date:
Instructor: Section:

8. Find the value of one share on the first trading day October. **8.** ______________

9. By how much did the value of one share increase from July to September? **9.** ______________

Objective 4 Read and understand a comparison line graph.

For extra help, see Example 4 on page 638 of your text and Section Lecture video for Section 9.3 and Exercise Solutions Clip 33.

The comparison line graph shows annual sales for two different stores from 2002 to 2006. Use the graph to solve Problems 10–12.

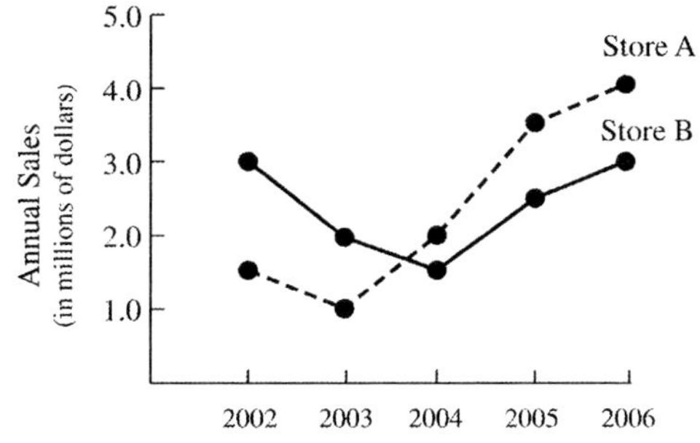

10. Find the annual sales for store A in 2005. **10.** ______________

11. Find the annual sales for store B in 2003. **11.** ______________

12. Find the amount of decrease and the percent of decrease in Store A's annual sales from 2002 to 2003. Round to the nearest whole percent, if necessary. **12.** ______________

Name: Date:
Instructor: Section:

Chapter 9 GRAPHS AND GRAPHING

9.4 The Rectangular Coordinate System

Learning Objectives

1. Plot a point, given the coordinates, and find the coordinates, given a point.
2. Identify the four quadrants and determine which points lie within each one.

Key Terms

Use the vocabulary terms listed below to complete each statement in exercises 1–8.

paired data	**horizontal axis**	**vertical axis**	
ordered pair	***x*-axis**	***y*-axis**	
coordinate system	**origin**	**coordinates**	**quadrants**

1. An ______________________ is the "address" of a point in a coordinate system.

2. In a rectangular coordinate system, the axis that goes "left and right" is called the ______________________ or the ___________________________.

3. In a rectangular coordinate system, the axis that goes "up and down" is called the ______________________ or the ___________________________.

4. When each number in a set of data is matched with another number by some rule of association, we call it ___________________________.

5. Together, the x-axis and the y-axis form a rectangular ______________________.

6. The x-axis and the y-axis divide the coordinate system into four regions called _______________.

7. The axis lines in a coordinate system intersect at the ____________________.

8. ___________________________ are the numbers in the ordered pair that specify the location of a point on a rectangular coordinate system.

Name: Date:
Instructor: Section:

Guided Examples

Review these examples for Objective 1:

1. Use the grid below to plot each point.

   a. (2, 5)

   Start at (0, 0). Move to the right along the horizontal axis until you reach 2. Then move up 5 units so that you are aligned with 5 on the vertical axis. Make a dot and label it (2, 5).

   b. (–5, 4)

   Start at (0, 0). Move to the left along the horizontal axis until you reach –5. Then move up 4 units so that you are aligned with 4 on the vertical axis. Make a dot and label it (–5, 4).

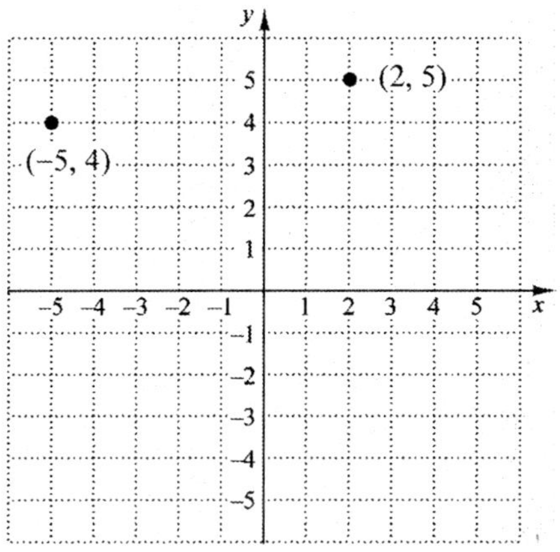

2. Use the grid below to plot each point.

   a. (1,–2)

   Move right 1 unit. Move down 2 units. Make a dot and label it (1,–2).

   b. (–4, 0)

   Move left 4 units. Make a dot and label it (–4, 0).

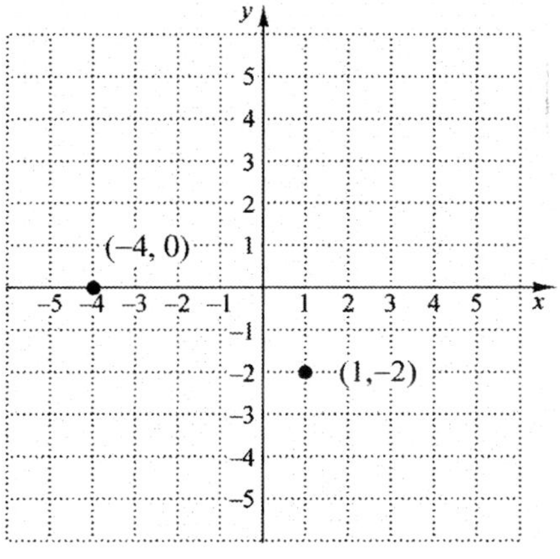

Now Try:

1. Use the grid to plot each point.

   a. (1, 3)

   b. (4,–1)

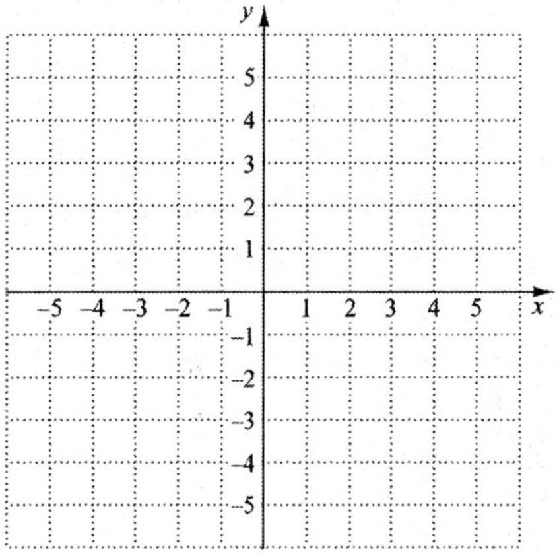

2. Use the grid to plot each point.

   a. (–2, 0)

   b. (0, 5)

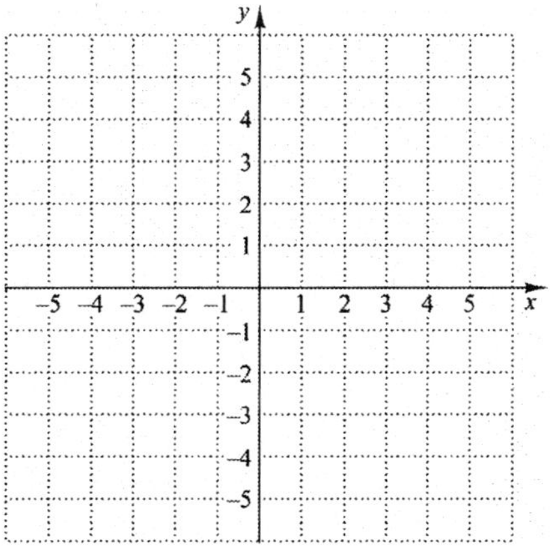

Name: Date:
Instructor: Section:

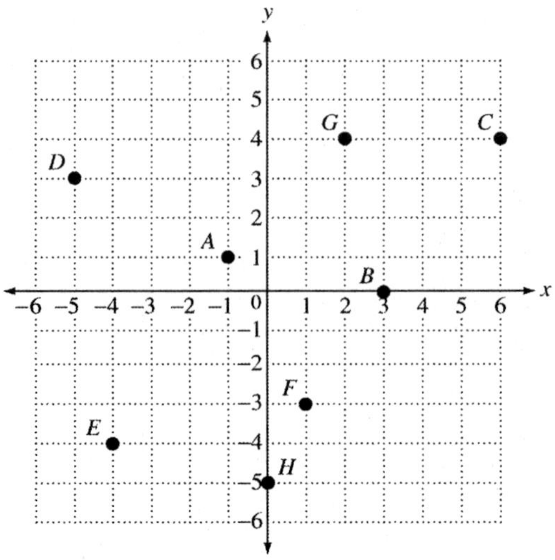

3. Give the coordinates of each point.	**3.** Give the coordinates of each point.
a. *A* To reach point *A* from the origin, move 1 unit to the left; then move up 1 unit. The coordinates are (–1, 1).	**a.** *E* ____________
b. *B* To reach point *B* from the origin, move 3 units to the left; do not move up or down. The coordinates are (3, 0).	**b.** *F* ____________
c. *C* To reach point *C* from the origin, move 6 units to the right; then move up 4 units. The coordinates are (6, 4).	**c.** *G* ____________
d. *D* To reach point *D* from the origin, move 5 units to the left; then move up 3 units. The coordinates are (–5, 3).	**d.** *H* ____________
Review these examples for Objective 2:	**Now Try:**
4. Identify the quadrant in which each point is located.	**4.** Identify the quadrant in which each point is located.
a. (5, 11) For (5, 11), the pattern is (+, +), so the point is in Quadrant I.	**a.** (4, 9) ____________
b. (3,–2) For (3, –2), the pattern is (+, –), so the point is in Quadrant IV.	**b.** (–6, 3) ____________

Name: Date:
Instructor: Section:

c. (–8, 0)

The point corresponding to (–8, 0) is on the x-axis, so it isn't in any quadrant.

c. (0, 9) __________

Objective 1 Plot a point, given the coordinates, and find the coordinates, given a point.

For extra help, see Examples 1–3 on pages 644–646 of your text and Section Lecture video for Section 9.4 and Exercise Solutions Clip 5 and 7.

1. *Plot each point on the rectangular coordinate system below. Label each point with its coordinates.*

a. (–3, 4)

b. (2, –1)

c. (0,–5)

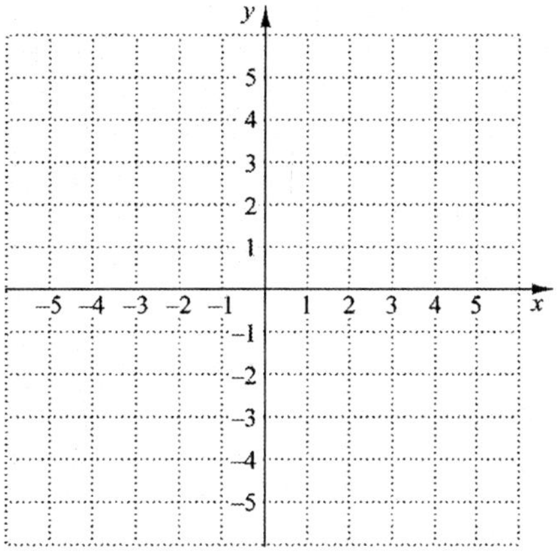

2. *Give the coordinates of each point.*

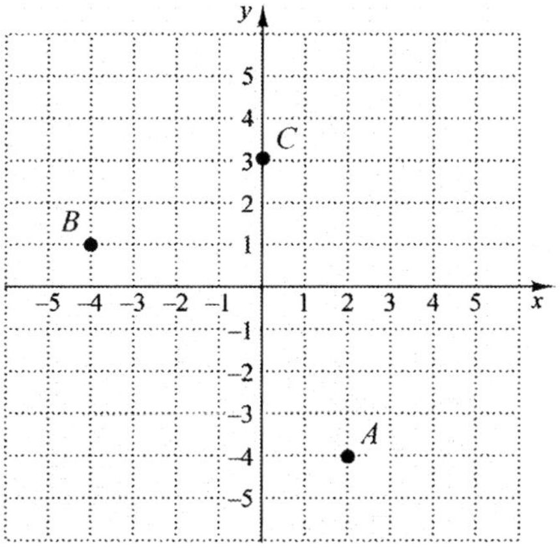

A __________

B __________

C __________

Objective 2 Identify the four quadrants and determine which points lie within each one.

For extra help, see Example 4 on page 646 of your text and Section Lecture video for Section 9.4 and Exercise Solutions Clip 9 and 11.

Identify the quadrant in which each point is located.

3. (3, –9) **3.** __________

4. (–2, –2) **4.** __________

5. (–3, 0) **5.** __________

Name: Date:
Instructor: Section:

Chapter 9 GRAPHS AND GRAPHING

9.5 Introduction to Graphing Linear Equations

Learning Objectives
1. Graph linear equations in two variables.
2. Identify the slope of a line as positive or negative.

Key Terms

Use the vocabulary terms listed below to complete each statement in exercises 1–2.

graph a linear equation **slope**

1. To __________________________, find at least three ordered pairs that satisfy the equation, and then, plot the points and connect them with a straight line.

2. As you move from left to right, if a line tilts downward, it has a negative __________________.

Guided Examples

Review these examples for Objective 1:

1. Graph $x+y=7$ by finding three solutions and plotting the ordered pairs. Then use the graph to find a fourth solution of the equation.

Set up a table to organize the information.

x	y	$x+y=7$	(x, y)
2	5	$2+5=7$	(2, 5)
3	4	$3+4=7$	(3, 4)
4	3	$4+3=7$	(4, 3)

Plot the ordered pairs and draw a line through the points, extending it in both directions.

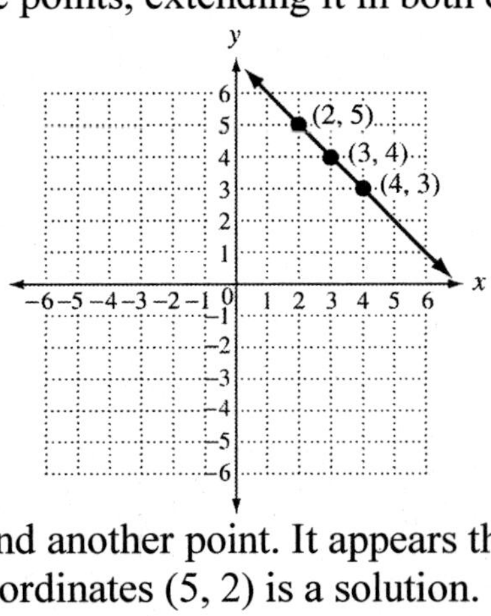

Find another point. It appears that a point with coordinates (5, 2) is a solution.

Now Try:

1. Graph $x-y=0$ by finding three solutions and plotting the ordered pairs. Then use the graph to find a fourth solution of the equation.

____

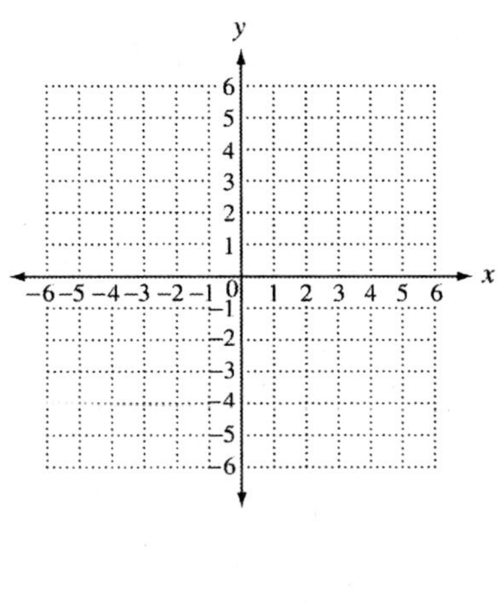

Name: Date:
Instructor: Section:

To check that (5, 2) is a solution, substitute 5 for x and 2 for y in the original equation.

$$x+y=7$$
$$5+2=7$$
$$7=7$$

The equation balances, so (5, 2) is another solution of $x+y=7$.

2. Graph $y=3x$ by finding three solutions and plotting the ordered pairs. Then use the graph to find a fourth solution of the equation.

Set up a table to organize the information.

x	y	$y=3x$	(x, y)
−1	−3	$-3=3(-1)$	$(-1,-3)$
0	0	$0=3(0)$	(0, 0)
1	3	$3=3(1)$	(1, 3)

Plot the ordered pairs and draw a line through the points, extending it in both directions.

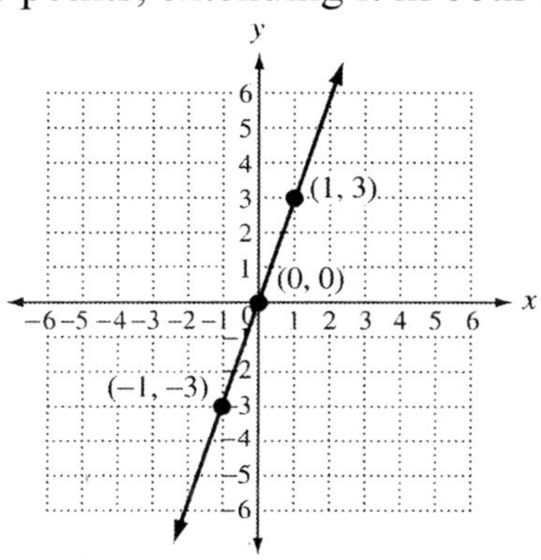

Find another point. It appears that a point with coordinates (2, 6) is a solution.
To check that (2, 6) is a solution, substitute 2 for x and 6 for y in the original equation.

$$y=3x$$
$$6=3(2)$$
$$6=6$$

The equation balances, so (2, 6) is another solution of $y=3x$.

2. Graph $y=4x-1$ by finding three solutions and plotting the ordered pairs. Then use the graph to find a fourth solution of the equation.

____

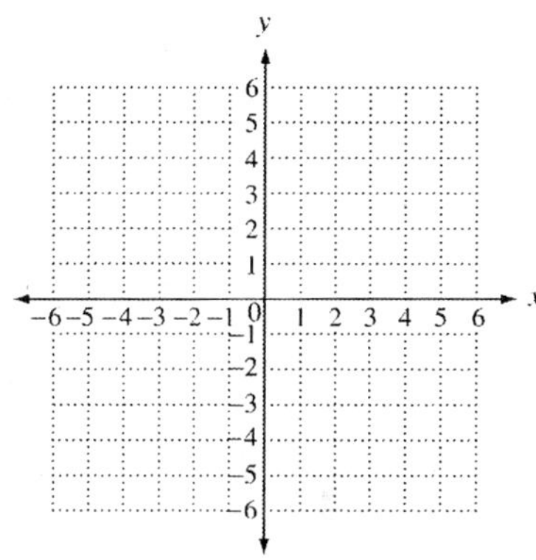

3. Graph $y=-\frac{1}{4}x$ by finding three solutions and plotting the ordered pairs. Then use the graph to find a fourth solution of the equation.

Set up a table to organize the information.

x	y	$y=-\frac{1}{4}x$	(x, y)
-4	1	$1=-\frac{1}{4}(-4)$	$(-4, 1)$
0	0	$0=-\frac{1}{4}(0)$	$(0, 0)$
4	-1	$-1=-\frac{1}{4}(4)$	$(4, -1)$

Plot the ordered pairs and draw a line through the points, extending it in both directions.

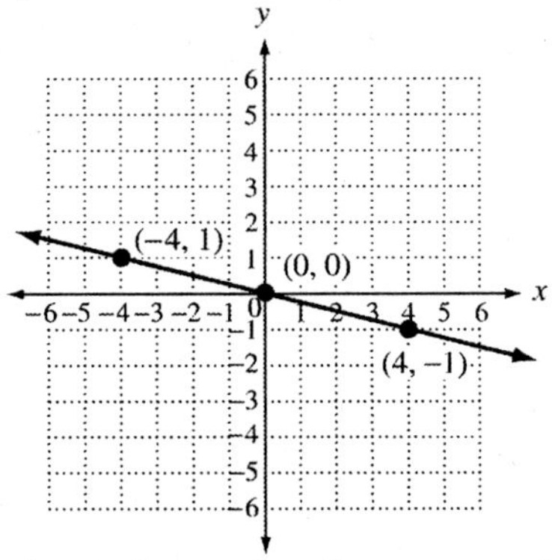

Find another point. It appears that a point with coordinates $\left(-2, \frac{1}{2}\right)$ is a solution.

To check that $\left(-2, \frac{1}{2}\right)$ is a solution, substitute -2 for x and $\frac{1}{2}$ for y in the original equation.

$$y=-\frac{1}{4}x$$

$$\frac{1}{2}=-\frac{1}{4}(-2)$$

$$\frac{1}{2}=\frac{1}{2}$$

The equation balances, so $\left(-2, \frac{1}{2}\right)$ is another solution of $y=-\frac{1}{4}x$.

3. Graph $y=\frac{1}{5}x$ by finding three solutions and plotting the ordered pairs. Then use the graph to find a fourth solution of the equation.

____

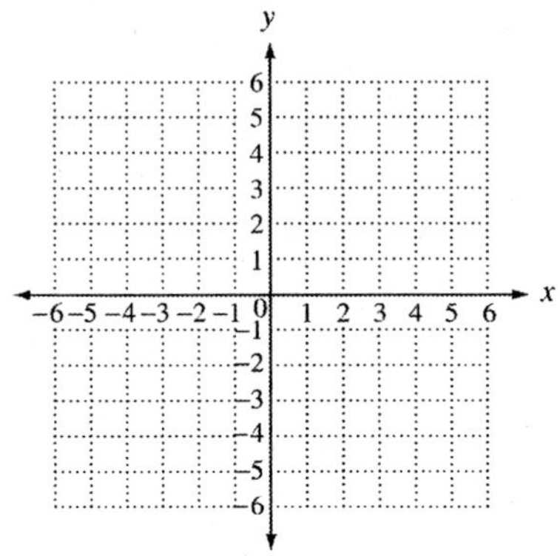

Name: Date:
Instructor: Section:

4. Graph $y=-3x+2$ by finding three solutions and plotting the ordered pairs. Then use the graph to find a fourth solution of the equation.

Set up a table to organize the information.

x	y	$y=-3x+2$	(x, y)
0	2	$2=-3(0)+2$	(0, 2)
1	−1	$-1=-3(1)+2$	(1, −1)
2	−4	$-4=-3(2)+2$	(2, −4)

Plot the ordered pairs and draw a line through the points, extending it in both directions.

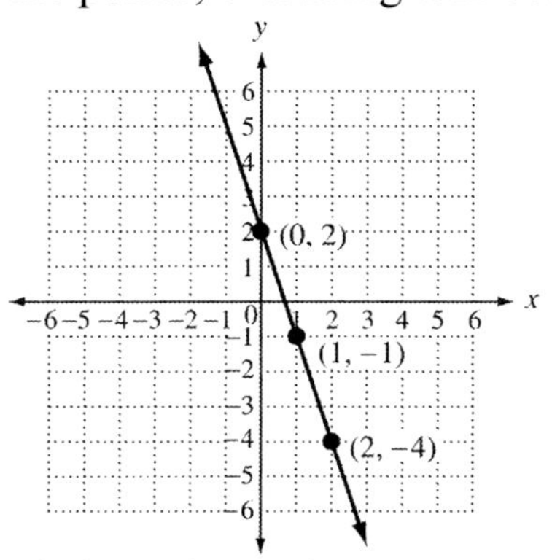

Find another point. It appears that a point with coordinates (−1, 5) is a solution.
To check that (−1, 5) is a solution, substitute −1 for x and 5 for y in the original equation.

$y=-3x+2$

$5=-3(-1)+2$

$5=5$

The equation balances, so (−1, 5) is another solution of $y=-3x+2$.

4. Graph $y=x+2$ by finding three solutions and plotting the ordered pairs. Then use the graph to find a fourth solution of the equation.

____

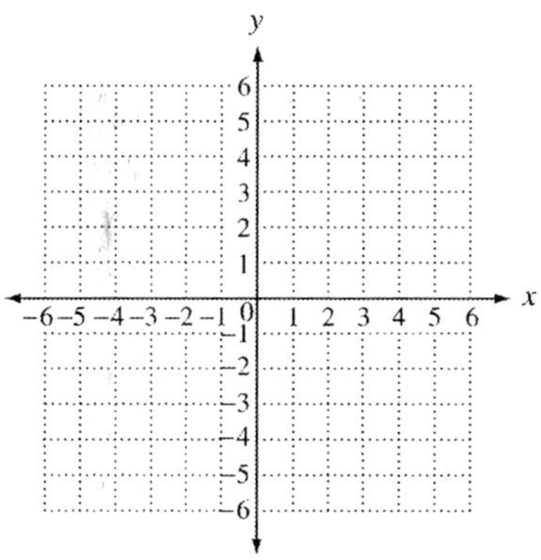

Review this example for Objective 2:

5. Look back at the graph of $y=-3x+2$ from Example 4. Then complete these sentences.
The graph of $y=-3x+2$ has a _____ slope.
As the value of x increases, the value of y ____.

The graph of $y=-3x+2$ has a negative slope (because it tilts downward).
As the value of x increases, the value of y decreases (does the opposite).

Now Try:

5. Look back at the graph of $y=x+2$ from Example 4. Then complete these sentences.
The graph of $y=x+2$ has a _____ slope. As the value of x increases, the value of y ____.

____

____

Name: Date:
Instructor: Section:

Objective 1 Graph linear equations in two variables.

For extra help, see Examples 1–4 on pages 650–654 of your text and Section Lecture video for Section 9.5 and Exercise Solutions Clip 5, 9, 13, and 17.

Graph each equation. Make your own table using the listed values of x.

1. $y = x - 4$
Use 0, 2, and 4 as the values of x.

1.

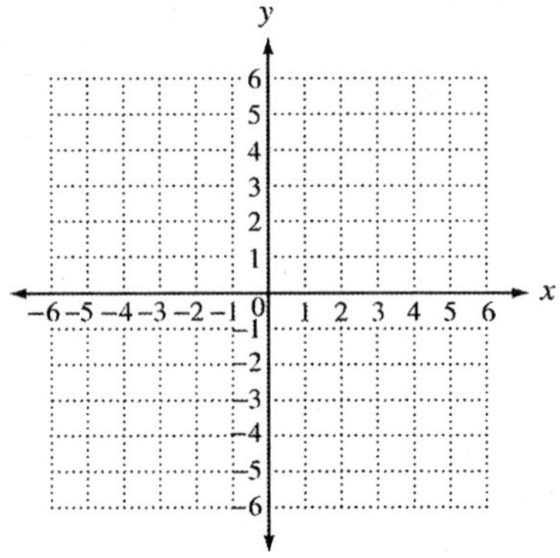

2. $x + y = -3$
Use –4, –3, and –2 as the values of x.

2.

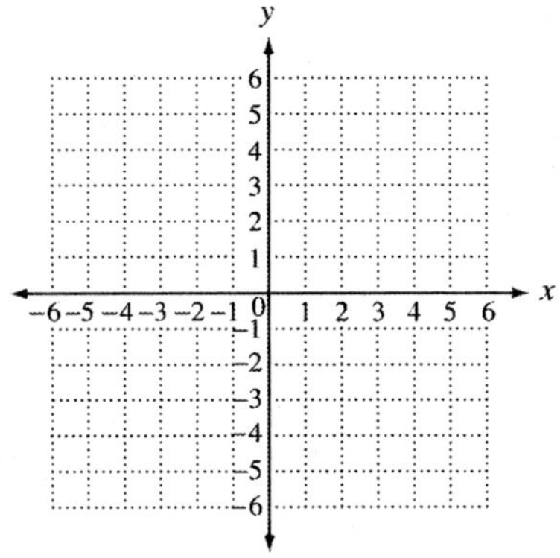

Objective 2 Identify the slope of a line as positive or negative.

For extra help, see Example 5 on page 656 of your text and Section Lecture video for Section 9.5.

Look back at the graphs in exercises 1–2. Then complete exercises 3–4.

3. The graph of $y = x - 4$ has a ____ slope. **3.** ________________

4. The graph of $x + y = -3$ has a ____ slope. **4.** ________________

Name: Date:
Instructor: Section:

Chapter 10 REAL NUMBERS, EQUATIONS, AND INEQUALITIES

10.1 Real Numbers and Expressions

Learning Objectives

1. Identify rational numbers, irrational numbers, and real numbers.
2. Use the symbols $\neq$, $<$, $\leq$, $>$, and $\geq$ to compare real numbers.
3. Reverse the direction of inequality statements.
4. Use the order of operations to simplify expressions with brackets.
5. Remove parentheses and simplify expressions using the distributive property.

Key Terms

Use the vocabulary terms listed below to complete each statement in exercises 1−6.

natural numbers	**whole numbers**	**integers**
rational number	**irrational number**	**real numbers**

1. The set $\{0, 1, 2, 3, \ldots\}$ is called the set of ______________________.
2. The set of ______________________ is $\{\ldots -3, -2, -1, 0, 1, 2, 3, \ldots\}$.
3. The set $\{1, 2, 3, \ldots\}$ is called the set of ______________________.
4. All numbers that can be represented by points on the number line make up the set of ______________________.
5. The decimal form of a(n) ______________________ does not terminate and does not repeat in a fixed block of digits.
6. A number that can be written as the quotient of two integers, with a denominator not 0, is a ______________________.

Guided Examples

Review these examples for Objective 1:

1. Graph these rational numbers on the number line, write them in the form of a mixed numbers or decimals. $-\frac{5}{4}, -\frac{3}{5}, \frac{1}{4}, \frac{6}{5}$

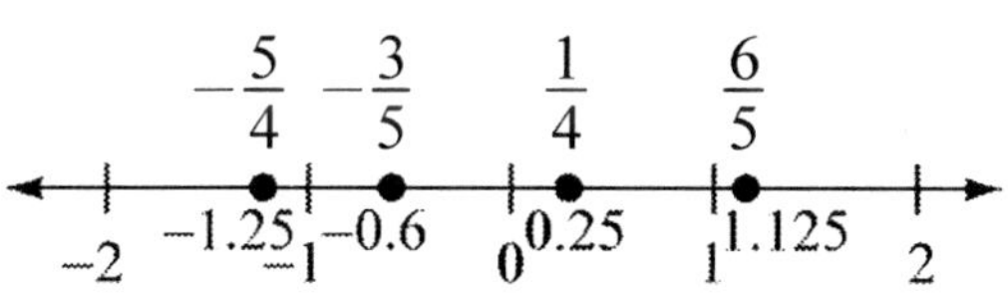

Now Try:

1. Graph these rational numbers on the number line, write them in the form of a mixed numbers or decimals. $-2\frac{1}{3}, -1.5, -\frac{2}{5}, \frac{5}{4}$

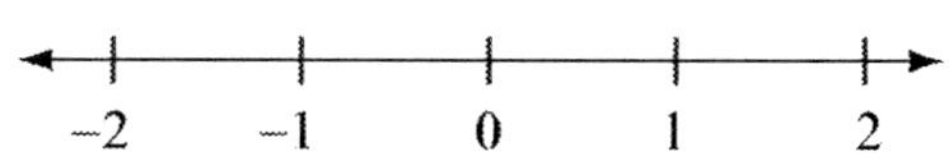

Name: Date:
Instructor: Section:

2. Identify each number as rational or irrational, and explain why. Use your calculator to find square roots.

a. 0.161616...

Rational, because the digits repeat in a fixed block.

b. 6.735

Rational, because the decimal terminates.

c. 0.10110111011110...

Irrational, because the digits do not repeat in a fixed block.

d. $\sqrt{13}$

Irrational, because the decimal value of $\sqrt{13}$ never terminates or repeats.

e. $\sqrt{25}$

Rational, because $\sqrt{25} = 5$.

f. $0.\overline{45}$

Rational, because the digits repeat in a fixed block.

2. Identify each number as rational or irrational. Use your calculator to find square roots.

a. 0.125125...

____

b. 3.78

____

c. 0.290299029990...

____

d. $\sqrt{17}$

____

e. $\sqrt{64}$

____

f. $0.\overline{82}$

____

Review these examples for Objective 2:

3. Label each statement as true or false and explain why.

a. $-5 \le -2$

True, because $-5 < -2$.

b. $\frac{1}{100} \ge \frac{1}{3}$

False, because $\frac{1}{100} < \frac{1}{3}$ and $\frac{1}{100} \ne \frac{1}{3}$.

c. $-9.7 \ge -9.7$

True, because $-9.7 = -9.7$.

d. $8 \div 2 \ne 2 \div 8$

True, because they are not equal. $8 \div 2$ is 4 and $2 \div 8$ is $\frac{1}{4}$ or 0.25.

Now Try:

3. Label each statement as true or false.

a. $-7 \le -6$

____

b. $\frac{1}{25} \ge \frac{1}{5}$

____

c. $-8.32 \ge -8.32$

____

d. $10 \div 5 \ne 5 \div 10$

____

Name: Date:
Instructor: Section:

Review these examples for Objective 3:

4. Write each statement with the inequality symbol reversed.

a. $5 < 19$

$5 < 19$ can be written as $19 > 5$.

b. $0 > -10$

$0 > -10$ becomes $-10 < 0$.

c. $\frac{1}{5} \le \frac{5}{6}$

$\frac{1}{5} \le \frac{5}{6}$ becomes $\frac{5}{6} \ge \frac{1}{5}$.

d. $-9 \ge -10$

$-9 \ge -10$ becomes $-10 \le -9$.

Now Try:

4. Write each statement with the inequality symbol reversed.

a. $\frac{2}{5} > \frac{2}{7}$ __________

b. $-4 < -3$ __________

c. $0 > -5$ __________

d. $8 < 11$ __________

Review this example for Objective 4:

5. Simplify $3[7+2(4+6)]$.

$3[7+2(4+6)]$
$3[7+2(10)]$
$3[7+20]$
$3[27]$
81

Now Try:

5. Simplify $4[6+5(3+2)]$. __________

Review these examples for Objective 5:

6. Write without parentheses.

a. $-4(5y+6)$

$-4(5y+6)$
$(-4\cdot 5y)+(-4\cdot 6)$
$-20y \; + \; -24$
$-20y \; - \; 24$

b. $-7(2a-8)$

$-7(2a-8)$
$(-7\cdot 2a)-(-7\cdot 8)$
$-14a \; - \; (-56)$
$-14a \; + \; 56$

Now Try:

6. Write without parentheses.

a. $-8(7y+2)$ __________

b. $-9(3a-4)$ __________

Name: Date:
Instructor: Section:

7. Write without parentheses.

a. $-(5x+9)$

$-1(5x+9)$
$(-1\cdot 5x)+(-1\cdot 9)$
$-5x \quad + (-9)$
$-5x \quad - \quad 9$

b. $-(-6r-4)$

$-1(-6r-4)$
$(-1\cdot -6r)-(-1\cdot 4)$
$6r \quad - (-4)$
$6r \quad + \quad 4$

8. Simplify each expression.

a. $15y+3(5+7y)$

$15y+3(5+7y)$
$15y+15+21y$
$36y+15$

b. $-(5-r)+6r$

$-1(5-r)+6r$
$-5+r+6r$
$-5+7r$

c. $7(3a^2-2a)-5(7a^2-8)$

$7(3a^2-2a)-5(7a^2-8)$
$21a^2-14a-35a^2+40$
$-14a^2-14a+40$

7. Write without parentheses.

a. $-(11x+3)$

____

b. $-(-15r-2)$

____

8. Simplify each expression.

a. $15y+3(8+4y)$

____

b. $-(9-r)+17r$

____

c. $4(9a^2-5a)-3(6a^2-2)$

____

Objective 1 Identify rational numbers, irrational numbers, and real numbers.

For extra help, see Examples 1–2 on pages 681–682 of your text and Section Lecture video for Section 10.1 and Exercise Solutions Clip 9.

Identify each number as **rational** *or* **irrational** and explain why.

1. $-0.37373737...$ **1.** ____________

2. $\sqrt{100}$ **2.** ____________

3. $\sqrt{37}$ **3.** ______________

Objective 2 Use the symbols ≠, <, ≤, >, and ≥ to compare real numbers.

For extra help, see Example 3 on page 683 of your text and Section Lecture video for Section 10.1 and Exercise Solutions Clip 15 and 27.

Label each statement as **true** *or* **false**.

4. $-76 < 45$ **4.** ______________

5. $\frac{2}{3} \neq \frac{12}{18}$ **5.** ______________

6. $0.75 > \frac{5}{8}$ **6.** ______________

Objective 3 Reverse the direction of inequality statements.

For extra help, see Example 4 on page 684 of your text and Section Lecture video for Section 10.1 and Exercise Solutions Clip 21.

Write each statement with the inequality symbol reversed.

7. $\frac{3}{4} > \frac{2}{3}$ **7.** ______________

8. $12 \geq 8$ **8.** ______________

9. $\frac{3}{8} \leq \frac{3}{7}$ **9.** ______________

Objective 4 Use the order of operations to simplify expression with brackets.

For extra help, see Example 5 on page 684 of your text and Section Lecture video for Section 10.1 and Exercise Solutions Clip 27.

Simplify each expression.

10. $19-3[8(5-2)+6]$ **10.** ______________

Name: Date:
Instructor: Section:

11. $4[5+2(8-6)]+12$ **11.** ________________

12. $2[4+2(5^2-3)]$ **12.** ________________

Objective 5 Remove parentheses and simplify expressions using the distributive property.

For extra help, see Examples 6–8 on pages 685–686 of your text and Section Lecture video for Section 10.1 and Exercise Solutions Clip 43 and 49.

Simplify each expression.

13. $11-(d-2)+(-6)$ **13.** ________________

14. $-4+s-s(12-21)$ **14.** ________________

15. $7(5n-2)-2(-6n-11)$ **15.** ________________

Name: Date:
Instructor: Section:

Chapter 10 REAL NUMBERS, EQUATIONS, AND INEQUALITIES

10.2 More on Solving Linear Equations

Learning Objectives	
1	Solve more difficult linear equations.
2	Solve equations that have no solution or infinitely many solutions.
3	Solve equations by first clearing fractions and decimals.

Key Terms

Use the vocabulary terms listed below to complete each statement in exercises 1–2.

distributive property **identity**

1. An equation that is true for all values of the variable is called a(n) ________________________.

2. We use the ______________________________ to remove the parentheses in the expression $8(y-5)$.

Guided Examples

Review these examples for Objective 1:

1. Solve this equation and check the solution:
$3x+4(x+3)=-20+3x$

Step 1 Use the distributive property.
$$3x+4(x+3)=-20+3x$$
$$3x+4x+12=-20+3x$$

Step 2 Combine like terms.
$$7x+12=-20+3x$$

Step 3 Subtract 12 from both sides.
$$\begin{aligned} 7x+12&=-20+3x \\ \underline{\quad -12} &\quad \underline{-12\quad} \\ 7x+0&=-32+3x \end{aligned}$$

Subtract $3x$ from both sides.
$$\begin{aligned} 7x\quad &=-32+3x \\ \underline{-3x\quad} &\quad \underline{\quad -3x} \\ 4x&=-32+0 \end{aligned}$$

Step 4 Divide both sides by 4.
$$\frac{4x}{4}=\frac{-32}{4}$$
$x=-8$ The solution is –8.

Now Try:

1. Solve this equation and check the solution:
$-1+4(3x+2)=-33+7x$

____

Name: Date:
Instructor: Section:

Step 5 Check.

$$3x+4(x+3)=-20+3x$$
$$3(-8)+4(-8+3)=-20+3(-8)$$
$$-24+4(-5)=-20+-24$$
$$-24+-20=-48$$
$$-48=-48$$

The correct solution is –8.

2. Solve this equation: $12a-5(a-2)=3a+5$

Step 1 Use the distributive property.

$$12a-5(a-2)=3a+5$$
$$12a-5a+10=3a+5$$

Step 2 Combine like terms.

$$7a+10=3a+5$$

Step 3 Subtract 10 from both sides.

$$\begin{array}{rcl} 7a+10 &=& 3a+5 \\ \underline{\quad -10} && \underline{\quad -10} \\ 7a+0 &=& 3a-5 \end{array}$$

Subtract $3a$ from both sides.

$$\begin{array}{rcl} 7a &=& 3a-5 \\ \underline{-3a} && \underline{-3a\quad} \\ 4a &=& 0-5 \end{array}$$

Step 4 Divide both sides by 4.

$$\frac{4a}{4}=\frac{-5}{4}$$
$$a=-\frac{5}{4}$$

The solution is $-\frac{5}{4}$. You can check the solution by going back to the original equation and replace each a with $-\frac{5}{4}$.

3. Solve: $-7(t-3)=3-4(t+3)$

$$-7(t-3)=3-4(t+3)$$
$$-7+21=3-4t-12$$
$$-7t+21=-9-4t$$

$$\begin{array}{rcl} -7t+21 &=& -9-4t \\ \underline{\quad -21} && \underline{-21\quad} \\ -7t+0 &=& -30-4t \end{array}$$

2. Solve this equation:
$8a-(3a-4)=11+2a$

____

3. Solve:
$23+2(t+2)=3-9(2-t)$

____

Name: Date:
Instructor: Section:

$$\begin{aligned} -7t &= -30 - 4t \\ +4t & \quad\quad +4t \\ \frac{-3t}{-3} &= \frac{-30}{-3} \\ t &= 10 \end{aligned}$$

The solution is 10.

Check

$$\begin{aligned} -7(t-3) &= 3-4(t+3) \\ -7(10-3) &= 3-4(10+3) \\ -7(7) &= 3-4(13) \\ -49 &= 3-52 \\ -49 &= -49 \end{aligned}$$

The equation balances, so 10 is the correct solution.

Review these examples for Objective 2:

4. Solve: $4x - 12 = 4(x-3)$

$$\begin{aligned} 4x-12 &= 4(x-3) \\ 4x-12 &= 4x-12 \\ +12 & \quad\quad +12 \\ 4x &= 4x \\ -4x & \quad\quad -4x \\ 0 &= 0 \end{aligned}$$

The variable has "disappeared." It is true that $0 = 0$. This is called an identity.
The solution for $4x - 12 = 4(x-3)$ is all real numbers.

5. Solve: $6w - 8(w+2) = -2w + 14$

$$\begin{aligned} 6w-8(w+2) &= -2w+14 \\ 6w-8w-16 &= -2w+14 \\ -2w-16 &= -2w+14 \\ +2w \quad & \quad +2w \\ -16 &= 14 \quad \text{False statement: no variable.} \end{aligned}$$

The variable has "disappeared." However, we are left with a false statement: $-16 \neq 14$. There is no solution.

Now Try:

4. Solve: $5(x-7) = 10 + 5x - 45$

____

5. Solve: $9w - 4(w+3) = 5w + 11$

____

Name: Date:
Instructor: Section:

Review these examples for Objective 3:

6. Solve: $\frac{3}{5}x-\frac{2}{3}x=\frac{1}{6}x-\frac{7}{15}$

The least common denominator is 30, so multiply both sides by 30.

$$\frac{3}{5}x-\frac{2}{3}x=\frac{1}{6}x-\frac{7}{15}$$

$$30\left(\frac{3}{5}x-\frac{2}{3}x\right)=30\left(\frac{1}{6}x-\frac{7}{15}\right)$$

$$30\left(\frac{3}{5}x\right)-30\left(\frac{2}{3}x\right)=30\left(\frac{1}{6}x\right)-30\left(\frac{7}{15}\right)$$

$$18x \quad - \quad 20x \quad = \quad 5x \quad - \quad 14$$

$$-2x=5x-14$$

$$\underline{-5x \quad -5x}$$

$$-7x=0-14$$

$$\frac{-7x}{-7}=\frac{-14}{-7}$$

$$x=2$$

The solution is 2. Check the solution by going back to the original equation and replacing each x with 2.

7. Solve: $-0.11t+0.15(t+6)=0.02(75)$

Clear decimals by multiplying both sides by 100.

$-0.11t+0.15(t+6)=0.02(75)$

$$-11t+15(t+6)=2(75)$$

$$-11t+15t+90=150$$

$$4t+90=150$$

$$\underline{-90 \quad -90}$$

$$4t=60$$

$$\frac{4t}{4}=\frac{60}{4}$$

$$t=15$$

The solution is 15. Check the solution by going back to the original equation and replacing each t with 15.

Now Try:

6. Solve: $\frac{3}{4}x-\frac{1}{8}x=-\frac{1}{2}x+\frac{9}{4}$

____

7. Solve:
$0.11t-0.06(t-7)=-0.03$

____

Name: Date:
Instructor: Section:

Objective 1 Solve more difficult linear equations.

For extra help, see Examples 1–3 on pages 689–691 of your text and Section Lecture video for Section 10.2 and Exercise Solutions Clip 17.

Solve each equation.

1. $4(z-2)-(3z-1)=2z-6$ **1.** ________________

2. $3(x+4)=6-2(x-8)$ **2.** ________________

3. $-(v+2)=3+v$ **3.** ________________

Objective 2 Solve equations that have no solution or infinitely many solutions.

For extra help, see Examples 4–5 on pages 692–693 of your text and Section Lecture video for Section 10.2 and Exercise Solutions Clip 27 and 29.

Solve each equation and check your solution.

4. $6y-3(y+2)=3(y-2)$ **4.** ________________

5. $1+3(2x+4)=15+6x$ **5.** ________________

6. $12+3(2k-1)=2(3k+1)$ **6.** ________________

Name: Date:
Instructor: Section:

Objective 3 Solve equations by first clearing fractions and decimals.

For extra help, see Examples 6–7 on pages 693–694 of your text and Section Lecture video for Section 10.2 and Exercise Solutions Clip 35 and 47.

Solve each equation and check your solution.

7. $\frac{3}{8}x-\frac{1}{3}x=\frac{1}{12}$ **7.** ______________

8. $\frac{1}{3}(2m-1)-\frac{3}{4}m=\frac{5}{6}$ **8.** ______________

9. $0.24x-0.38(x+2)=-0.34(x+4)$ **9.** ______________

Name: Date:
Instructor: Section:

Chapter 10 REAL NUMBERS, EQUATIONS, AND INEQUALITIES

10.3 Formulas and Solving for a Specified Variable

Learning Objectives
1 Solve a formula for one variable, given the values of the other variables.
2 Solve a formula for a specified variable.

Key Terms

Use the vocabulary terms listed below to complete each statement in exercises 1–3.

formula **area** **perimeter**

1. An equation in which variables are used to describe a relationship is called a(n) ____________________.

2. The distance around a figure is called its ____________________.

3. A measure of the surface covered by a figure is called its ____________________.

Guided Examples

Review these examples for Objective 1:

1. Find the value of the remaining variable in each formula.

a. $I = prt$; $I = 288$, $r = 0.04$, $t = 3$

Substitute the given values into the formula and then solve for p.

$$I = prt$$
$$288 = p(0.04)(3)$$
$$\frac{288}{0.12} = \frac{0.12p}{0.12}$$
$$2400 = p$$

Check

$$I = prt$$
$$288 = 2400(0.04)(3)$$
$$288 = 288$$

The solution for p is 2400.

b. $P = 2L + 2W$; $P = 42$, $W = 6$

$$P = 2L + 2W$$
$$42 = 2L + 2(6)$$

Now Try:

1. Find the value of the remaining variable in each formula.

a. $V = LWH$; $V = 24$, $L = 2$, $H = 3$

____

b. $m = \frac{y_1 - y_2}{x_1 - x_2}$; $m = \frac{1}{6}$, $x_1 = -5$, $x_2 = 7$, $y_2 = -5$

____

$$42 = 2L + 12$$
$$\underline{-12} \quad \underline{\quad -12}$$
$$30 = 2L$$
$$\frac{30}{2} = \frac{2L}{2}$$
$$15 = L$$

Check

$$P = 2L + 2W$$
$$42 = 2(15) + 2(6)$$
$$42 = 30 + 12$$
$$42 = 42$$

The solution for L is 15.

2. Find the height of a triangular banner whose area is 48 square inches and base is 12 inches.

The formula for the area of a triangle is $A = \frac{1}{2}bh.$ Find the height by substituting 48 for A and 12 for b. Then solve for h.

$$A = \frac{1}{2}bh$$
$$48 = \frac{1}{2}(12)h$$
$$48 = 6h$$
$$\frac{48}{6} = \frac{6h}{6}$$
$$8 = h$$

The height is 8 inches.

Check
If the base is 12 inches, and the height is 8 inches, then $A = \frac{1}{2}(12)(8) = 48$ square inches.

This matches the area given in the problem.

2. Linda invests \$5000 at 6% simple interest and earns \$450. How long did Linda invest her money?

____

Review these examples for Objective 2:

3. Solve $V = LWH$ for H.

$$V = LWH$$
$$\frac{V}{LW} = \frac{LWH}{LW}$$
$$\frac{V}{LW} = H$$

Now Try:

3. Solve $A = \frac{1}{2}bh$ for b.

____

Name: Date:
Instructor: Section:

4.

a. Solve $S=\frac{a}{1-r}$ for r.

Start by multiplying both sides by the LCD, $1-r$.

$$S=\frac{a}{1-r}$$

$$(1-r)S=(1-r)\frac{a}{1-r}$$

$$S(1-r)=a$$

$$S-Sr=a$$

$$\underline{-S \qquad -S}$$

$$-Sr=a-S$$

$$\frac{-Sr}{-S}=\frac{a-S}{-S}$$

$$r=\frac{a-S}{-S}, \text{ or } \frac{S-a}{S}$$

b. Solve $d=\frac{a}{2}(2t-1)$ for t.

$$d=\frac{a}{2}(2t-1)$$

$$2d=2\cdot\frac{a}{2}(2t-1)$$

$$2d=a(2t-1)$$

$$2d=2at-a$$

$$\underline{+a \qquad +a}$$

$$2d+a=2at$$

$$\frac{2d+a}{2a}=\frac{2at}{2a}$$

$$\frac{2d+a}{2a}=t, \text{ or } t=\frac{d}{a}+\frac{1}{2}$$

4.

a. Solve $m=\frac{y_1-y_2}{x_1-x_2}$ for y_1.

____

b. Solve $S=(n-2)180$ for n.

____

Objective 1 Solve a formula for one variable, given the values of the other variables.

For extra help, see Examples 1–2 on pages 699–700 of your text and Section Lecture video for Section 10.3 and Exercise Solutions Clip 3 and 15.

In the following exercises, a formula is given, along with the values of all but one of the variables in the formula. Find the value of the variable that is not given.

1. $S=\frac{a}{1-r}; S=60,\ r=0.4$

1. ____________

Name: Date:
Instructor: Section:

2. $C = \frac{5}{9}(F - 32);\ F = 104$ **2.** ________________

Use a formula to write an equation for the problem, then solve it.

3. Ruth has 42 feet of binding for a rectangular rug that she is weaving. If the rug is 9 feet wide, how long can she make the rug if she wishes to use all the binding on the perimeter of the rug? **3.** ________________

Objective 2 Solve a formula for a specified variable.

For extra help, see Examples 3–4 on pages 701–702 of your text and Section Lecture video for Section 10.3 and Exercise Solutions Clip 19 and 29.

Solve each formula for the specified variable.

4. $a_n = a_1 + (n-1)d$ for d **4.** ________________

5. $P = A - Art$ for r **5.** ________________

6. $V = \frac{1}{3}\pi r^2 h$ for h **6.** ________________

Name: Date:
Instructor: Section:

Chapter 10 REAL NUMBERS, EQUATIONS, AND INEQUALITIES

10.4 Solving Linear Inequalities

Learning Objectives

1	Graph the solutions of inequalities on a number line.
2	Use the addition property of inequality.
3	Use the multiplication property of inequality.
4	Solve linear inequalities using both properties of inequality.
5	Use inequalities to solve application problems.

Key Terms

Use the vocabulary terms listed below to complete each statement in exercises 1–2.

inequalities **linear inequality**

1. A(n) ____________________________ can be written in the form $Ax + B > C$, $Ax + B \geq C$, $Ax + B < C$, or $Ax + B \leq C$, where A, B, and C are real numbers with $A \neq 0$.

2. Algebraic expressions related by $>$, $\geq$, $<$, or $\leq$ are called ______________________.

Guided Examples

Review these examples for Objective 1:

1.

a. Graph $3 < x$.

$3 < x$ can be written as $x > 3$.
The statement $x > 3$ says that *a* can take any value greater than 3, but x cannot equal 3 itself. We show this on the graph by placing an open circle at 3, and drawing an arrow to the right.

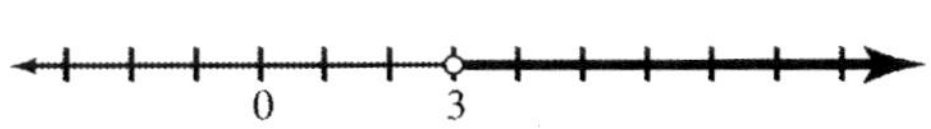

b. Graph $y \geq -2$.

The statement $y \geq -2$ means that y can take any value greater than or equal to –2. We use a closed circle at –2 to indicate that it is part of the graph, and drawing an arrow to the right.

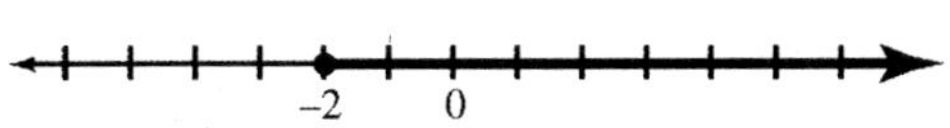

Now Try:

1.

a. Graph $x > 2$.

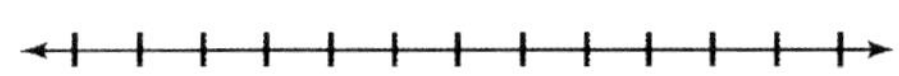

b. Graph $y < -2$.

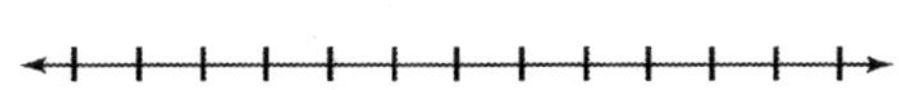

Name: Date:
Instructor: Section:

2. Graph $-3 \le y < 0$.

We graph the solutions of this inequality by placing a closed circle at –3 and an open circle at 0, then drawing a line segment between the two circles.

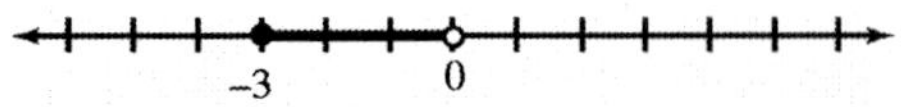

2. Graph $-3 < a \le 2$.

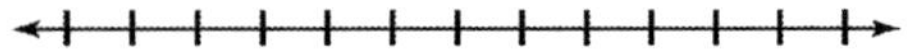

Review this example for Objective 2:

3. Solve $5a + 3 \le 6a$.

$$\begin{array}{rcl} 5a + 3 & \le & 6a \\ \underline{-5a\quad\ } & & \underline{-5a} \\ 3 & \le & a \end{array}$$

The graph of the solutions, $a \ge 3$, is shown below.

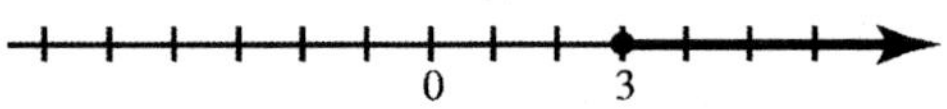

Now Try:

3. Solve $6 + 3x < 4x + 4$.

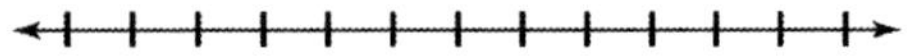

Review these examples for Objective 3:

4. Solve $4k \ge -16$.

Using the multiplication property, we divide both sides by 4. The inequality symbol does not change.

$$4k \ge -16$$

$$\frac{4k}{4} \ge \frac{-16}{4}$$

$$k \ge -4$$

The graph of the solutions is shown below.

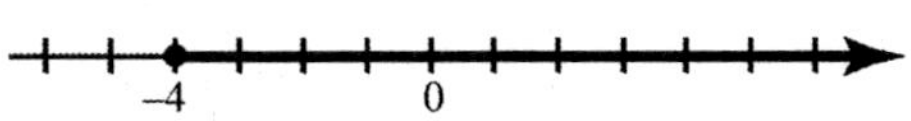

Now Try:

4. Solve $-3 > \frac{3}{4}n$.

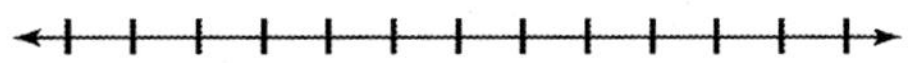

5. Solve $-5t \le -35$.

Here, divide both sides of the inequality by –5. The inequality symbol changes direction because we are dividing by a negative number.

$$-5t \le -35$$

$$\frac{-5t}{-5} \ge \frac{-35}{-5}$$

$$t \ge 7$$

The graph of the solutions is shown below.

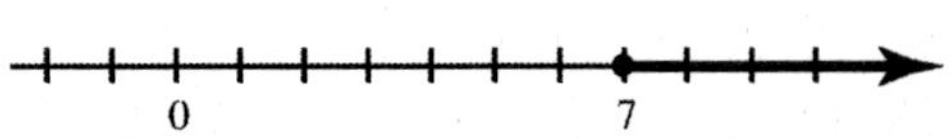

5. Solve $-2s < 4$.

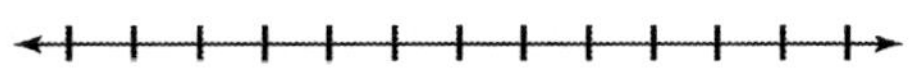

Name: Date:
Instructor: Section:

Review this example for Objective 4:

6. Solve $7(2-x)-3 \le -2(x-4)-x$.

Step 1 Use the distributive property to remove parentheses; then combine like terms.

$$7(2-x)-3 \le -2(x-4)-x$$
$$14-7x-3 \le -2x+8-x$$
$$-7x+11 \le -3x+8$$

Step 2 Use the addition property.

$$\begin{aligned} -7x+11 &\le -3x+8 \\ +3x \quad &\quad\ +3x \\ \hline -4x+11 &\le 8 \\ -11 \quad &\quad -11 \\ \hline -4x \quad &\le -3 \end{aligned}$$

Step 3 Divide both sides by –4, a negative number, and reverse the direction of the inequality symbol.

$$\frac{-4x}{-4} \ge \frac{-3}{-4}$$
$$x \ge \frac{3}{4}$$

A graph of the solutions is shown below.

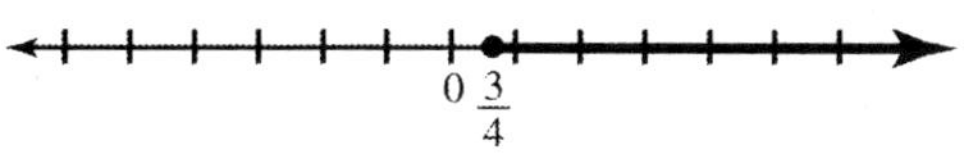

Now Try:

6. Solve
$5(2z+2)-2(z-3) > 3(2z+5)+z$.

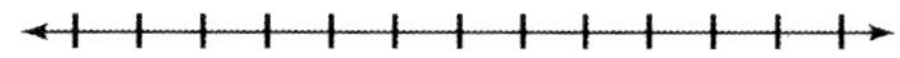

Review this example for Objective 5:

7. Lauren has grades of 98 and 86 on her first two chemistry quizzes. What must she score on her third quiz to have an average of at least 91 on the three quizzes?

Let x represent the score on the third quiz. To find the average of the three scores, add them and divide the sum by 3.

$$\frac{98+86+x}{3} \ge 91$$
$$3\left(\frac{184+x}{3}\right) \ge 3(91)$$
$$184+x \ge 273$$
$$\begin{aligned} -184 \quad &\quad -184 \\ \hline x &\ge 89 \end{aligned}$$

She must score 89 or more on the third quiz to have an average of at least 91.

Now Try:

7. Ruth tutors mathematics in the evenings in an office for which she pays \$600 per month rent. If rent is her only expense and she charges each student \$40 per month, how many students must she teach to make a profit of at least \$1600 per month?

____

Name: Date:
Instructor: Section:

Objective 1 Graph the solutions of inequalities on a number line.

For extra help, see Examples 1–2 on pages 705–706 of your text and Section Lecture video for Section 10.4 and Exercise Solutions Clip 7 and 9.

Graph each inequality on the given number line.

1. $x > 1$

1.

2. $-1 < x < 3$

2.

3. $-4 \geq a > -8$

3.

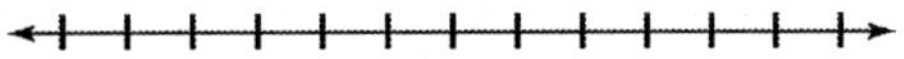

Objective 2 Use the addition property of inequality.

For extra help, see Example 3 on page 707 of your text and Section Lecture video for Section 10.4 and Exercise Solutions Clip 15.

Solve each inequality and then graph the solution.

4. $3 + 5p \leq 4p + 3$

4.

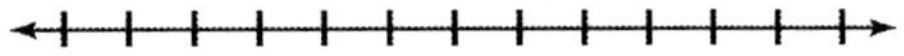

5. $9 + 8b > 9b + 11$

5.

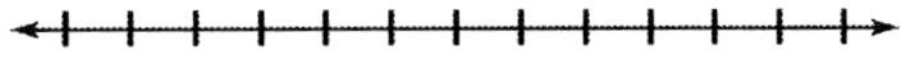

6. $3x + 5 \geq 2x - 5$

6.

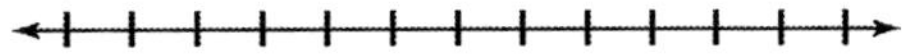

Name: Date:
Instructor: Section:

Objective 3 Use the multiplication property of inequality.

For extra help, see Examples 4–5 on pages 708–709 of your text and Section Lecture video for Section 10.4 and Exercise Solutions Clip 25.

Solve each inequality and then graph the solution.

7. $\frac{3}{5}n \geq 0$ **7.**

8. $-9m > -36$ **8.**

9. $-4x \leq -4$ **9.**

Objective 4 Solve linear inequalities using both properties of inequality.

For extra help, see Example 6 on pages 709–710 of your text and Section Lecture video for Section 10.4 and Exercise Solutions Clip 39.

Solve each inequality and then graph the solution.

10. $4(y-3)+2 > 3(y-2)$ **10.**

11. $-3(m+4)+1 \leq -4(m-2)$ **11.**

12. $3x - \frac{3}{4} \geq 2x + \frac{1}{3}$ **12.**

Name: Date:
Instructor: Section:

Objective 5 Use inequalities to solve application problems.

For extra help, see Example 7 on page 710 of your text and Section Lecture video for Section 10.4 and Exercise Solutions Clip 49.

Solve each problem.

13. Nina has a budget of $230 for gifts for this year. So far she has bought gifts costing $47.52, $38.98, and $26.98. If she has three more gifts to buy, find the average amount she can spend on each gift and still stay within her budget.

13. ________________

14. Two sides of a triangle are equal in length, with the third side 8 feet longer than one of the equal sides. The perimeter of the triangle cannot be more than 38 feet. Find the largest possible value for the length of the equal sides.

14. ________________

15. A roll has 10 fewer calories than twice the number of calories in a slice of bread. Together they contain at least 185 calories. Find the least number of calories in the slice of bread.

15. ________________

Name: Date:
Instructor: Section:

Chapter 11 GRAPHS OF LINEAR EQUATIONS AND INEQUALITIES IN TWO VARIABLES

11.1 Linear Equations in Two Variables; The Rectangular Coordinate System

Learning Objectives

1. Interpret graphs.
2. Write a solution as an ordered pair.
3. Decide whether a given ordered pair is a solution of a given equation.
4. Complete ordered pairs for a given equation.
5. Complete a table of values.
6. Plot ordered pairs.

Key Terms

Use the vocabulary terms listed below to complete each statement in exercises 1−14.

bar graph	**line graph**	**linear equation in two variables**	
ordered pair		**table of values**	***x*-axis**
***y*-axis**	**rectangular (Cartesian) coordinate system**		
origin	**quadrants**	**plane**	**coordinates**
plot	**scatter diagram**		

1. A ______________________________ uses dots connected by lines to show trends.

2. An equation that can be written in the form $Ax + By = C$, where A, B, and C are real numbers and $A, B \neq 0$, is called a ______________________________.

3. A ______________________________ uses bars of various heights or lengths to show quantity or frequency.

4. ______________________________ are the numbers in the ordered pair that specify the location of a point on a rectangular coordinate system.

5. In a coordinate system, the horizontal axis is called the ________________________.

6. In a coordinate system, the vertical axis is called the ________________________.

7. A pair of numbers written between parentheses in which order is important is called a(n) ______________________________.

8. Together, the x-axis and the y-axis form a ______________________.

9. A coordinate system divides the plane into four regions called ________________.

Name: Date:
Instructor: Section:

10. The axis lines in a coordinate system intersect at the ____________________.

11. To ____________________ an ordered pair is to find the corresponding point on a coordinate system.

12. A graph of ordered pairs is called a ______________________________.

13. A table showing selected ordered pairs of numbers that satisfy an equation is called a ______________________.

14. A flat surface determined by two intersecting lines is a __________________.

Guided Examples

The bar graph shows the enrollment for the fall semester at a small college for the years 2002 to 2006.

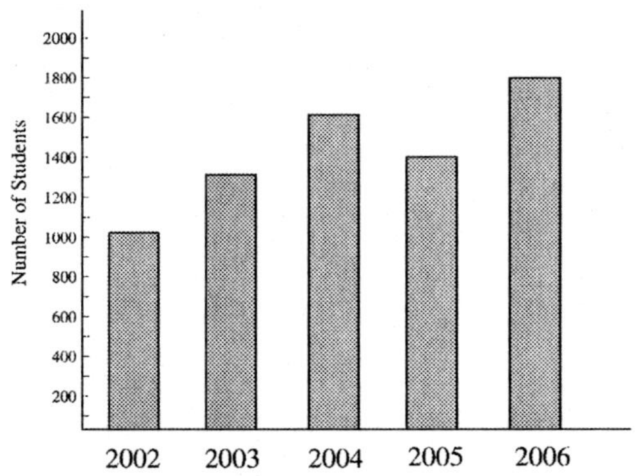

Review these examples for Objective 1:

1.

a. In what years was enrollment greater than 1500?

Locate 1500 on the vertical axis and follow the line across to the right. Two years—2004 and 2006—have bars that extend above the line for 1500, so enrollment was greater than 1500 in those years.

b. Estimate the enrollment in 2003.

Locate the top of the bar for 2003 and move horizontally across to the vertical scale to see that it is about 1300. So enrollment for 2003 was about 1300 students.

c. Describe the change in enrollment from 2003 to 2004.

Enrollment increased by about 300 students from 2003 to 2004.

Now Try:

1.

a. In what years was enrollment less than 1200?

____

b. Estimate the enrollment in 2005.

____

c. Describe the change in enrollment from 2004 to 2005.

____

____

Name: Date:
Instructor: Section:

The line graph shows the number of degrees awarded by a university for the years 2000–2005.

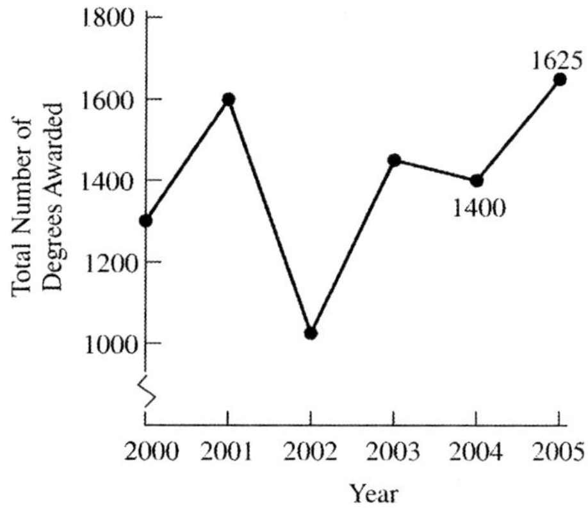

2.

a. Between which years did the number of degrees awarded decrease?

The line between 2001 and 2002 and between 2003 and 2004 falls, so the number of degrees awarded decreased between 2001-2002 and 2003-2004.

b. Between which two years did the total number of degrees awarded show the smallest change?

The line between 2003 and 2004 falls the least, so the total number of degrees awarded declined the least from 2003 to 2004.

c. Estimate the total number of degrees awarded in 2002 and in 2005. About how many more degrees were awarded in 2005?

Move up from 2002 on the horizontal scale to the point plotted for 2002. This point is about 1000. So about 1000 degrees were awarded in 2002.
Similarly, locate the point plotted for 2005. Moving across to the vertical scale, the graph indicates that the number of degrees awarded in 2005 was 1625.
Between 2002 and 2005, the increase was
$1625 - 1000 = 625$.

2.

a. Between which years did the number of degrees awarded increase?

____

b. Between which two years did the total number of degrees awarded show the greatest decline?

____

c. Estimate the total number of degrees awarded in 2000 and in 2001. About how many more degrees were awarded in 2001?

____

Name: Date:
Instructor: Section:

Review these examples for Objective 3:

3. Decide whether each ordered pair is a solution of the equation $4x+5y=40$.

a. $(5,\ 4)$

Substitute 5 for *x* and 4 for *y* in the given equation.

$$4x+5y=40$$
$$4(5)+5(4)\stackrel{?}{=}40$$
$$20+20\stackrel{?}{=}40$$
$$40=40 \quad \text{True}$$

This result is true, so $(5,\ 4)$ is a solution of $4x+5y=40$.

b. $(-3,\ 6)$

Substitute –3 for *x* and 6 for *y* in the given equation.

$$4x+5y=40$$
$$4(-3)+5(6)\stackrel{?}{=}40$$
$$-12+30\stackrel{?}{=}40$$
$$18=40 \quad \text{False}$$

This result is false, so $(-3,\ 6)$ is not a solution of $4x+5y=40$.

Now Try:

3. Decide whether each ordered pair is a solution of the equation $3x-4y=12$.

a. $(8,\ 3)$

____

b. $(5,-4)$

____

Review these examples for Objective 4:

4. Complete each ordered pair for the equation $y=5x+8$.

a. (3,____)

Replace *x* with 3.

$$y=5x+8$$
$$y=5(3)+8$$
$$y=15+8$$
$$y=23$$

The ordered pair is (3, 23).

Now Try:

4. Complete each ordered pair for the equation $y=4x-7$.

a. (5,____)

____

b. $(\_\_\_\_, -12)$

Replace y with –12.

$$y = 5x + 8$$
$$-12 = 5x + 8$$
$$-20 = 5x$$
$$-4 = x$$

The ordered pair is (–4, –12).

b. $(\_\_\_\_, -15)$

____

Review these examples for Objective 5:

5. Complete the table of values for each equation. Then write the results as ordered pairs.

a. $2x - 3y = 6$

x	y
9	
6	
	–2
	8

From the table, we can write the ordered pairs: (9, ____), (6, ____), (____, –2), (____, 8).
From the first row of the table, let $x = 9$ in the equation. From the second row of the table, let $x = 6$.

If $x = 9$,	If $x = 6$,
$2x - 3y = 6$	$2x - 3y = 6$
$2(9) - 3y = 6$	$2(6) - 3y = 6$
$18 - 3y = 6$	$12 - 3y = 6$
$-3y = -12$	$-3y = -6$
$y = 4$	$y = 2$

The first two ordered pairs are (9, 4) and (6, 2).
From the third and fourth rows of the table, let $y = -2$ and $y = 8$, respectively.

If $y = -2$,	If $y = 8$,
$2x - 3y = 6$	$2x - 3y = 6$
$2x - 3(-2) = 6$	$2x - 3(8) = 6$
$2x + 6 = 6$	$2x - 24 = 6$
$2x = 0$	$2x = 30$
$x = 0$	$x = 15$

The last two ordered pairs are (0, –2) and (15, 8).
The completed table and corresponding ordered

Now Try:

5. Complete the table of values for each equation. Then write the results as ordered pairs.

a. $4x - y = 8$

x	y
1	
5	
	0
	4

____

pairs follow.

x	y	Ordered pairs
9	4	$\rightarrow$ (9, 4)
6	2	$\rightarrow$ (6, 2)
0	-2	$\rightarrow$ (0, -2)
15	8	$\rightarrow$ (15, 8)

b. $x = 2$

x	y
	-7
	4
	2

The given equation is $x = 2$. No matter which value of y is chosen, the value of x is always 2.

x	y	Ordered pairs
2	-7	$\rightarrow$ (2, -7)
2	4	$\rightarrow$ (2, 4)
2	2	$\rightarrow$ (2, 2)

b. $x = 6$

x	y
	-5
	0
	3

____

Review these examples for Objective 6:

6. Plot the given points in a coordinate system.

(5, 4) (–2,–1) (–3, 5) (4,–2)
(1,–2.5) (3, 0) (0, 4)

Step 1 Move right or left the number of units that correspond to the x-coordinate in the ordered pair—right if the x-coordinate is positive and left if it is negative.

Step 2 Then turn and move up or down the number of units that corresponds to the y-coordinate—up if the y-coordinate is positive or down if it is negative.

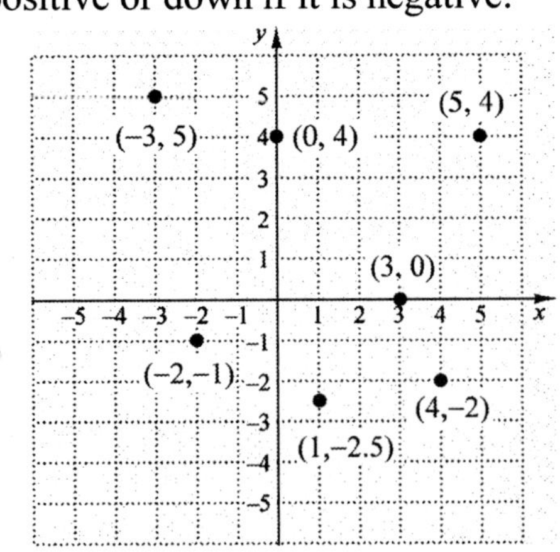

Now Try:

6. Plot the given points in a coordinate system.

(2, 4) (–5, 1) (–4,–2)
(3,–5) (2,–1.5) (6, 0)
(0,–6)

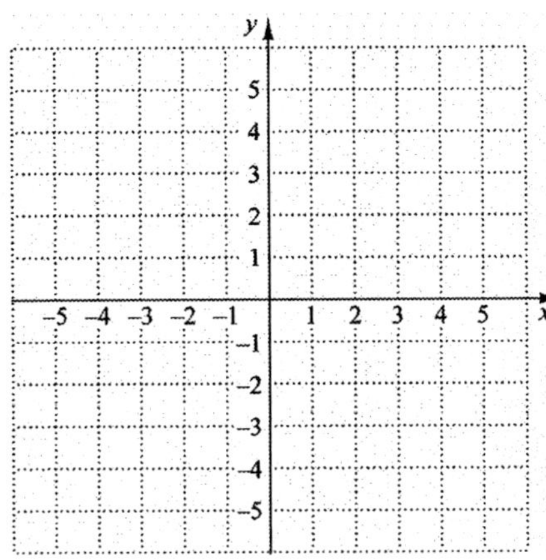

Name: Date:
Instructor: Section:

An accountant keeps track of the balance for a business loan given by the linear equation

$$y = 137.24x + 9879,$$

where x is the number of months, and y is the balance due. For this loan, payment is expected in full at the end of the term.

7.

a. Complete the table of values for the given linear equation.

x (months)	y (balance due)
1	
3	
10	

To find y for $x = 1$, substitute into the equation.

$y = 137.24x + 9879$

$y = 137.24(1) + 9879$

$y = 10,016.24$

x (months)	y (balance due)
1	10,016.24
3	10,290.72
10	11,251.40

So the ordered pairs are (1, 10,016.24), (3, 10,290.72), and (10, 11,251.40).

b. Graph the ordered pairs found in part (a).

A graph of ordered pairs of data is a scatter diagram.

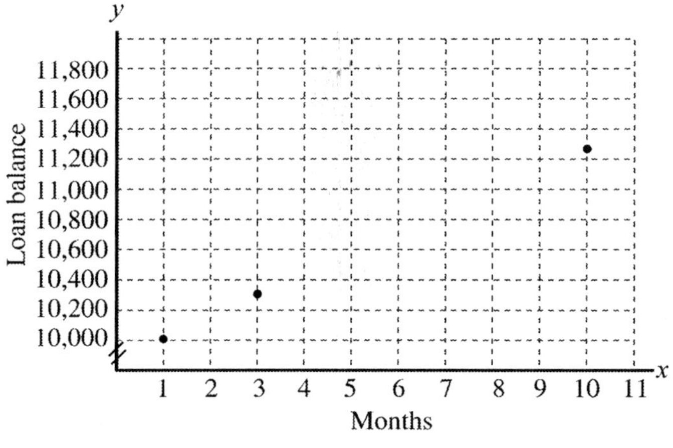

7.

a. Complete the table of values for the given linear equation.

x (months)	y (balance due)
2	
5	
11	

b. Graph the ordered pairs found in part (a).

Name: Date:
Instructor: Section:

Objective 1 Interpret graphs.

For extra help, see Examples 1–2 on pages 726–727 of your text and Section Lecture video for Section 11.1.

The bar graph shows the enrollment for the fall semester at a small college for the years 2002 to 2006. Use this graph for problems 1–2.

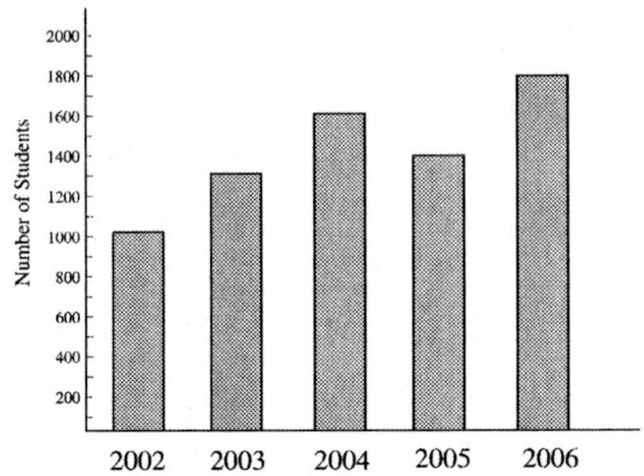

1. In what years was enrollment less than 1500 students? **1.** ________________

2. Estimate the enrollment in 2004. **2.** ________________

The line graph shows the number of degrees awarded by a university for the years 2000–2005. Use this graph to answer exercise 3.

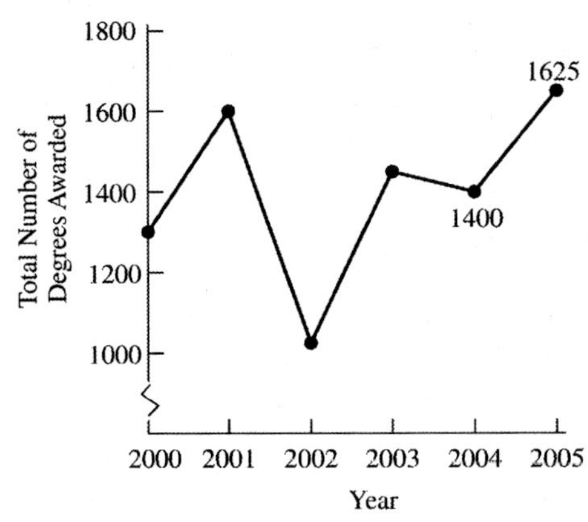

3. If 20% of the degrees awarded in 2005 were MBA degrees, how many MBAs were awarded in 2005? **3.** ________________

Name: Date:
Instructor: Section:

Objective 2 Write a solution as an ordered pair.

For extra help, see page 728 of your text and Section Lecture video for Section 11.1.

Write each solution as an ordered pair.

4. $x = 4$ and $y = 7$ **4.** ________

5. $y = \frac{1}{3}$ and $x = 0$ **5.** ________

6. $x = 0.2$ and $y = 0.3$ **6.** ________

Objective 3 Decide whether a given ordered pair is a solution of a given equation.

For extra help, see Example 3 on page 729 of your text and Section Lecture video for Section 11.1 and Exercise Solutions Clip 22 and 26.

Decide whether the given ordered pair is a solution of the given equation.

7. $4x - 3y = 10; (1, 2)$ **7.** ________

8. $2x - 3y = 1; \left(0, \frac{1}{3}\right)$ **8.** ________

9. $x = -7; (-7, 9)$ **9.** ________

Name: Date:
Instructor: Section:

Objective 4 Complete ordered pairs for a given equation.

For extra help, see Example 4 on page 729 of your text and Section Lecture video for Section 11.1.

For each of the given equations, complete the ordered pairs beneath it.

10. $y = 2x - 5$

(a) $(2, \ )$

(b) $(0, \ )$

(c) $(\ ,3)$

(d) $(\ ,-7)$

(e) $(\ ,9)$

10.
(a) ______________
(b) ______________
(c) ______________
(d) ______________
(e) ______________

11. $y = 3 + 2x$

(a) $(-4, \ )$

(b) $(2, \ )$

(c) $(\ ,0)$

(d) $(-2, \ )$

(e) $(\ ,-7)$

11.
(a) ______________
(b) ______________
(c) ______________
(d) ______________
(e) ______________

Objective 5 Complete a table of values.

For extra help, see Example 5 on page 730 of your text and Section Lecture video for Section 11.1 and Exercise Solutions Clip 42.

Complete each table of values. Write the results as ordered pairs.

12. $2x + 5 = 7$

x	y
	-3
	0
	5

12. ______________

Name: Date:
Instructor: Section:

13. $y-4=0$ **13.** ________________

x	y
-4	
0	
6	

14. $4x+3y=12$ **14.** ________________

x	y
0	
	0
	-1

Objective 6 Plot ordered pairs.

For extra help, see Examples 6–7 on pages 732–733 of your text and Section Lecture video for Section 11.1 and Exercise Solutions Clip 61 and 65.

Plot the each ordered pair on a coordinate system.

15. $(0,-2)$ **15.** ________________

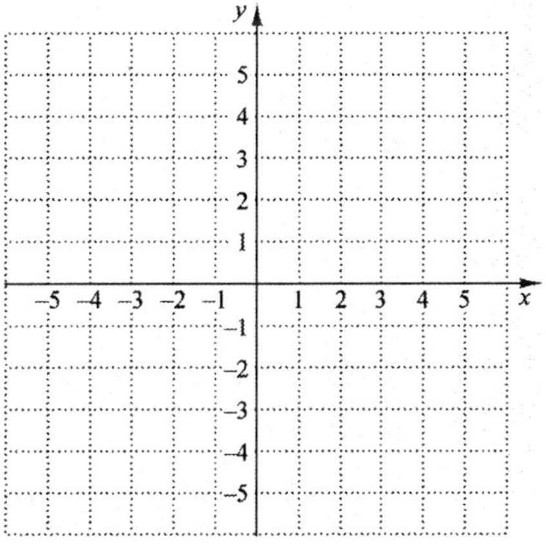

Name: Date:
Instructor: Section:

16. $(-3,\ 4)$

16. ________________

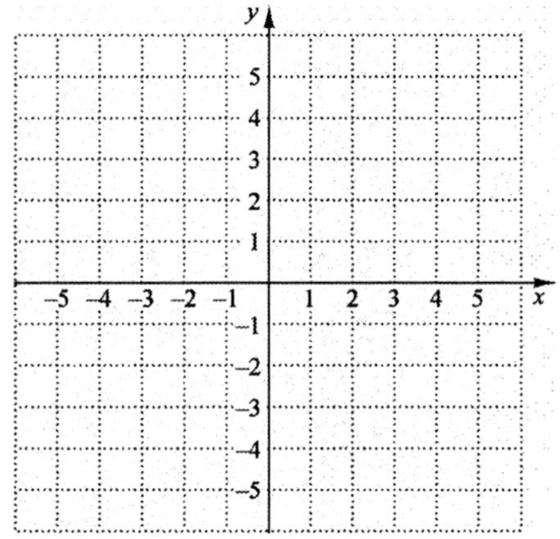

17. $(2,-5)$

17. ________________

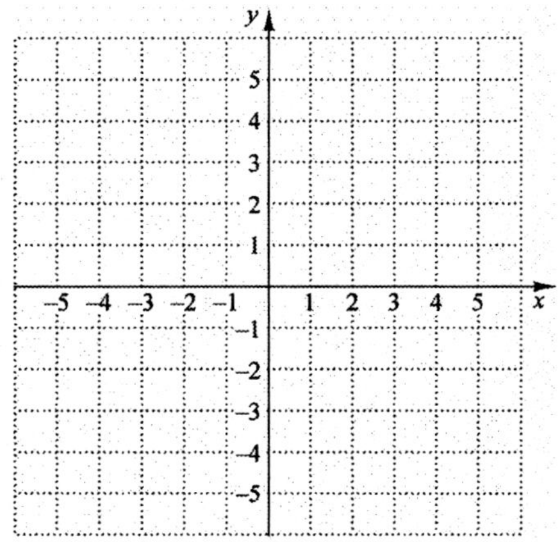

Name: Date:
Instructor: Section:

Chapter 11 GRAPHS OF LINEAR EQUATIONS AND INEQUALITIES IN TWO VARIABLES

11.2 Graphing Linear Equations in Two Variables

Learning Objectives

1	Graph linear equations by plotting ordered pairs.
2	Find intercepts.
3	Graph linear equations of the form $Ax + By = 0$.
4	Graph linear equations of the form $y = k$ or $x = k$.
5	Use a linear equation to model data.

Key Terms

Use the vocabulary terms listed below to complete each statement in exercises 1–4.

graph **graphing** ***y*-intercept** ***x*-intercept**

1. If a graph intersects the y-axis at k, then the ___________________ is $(0, k)$.

2. If a graph intersects the x-axis at k, then the ___________________ is $(k, 0)$.

3. The process of plotting the ordered pairs that satisfy a linear equation and drawing a line through them is called ________________________.

4. The set of all points that correspond to the ordered pairs that satisfy the equation is called the ______________________ of the equation.

Guided Examples

Review this example for Objective 1:

1. Graph $2x+3y=6$.

First let $x = 0$ and then let $y = 0$ to determine two ordered pairs.

$2(0)+3y=6$	$2x+3(0)=6$
$0+3y=6$	$2x+0=6$
$3y=6$	$2x=6$
$y=2$	$x=3$

The ordered pairs are (0, 2) and (3, 0). Find a third ordered pair by choosing a number other than 0 for x or y. We choose $y = 4$.

$$2x+3(4)=6$$
$$2x+12=6$$
$$2x=-6$$
$$x=-3$$

This gives the ordered pair (–3, 4). We plot the

Now Try:

1. Graph $x+y=3$.

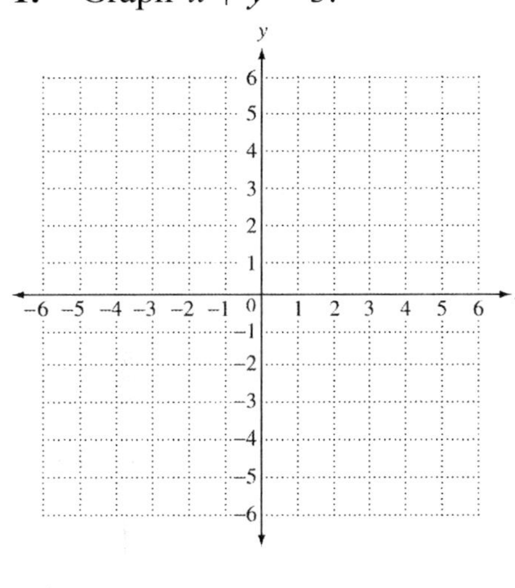

three ordered pairs (0, 2), (3, 0), and (–3, 4) and draw a line through them.

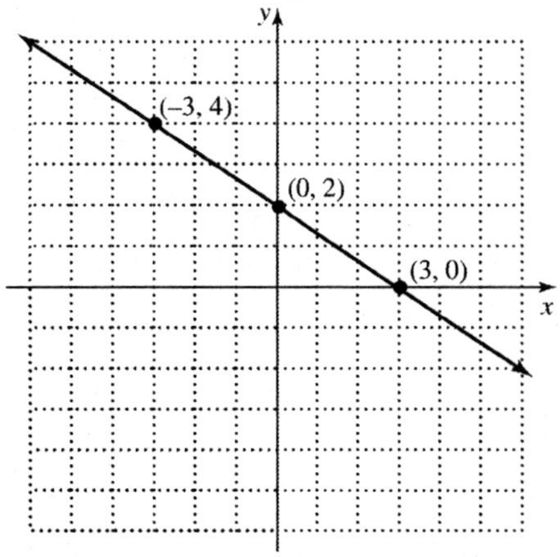

Review this example for Objective 2:

2. Graph $3x + y = 6$ using intercepts.

To find the *y*-intercept, let *x* = 0.
To find the *x*-intercept, let *y* = 0.

$$\begin{array}{r|r} 3(0)+y=6 & 3x+0=6 \\ 0+y=6 & 3x=6 \\ y=6 & x=2 \end{array}$$

The intercepts are (0, 6) and (2, 0). To find a third point, as a check, we let *x* = 1.

$$3(1)+y=6$$
$$3+y=6$$
$$y=3$$

This gives the ordered pair (1, 3).

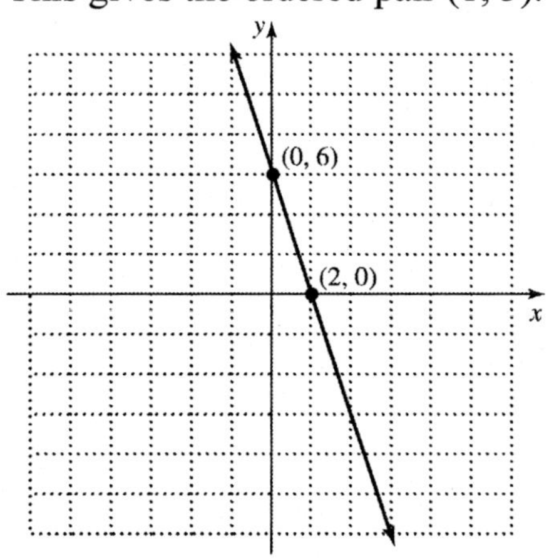

Now Try:

2. Graph $5x - 2y = -10$ using intercepts.

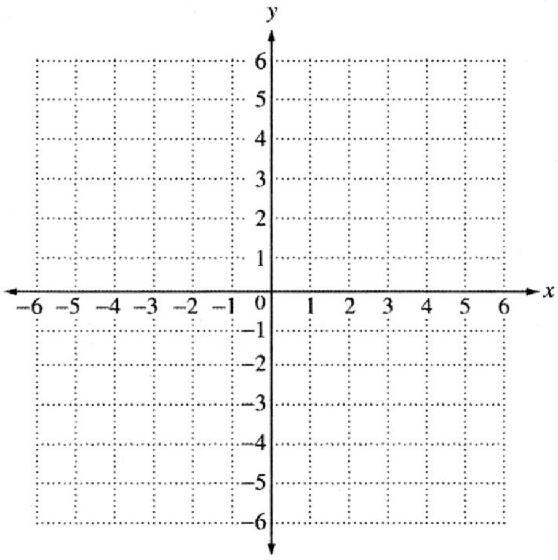

3. Graph $y = \frac{3}{2}x - 4$.

To find the y-intercept, let $x = 0$.
To find the x-intercept, let $y = 0$.

$y = \frac{3}{2}(0) - 4$	$0 = \frac{3}{2}x - 4$
$y = 0 - 4$	$4 = \frac{3}{2}x$
$y = -4$	$\frac{8}{3} = x$

The intercepts are (0, –4) and $\left(\frac{8}{3}, 0\right)$. To find a third point, as a check, we let $x = 2$.

$y = \frac{3}{2}(2) - 4$

$y = 3 - 4$

$y = -1$

This gives the ordered pair (2, –1).

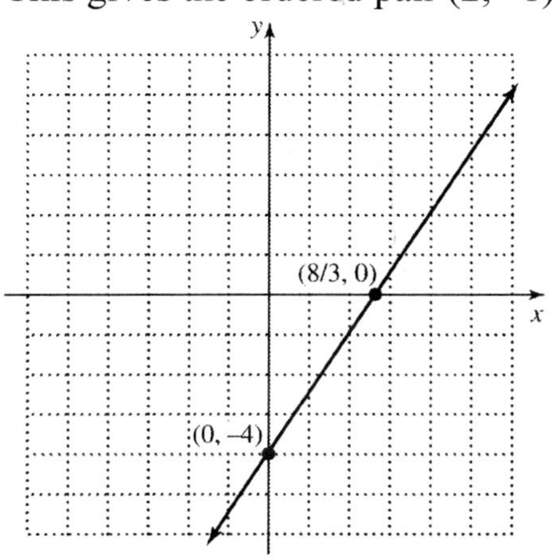

3. Graph $y = 4x - 4$.

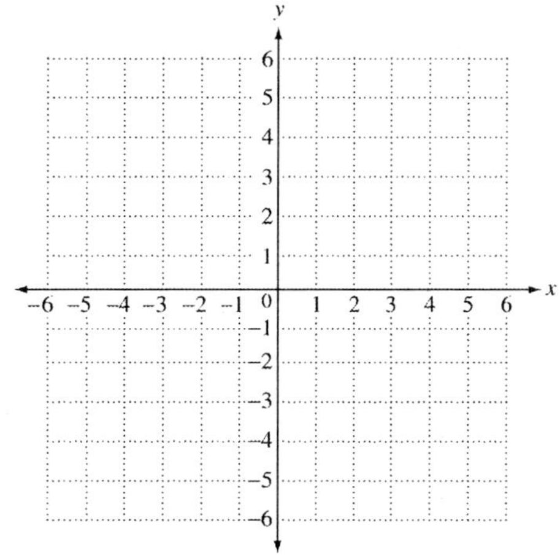

Review this example for Objective 3:

4. Graph $x + 5y = 0$.

To find the y-intercept, let $x = 0$.
To find the x-intercept, let $y = 0$.

$0 + 5y = 0$	$x + 5(0) = 0$
$5y = 0$	$x + 0 = 0$
$y = 0$	$x = 0$

The x- and y-intercepts are the same point (0, 0). We must select two other values for x or y to find two other points. We choose $y = 1$ and $y = -1$.

Now Try:

4. Graph $3x - y = 0$.

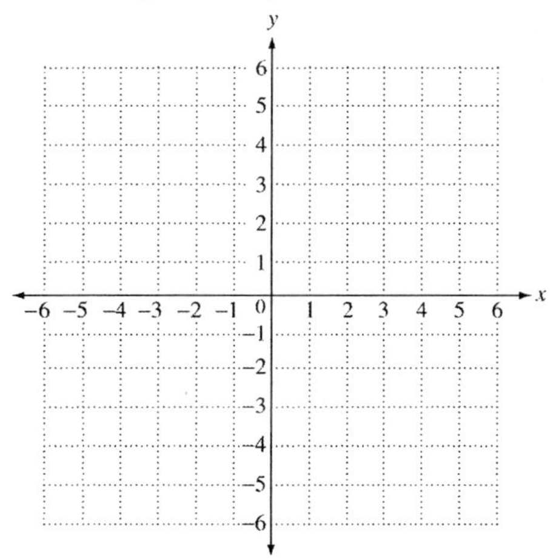

$$x+5(1)=0 \qquad x+5(-1)=0$$
$$x+5=0 \qquad x-5=0$$
$$x=-5 \qquad x=5$$

We use (–5, 1), (0, 0), and (5, –1) to draw the graph.

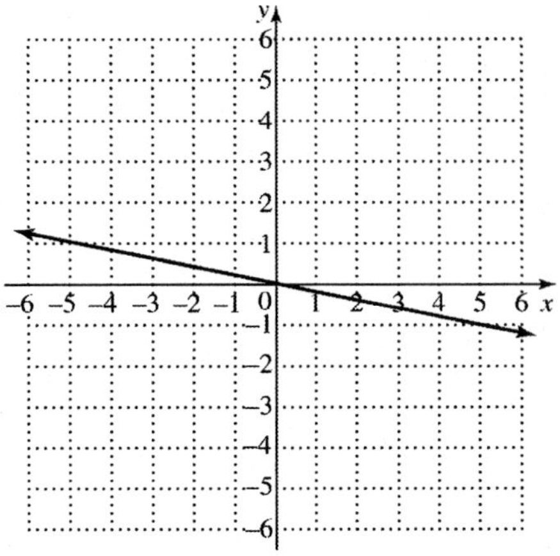

Review these examples for Objective 4:

5. Graph $y=-2$.

For any value of x, y is always –2. Three ordered pairs that satisfy the equation are (–4, –2), (0, –2) and (2, –2). Drawing a line through these points gives the horizontal line. The y-intercept is (0, –2). There is no x-intercept.

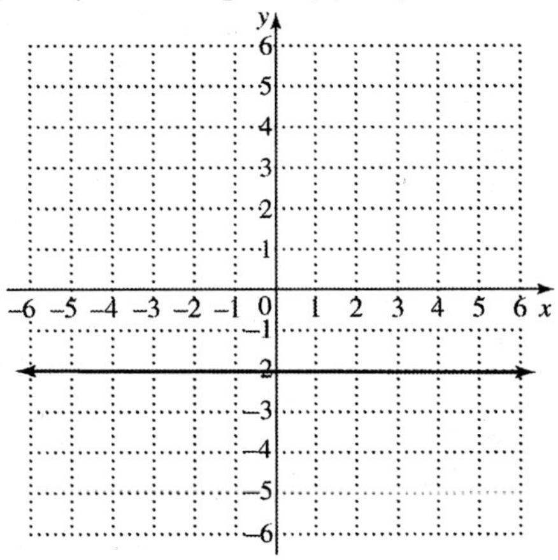

Now Try:

5. Graph $y=4$.

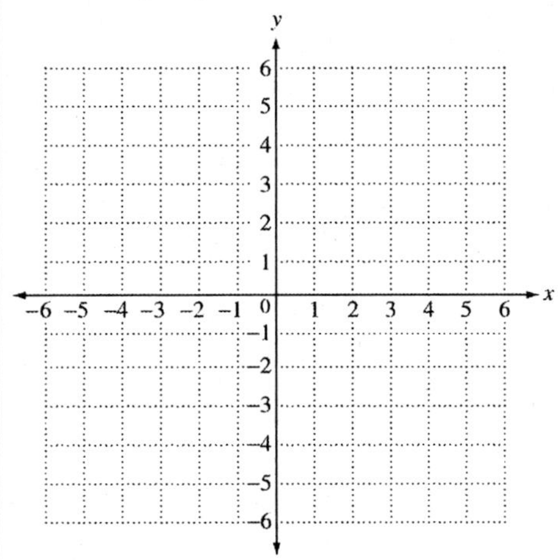

Name: Date:
Instructor: Section:

6. Graph $x+4=0$.

First we subtract 4 from each side of the equation to get the equivalent equation $x = -4$. All ordered-pair solutions of this equation have x-coordinate -4.
Three ordered pairs that satisfy the equation are $(-4, -1)$, $(-4, 0)$, and $(-4, 3)$. The graph is a vertical line. The x-intercept is $(-4, 0)$. There is no y-intercept.

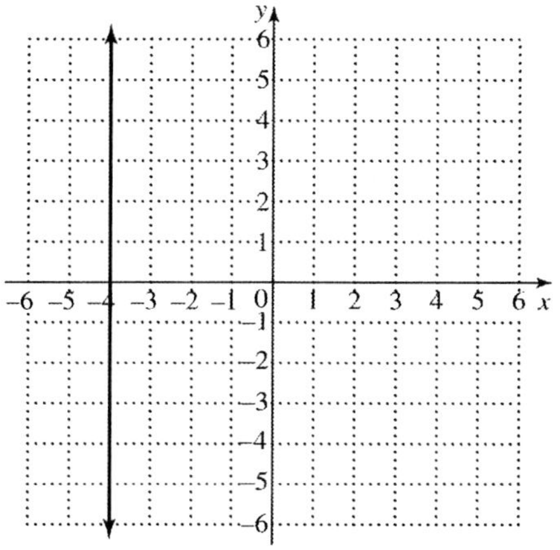

6. Graph $x=0$.

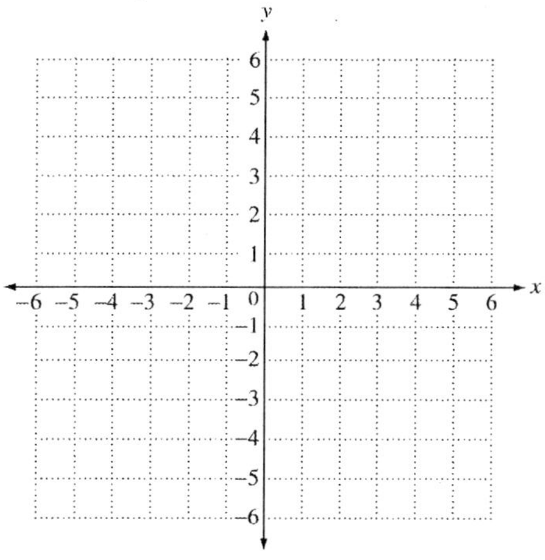

Review these examples for Objective 5:

7. Every year sea turtles return to a certain group of islands to lay eggs. The number of turtle eggs that hatch can be approximated by the equation $y=-70x+3260$, where y is the number of eggs that hatch and $x=0$ representing 1990.

a. Use this equation to find the number of eggs that hatched in 1995, 2000, and 2005, and 2015.

Substitute the appropriate value for each year x to find the number of eggs hatched in that year.

For 1995:

$y=-70(5)+3260$ $\quad$ $1995-1990=5$

$y=2910$ eggs $\quad$ Replace x with 5.

For 2000:

$y=-70(10)+3260$ $\quad$ $2000-1990=10$

$y=2560$ eggs $\quad$ Replace x with 10.

Now Try:

7. Suppose that the demand and price for a certain model of calculator are related by the equation $y=45-\frac{3}{5}x$, where y is the price (in dollars) and x is the demand (in thousands of calculators).

a. Assuming that this model is valid for a demand up to 50,000 calculators, use this equation to find the price of calculators at each level of demand.

0 calculators ____________

5000 calculators ____________

20,000 calculators ____________

45,000 calculators ____________

For 2005:

$y = -70(15) + 3260$ $\quad 2005 - 1990 = 15$

$y = 2210$ eggs $\quad$ Replace x with 15.

For 2015:

$y = -70(25) + 3260$ $\quad 2015 - 1990 = 25$

$y = 1510$ eggs $\quad$ Replace x with 25.

b. Write the information from part (a) as four ordered pairs, and use them to graph the given linear equation.

Since x represents the year and y represents the number of eggs, the ordered pairs are (5, 2910), (10, 2560), (15, 2210), and (25, 1510).

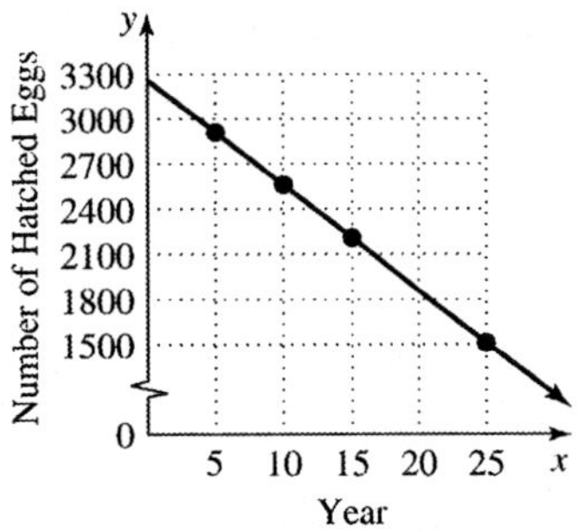

c. Use this graph and the equation to estimate the number of eggs that will hatch in 2010.

For 2010, $x = 20$. On the graph, find 20 on the horizontal axis, move up to the graphed line and then across to the vertical axis. It appears that in 2010, there were about 1900 eggs.

To use the equation, substitute 20 for x.

$y = -70(20) + 3260$

$y = 1860$ eggs

This result for 2020 is close to our estimate of 1900 eggs from the graph.

b. Write the information from part (a) as four ordered pairs, and use them to graph the given linear equation.

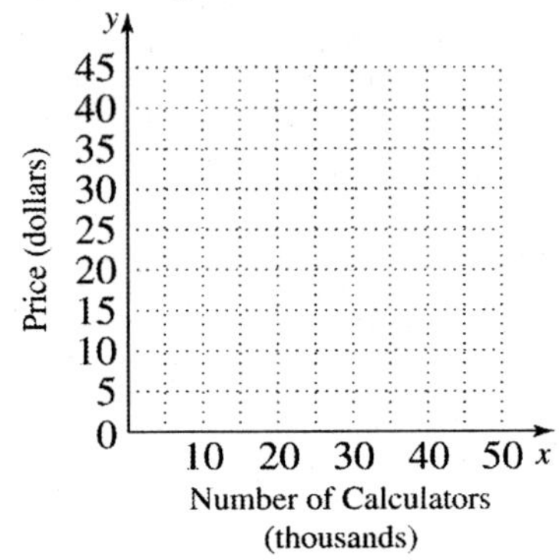

c. Use this graph and the equation to estimate the price of 30,000 calculators.

____

Name: Date:
Instructor: Section:

Objective 1 Graph linear equations by plotting ordered pairs.

For extra help, see Example 1 on page 740 of your text and Section Lecture video for Section 11.2 and Exercise Solutions Clip 9 and 11.

Complete the ordered pairs for each equation. Then graph the equation by plotting the points and drawing a line through them.

1. $y = 3x - 2$

$(0, \ )$

$(\ ,0)$

$(2, \ )$

1.

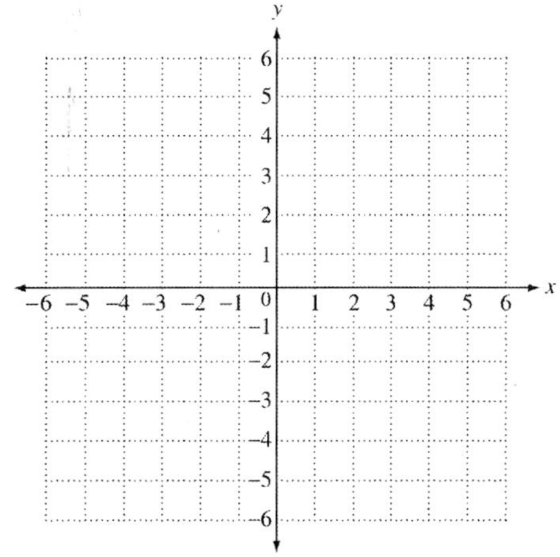

2. $x - y = 4$

$(0, \ )$

$(\ ,0)$

$(-2, \ )$

2.

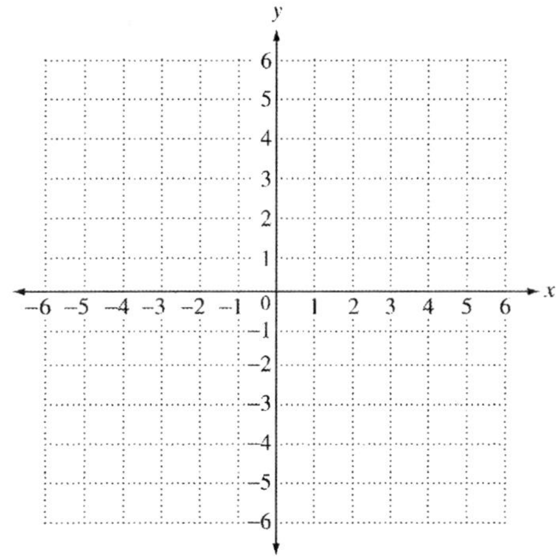

Name: Date:
Instructor: Section:

3. $x = 2y + 1$

$(0,\ \ )$

$(\ \ ,0)$

$(\ \ ,-2)$

3.

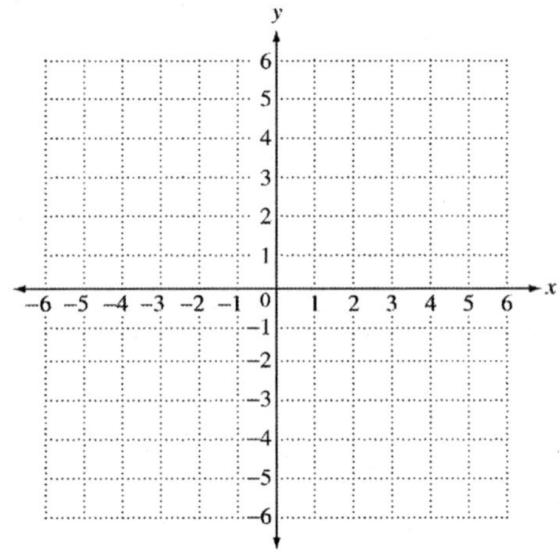

Objective 2 Find intercepts.

For extra help, see Examples 2–3 on pages 741–742 of your text and Section Lecture video for Section 11.2 and Exercise Solutions Clip 9 and 11.

Find the intercepts for each equation. Then graph the equation.

4. $y = \frac{2}{3}x - 2$

4.

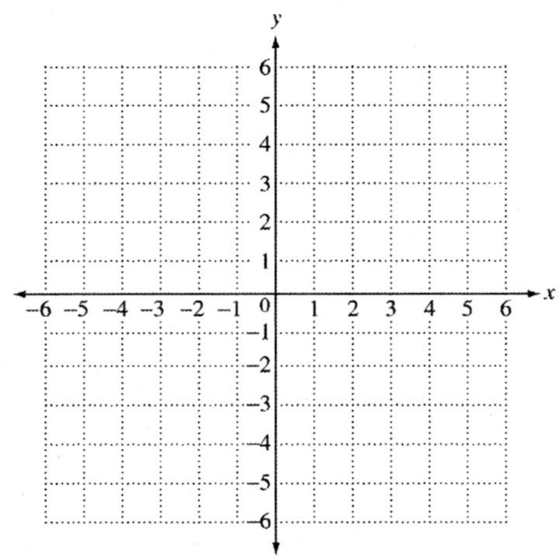

5. $4x - 7y = -8$

5.

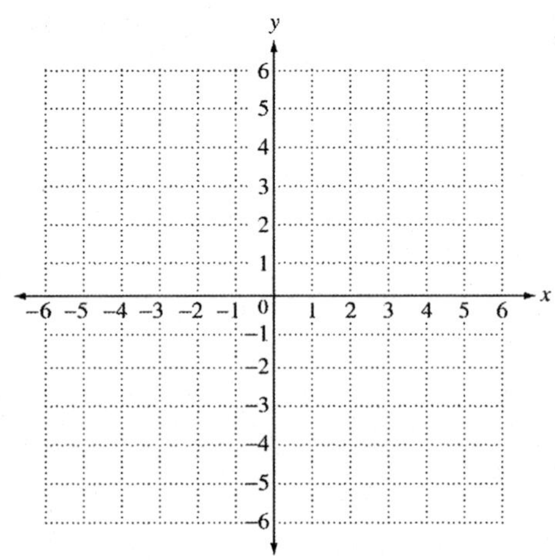

Name: Date:
Instructor: Section:

Objective 3 Graph linear equations of the form $Ax + By = 0$.

For extra help, see Example 4 on page 742 of your text and Section Lecture video for Section 11.2 and Exercise Solutions Clip 31 and 33.

Graph each equation.

6. $-3x - 2y = 0$

6.

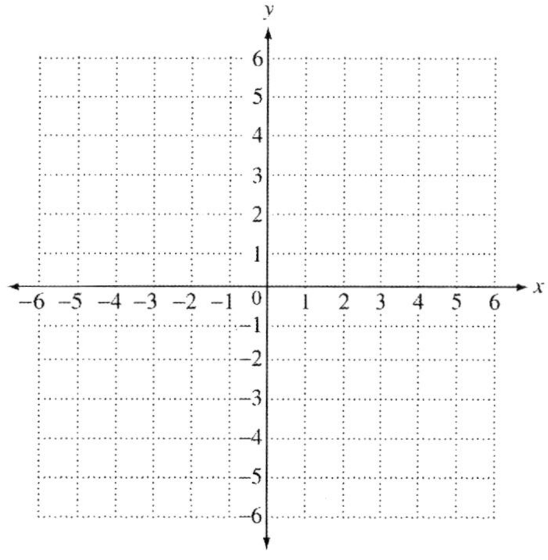

7. $x + y = 0$

7.

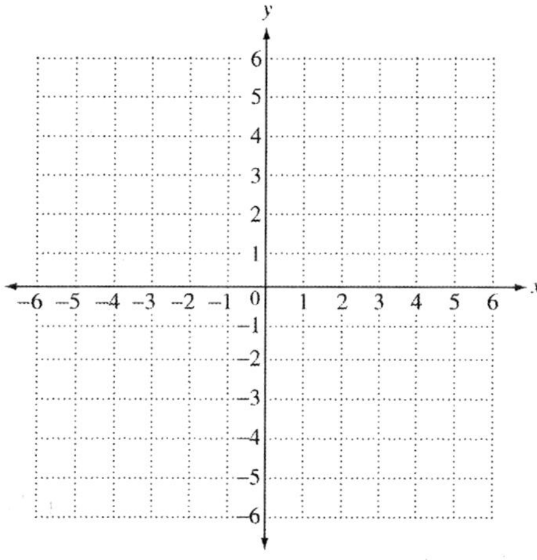

8. $y = 2x$

8.

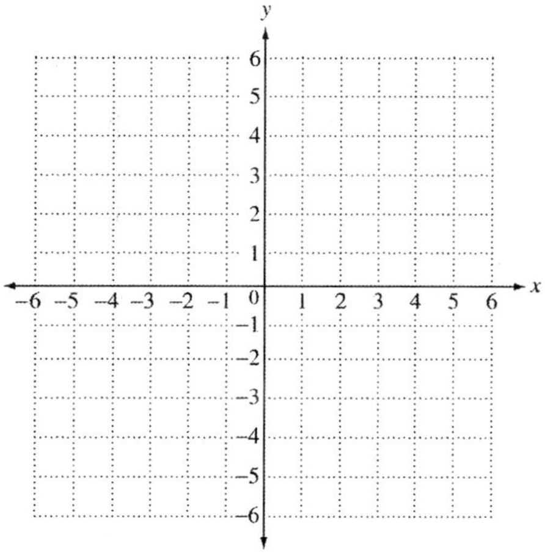

Name: Date:
Instructor: Section:

Objective 4 Graph linear equations of the form *y* = *k* or *x* = *k*.

For extra help, see Examples 5–6 on page 743 of your text and Section Lecture video for Section 11.2 and Exercise Solutions Clip 35.

Graph each equation.

9. $x-1=0$

9.

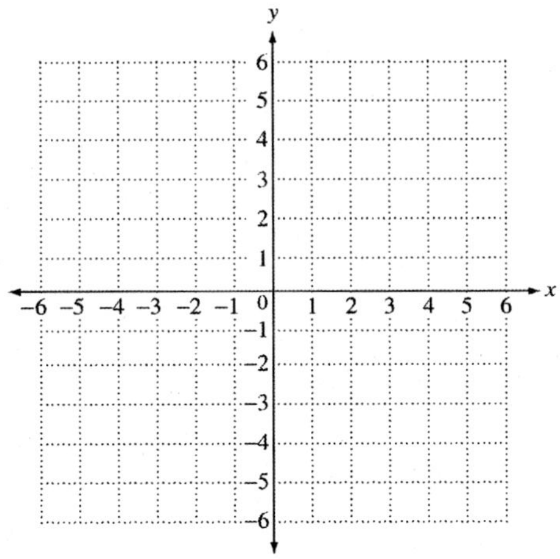

10. $y+3=0$

10.

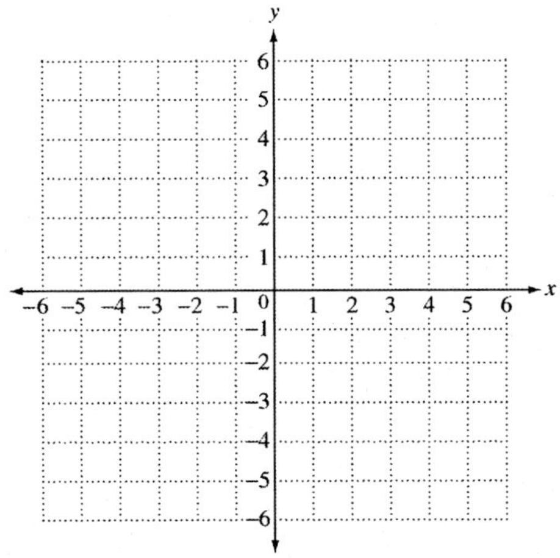

Name: Date:
Instructor: Section:

Objective 5 Use a linear equation to model data.

For extra help, see Example 7 on page 745 of your text and Section Lecture video for Section 11.2 and Exercise Solutions Clip 47 and 48.

Solve each problem. Then graph the equation.

11. The profit y in millions of dollars earned by a small computer company can be approximated by the linear equation $y = .63x + 4.9$, where $x = 0$ corresponds to 2004, $x = 1$ corresponds to 2005, and so on. Use this equation to approximate the profit in each year from 2004 through 2007.

11. 2004 ___________

2005 ___________

2006 ___________

2007 ___________

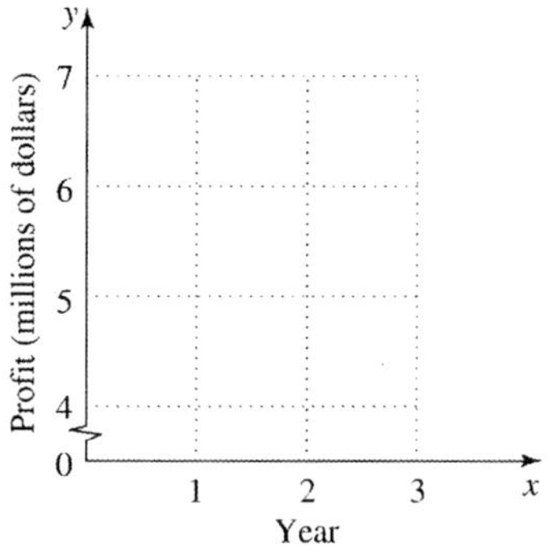

12. The number of band instruments sold by Elmer's Music Shop can be approximated by the equation $y = 325 + 42x$, where y is the number of instruments sold and x is the time in years, with $x = 0$ representing 2003. Use this equation to approximate the number of instruments sold in each year from 2003 through 2006.

12. 2003 ___________

2004 ___________

2005 ___________

2006 ___________

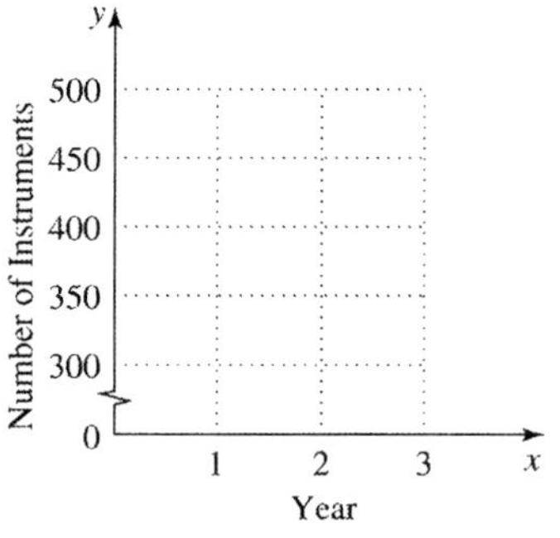

Name: Date:
Instructor: Section:

13. According to *The Old Farmer's Almanac*, the temperature in degrees Celsius can be determined by the equation $y = \frac{1}{3}x + 4$, where x is the number of cricket chirps in 25 seconds and y is the temperature in degrees Celsius. Use this equation to find the temperature when there are 48 chirps, 54 chirps, 60 chirps, and 66 chirps.

13. 48 ______________

54 ______________

60 ______________

66 ______________

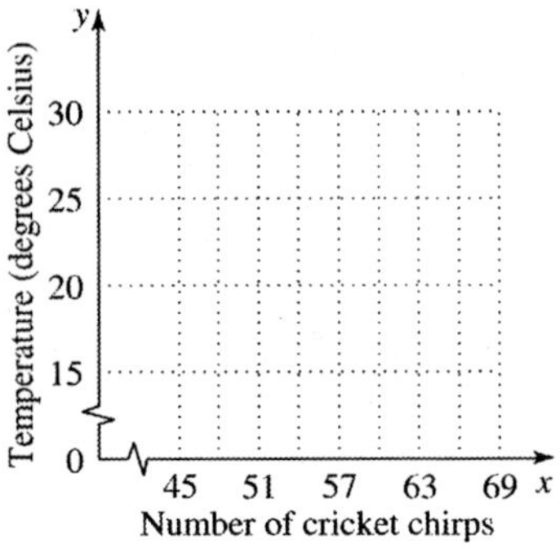

Name: Date:
Instructor: Section:

Chapter 11 GRAPHS OF LINEAR EQUATIONS AND INEQUALITIES IN TWO VARIABLES

11.3 Slope of a Line

Learning Objectives
1. Find the slope of a line given two points.
2. Find the slope from the equation of a line.
3. Use slope to determine whether two lines are parallel, perpendicular, or neither.

Key Terms

Use the vocabulary terms listed below to complete each statement in exercises 1–5.

rise **run** **slope** **parallel lines**

perpendicular lines

1. Two lines that intersect in a 90° angle are called ______________________________.
2. The __________________ of a line is the ratio of the change in y compared to the change in x when moving along the line from one point to another.
3. The vertical change between two different points on a line is called the __________________________.
4. Two lines in a plane that never intersect are called __________________________.
5. The horizontal change between two different points on a line is called the __________________________.

Guided Examples

Review these examples for Objective 1:

1. Find the slope of the line.

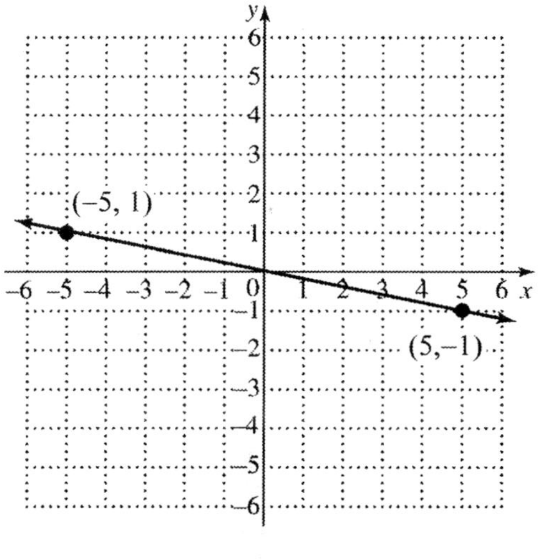

Now Try:

1. Find the slope of the line.

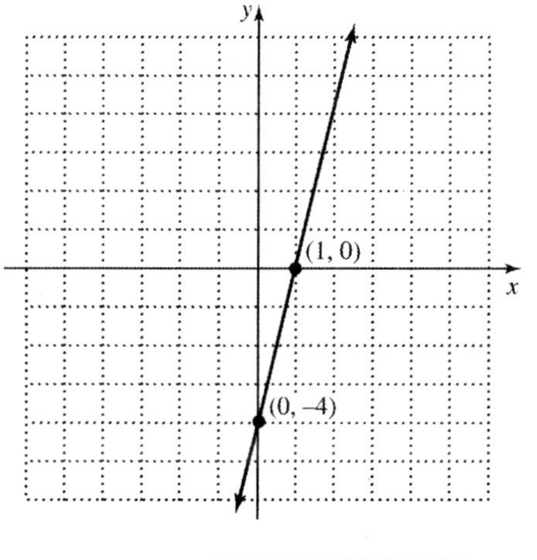

____

We use the two points shown on the line. The vertical change is the difference in the y-values, or –1 – 1 = –2, and the horizontal change is the difference in the x-values or 5 – (–5) = 10. Thus, the line has

$$\text{slope} = \frac{-2}{10}, \text{ or } -\frac{1}{5}.$$

2. Find the slope of each line.

a. The line passing through (–5, 4) and (2,–6)

Apply the slope formula.

$(x_1, y_1) = (-5, 4)$ and $(x_2, y_2) = (2, -6)$

$$\text{slope } m = \frac{y_2 - y_1}{x_2 - x_1} = \frac{-6-4}{2-(-5)}$$
$$= \frac{-10}{7}, \text{or } -\frac{10}{7}$$

b. The line passing through (–7,–3) and (11, 8)

Apply the slope formula.

$(x_1, y_1) = (-7, -3)$ and $(x_2, y_2) = (11, 8)$

$$\text{slope } m = \frac{y_2 - y_1}{x_2 - x_1} = \frac{8-(-3)}{11-(-7)} = \frac{11}{18}$$

3. Find the slope of the line passing through (–9, 3) and (4, 3).

$(x_1, y_1) = (-9, 3)$ and $(x_2, y_2) = (4, 3)$

$$m = \frac{y_2 - y_1}{x_2 - x_1} = \frac{3-3}{4-(-9)} = \frac{0}{13} = 0$$

4. Find the slope of the line passing through (–5, 3) and (–5, 8).

$(x_1, y_1) = (-5, 3)$ and $(x_2, y_2) = (-5, 8)$

$$m = \frac{y_2 - y_1}{x_2 - x_1} = \frac{8-3}{-5-(-5)} = \frac{5}{0} \text{ undefined slope}$$

2. Find the slope of each line.

a. The line passing through (–6, 7) and (3, –9)

____

b. The line passing through (–5, –2) and (12, 7)

____

3. Find the slope of the line passing through (8,–5) and (–7,–5).

____

4. Find the slope of the line passing through (9, 11) and (9,–7).

____

Name: Date:
Instructor: Section:

Review these examples for Objective 2:

5. Find the slope of each line.

a. $4x-3y=7$

Step 1 Solve the equation for y.

$$4x-3y=7$$
$$-3y=-4x+7$$
$$y=\frac{4}{3}x-\frac{7}{3}$$

Step 2 The slope is given by the coefficient of x, so the slope is $\frac{4}{3}$.

b. $6x+3y=1$

$$3y=-6x+1$$
$$y=-2x+\frac{1}{3}$$

The slope of this line is given by the coefficient of x, which is –2.

c. $5y+x=5$

$$5y+x=5$$
$$5y=-x+5$$
$$y=-\frac{x}{5}+1$$
$$y=-\frac{1}{5}x+1$$

The coefficient of x is $-\frac{1}{5}$, so the slope of the line is $-\frac{1}{5}$.

Now Try:

5. Find the slope of each line.

a. $7x-4y=8$ ____________

b. $9x+6y=4$ ____________

c. $8y+x=16$ ____________

Review these examples for Objective 3:

6. Decide whether each pair of lines is parallel, perpendicular, or neither.

a. $x+3y=8$ and $-3x+y=5$

Find the slope of each line by first solving each equation for y.

$3y=-x+8$	$y=3x+5$
$y=-\frac{1}{3}x+\frac{8}{3}$	
The slope is $-\frac{1}{3}$.	The slope is 3.

Now Try:

6. Decide whether each pair of lines is parallel, perpendicular, or neither.

a. $9x-y=7$ and $x+9y=11$ ____________

Because the slopes are not equal, the lines are not parallel.

Check the product of the slopes: $-\frac{1}{3}(3)=-1$.

The two lines are perpendicular because the product of their slopes is –1.

b. $5x-y=3$

$15x-3y=12$

Solve each equation for y.

$y=5x-3$

$y=5x-4$

Both lines have slope 5, so the lines are parallel.

c. $6x+5y=10$

$5x-3y=4$

Solve each equation for y.

$y=-\frac{6}{5}x+2$

$y=\frac{5}{3}x-\frac{4}{3}$

The slopes are $-\frac{6}{5}$ and $\frac{5}{3}$. Because $-\frac{6}{5}\neq\frac{5}{3}$ and $-\frac{6}{5}\left(\frac{5}{3}\right)\neq-1$, these lines are neither parallel nor perpendicular.

d. $x-6y=2$

$6x-y=9$

Solve each equation for y.

$y=\frac{1}{6}x-\frac{1}{3}$

$y=6x-9$

Here the slopes are $\frac{1}{6}$ and 6. The lines are neither parallel nor perpendicular.

b. $2x-4y=7$

$3x-6y=8$

____

c. $5x-8y=20$

$8x-5y=-9$

____

d. $x-6y=9$

$x+6y=-3$

____

Name: Date:
Instructor: Section:

Objective 1 Find the slope of a line given two points.

For extra help, see Examples 1–4 on pages 753–756 of your text and Section Lecture video for Section 11.3 and Exercise Solutions Clip 5, 23, 29, 31, and 35.

Find the slope of the line through the given points.

1. (4, 3) and (3, 5) **1.** ________________

2. (−4, 6) and (−4, −1) **2.** ________________

3. (−3, 3) and (6, 3) **3.** ________________

Objective 2 Find the slope from the equation of a line.

For extra help, see Example 5 on page 757 of your text and Section Lecture video for Section 11.3 and Exercise Solutions Clip 41.

Find the slope of each line.

4. $7y-4x=11$ **4.** ________________

5. $3y=2x-1$ **5.** ________________

6. $y=-\frac{2}{5}x-4$ **6.** ________________

Name: Date:
Instructor: Section:

Objective 3 Use slope to determine whether two lines are parallel, perpendicular, or neither.

For extra help, see Example 6 on page 759 of your text and Section Lecture video for Section 11.3 and Exercise Solutions Clip 51 and 55.

In each pair of equations, give the slope of each line, and then determine whether the two lines are **parallel**, **perpendicular**, *or* **neither**.

7. $-x + y = -7$

$x - y = -3$

7. ________________

8. $4x + 2y = 8$

$x + 4y = -3$

8. ________________

9. $9x + 3y = 2$

$x - 3y = 5$

9. ________________

Name: Date:
Instructor: Section:

Chapter 11 GRAPHS OF LINEAR EQUATIONS AND INEQUALITIES IN TWO VARIABLES

11.4 Writing and Graphing Equations of Lines

Learning Objectives

1	Use the slope-intercept form of the equation of a line.
2	Graph a line by using its slope and a point on the line.
3	Write an equation of a line by using its slope and any point on the line.
4	Write an equation of a line by using two points on the line.
5	Find an equation of a line that fits a data set.

Key Terms

Use the vocabulary terms listed below to complete each statement in exercises 1–3.

slope-intercept form **point-slope form** **standard form**

1. A linear equation in the form $y - y_1 = m(x - x_1)$ is written in ______________________.

2. A linear equation in the form $Ax + By = C$ is written in ______________________.

3. A linear equation in the form $y = mx + b$ is written in ______________________.

Guided Examples

Review these examples for Objective 1:

1. Identify the slope and y-intercept of the line with each equation.

a. $y = -8x + 7$

The slope is –8, and the y-intercept is (0, 7).

b. $y = -x - 6$

The equation can be written as
$y = (-1)x + (-6)$.
The slope is –1, and the y-intercept is (0, –6).

c. $y = 11x$

The equation can be written as
$y = 11x + 0$.
The slope is 11, and the y-intercept is (0, 0).

Now Try:

1. Identify the slope and y-intercept of the line with each equation.

a. $y = -12x + 6$

____

b. $y = -x - 9$

____

c. $y = 23x$

____

Name: Date:
Instructor: Section:

d. $y = \frac{x}{9} - \frac{5}{4}$

The equation can be written as

$$y = \frac{1}{9}x + \left(-\frac{5}{4}\right).$$

The slope is $\frac{1}{9}$, and the y-intercept is $\left(0, -\frac{5}{4}\right)$.

d. $y = -\frac{x}{7} - \frac{7}{5}$

____

2. Write an equation of the line with slope $\frac{5}{7}$ and y-intercept $(0, -6)$.

Here, $m = \frac{5}{7}$ and $b = -6$, so we can write the following equation.

$$y = mx + b$$

$$y = \frac{5}{7}x + (-6), \text{ or } y = \frac{5}{7}x - 6$$

2. Write an equation of the line with slope $\frac{7}{9}$ and y-intercept $(0, 8)$.

____

Review these examples for Objective 2:

3. Graph each equation by using the slope and y-intercept.

a. $y = -3x + 3$

Step 1 The equation is in slope-intercept form.

Step 2 The y-intercept is $(0, 3)$. Graph this point.

Step 3 The slope is -3. By definition,

$$\text{slope } m = \frac{\text{change in } y \text{ (rise)}}{\text{change in } x \text{ (run)}} = \frac{-3}{1}$$

From the y-intercept, count down 3 units and to the right 1 unit to obtain the point $(1, 0)$.

Step 4 Draw the line through the points $(0, 3)$ and $(1, 0)$ to obtain the graph.

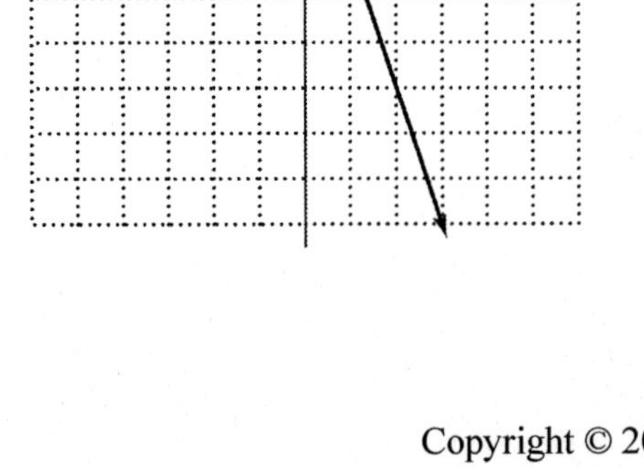

Now Try:

3. Graph each equation by using the slope and y-intercept.

a. $y = -x + 2$

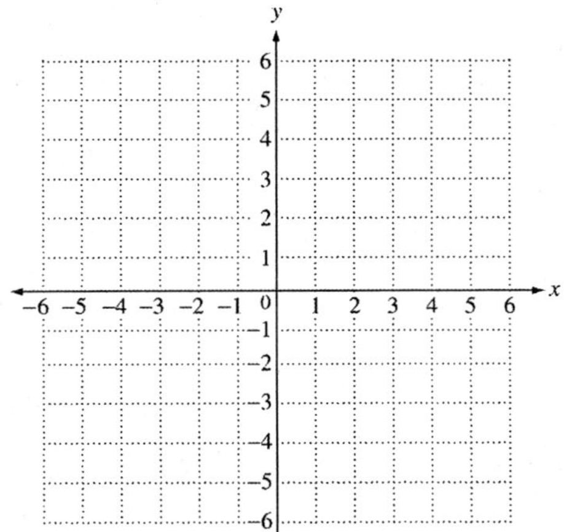

Name: Date:
Instructor: Section:

b. $2x - 3y = 6$

Step 1 Solve for y to write the equation in slope-intercept form.

$$2x - 3y = 6$$
$$-3y = -2x + 6$$
$$y = \frac{2}{3}x - 2$$

Step 2 The y-intercept is (0, –2). Graph this point.

Step 3 The slope is $\frac{2}{3}$. By definition,

$$\text{slope } m = \frac{\text{change in } y \text{ (rise)}}{\text{change in } x \text{ (run)}} = \frac{2}{3}$$

From the y-intercept, count up 2 units and to the right 3 units to obtain the point (3, 0).

Step 4 Draw the line through the points (0, –2) and (3, 0) to obtain the graph.

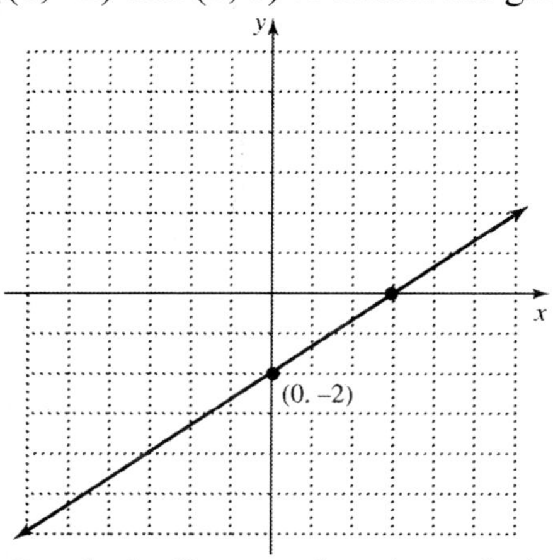

4. Graph the line passing through the point (1, –3), with slope $-\frac{5}{2}$.

First, locate the point (1, –3). Then write the slope $-\frac{5}{2}$ as

$$\text{slope } m = \frac{\text{change in } y \text{ (rise)}}{\text{change in } x \text{ (run)}} = \frac{5}{-2}.$$

Locate another point on the line by counting up 5 units from (1, –3), and then to the left 2 units. Finally, draw the line through this new point, (–1, 2).

b. $y = \frac{2}{3}x$

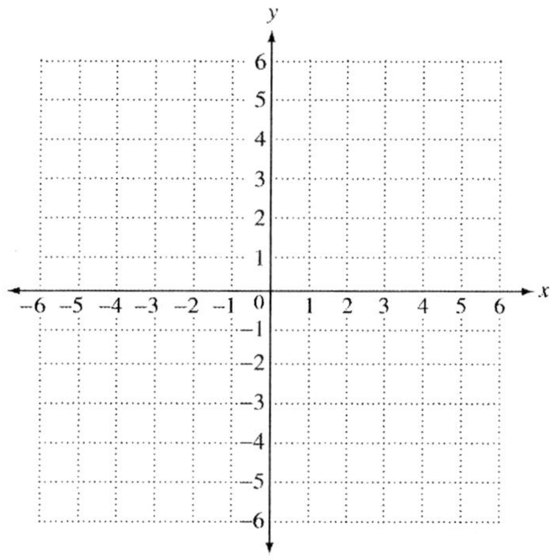

4. Graph the line passing through the point (2, 2), with slope $\frac{1}{3}$.

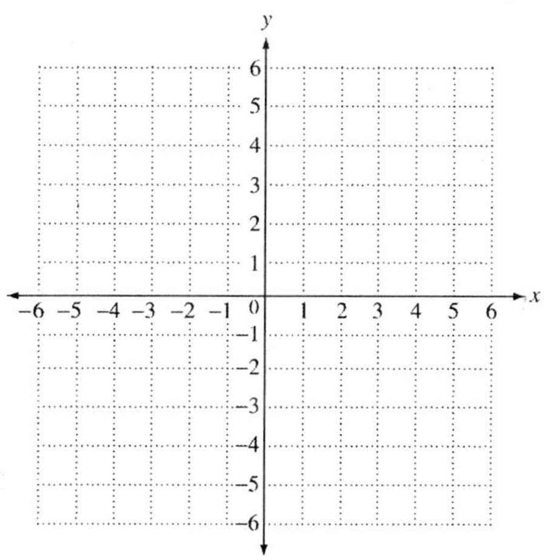

Name: Date:
Instructor: Section:

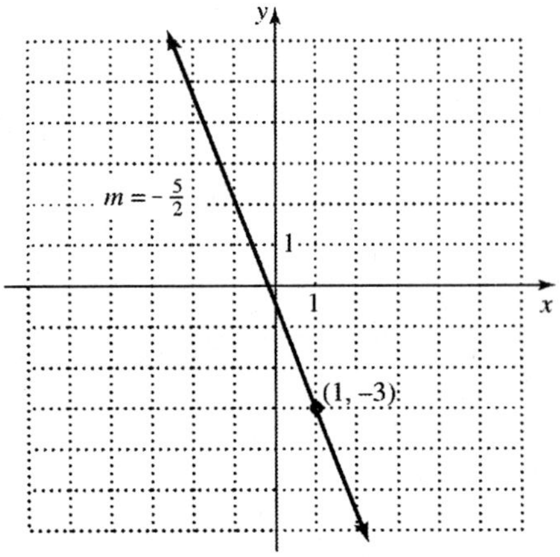

Review these examples for Objective 3:

5. Write an equation, in slope-intercept form, of the line having slope 5 passing through the point (2, 9).

Since the line passes through the point (2, 9), we can substitute $x = 2$, $y = 9$, and slope $m = 5$ into $y = mx + b$ and solve for b.

$$y = mx + b$$
$$9 = 5(2) + b$$
$$-1 = b$$

Now substitute the values of m and b into slope-intercept form.

$$y = mx + b$$
$$y = 5x - 1$$

6. Write an equation of each line. Give the final answer in slope-intercept form.

a. The line passing through (–3, 6) with slope –4.

The given point is (–3, 6) so $x_1 = -3$ and $y_1 = 6$. Also, $m = -4$. Substitute these values into the point-slope form.

$$y - y_1 = m(x - x_1)$$
$$y - 6 = -4[x - (-3)]$$
$$y - 6 = -4(x + 3)$$
$$y - 6 = -4x - 12$$
$$y = -4x - 6$$

Now Try:

5. Write an equation, in slope-intercept form, of the line having slope 6 passing through the point (–1, 4).

____

6. Write an equation of each line. Give the final answer in slope-intercept form.

a. The line passing through (8, –7) with slope $\frac{1}{4}$.

____

Name: Date:
Instructor: Section:

b. The line passing through (5, –3) with slope $\frac{5}{4}$.

$$y - y_1 = m(x - x_1)$$
$$y - (-3) = \frac{5}{4}(x - 5)$$
$$y + 3 = \frac{5}{4}x - \frac{25}{4}$$
$$y = \frac{5}{4}x - \frac{37}{4}$$

b. The line passing through (6, 11) with slope $-\frac{2}{3}$.

____

Review this example for Objective 4:

7. Write the equation of the line passing through the point (6, 8) and (–3, 5). Give the final answer in slope-intercept form and then in standard form.

First, find the slope of the line.
$(x_1, y_1) = (6,\ 8)$ and $(x_2, y_2) = (-3,\ 5)$

$$\text{slope } m = \frac{y_2 - y_1}{x_2 - x_1} = \frac{5 - 8}{-3 - 6} = \frac{-3}{-9} = \frac{1}{3}$$

Now use (x_1, y_1), here (6, 8) and point-slope form.

$$y - y_1 = m(x - x_1)$$
$$y - 8 = \frac{1}{3}(x - 6)$$
$$y - 8 = \frac{1}{3}x - 2$$
$$y = \frac{1}{3}x + 6 \quad \text{Slope-intercept form}$$
$$3y = x + 18$$
$$-x + 3y = 18$$
$$x - 3y = -18 \quad \text{Standard form}$$

Now Try:

7. Write the equation of the line passing through the point (7, 15) and (15, 9). Give the final answer in slope-intercept form and then in standard form.

____

Name: Date:
Instructor: Section:

Review this example for Objective 5:

8. The table shows the number of internet users in the world from 1998 to 2005, where year 0 represents 1998.

Year	Number of Internet Users (millions)
0	147
2	361
4	587
6	817
8	1093

Plot the data and find an equation that approximates it.

Letting y represent the number of internet users in year x, we plot the data.

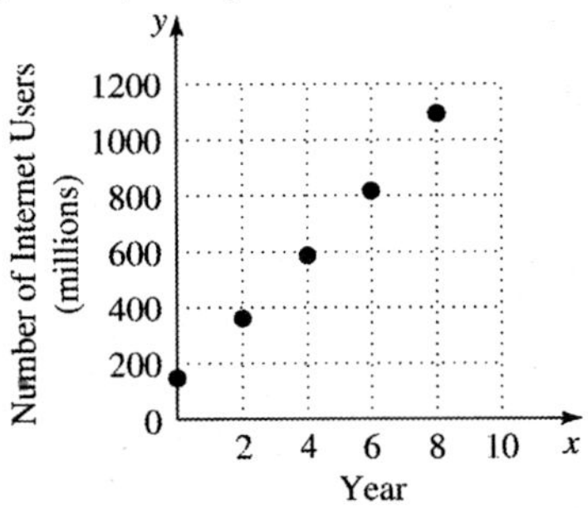

The points appear to lie approximately in a straight line. To find an equation of the line, we choose the ordered pairs (0, 147) and (8, 1093) from the table and find the slope of the line through these points.

$$(x_1, y_1) = (0, 147) \text{ and } (x_2, y_2) = (8, 1093)$$

$$\text{slope } m = \frac{y_2 - y_1}{x_2 - x_1} = \frac{1093 - 147}{8 - 0} = \frac{946}{8}$$

$$= 118.25$$

Use the slope, 118.25, and the point (0, 147) in slope-intercept form.

$$y = mx + b$$

$$147 = 118.25(0) + b$$

$$147 = b$$

Thus, $m = 118.25$ and $b = 147$, so the equation of the line is $y = 118.25x + 147$.

Now Try:

8. The table shows the average annual telephone expenditures for residential and pay telephones from 2001 to 2006, where year 0 represents 2001.

Year	Annual Telephone Expenditures
0	\$686
2	\$620
3	\$592
4	\$570
5	\$542

Plot the data and find an equation that approximates it.

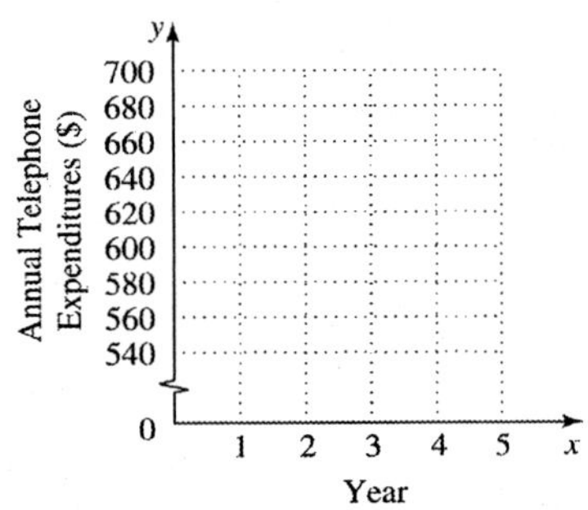

____

Name: Date:
Instructor: Section:

Objective 1 Use the slope-intercept form of the equation of a line.

For extra help, see Examples 1–2 on pages 764–765 of your text and Section Lecture video for Section 11.4 and Exercise Solutions Clip 11.

Identify the slope and y-intercept of the line with each equation.

1. $y = \frac{3}{2}x - \frac{2}{3}$ **1.** ______________

____

2. $y = -4x$ **2.** ______________

____

Write an equation of the line with the given slope and y-intercept.

3. slope –3; y-intercept (0, 3) **3.** ______________

Objective 2 Graph a line by using its slope and a point on the line.

For extra help, see Examples 3–4 on pages 765–766 of your text and Section Lecture video for Section 11.4 and Exercise Solutions Clip 29 and 35.

Graph each equation by using the slope and y-intercept.

4. $4x - y = 4$ **4.**

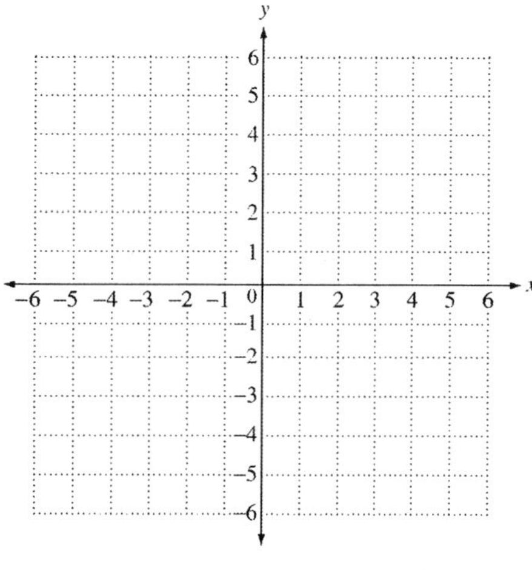

Name: Date:
Instructor: Section:

5. $y = -3x + 6$

5.

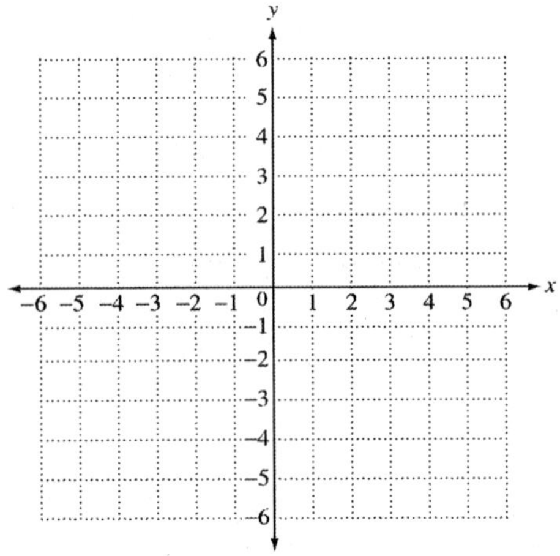

Graph the line passing through the given point and having the given slope.

6. $(-2, -2)$; $m = 0$

6.

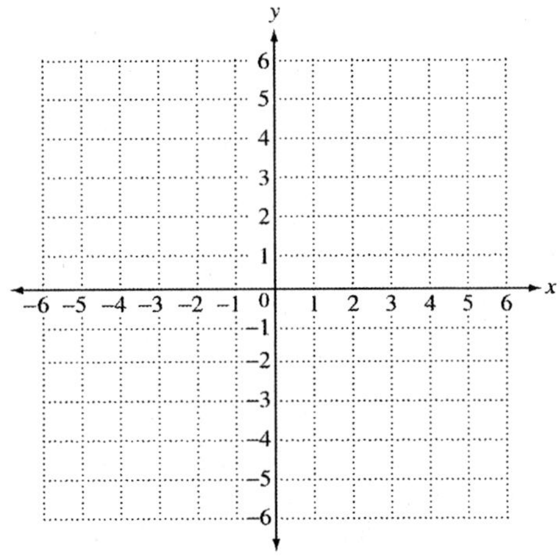

Objective 3 Write an equation of a line by using its slope and any point on the line.

For extra help, see Examples 5–6 on pages 767–768 of your text and Section Lecture video for Section 11.4 and Exercise Solutions Clip 35.

Write an equation for the line passing through the given point and having the given slope. Write the equations in slope-intercept form, if possible.

7. $(-3, 4)$; $m = -\frac{3}{5}$

7. ________________

8. $(-4,-3)$; $m=-2$ **8.** ________________

9. $(2,\ 2)$; $m=-\frac{3}{2}$ **9.** ________________

Objective 4 Write an equation of a line by using two points on the line.

For extra help, see Example 7 on page 769 of your text and Section Lecture video for Section 11.4 and Exercise Solutions Clip 50.

Write an equation for the line passing through each pair of points. Write the equations in standard form.

10. $(-2,\ 1)$ and $(3,\ 11)$ **10.** ________________

11. $(2,\ 3)$ and $(-2,-3)$ **11.** ________________

12. $(3,-4)$ and $(2,\ 7)$. **12.** ________________

Name: Date:
Instructor: Section:

Objective 5 Find an equation of a line that fits a data set.

For extra help, see Example 8 on page 770 of your text and Section Lecture video for Section 11.4.

Plot the data and find an equation that approximates it.

13. The table shows the U.S. municipal solid waste recycling percents since 1985, where year 0 represents 1985.

Year	Recycling Percent
0	10.1
5	16.2
10	26.0
15	29.1
20	32.5

13.

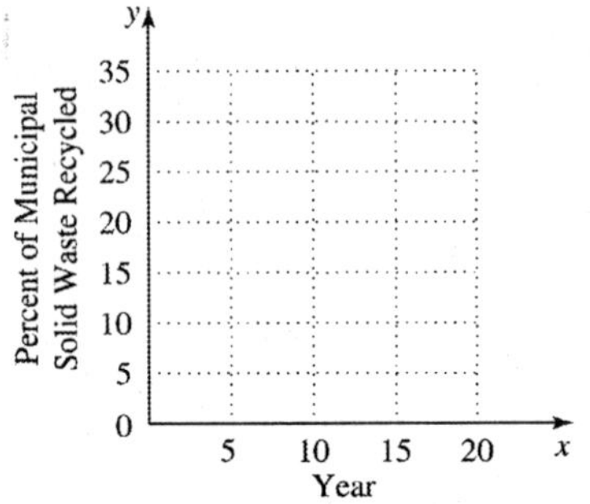

____

14. The table shows the approximate consumer expenditures for food in the U.S. in billions of dollars for selected years, where year 0 represents 1985.

Year	Food Expenditures (billions of dollars)
0	233
5	298
10	343
15	417
20	515

14.

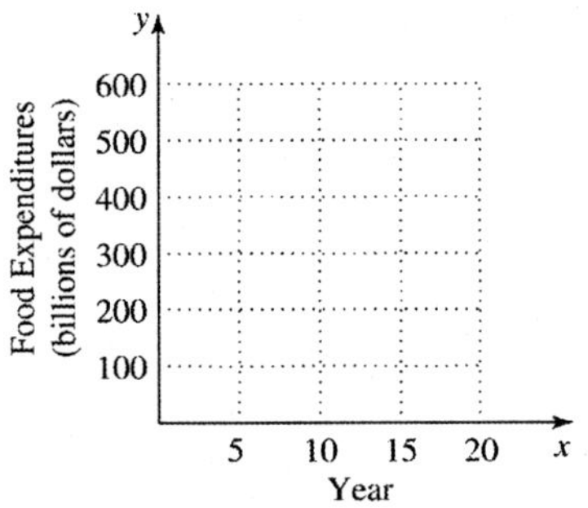

____

Name: Date:
Instructor: Section:

Chapter 11 GRAPHS OF LINEAR EQUATIONS AND INEQUALITIES IN TWO VARIABLES

11.5 Graphing Linear Inequalities in Two Variables

Learning Objectives

1. Graph linear inequalities in two variables.
2. Graph an inequality with a boundary line through the origin.

Key Terms

Use the vocabulary terms listed below to complete each statement in exercises 1–2.

linear inequality in two variables **boundary line**

1. In the graph of a linear inequality, the ____________________________ separates the region that satisfies the inequality from the region that does not satisfy the inequality.

2. An inequality that can be written in the form $Ax + By < C$, $Ax + By > C$, $Ax + By \le C$, or $Ax + By \ge C$ is called a ________________________________.

Guided Examples

Review these examples for Objective 1:

1. Graph $3x - 2y \le 6$.

The inequality $3x - 2y \le 6$ means that
$3x - 2y < 6$ or $3x - 2y = 6$.
We begin by graphing the line $3x - 2y = 6$ with intercepts (0, –3) and (2, 0). This boundary line divides the plane into two regions, one of which satisfies the inequality. We use the test point (0, 0) to see whether the resulting statement is true or false, thereby determining whether the point is in the shaded region or not.

$$3x - 2y \le 6$$
$$3(0) - 2(0) \overset{?}{\le} 6$$
$$0 - 0 \overset{?}{\le} 6$$
$$0 \le 6 \quad \text{True}$$

Since the last statement is true, we shade the region that includes the test point (0, 0). The shaded region, along with the boundary line, is the desired graph.

Now Try:

1. Graph $2x + 5y \le -8$.

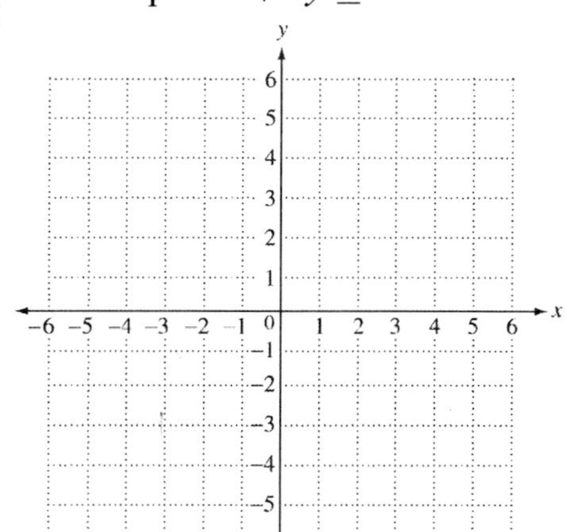

Name: Date:
Instructor: Section:

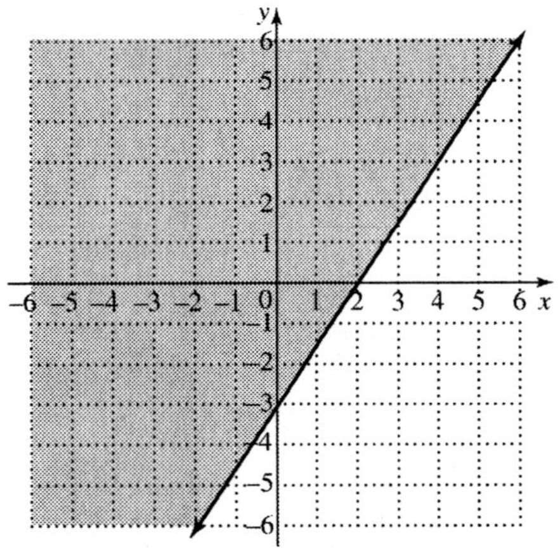

2. Graph $2x+5y>-10$.

This equation does not include equality. Therefore, the points on the line $2x+5y=-10$ do not belong to the graph. However, the line still serves as a boundary for two regions.

To graph the inequality, first graph the equation $2x+5y>-10$. Use a dashed line to show that the points on the line are not solutions of the inequality $2x+5y>-10$. Then choose a test point to see which region satisfies the inequality.

$$2x+5y>-10$$
$$2(0)+5(0)\overset{?}{>}-10$$
$$0+0\overset{?}{>}-10$$
$$0>-10 \text{ True}$$

Since 0 > –10 is true, the graph of the inequality is the region that contains (0, 0). Shade that region.

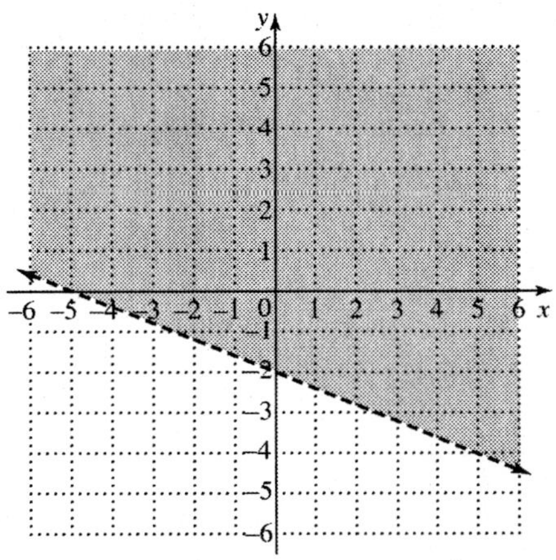

2. Graph $5x+4y>20$.

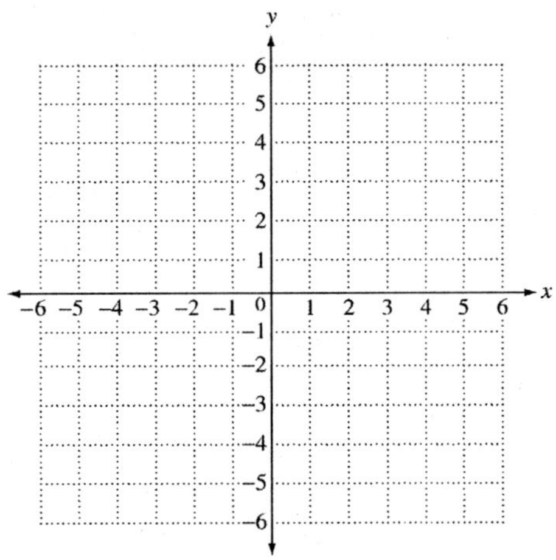

Name: Date:
Instructor: Section:

3. Graph $x - 4 \leq -1$.

First, solve the inequality for x.

$x \leq 3$

Now graph the line $x = 3$, a vertical line through the point (3, 0). Use a solid line, and choose (0, 0) as a test point.

$0 \leq 3$ True

Because $0 \leq 3$ is true, we shade the region containing (0, 0).

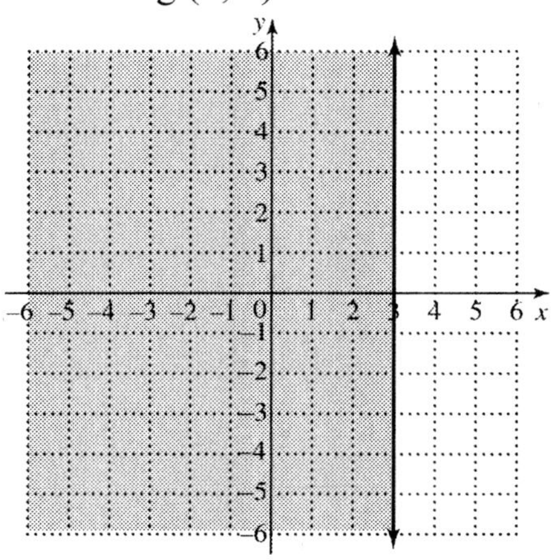

3. Graph $y \geq -1$.

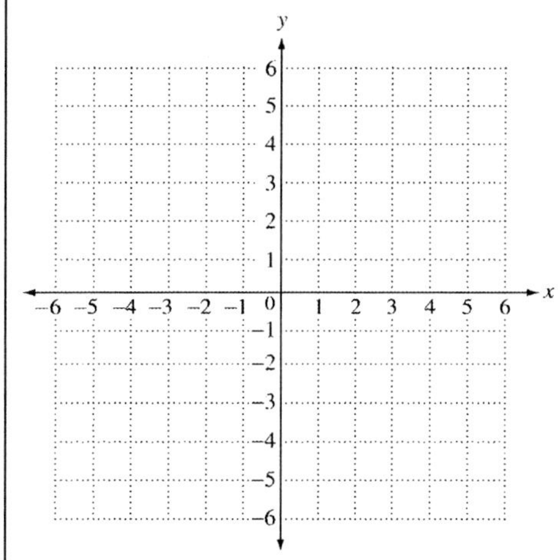

Review this example for Objective 2:

4. Graph $y \geq 3x$.

We graph $y = 3x$ using a solid line through (0, 0), (1, 3) and (2, 6). Because (0, 0) is on the line $y \geq 3x$, it cannot be used as a test point. Instead we choose a test point off the line, say (3, 0).

$0 \overset{?}{\geq} 3(3)$

$0 \geq 9$ False

Because $0 \geq 9$ is false, shade the other region.

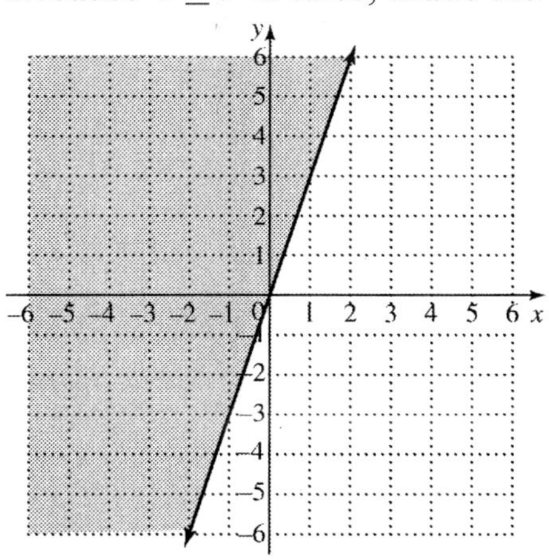

Now Try:

4. Graph $y \geq x$.

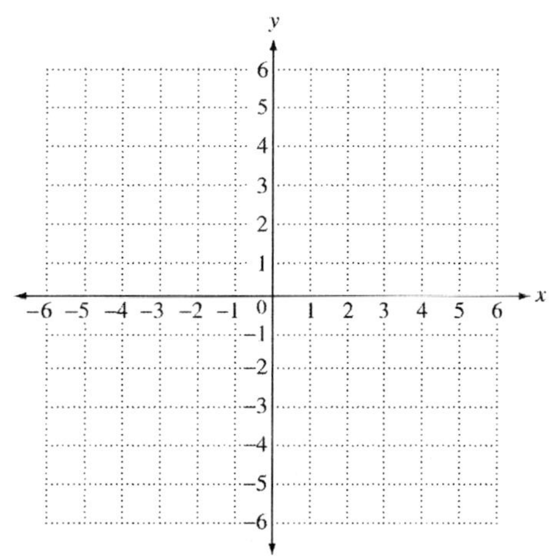

Name: Date:
Instructor: Section:

Objective 1 Graph linear inequalities in two variables.

For extra help, see Examples 1–3 on pages 780–782 of your text and Section Lecture video for Section 11.5 and Exercise Solutions Clip 27 and 31.

Graph each linear inequality.

1. $y \geq x - 1$

1.

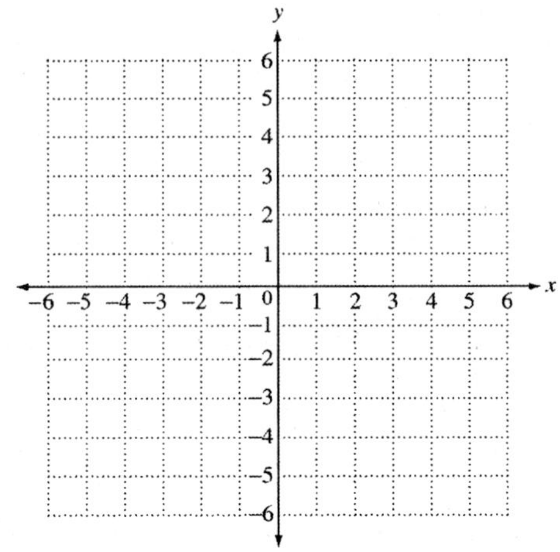

2. $y > -x + 2$

2.

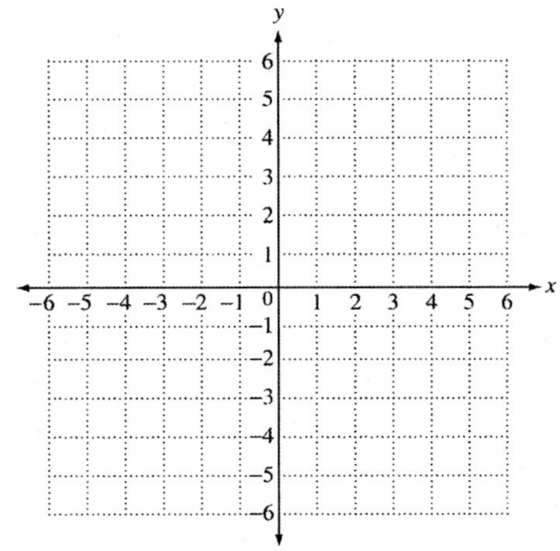

3. $3x - 4y - 12 > 0$

3.

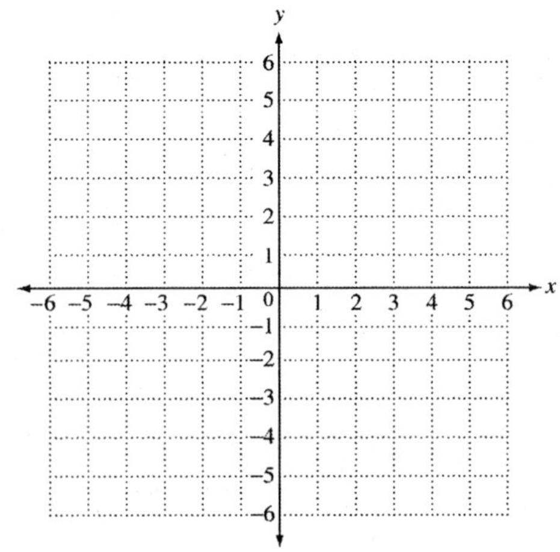

Name: Date:
Instructor: Section:

Objective 2 Graph an inequality with a boundary line through the origin.

For extra help, see Example 4 on page 782 of your text and Section Lecture video for Section 11.5 and Exercise Solutions Clip 35.

Graph each linear inequality.

4. $y \leq \frac{2}{5}x$

4.

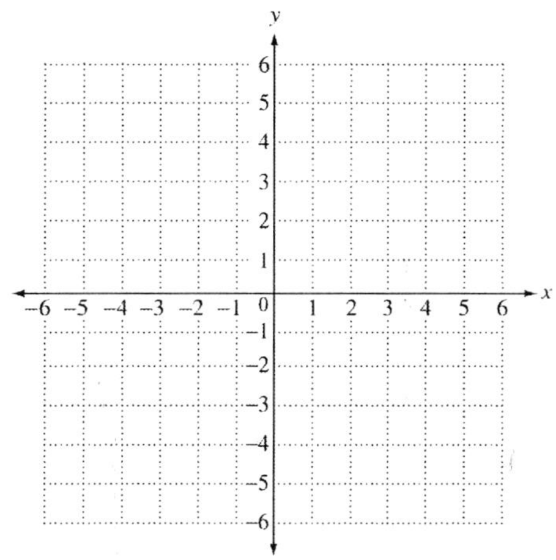

5. $y \geq \frac{1}{3}x$

5.

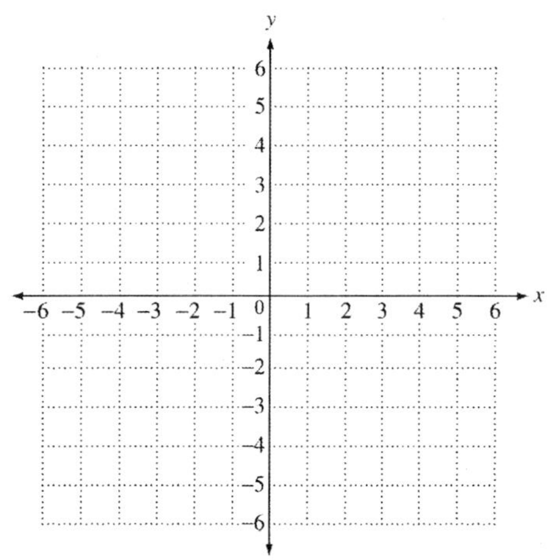

6. $x < -2y$

6.

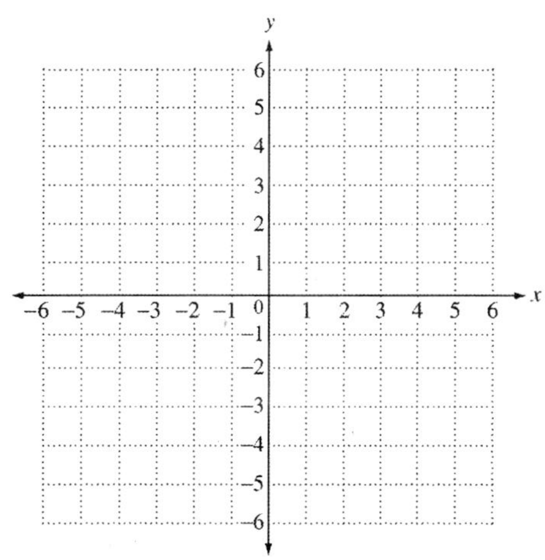

Name: Date:
Instructor: Section:

Chapter 12 SYSTEMS OF LINEAR EQUATIONS AND INEQUALITIES

12.1 Solving Systems of Linear Equations by Graphing

Learning Objectives

1. Decide whether a given ordered pair is a solution of a system.
2. Solve linear systems by graphing.
3. Solve special systems by graphing.
4. Identify special systems without graphing.

Key Terms

Use the vocabulary terms listed below to complete each statement in exercises 1–7.

system of linear equations	**solution of the system**
solution set of the system	**consistent system**
inconsistent system	**independent equations**
dependent equations	

1. Equations of a system that have different graphs are called ______________________________.

2. A system of equations with at least one solution is a ______________________________.

3. The set of all ordered pairs that are solutions of a system is the ______________________________.

4. The ______________________________ of linear equations is an ordered pair that makes all the equations of the system true at the same time.

5. Equations of a system that have the same graph (because they are different forms of the same equation) are called ______________________________.

6. A system with no solution is called a(n) ______________________________.

7. A(n) ______________________________ consists of two or more linear equations with the same variables.

Name: Date:
Instructor: Section:

Guided Examples

Review these examples for Objective 1:

1. Determine whether the ordered pair (5, –2) is a solution of each system.

a. $x+3y=-1$
$2x-4y=18$

To decide whether (5, –2) is a solution of the system, substitute 5 for x and –2 for y in each equation.

$x+3y=-1$	$2x-4y=18$
$5+3(-2)\stackrel{?}{=}-1$	$2(5)-4(-2)\stackrel{?}{=}18$
$5+(-6)\stackrel{?}{=}-1$	$10+8\stackrel{?}{=}18$
$-1=-1$ True	True $18=18$

Because (5, –2) satisfies both equations it is a solution of the system.

b. $4x+5y=10$
$3x+8y=6$

Again, substitute 5 for x and –2 for y in each equation.

$4x+5y=10$	$3x+8y=6$
$4(5)+5(-2)\stackrel{?}{=}10$	$3(5)+8(-2)\stackrel{?}{=}6$
$20-10\stackrel{?}{=}10$	$15-16\stackrel{?}{=}6$
$10=10$ True	False $-1=6$

The ordered pair (5, –2) is not a solution of this system because it does not satisfy the second equation.

Now Try:

1. Determine whether the ordered pair (6, 5) is a solution of each system.

a. $2x-3y=-3$
$4x+y=29$

____

b. $5x-6y=0$
$6x+5y=50$

____

Name: Date:
Instructor: Section:

Review this example for Objective 2:

2. Solve the system of equation by graphing both equations on the same axes.

$$6x - 5y = 4$$
$$2x - 5y = 8$$

Graph these equations by plotting several points for each line. To find the x-intercept, let $y = 0$. To find the y-intercept, let $x = 0$.
The tables show the intercepts and a check point for each graph.

$6x - 5y = 4$

x	y
0	$-\frac{4}{5}$
$\frac{2}{3}$	0
4	4

$2x - 5y = 8$

x	y
0	$\frac{8}{5}$
4	0
-6	-4

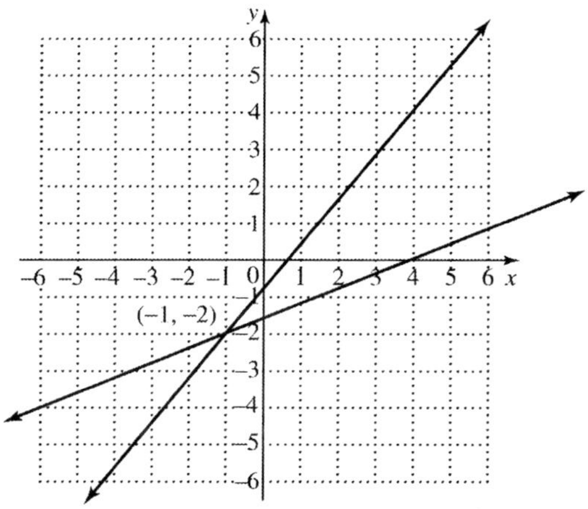

The lines suggest that the graphs intersect at the point (–1, –2). We check by substituting –1 for x and –2 for y in both equations.

$$6x - 5y = 4$$
$$6(-1) - 5(-2) \stackrel{?}{=} 4$$
$$-6 + 10 \stackrel{?}{=} 4$$
$$4 = 4 \quad \text{True}$$

$$2x - 5y = 8$$
$$2(-1) - 5(-2) \stackrel{?}{=} 8$$
$$-2 + 10 \stackrel{?}{=} 8$$
$$\text{True} \quad 8 = 8$$

Because (–1, –2) satisfies both equations, the solution set of this system is {(–1, –2)}.

Now Try:

2. Solve the system of equation by graphing both equations on the same axes.

$$3x - y = -7$$
$$2x + y = -3$$

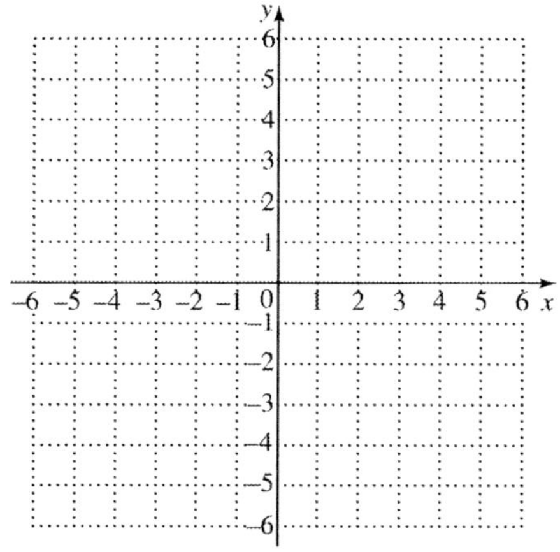

Name: Date:
Instructor: Section:

Review these examples for Objective 3:

3. Solve each system by graphing.

a. $x - y = 1$

$x - y = -1$

The graphs of these two equations are parallel and have no points in common. There is no solution for this system. The solution set is $\varnothing$.

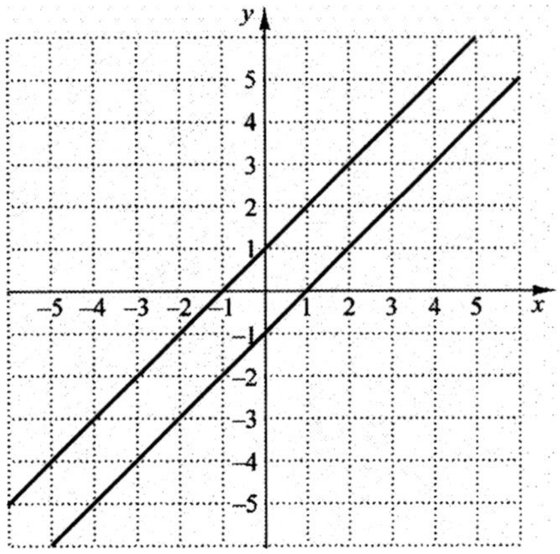

b. $3x - y = 0$

$2y = 6x$

The graphs of these two equations are the same line.

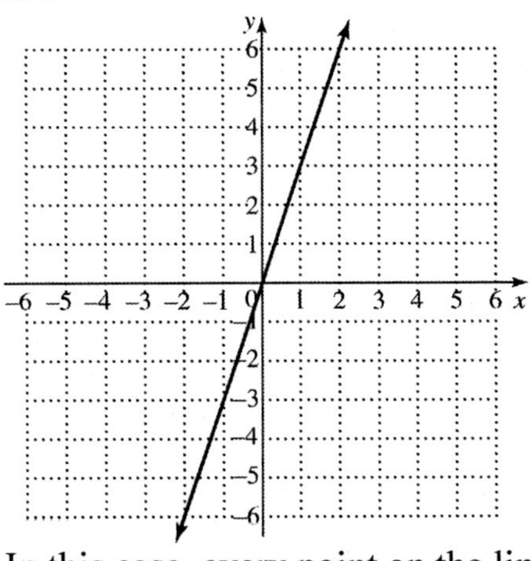

In this case, every point on the line is a solution of the system, and the solution set contains an infinite number of ordered pairs, each of which satisfies both equations of the system. We write the solution set as

$\{(x, y) | \ 3x - y = 0\}$.

Now Try:

3. Solve each system by graphing.

a. $4x - 2y = 8$

$6x - 3y = 12$

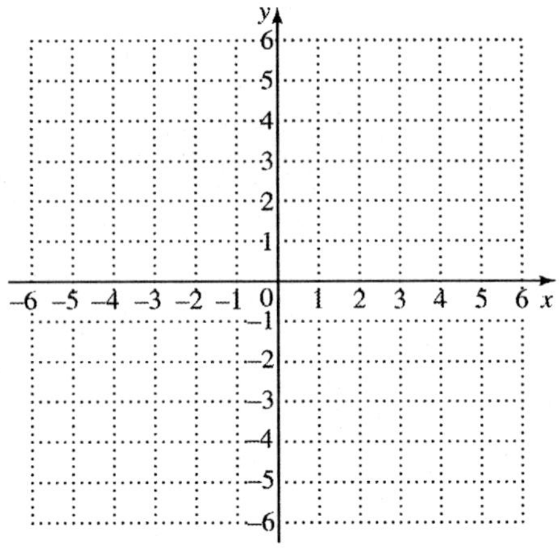

____

b. $x - 3y = 6$

$x - 3y = 4$

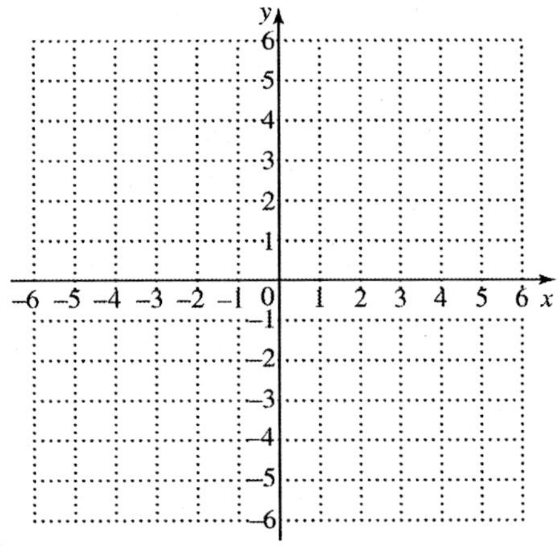

____

Name: Date:
Instructor: Section:

Review these examples for Objective 4:

4. Describe each system without graphing. State the number of solutions.

a. $5x+4y=20$
$-4y=5x+4$

Write each equation in slope-intercept form.

$5x+4y=20$	$-4y=5x+4$
$4y=-5x+20$	$y=-\frac{5}{4}x-1$
$y=-\frac{5}{4}x+5$	

Both equations have slope $-\frac{5}{4}$ but they have different y-intercepts, (0, 5) and (0, –1). These equations have graphs that are parallel lines. Thus, the system has no solution.

b. $3x-y=6$
$x=\frac{y}{3}+2$

Write each equation in slope-intercept form.

$3x-y=6$	$x=\frac{y}{3}+2$
$-y=-3x+6$	$\frac{y}{3}+2=x$
$y=3x-6$	$y=3x-6$

The equations are exactly the same – the graphs are the same line. Thus, the system has an infinite number of solutions.

c. $3x-4y=12$
$2x-3y=6$

Write each equation in slope-intercept form.

$3x-4y=12$	$2x-3y=6$
$-4y=-3x+12$	$-3y=-2x+6$
$y=\frac{3}{4}x-3$	$y=\frac{2}{3}x-2$

The graphs are neither parallel nor the same line, since the slopes are different. This system has exactly one solution.

Now Try:

4. Describe each system without graphing. State the number of solutions.

a. $7x-14y=35$
$3x-6y=25$

____

b. $6x+3y=18$
$10x+5y=30$

____

c. $x-6y=4$
$6x-y=9$

____

Name: Date:
Instructor: Section:

Objective 1 Decide whether a given ordered pair is a solution of a system.

For extra help, see Example 1 on page 798 of your text and Section Lecture video for Section 12.1 and Exercise Solutions Clip 11 and 13.

Decide whether the given ordered pair is a solution of the given system.

1. $(2,-4)$

$2x+3y=6$

$3x-2y=14$

1. ______________

2. $(-3,-1)$

$5x-3y=-12$

$2x+3y=-9$

2. ______________

3. $(4,\ 0)$

$4x+3y=16$

$x-4y=-4$

3. ______________

Objective 2 Solve linear systems by graphing.

For extra help, see Example 2 on page 799 of your text and Section Lecture video for Section 12.1 and Exercise Solutions Clip 19.

Solve each system by graphing both equations on the same axes.

4. $x-2y=6$

$2x+y=2$

4.

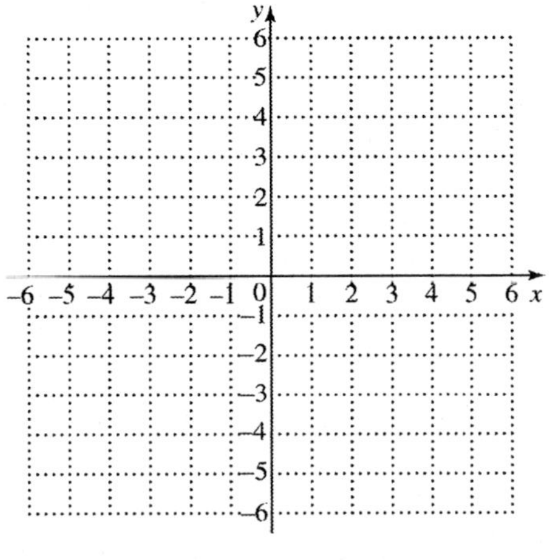

____

Name: Date:
Instructor: Section:

5. $2x = y$

$5x + 3y = 0$

5.

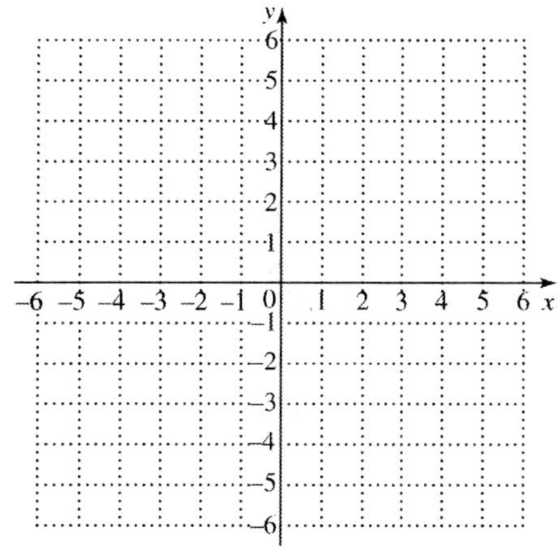

6. $3x + 2 = y$

$2x - y = 0$

6.

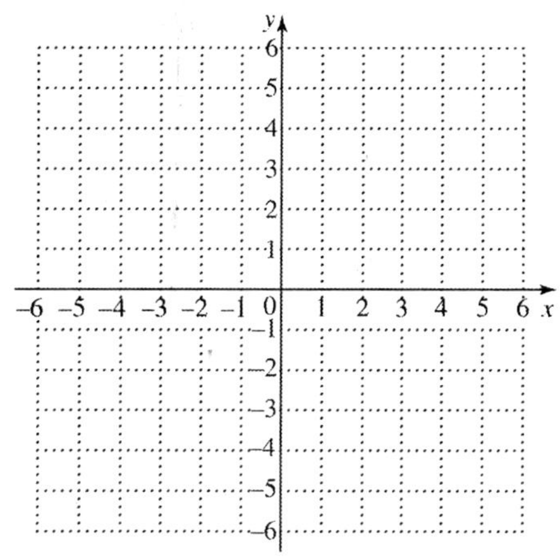

Name: Date:
Instructor: Section:

Objective 3 Solve special systems by graphing.

For extra help, see Example 3 on pages 800–801 of your text and Section Lecture video for Section 12.1 and Exercise Solutions Clip 29.

Solve each system of equations by graphing both equations on the same axes. If the two equations produce parallel lines, write ***no solution****. If the two equations produce the same line, write* ***infinite number of solutions****.*

7. $8x + 4y = -1$

$4x + 2y = 3$

7.

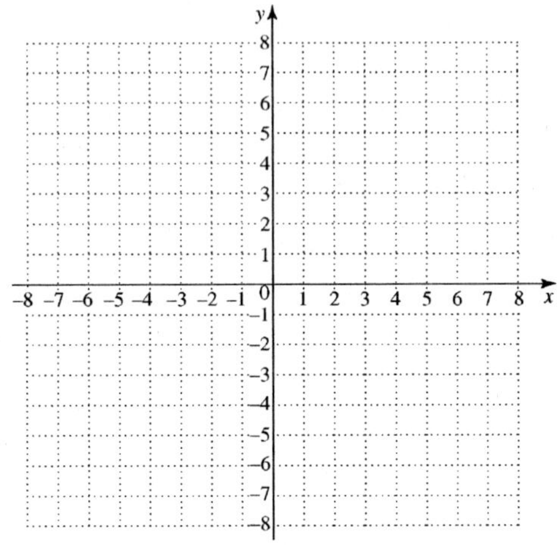

8. $-3x + 2y = 6$

$-6x + 4y = 12$

8.

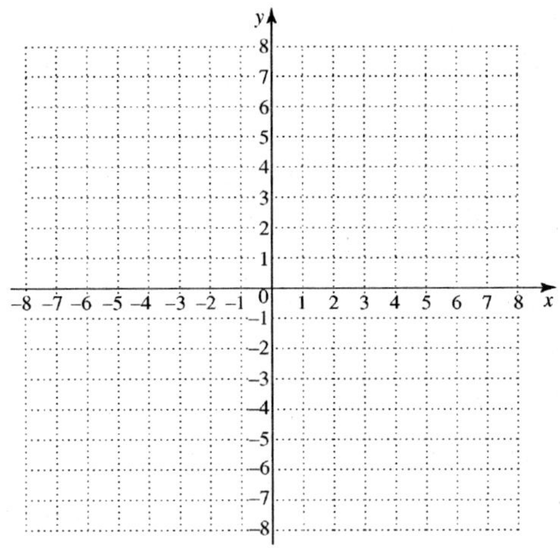

Name: Date:
Instructor: Section:

Objective 4 Identify special systems without graphing.

For extra help, see Example 4 on page 802 of your text and Section Lecture video for Section 12.1 and Exercise Solutions Clip 39.

Without graphing, answer the following equations for each linear system.
(*a*) *Is the system inconsistent, are the equations dependent, or neither?*
(*b*) *Is the graph a pair of intersecting lines, a pair of parallel lines, or one line?*
(*c*) *Does the system have one solution, no solution, or an infinite number of solutions?*

9. $y = 2x + 1$
$3x - y = 7$

9. (a)_______________

(b)_______________

(c)_______________

10. $-2x + y = 4$
$-4x + 2y = -2$

10. (a)_______________

(b)_______________

(c)_______________

11. $4x + 3y = 12$
$-12x = -36 + 9y$

11. (a)_______________

(b)_______________

(c)_______________

Name: Date:
Instructor: Section:

Chapter 12 SYSTEMS OF LINEAR EQUATIONS AND INEQUALITIES

12.2 Solving Systems of Linear Equations by Substitution

Learning Objectives	
1	Solve linear systems by substitution.
2	Solve special systems by substitution.
3	Solve linear systems with fractions and decimals.

Key Terms

Use the vocabulary terms listed below to complete each statement in exercises 1–4.

substitution **ordered pair** **inconsistent system**

dependent system

1. The solution of a linear system of equations is written as a(n) ________________.

2. When one expression is replaced by another, ____________________ is being used.

3. A system of equations in which all solutions of the first equation are also solutions of the second equation is a(n) ______________________________.

4. A system of equations that has no common solution is called a(n) ________________________.

Guided Examples

Review these examples for Objective 1:

1. Solve the system by the substitution method.

$$2x + 5y = 22 \quad (1)$$
$$y = 4x \quad (2)$$

Equation (2) is already solved for y. We substitute $4x$ for y in equation (1).

$$2x + 5y = 22$$
$$2x + 5(4x) = 22$$
$$2x + 20x = 22$$
$$22x = 22$$
$$x = 1$$

Find the value of y by substituting 1 for x in either equation. We use equation (2).

$$y = 4x$$
$$y = 4(1) = 4$$

Now Try:

1. Solve the system by the substitution method.

$$x + y = 7$$
$$y = 6x$$

____

We check the solution (1, 4) by substituting 1 for x and 4 for y in both equations.

$2x+5y=22$	$y=4x$
$2(1)+5(4)\stackrel{?}{=}22$	$4\stackrel{?}{=}4(1)$
$2+20\stackrel{?}{=}22$	True $4=4$
$22=22$ True	

Since (1, 4) satisfies both equations, the solution set of the system is $\{(1, 4)\}$.

2. Solve the system by the substitution method.

$$4x+5y=13 \quad (1)$$
$$x=-y+2 \quad (2)$$

Equation (2) gives x in terms of y. We substitute $-y+2$ for x in equation (1).

$$4x+5y=13$$
$$4(-y+2)+5y=13$$
$$-4y+8+5y=13$$
$$y+8=13$$
$$y=5$$

Find the value of x by substituting 5 for y in either equation. We use equation (2).

$$x=-y+2$$
$$x=-5+2=-3$$

We check the solution (–3, 5) by substituting –3 for x and 5 for y in both equations.

$4x+5y=13$	$x=-y+2$
$4(-3)+5(5)\stackrel{?}{=}13$	$-3\stackrel{?}{=}-5+2$
$-12+25\stackrel{?}{=}13$	True $-3=-3$
$13=13$ True	

Both results are true, so the solution set of the system is $\{(-3, 5)\}$.

3. Solve the system by the substitution method.

$$4x=5-y \quad (1)$$
$$7x+3y=15 \quad (2)$$

Step 1 Solve one of the equations for x or y. Solve equation (1) for y to avoid fractions.

$$4x=5-y$$
$$y+4x=5$$
$$y=-4x+5$$

2. Solve the system by the substitution method.

$$2x+3y=6$$
$$x=5-y$$

____

3. Solve the system by the substitution method.

$$2x+7y=2$$
$$3y=2-x$$

____

Step 2 Now substitute $-4x+5$ for y in equation (2).

$$7x+3y=15$$
$$7x+3(-4x+5)=15$$

Step 3 Solve the equation from Step 2.

$$7x-12x+15=15$$
$$-5x+15=15$$
$$-5x=0$$
$$x=0$$

Step 4 Equation (1) solved for y is $y=-4x+5$. Substitute 0 for x.

$$y=-4(0)+5=5$$

Step 5 Check that (0, 5) is the solution.

$4x=5-y$	$7x+3y=15$
$4(0)\stackrel{?}{=}5-5$	$7(0)+3(5)\stackrel{?}{=}15$
$0=0$ True	True $15=15$

Since both results are true, the solution set of the system is $\{(0, 5)\}$.

Review these examples for Objective 2:

4. Use substitution to solve the system.

$$x=7-3y \quad (1)$$
$$5x+15y=1 \quad (2)$$

Because equation (1) is already solved for x, we substitute $7-3y$ for x in equation (2).

$$5x+15y=1$$
$$5(7-3y)+15y=1$$
$$35-15y+15y=1$$
$$35=1 \quad \text{False}$$

A false result, here 35 = 1, means that the equations in the system have graphs that are parallel lines. The system is inconsistent and has no solution, so the solution set is ∅.

5. Use substitution to solve the system.

$$14x-7y=21 \quad (1)$$
$$-2x+y=-3 \quad (2)$$

Begin by solving equation (2) for y to get $y=2x-3$. Substitute $2x-3$ for y in

Now Try:

4. Use substitution to solve the system.

$$5x-10y=8$$
$$x=2y+5$$

____

5. Use substitution to solve the system.

$$5x+4y=20$$
$$-10x+40=8y$$

____

equation (1).

$$14x - 7y = 21$$
$$14x - 7(2x - 3) = 21$$
$$14x - 14x + 21 = 21$$
$$0 = 0$$

This true result means that every solution of one equation is also a solution of the other, so the system has an infinite number of solutions. The solution set is $\{(x, y) | 14x - 7y = 21\}$.

Review these examples for Objective 3:

6. Solve the system by the substitution method.

$$\frac{1}{2}x - y = 3 \quad (1)$$
$$\frac{1}{5}x + \frac{1}{2}y = \frac{3}{10} \quad (2)$$

Clear equation (1) of fractions by multiplying each side by 2.

$$2\left(\frac{1}{2}x - y\right) = 2(3)$$
$$2\left(\frac{1}{2}x\right) - 2y = 2(3)$$
$$x - 2y = 6$$

Clear equation (2) of fractions by multiplying each side by 10.

$$10\left(\frac{1}{5}x + \frac{1}{2}y\right) = 10\left(\frac{3}{10}\right)$$
$$10\left(\frac{1}{5}x\right) + 10\left(\frac{1}{2}y\right) = 10\left(\frac{3}{10}\right)$$
$$2x + 5y = 3$$

The given system of equations has been simplified to an equivalent system.

$$x - 2y = 6 \quad (3)$$
$$2x + 5y = 3 \quad (4)$$

To solve the system by substitution, solve equation (3) for x.

$$x - 2y = 6$$
$$x = 2y + 6$$

Now substitute the result for x in equation (4).

Now Try:

6. Solve the system by the substitution method.

$$x + \frac{1}{2}y = \frac{1}{2}$$
$$\frac{1}{2}x + \frac{1}{5}y = 0$$

____

$$2x+5y=3$$
$$2(2y+6)+5y=3$$
$$4y+12+5y=3$$
$$9y+12=3$$
$$9y=-9$$
$$y=-1$$

Substitute –1 for y in $x=2y+6$ (equation (3) solved for x).

$$x=2(-1)+6=4$$

Check (4, –1) in both of the original equations. The solution set is $\{(4,-1)\}$.

7. Solve the system by the substitution method.

$$0.4x+2.5y=8 \quad (1)$$
$$-0.1x+3.5y=14.5 \quad (2)$$

Clear each equation of decimals, by multiplying by 10.

$$0.4x+2.5y=8$$
$$10(0.4x+2.5y)=10(8)$$
$$10(0.4x)+10(2.5y)=10(8)$$
$$4x+25y=80$$

$$-0.1x+3.5y=14.5$$
$$10(-0.1x+3.5y)=10(14.5)$$
$$10(-0.1x)+10(3.5y)=10(14.5)$$
$$-x+35y=145$$

Now solve the equivalent system of equations by substitution.

$$4x+25y=80 \quad (3)$$
$$-x+35y=145 \quad (4)$$

Equation (4) can be solved for x.

$$x=35y-145$$

Substitute this result for x in equation (3).

$$4x+25y=80$$
$$4(35y-145)+25y=80$$
$$140y-580+25y=80$$
$$165y-580=80$$
$$165y=660$$
$$y=4$$

7. Solve the system by the substitution method.

$$0.6x+0.8y=-2.2$$
$$0.7x-0.1y=2.6$$

____

Name: Date:
Instructor: Section:

Since equation (4) solved for x is $x = 35y - 145$, substitute 4 for y.

$x = 35(4) - 145 = -5$

Check (–5, 4) in both of the original equations.
The solution set is $\{(-5, 4)\}$.

Objective 1 Solve linear systems by substitution.

For extra help, see Examples 1–3 on pages 807–809 of your text and Section Lecture video for Section 12.2 and Exercise Solutions Clip 5, 7, and 9.

Solve each system by the substitution method. Check each solution.

1. $3x + 2y = 14$
$y = x + 2$

1. ________________

2. $x + y = 9$
$5x - 2y = -4$

2. ________________

3. $3x - 21 = y$
$y + 2x = -1$

3. ________________

Objective 2 Solve special systems by substitution.

For extra help, see Examples 4–5 on pages 809–810 of your text and Section Lecture video for Section 12.2 and Exercise Solutions Clip 17 and 19.

Solve each system by the substitution method. Use set-builder notation for dependent equations.

4. $y = -\frac{1}{3}x + 5$
$3y + x = -9$

4. ________________

Name: Date:
Instructor: Section:

5. $\frac{1}{2}x+3=y$

$6=-x+2y$

5. ________________

6. $4x+3y=2$

$8x+6y=6$

6. ________________

Objective 3 Solve linear systems with fractions and decimals.

For extra help, see Examples 6–7 on pages 811–812 of your text and Section Lecture video for Section 12.2 and Exercise Solutions Clip 23 and 25.

Solve each system by the substitution method. Check each solution.

7. $\frac{5}{4}x-y=-\frac{1}{4}$

$-\frac{7}{8}x+\frac{5}{8}y=1$

7. ________________

8. $\frac{1}{4}x+\frac{3}{8}y=-3$

$\frac{5}{6}x-\frac{3}{7}y=-10$

8. ________________

9. $0.6x+0.8y=1$

$0.4y=0.5-0.3x$

9. ________________

Name: Date:
Instructor: Section:

Chapter 12 SYSTEMS OF LINEAR EQUATIONS AND INEQUALITIES

12.3 Solving Systems of Linear Equations by Elimination

Learning Objectives

1. Solve linear systems by elimination.
2. Multiply when using the elimination method.
3. Use an alternative method to find the second value in a solution.
4. Solve special systems by elimination.

Key Terms

Use the vocabulary terms listed below to complete each statement in exercises 1–3.

addition property of equality **elimination method** **substitution**

1. Using the addition property to solve a system of equations is called the ____________________________.

2. The ______________________________ states that the same added quantity to each side of an equation results in equal sums.

3. ________________________ is being used when one expression is replaced by another.

Guided Examples

Review these examples for Objective 1:

1. Use the elimination method to solve the system.

$$x + y = 6 \quad (1)$$
$$-x + y = 4 \quad (2)$$

Add the equations vertically.

$$\begin{array}{rl} x + y = 6 & (1) \\ \underline{-x + y = 4} & (2) \\ 2y = 10 & \\ y = 5 & \end{array}$$

To find the x-value, substitute 5 for y in either of the two equations of the system. We choose equation (1).

$$x + y = 6$$
$$x + 5 = 6$$
$$x = 1$$

Check the solution (1, 5), by substituting 1 for x

Now Try:

1. Use the elimination method to solve the system.

$$x + y = 11$$
$$x - y = 5$$

____

and 5 for y in both equations of the given system.

$x+y=6$	$-x+y=4$
$1+5\stackrel{?}{=}6$	$-1+5\stackrel{?}{=}4$
$6=6$ True	True $4=4$

Since both results are true, the solution set of the system is $\{(1, 5)\}$.

2. Solve the system.

$$x-11=2y \quad (1)$$
$$-5y=x+3 \quad (2)$$

Step 1 Write both equations in standard form.

$$x-2y=11$$
$$-x-5y=3$$

Step 2 Because the coefficients of x are 1 and –1, adding will eliminated x. It is not necessary to multiply either equation by a number.

Step 3 Add the two equations.

$$\begin{array}{rl} x-2y=11 & (1) \\ -x-5y=3 & (2) \\ \hline -7y=14 & \end{array}$$

Step 4 Solve.

$$y=-2$$

Step 5 Find the value of x by substituting –2 for y in either of the original equations.

$$x-11=2y$$
$$x-11=2(-2)$$
$$x-11=-4$$
$$x=7$$

Step 6 Check the ordered-pair solution (7, –2) in both of the original equations.

$x-11=2y$	$-5y=x+3$
$7-11\stackrel{?}{=}2(-2)$	$-5(-2)\stackrel{?}{=}7+3$
$-4=-4$ True	True $10=10$

Since (7, –2) is a solution of both equations, the solution set is $\{(7, -2)\}$.

2. Solve the system.

$$5x+4y=50$$
$$3x-4y=-2$$

____

Name: Date:
Instructor: Section:

Review this example for Objective 2:

3. Solve the system.

$$3x+8y=-2 \quad (1)$$
$$2x+7y=2 \quad (2)$$

To eliminate x, multiply equation (1) by 2 and multiply equation (2) by –3. Then add.

$$\begin{aligned} 6x+16y&=-4 \\ -6x-21y&=-6 \\ \hline -5y&=-10 \\ y&=2 \end{aligned}$$

Substituting 2 for y in either equation (1) or (2) gives $x = -6$. Check that the solution set of the system is $\{(-6, 2)\}$.

Now Try:

3. Solve the system.

$$3x+4y=24$$
$$4x+3y=11$$

____

Review this example for Objective 3:

4. Solve the system.

$$6x=7-3y \quad (1)$$
$$8x-5y=5 \quad (2)$$

Write equation (1) in standard form.

$$6x+3y=7 \quad (3)$$
$$8x-5y=5 \quad (4)$$

One way to proceed is to eliminate y by multiplying each side of equation (3) by 5 and each side of equation (4) by 3, and then adding.

$$\begin{aligned} 30x+15y&=35 \\ 24x-15y&=15 \\ \hline 54x\quad\quad&=50 \end{aligned}$$

$$x=\frac{50}{54} \text{ or } \frac{25}{27}$$

Substituting $\frac{25}{27}$ for x in one of the given equations would give y, but the arithmetic would be complicated. Instead, solve for y by starting again with the original equations in standard form, and eliminating x. Multiply equation (3) by 4 and equation (4) by –3.

$$\begin{aligned} 24x+12y&=28 \\ -24x+15y&=-15 \\ \hline 27y&=13 \\ y&=\frac{13}{27} \end{aligned}$$

The solution set is $\left\{\left(\frac{25}{27}, \frac{13}{27}\right)\right\}$.

Now Try:

4. Solve the system.

$$8x=5y+1$$
$$6x-8y=-2$$

____

Name: Date:
Instructor: Section:

Review these examples for Objective 4:

5. Solve each system by the elimination method.

a. $5x+10y=9$ (1)

$3x+6y=8$ (2)

Multiply each side of equation (1) by 3 and each side of equation (2) by –5.

$$\begin{aligned} 15x+30y&=27 \\ -15x-30y&=-40 \\ \hline 0&=-13 \quad \text{False} \end{aligned}$$

The false statement 0 = –13 indicates that the system has solution set ∅.

b. $7x+y=9$ (1)

$-14x-2y=-18$ (2)

Multiply each side of equation (1) by 2.

$$\begin{aligned} 14x+2y&=18 \\ -14x-2y&=-18 \\ \hline 0&=0 \quad \text{True} \end{aligned}$$

A true statement occurs when the equations are equivalent. This indicates that every solution of one equation is also a solution of the other. The solution set is $\{(x, y) \mid 7x+y=9\}$.

Now Try:

5. Solve each system by the elimination method.

a. $2x+6y=5$

$5x+15y=8$

____

b. $9x-7y=5$

$18x=14y+10$

____

Objective 1 Solve linear systems by elimination.

For extra help, see Examples 1–2 on pages 815–816 of your text and Section Lecture video for Section 12.3 and Exercise Solutions Clip 5.

Solve each system by the elimination method. Check your answers.

1. $x-4y=-4$

$-x+y=-5$

1. _______________

2. $2x-y=10$

$3x+y=10$

2. _______________

Name: Date:
Instructor: Section:

3. $x - 3y = 5$
$-x + 4y = -5$

3. ________________

Objective 2 Multiply when using the elimination method.

For extra help, see Example 3 on page 817 of your text and Section Lecture video for Section 12.3 and Exercise Solutions Clip 13 and 17.

Solve each system by the elimination method. Check your answers.

4. $6x + 7y = 10$
$2x - 3y = 14$

4. ________________

5. $8x + 6y = 10$
$4x - y = 1$

5. ________________

6. $6x + y = 1$
$3x - 4y = 23$

6. ________________

Objective 3 Use an alternative method to find the second value in a solution.

For extra help, see Example 4 on pages 817–818 of your text and Section Lecture video for Section 12.3 and Exercise Solutions Clip 29 and 32.

Solve each system by the elimination method. Check your answers.

7. $4x - 3y - 20 = 0$
$6x + 5y + 8 = 0$

7. ________________

Name: Date:
Instructor: Section:

8. $6x = 16 - 7y$
$4x = 3y + 26$

8. ________________

9. $2x = 14 + 4y$
$6y = -5x + 3$

9. ________________

Objective 4 Solve special systems by elimination.

For extra help, see Example 5 on page 818 of your text and Section Lecture video for Section 12.3 and Exercise Solutions Clip 25.

Solve each system by the elimination method. Use set-builder notation for dependent equations. Check your answers.

10. $12x - 8y = 3$
$6x - 4y = 6$

10. ________________

11. $2x + 4y = -6$
$-x - 2y = 3$

11. ________________

12. $15x + 6y = 9$
$10x + 4y = 18$

12. ________________

Name: Date:
Instructor: Section:

Chapter 12 SYSTEMS OF LINEAR EQUATIONS AND INEQUALITIES

12.4 Applications of Linear Systems

Learning Objectives	
1	Solve problems about unknown numbers.
2	Solve problems about quantities and their costs.
3	Solve problems about mixtures.
4	Solve problems about distance, rate (or speed), and time.

Key Terms

Use the vocabulary terms listed below to complete each statement in exercises 1−2.

system of linear equations $d = rt$

1. The formula that relates distance, rate, and time is ____________________.

2. A ____________________ consists of at least two linear equations with different variables.

Guided Examples

Review this example for Objective 1:

1. Two towns have a combined population of 9045. There are 2249 more people living in one than in the other. Find the population in each town.

Step1 Read the problem carefully. We are to find the population of each town.

Step 2 Assign variables. Let x = the population of the larger town, and y = the population of smaller town.

Step 3 Write two equations. There are 2249 more people living in one town. The total population is 9045.

$$x = y + 2249 \quad (1)$$

$$x + y = 9045 \quad (2)$$

Step 4 Solve the system. We use substitution. Substitute y + 2249 for x in equation (2).

Now Try:

1. A rope 82 centimeters long is cut into two pieces with one piece four more than twice as long as the other. Find the length of each piece.

____

$$x + y = 9045$$
$$(y + 2249) + y = 9045$$
$$2249 + 2y = 9045$$
$$2y = 6796$$
$$y = 3398$$

To find x, substitute 3398 into either original equation. We use equation (1).

$$x = 3398 + 2249 = 5647$$

Step 5 State the answer. The population of the larger town is 5647. The population of the smaller town is 3398.

Step 6 Check the answer in the original problem. $3398 + 2249 = 5647$ and $5647 + 3398 = 9045$ The answers check.

Review this example for Objective 2:

2. The total receipts for a basketball game were \$4690.50. There were 723 tickets sold, some for children and some for adults. If the adult tickets cost \$9.50 and the children's tickets cost \$4, how many of each type were there?

Step1 Read the problem.

Step 2 Assign variables. Let x = the number of adult tickets and y = the number of children's tickets.

Step 3 Write two equations. The total number of tickets is 723. The total value of the tickets is \$4690.50.

$$x + y = 723 \quad (1)$$
$$9.50x + 4y = 4690.5 \quad (2)$$

Step 4 Solve the system. Solve using elimination. Multiply equation (1) by –4.

$$-4x - 4y = -2892$$
$$\underline{9.50x + 4y = 4690.5}$$
$$5.50x \qquad = 1798.50$$
$$x = 327$$

Substitute 327 for x in equation (1) to find y.

$$x + y = 723$$
$$327 + y = 723$$
$$y = 396$$

Now Try:

2. Twice as many general admission tickets to a basketball game were sold as reserved seat tickets. General admission tickets cost \$10 and reserved seat tickets cost \$15. If the total value of both kinds of tickets was \$26,250, how many tickets of each kind were sold?

____

Step 5 State the answer. The number of adult tickets is 327 and the number of children's tickets is 396.

Step 6 Check. The sum of the tickets is 723. The value of the tickets is

$$9.50(327)+4(396)=4690.50$$

This checks.

Review this example for Objective 3:

3. A mixture of 75% solution should be mixed with a 55% solution to get 70 liters of 63% solution. Determine the number of liters required of the 55% and 75% solutions.

Step1 Read the problem carefully.

Step 2 Assign variables. Let x = the number of liters of the 75% liquid and y = the number of liters of the 55% liquid.

Step 3 Write two equations. The total amount of liquid of the final mixture is 70 liters. The amount of 75% solution mixed with 55% solution will equal the 70 liters of 63% solution.

$$x+y=70 \quad (1)$$
$$0.75x+0.55y=0.63(70) \quad (2)$$

Step 4 Solve the system. Solve by substitution. Solving equation (1) for x results in $70-y$.

$$0.75x+0.55y=0.63(70)$$
$$0.75(70-y)+0.55y=44.1$$
$$52.5-0.75y+0.55y=44.1$$
$$-0.20y+52.5=44.1$$
$$-0.20y=-8.4$$
$$y=42$$

Substitute 42 for y in equation (1).

$$x=70-42=28$$

Step 5 State the answer. There should be 28 liters of 75% solution and 42 liters of 55% solution.

Step 6 Check the answer in the original problems. 28 + 42 = 70 and $0.75(28)+0.55(42)=44.1$ The answer checks.

Now Try:

3. A pharmacist wants to add water to a solution that contains 80% medicine. She wants to obtain 12 oz. of a solution that is 20% medicine. How much water and how much of the 80% solution should she use?

____

Name: Date:
Instructor: Section:

Review these examples for Objective 4:

4. Bill and Hillary start in Washington and fly in opposite directions. At the end of 4 hours, they are 4896 kilometers apart. If Bill flies 60 kilometers per hour faster than Hillary, what are their speeds?

Step1 Read the problem carefully.

Step 2 Assign variables. Let x = Bill's rate of speed and y = Hillary's rate of speed.

Step 3 Write two equations.

$$x = 60 + y \quad (1)$$
$$4x + 4y = 4896 \quad (2)$$

Step 4 Solve the system. . Solve by substitution.

$$4(60 + y) + 4y = 4896$$
$$240 + 4y + 4y = 4896$$
$$240 + 8y = 4896$$
$$8y = 4656$$
$$y = 582$$

Substitute 582 for y in equation (1).

$$x = 60 + 582 = 642$$

Step 5 State the answer. Bill's rate is 642 kmh and Hillary's rate is 582 kmh.

Step 6 Check. Since $4(642) + 4(582) = 4896$ and $642 = 582 + 60$ the answers check.

5. It takes a kayak $1\frac{1}{2}$ hours to go 24 miles downstream and 4 hours to return. Find the speed of the current and the speed of the kayak in still water.

Step1 Read the problem carefully.

Step 2 Assign variables. Let x = the speed of the kayak and y = the speed of the current.

Step 3 Write two equations. When going downstream, the speed of the current is added to the speed of the kayak. When going upstream (the return), the speed of the current is subtracted from the speed of the kayak. Also, rate times time equals distance.

Now Try:

4. Enid leaves Cherry Hill, driving by car toward New York, which is 90 miles away. At the same time, Jerry, riding his bicycle, leaves New York cycling toward Cherry Hill. Enid is traveling 28 miles per hour faster than Jerry. They pass each other $1\frac{1}{2}$ hours later. What are their speeds?

____

5. A plane can travel 300 miles per hour with the wind and 230 miles per hour against the wind. Find the speed of the wind and the speed of the plane in still air.

____

Name: Date:
Instructor: Section:

$$\frac{3}{2}(x+y)=24$$
$$4(x-y)=24$$

Dividing the first equation by $\frac{3}{2}$ and the second by 4, we have our system of equations.

$$x+y=16 \quad (1)$$
$$x-y=6 \quad (2)$$

Step 4 Solve the system. We solve using elimination.

$$\begin{aligned} x+y&=16 \\ x-y&=6 \\ \hline 2x\quad&=22 \\ x&=11 \end{aligned}$$

Since $x+y=16$ and $x=11$, it follows that $y=5$.

Step 5 State the answer. The speed of the kayak is 11 mph and the speed of the current is 5 mph.

Step 6 Check. The answers seem reasonable, and true statements result when the values are substituted in to the equations of the system.

Objective 1 Solve problems about unknown numbers.

For extra help, see Example 1 on page 824 of your text and Section Lecture video for Section 12.4.

Write a system of equations for each problem, then solve the problem.

1. The difference between two numbers is 14. If two times the smaller is added to one-half the larger, the result is 52. Find the numbers.

1.
larger number__________
smaller number__________

2. There are a total of 49 students in the two second grade classes at Jefferson School. If Carla has 7 more students in her class than Linda, find the number of students in each class.

2.
Carla's class __________
Linda's class__________

Name: Date:
Instructor: Section:

3. The perimeter of a rectangular room is 50 feet. The length is three feet greater than the width. Find the dimensions of the rectangle.

3.
length ______________
width______________

Objective 2 Solve problems about quantities and their costs.

For extra help, see Example 2 on pages 825–826 of your text and Section Lecture video for Section 12.4 and Exercise Solutions Clip 21.

Write a system of equations for each problem, then solve the problem.

4. There were 411 tickets sold for a soccer game, some for students and some for nonstudents. Student tickets cost \$4.25 and nonstudent tickets cost \$8.50 each. The total receipts were \$3021.75. How many of each type were sold?

4.
student tix____________
nonstudent tix_________

5. A cashier has some \$5 bills and some \$10 bills. The total value of the money is \$750. If the number of tens is equal to twice the number of fives, how many of each type are there?

5.
\$5 bills ______________
\$10 bills _____________

6. Luke plans to buy 10 ties with exactly \$162. If some ties cost \$14, and the others cost \$25, how many ties of each price should he buy?

6.
\$14 ties______________
\$25 ties______________

Name: Date:
Instructor: Section:

Objective 3 Solve problems about mixtures.

For extra help, see Example 3 on pages 826–827 of your text and Section Lecture video for Section 12.4 and Exercise Solutions Clip 29.

Write a system of equations for each problem, then solve the problem.

7. Jorge wishes to make 150 pounds of coffee blend that can be sold for \$8 per pound. The blend will be a mixture of coffee worth \$6 per pound and coffee worth \$12 per pound. How many pounds of each kind of coffee should be used in the mixture?

7.
\$6 coffee____________
\$12 coffee____________

8. How many liters of water should be added to 25% antifreeze solution to get 30 liters of a 20% solution? How many liters of 25% solution are needed?

8.
water____________
25% solution ____________

9. Ben wishes to blend candy selling for \$1.60 a pound with candy selling for \$2.50 a pound to get a mixture that will be sold for \$1.90 a pound. How many pounds of the \$1.60 and the \$2.50 candy should be used to get 30 pounds of the mixture?

9.
\$1.60 candy ____________
\$2.50 candy ____________

Name: Date:
Instructor: Section:

Objective 4 Solve problems about distance, rate (or speed), and time.

For extra help, see Examples 4–5 on pages 828–829 of your text and Section Lecture video for Section 12.4 and Exercise Solutions Clip 35 and 39.

Write a system of equations for each problem, and then solve the problem.

10. It takes Carla's boat $\frac{1}{2}$ hour to go 8 miles downstream and 1 hour to make the return trip upstream. Find the speed of the current and the speed of Carla's boat in still water.

10.
boat speed____________
current speed __________

11. Two planes left Philadelphia traveling in opposite directions. Plane A left 15 minutes before plane B. After plane B had been flying for 1 hour, the planes were 860 miles apart. What were the speeds of the two planes if plane A was flying 40 miles per hour faster than plane B?

11.
plane A____________
plane B ____________

12. At the beginning of a fund-raising walk, Steve and Vic are 30 miles apart. If they leave at the same time and walk in the same direction, Steve would overtake Vic in 15 hours. If they walked toward each other, they would meet in 3 hours. What are their speeds?

12.
Steve____________
Vic ____________

Name: Date:
Instructor: Section:

Chapter 12 SYSTEMS OF LINEAR EQUATIONS AND INEQUALITIES

12.5 Solving Systems of Linear Inequalities

Learning Objectives
1 Solve systems of linear inequalities by graphing.

Key Terms

Use the vocabulary terms listed below to complete each statement in exercises 1–2.

system of linear inequalities

solution set of a system of linear inequalities

1. All ordered pairs that make all inequalities of the system true at the same time is called the ______________________________.

2. A ______________________________ contains two or more linear inequalities (and no other kinds of inequalities).

Guided Examples

Review this example for Objective 1:

1. Graph the solution set of the system.

$$x + y \leq 3$$
$$5x - y \geq 5$$

To graph $x + y \leq 3$, graph the solid boundary line $x + y = 3$ using the intercepts (0, 3) and (3, 0). Determine the region to shade using (0, 0) as a test point.

$$x + y \leq 3$$
$$0 + 0 \overset{?}{\leq} 3$$
$$0 \leq 3 \quad \text{True}$$

Shade the region containing (0, 0).

To graph $5x - y \geq 5$, graph the solid boundary line $5x - y = 5$ using the intercepts (0, –5) and (1, 0). Determine the region to shade using (0, 0) as a test point.

$$5x - y \geq 5$$
$$5(0) - 0 \overset{?}{\geq} 5$$
$$0 \geq 5 \quad \text{False}$$

Now Try:

1. Graph the solution set of the system.

$$3x - y \leq 3$$
$$x + y \leq 0$$

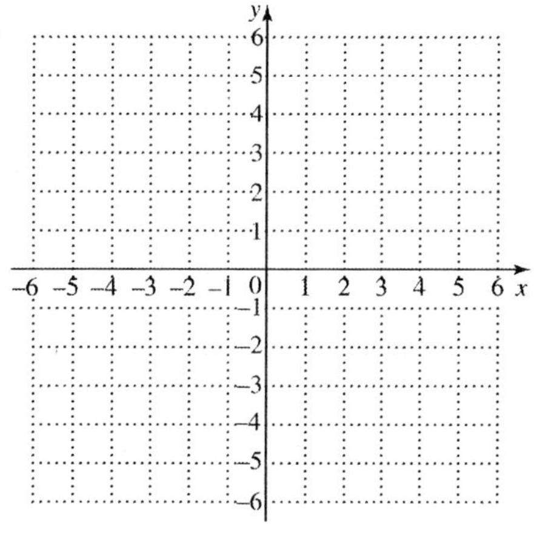

Name: Date:
Instructor: Section:

Shade the region that does not contain (0, 0).

The solution set of this system includes all points in the intersection (overlap) of the graph of the two inequalities. This intersection is the gray shaded region and portions of the two boundary lines that surround it.

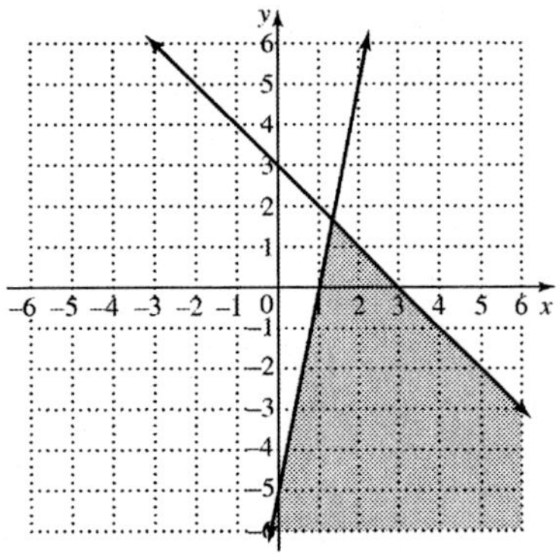

2. Graph the solution set of the system.

$$4x - y > 2$$
$$x + y > -2$$

Graph $4x - y > 2$ using a dashed line for $4x - y = 2$ and intercepts (0, –2) and $\left(\frac{1}{2}, 0\right)$. Using the test point (0, 0), we shade the region that does not contain the point (0, 0).

Graph $x + y > -2$ using a dashed line for $x + y = -2$ and intercepts (–2, 0) and (0, –2). Using the test point (0, 0), we shade the region that does contain the point (0, 0).

The solution set is marked in gray. The solution set does not include either boundary line.

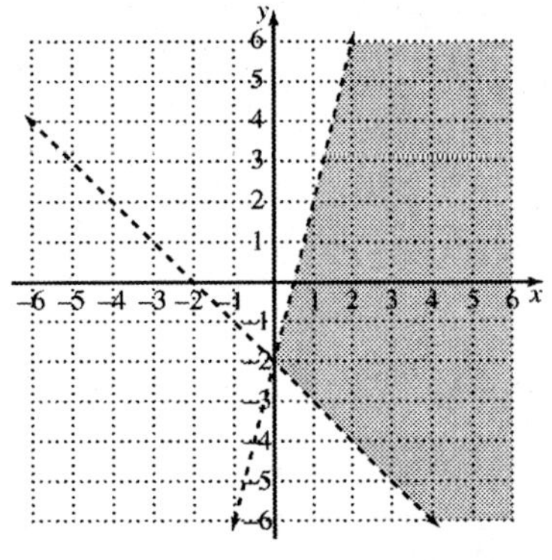

2. Graph the solution set of the system.

$$6x - y > 6$$
$$2x + 5y < 10$$

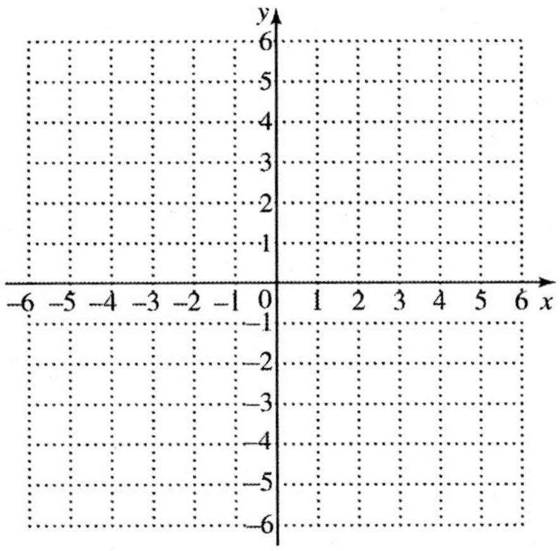

Name: Date:
Instructor: Section:

3. Graph the solution set of the system.

$$y \geq -1$$
$$2x - y > -1$$

Recall that $y = -1$ is a horizontal line through the point (0, –1). Use a solid line, and shade the region with the point (0, 0).

The graph of $2x - y > -1$ is created using a dashed line through the intercepts (0, 1) and $\left(-\frac{1}{2}, 0\right)$. Shade the region with the point (0, 0).

The solution set is marked in gray. The solution set includes the boundary line $y \geq -1$.

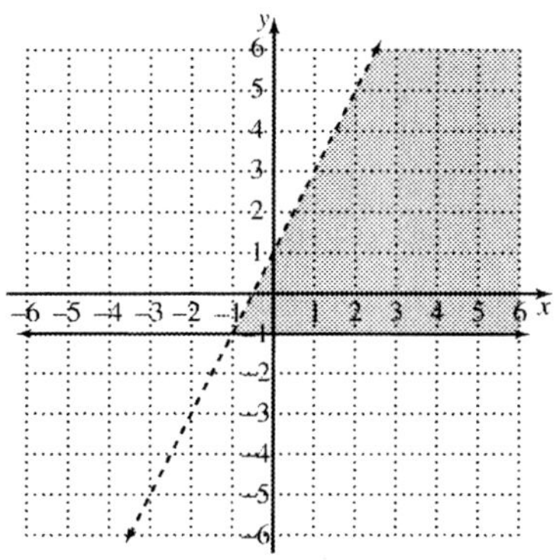

3. Graph the solution set of the system.

$$x - 3y \leq -7$$
$$x < 2$$

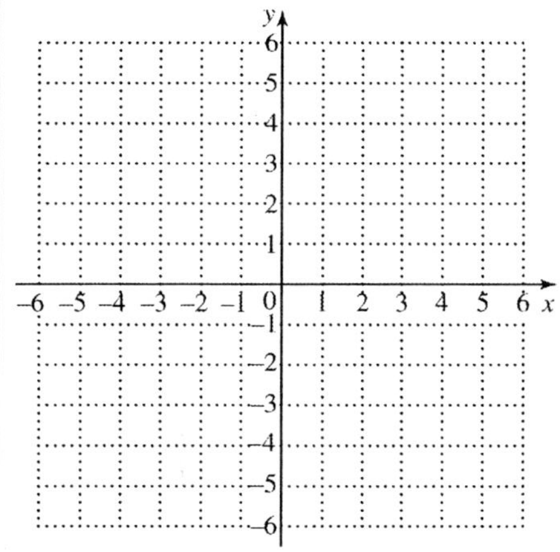

Objective 1 Solve systems of linear inequalities by graphing.

For extra help, see Examples 1–3 on pages 836–838 of your text and Section Lecture video for Section 12.5 and Exercise Solutions Clip 9, 13, and 21.

Graph the solution of each system of linear inequalities.

1. $4x + 5y \leq 20$

$$y \leq x + 3$$

1.

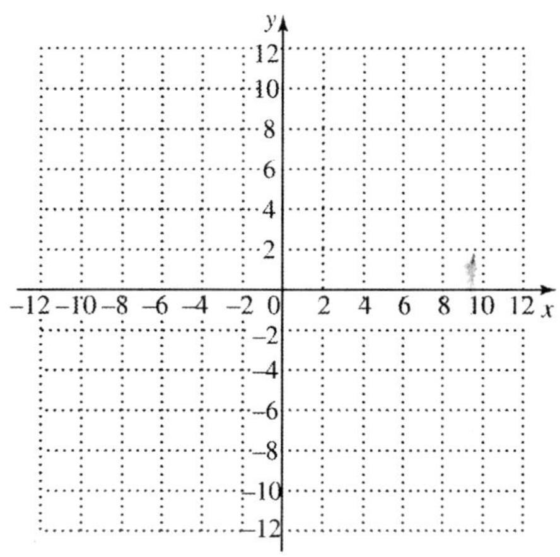

Name: Date:
Instructor: Section:

2. $x < 2y + 3$
$0 < x + y$

2.

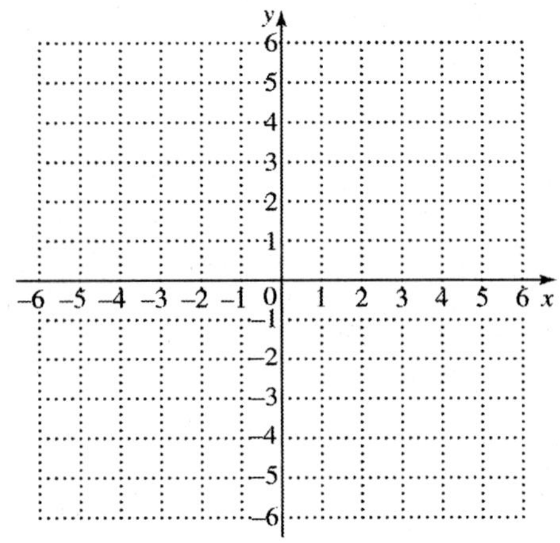

3. $y < 4$
$x \geq -3$

3.

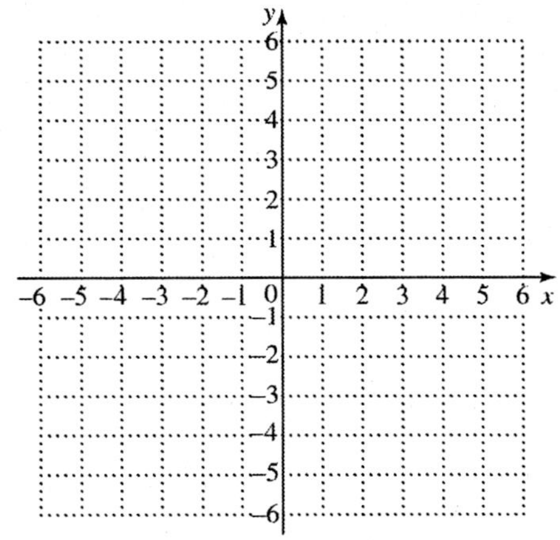

Name: Date:
Instructor: Section:

Chapter 13 EXPONENTS AND POLYNOMIALS

13.1 Adding and Subtracting Polynomials

Learning Objectives	
1	Review combining like terms.
2	Know the vocabulary for polynomials.
3	Evaluate polynomials.
4	Add polynomials.
5	Subtract polynomials.
6	Add and subtract polynomials with more than one variable.

Key Terms

Use the vocabulary terms listed below to complete each statement in exercises 1–9.

term	**like terms**	**polynomial**
descending powers		**degree of a term**
degree of a polynomial		**monomial**
binomial		**trinomial**

1. The ____________________________ is the sum of the exponents on the variables in that term.

2. A polynomial in x is written in ______________________________ if the exponents on x in its terms are decreasing order.

3. A _________________ is a number, a variable, or a product or quotient of a number and one or more variables raised to powers.

4. A polynomial with exactly three terms is called a __________________________.

5. A ______________________ is a term, or the sum of a finite number of terms with whole number exponents.

6. A polynomial with exactly one term is called a ________________________.

7. The _______________________________ is the greatest degree of any term of the polynomial.

8. A ______________________ is a polynomial with exactly two terms.

9. Terms with exactly the same variables (including the same exponents) are called ______________________________.

Name: Date:
Instructor: Section:

Guided Examples

Review these examples for Objective 1:

1. Simplify each expression by combining like terms.

a. $-7x^2+10x^2$

$$\begin{aligned}-7x^2+10x^2 &= (-7+10)x^2\\ &= 3x^2\end{aligned}$$

b. $10rs-14rs+9rs$

$$\begin{aligned}10rs-14rs+9rs &= (10-14+9)rs\\ &= 5rs\end{aligned}$$

c. $4x^5-15x^5+x^5$

$$\begin{aligned}4x^5-15x^5+x^5 &= (4-15+1)x^5\\ &= -10x^5\end{aligned}$$

d. $y+\frac{3}{4}y$

$$\begin{aligned}y+\frac{3}{4}y &= 1y+\frac{3}{4}y\\ &= \left(\frac{4}{4}+\frac{3}{4}\right)y\\ &= \frac{7}{4}y\end{aligned}$$

e. $19m^3+6m+5m^3$

$$\begin{aligned}19m^3+6m+5m^3 &= (19+5)m^3+6m\\ &= 24m^3+6m\end{aligned}$$

f. $8x+17y$

These are unlike terms. They cannot be combined.

Now Try:

1. Simplify each expression by combining like terms.

a. $-6x^4+11x^4$

____

b. $19pq-15pq+3pq$

____

c. $9x^7-18x^7+x^7$

____

d. $x+\frac{3}{5}x$

____

e. $22m^2+15m^3+7m^2$

____

f. $8p+9q$

____

Name: Date:
Instructor: Section:

Review these examples for Objective 2:

2. Simplify each polynomial if possible. Then give the degree and tell whether the polynomial is a monomial, a binomial, a trinomial, or none of these.

a. $5x^4 + 7x$

We cannot simplify further. This is a binomial of degree 4.

b. $9x - 7x + 3x$

$9x - 7x + 3x = 5x$

The degree is 1. The simplified polynomial is a monomial.

Now Try:

2. Simplify each polynomial if possible. Then give the degree and tell whether the polynomial is a monomial, a binomial, a trinomial, or none of these.

a. $8x^3 + 4x^2 + 6$

____

b. $x^5 + 3x^5$

____

Review these examples for Objective 3:

3. Find the value of $4x^3 + 6x^2 - 5x - 5$ for

a. $x = -3$

$$\begin{aligned} &4x^3 + 6x^2 - 5x - 5 \\ &= 4(-3)^3 + 6(-3)^2 - 5(-3) - 5 \\ &= 4(-27) + 6(9) - 5(-3) - 5 \\ &= -108 + 54 + 15 - 5 \\ &= -44 \end{aligned}$$

b. $x = 2$

$$\begin{aligned} &4x^3 + 6x^2 - 5x - 5 \\ &= 4(2)^3 + 6(2)^2 - 5(2) - 5 \\ &= 4(8) + 6(4) - 5(2) - 5 \\ &= 32 + 24 - 10 - 5 \\ &= 41 \end{aligned}$$

Now Try:

3. Find the value of $5x^4 + 3x^2 - 9x - 7$ for

a. $x = 4$

____

b. $x = -4$

____

Review these examples for Objective 4:

4. Add vertically.

a. $5x^4 - 7x^3 + 9$ and $-3x^4 + 8x^3 - 7$

Write like terms in columns.

$$\begin{array}{r} 5x^4 - 7x^3 + 9 \\ \underline{-3x^4 + 8x^3 - 7} \end{array}$$

Now add, column by column.

Now Try:

4. Add vertically.

a. $8x^3 - 9x^2 + x$ and $-3x^3 + 4x^2 + 3x$

____

Name: Date:
Instructor: Section:

$$\begin{array}{ccc} 5x^4 & -7x^3 & 9 \\ \underline{-3x^4} & \underline{8x^3} & \underline{-7} \\ 2x^4 & x^3 & 2 \end{array}$$

Add the three sums together to obtain the answer.

$2x^4 + x^3 + 2$

b. $3x^3 + 7x + 5$ and $x^4 - 6x$

Write like terms and add column by column.

$$\begin{array}{llll} & 3x^3 & +7x & +5 \\ x^4 & & -6x & \\ \hline x^4 & +3x^3 & +x & +5 \end{array}$$

b. $9x^4 + 3x - 6$ and $7x^2 + 4x$

____

5. Find each sum.

a. Add $5x^4 - 7x^3 + 9$ and $-3x^4 + 8x^3 - 7$

$(5x^4 - 7x^3 + 9) + (-3x^4 + 8x^3 - 7)$
$= 5x^4 - 3x^4 - 7x^3 + 8x^3 + 9 - 7$
$= 2x^4 + x^3 + 2$

b. $(5x^4 - 7x^2 + 6x) + (-3x^3 + 4x^2 - 7)$

$(5x^4 - 7x^2 + 6x) + (-3x^3 + 4x^2 - 7)$
$= 5x^4 - 3x^3 - 7x^2 + 4x^2 + 6x - 7$
$= 5x^4 - 3x^3 - 3x^2 + 6x - 7$

5. Find each sum.

a. Add $15x^3 - 5x + 3$ and $-11x^3 + 6x + 9$

____

b. $(8x^2 - 6x + 4) + (7x^3 - 8x - 5)$

____

Review these examples for Objective 5:

6. Perform each subtraction.

a. $(8x - 3) - (4x - 7)$

Use the definition of subtraction and the distributive property.

$$\begin{aligned} (8x-3)-(4x-7) &= (8x-3)+[-(4x-7)] \\ &= (8x-3)+[-1(4x-7)] \\ &= (8x-3)+(-4x+7) \\ &= 4x+4 \end{aligned}$$

Now Try:

6. Perform each subtraction.

a. $(7x - 6) - (5x - 8)$

____

Name: Date:
Instructor: Section:

b. Subtract $8x^3 - 5x^2 + 8$ from $9x^3 + 6x^2 - 7$.

$$(9x^3 + 6x^2 - 7) - (8x^3 - 5x^2 + 8)$$
$$= (9x^3 + 6x^2 - 7) + (-8x^3 + 5x^2 - 8)$$
$$= x^3 + 11x^2 - 15$$

7. Subtract by columns:
$(15x^3 - 8x^2 + 3x - 7) - (9x^3 - 5x^2 - 6x + 4)$.

Arrange like terms in columns.

$$\begin{array}{rrrr} 15x^3 & -8x^2 & +3x & -7 \\ \underline{9x^3} & \underline{-5x^2} & \underline{-6x} & \underline{+4} \end{array}$$

Change all the signs in the second row, and then add.

$$\begin{array}{rrrr} 15x^3 & -8x^2 & +3x & -7 \\ -9x^3 & +5x^2 & +6x & -4 \\ \hline 6x^3 & -3x^2 & +9x & -11 \end{array}$$

8. Perform the indicated operations to simplify the expression
$(5 - 2x + 9x^2) - (7 - 5x + 8x^2) + (6 + 3x - 5x^2)$

Rewrite, changing the subtraction to adding the opposite.

$$(5-2x+9x^2)-(7-5x+8x^2)+(6+3x-5x^2)$$
$$=(5-2x+9x^2)+(-7+5x-8x^2)+(6+3x-5x^2)$$
$$=(-2+3x+x^2)+(6+3x-5x^2)$$
$$=4+6x-4x^2$$

b. $(7x^3-3x-5)-(18x^3+4x-6)$

____

7. Subtract by columns:
$(16x^3 - 7x^2 - 5x + 3)$
$-(7x^3 - 8x^2 + 6x - 4)$

____

8. Perform the indicated operations to simplify the expression
$(10-7x+6x^2)-(5-11x+3x^2)$
$+(2+4x-7x^2)$

____

Review these examples for Objective 6:

9. Add or subtract as indicated.

a. $(5a + 3ab + b) + (7a - 2ab - b)$

$$(5a + 3ab + b) + (7a - 2ab - b)$$
$$= 5a + 3ab + b + 7a - 2ab - b$$
$$= 12a + ab$$

b. $(3x^2y + 5xy + y^2) - (4x^2y + xy - 3y^2)$

$$(3x^2y + 5xy + y^2) - (4x^2y + xy - 3y^2)$$
$$= 3x^2y + 5xy + y^2 - 4x^2y - xy + 3y^2$$
$$= -x^2y + 4xy + 4y^2$$

Now Try:

9. Add or subtract as indicated.

a. $(10a + 6ab + 2b)$
$+(3a - 5ab - 2b)$

____

b. $(7x^2y + 3xy + 4y^2)$
$-(6x^2y - xy + 4y^2)$

____

Name: Date:
Instructor: Section:

Objective 1 Review combining like terms.

For extra help, see Example 1 on page 850 of your text and Section Lecture video for Section 13.1 and Exercise Solutions Clip 17, 27, and 29.

In each polynomial, combine like terms whenever possible. Write the result with descending powers.

1. $7z^3 - 4z^3 + 5z^3 - 11z^3$ **1.** ________________

2. $-1.3z^7 + 0.4z^7 + 2.6z^8$ **2.** ________________

3. $6c^3 - 9c^2 - 2c^2 + 14 + 3c^2 - 6c - 8 + 2c^3$ **3.** ________________

Objective 2 Know the vocabulary for polynomials.

For extra help, see Example 2 on page 851 of your text and Section Lecture video for Section 13.1 and Exercise Solutions Clip 7, 31, and 33.

For each polynomial, first simplify, if possible, and write the resulting polynomial in descending powers of the variable. Then give the degree of this polynomial, and tell whether it is a monomial, *a* binomial, *a* trinomial, *or* none of these.

4. $3n^8 - n^2 - 2n^8$ **4.** ________________

degree: __________

type: ____________

5. $-d^2 + 3.2d^3 - 5.7d^8 - 1.1d^5$ **5.** ________________

degree: __________

type: ____________

6. $-6c^4 - 6c^2 + 9c^4 - 4c^2 + 5c^5$ **6.** ________________

degree: __________

type: ____________

Name: Date:
Instructor: Section:

Objective 3 Evaluate polynomials.

For extra help, see Example 3 on page 852 of your text and Section Lecture video for Section 13.1 and Exercise Solutions Clip 43 and 45.

Find the value of each polynomial (a) *when* $x = -2$ *and* (b) *when* $x = 3$.

7. $3x^3 + 4x - 19$

7. a._______________

b._______________

8. $-4x^3 + 10x^2 - 1$

8. a._______________

b._______________

9. $x^4 - 3x^2 - 8x + 9$

9. a._______________

b._______________

Objective 4 Add polynomials.

For extra help, see Examples 4–5 on pages 852–853 of your text and Section Lecture video for Section 13.1 and Exercise Solutions Clip 63.

Add.

10.
$$\begin{array}{r} 9m^3 + 4m^2 - 2m + 3 \\ \underline{-4m^3 - 6m^2 - 2m + 1} \end{array}$$

10. _______________

11. $(x^2 + 6x - 8) + (3x^2 - 10)$

11. _______________

12. $(3r^3 + 5r^2 - 6) + (2r^2 - 5r + 4)$

12. _______________

Name: Date:
Instructor: Section:

Objective 5 Subtract polynomials.

For extra help, see Examples 6–8 on pages 853–854 of your text and Section Lecture video for Section 13.1 and Exercise Solutions Clip 61 and 69.

Subtract.

13. $(-8w^3+11w^2-12)-(-10w^2+3)$ **13.** ______________

14. $(8b^4-4b^3+7)-(2b^2+b+9)$ **14.** ______________

15. $(9x^3+7x^2-6x+3)-(6x^3-6x+1)$ **15.** ______________

Objective 6 Add and subtract polynomials with more than one variable.

For extra help, see Example 9 on page 854 of your text and Section Lecture video for Section 13.1 and Exercise Solutions Clip 79.

Add or subtract as indicated.

16. $(-2a^6+8a^4b-b^2)-(a^6+7a^4b+2b^2)$ **16.** ______________

17. $(4ab+2bc-9ac)+(3ca-2cb-9ba)$ **17.** ______________

18. $(2x^2y+2xy-4xy^2)+(6xy+9xy^2)-(9x^2y+5xy)$ **18.** ______________

Name: Date:
Instructor: Section:

Chapter 13 EXPONENTS AND POLYNOMIALS

13.2 The Product Rule and Power Rules for Exponents

Learning Objectives

1. Use exponents.
2. Use the product rule for exponents.
3. Use the rule $(a^m)^n = a^{mn}$.
4. Use the rule $(ab)^m = a^m b^m$.
5. Use the rule $\left(\frac{a}{b}\right)^m = \frac{a^m}{b^m}$.
6. Use combinations of the rules for exponents.
7. Use the rules for exponents in a geometry problem.

Key Terms

Use the vocabulary terms listed below to complete each statement in exercises 1–3.

exponential expression **base** **power**

1. 2^5 is read "2 to the fifth ______________________".
2. A number written with an exponent is called a(n) ____________________________.
3. The ________________ is the number being multiplied repeatedly.

Guided Examples

Review this example for Objective 1:

1. Write $5\cdot5\cdot5$ in exponential form and evaluate.

   Since 5 occurs as a factor three times, the base is 5 and the exponent is 3.

   $5\cdot5\cdot5 = 5^3 = 125$

2. Name the base and exponent of each expression. Then evaluate.

   a. 3^4

   Base: 3
   Exponent: 4
   Value: $3^4 = 3\cdot3\cdot3\cdot3 = 81$

Now Try:

1. Write $4\cdot4\cdot4\cdot4\cdot4$ in exponential form and evaluate.

   ____

2. Name the base and exponent of each expression. Then evaluate.

   a. 2^6

   ____

b. -3^4

Base: 3
Exponent: 4
Value: $-3^4 = -1\cdot(3\cdot3\cdot3\cdot3) = -81$

c. $(-3)^4$

Base: –3
Exponent: 4
Value: $(-3)^4 = (-3)(-3)(-3)(-3) = 81$

b. -2^6

____

c. $(-2)^6$

____

Review these examples for Objective 2:

3. Use the product rule for exponents to simplify, if possible.

a. $8^4 \cdot 8^5$

$$8^4 \cdot 8^5 = 8^{4+5} = 8^9$$

b. $(-5)^8(-5)^3$

$$(-5)^8(-5)^3 = (-5)^{8+3} = (-5)^{11}$$

c. $x^3 \cdot x$

$$x^3 \cdot x = x^3 \cdot x^1 = x^{3+1} = x^4$$

d. $m^7m^8m^9$

$$m^7m^8m^9 = m^{7+8+9} = m^{24}$$

e. $5^2 \cdot 4^3$

$$5^2 \cdot 4^3 = 25\cdot64 = 1600$$

f. $5^2 + 5^3$

$$5^2 + 5^3 = 25 + 125 = 150$$

Now Try:

3. Use the product rule for exponents to simplify, if possible.

a. $9^6 \cdot 9^7$

____

b. $(-6)^2(-6)^4$

____

c. $x^5 \cdot x$

____

d. $m^{11}m^9m^7$

____

e. $5^3 \cdot 2^2$

____

f. $3^4 + 3^3$

____

4. Multiply $5x^4$ and $6x^9$.

$$5x^4 \cdot 6x^9 = (5 \cdot 6) \cdot (x^4 \cdot x^9)$$
$$= 30x^{4+9}$$
$$= 30x^{13}$$

4. Multiply $6x^5$ and $3x^6$.

____

Review these examples for Objective 3:

5. Use power rule (a) for exponents to simplify.

a. $(3^4)^5$

$$(3^4)^5 = 3^{4 \cdot 5}$$
$$= 3^{20}$$

b. $(5^6)^3$

$$(5^6)^3 = 5^{6 \cdot 3}$$
$$= 5^{18}$$

c. $(x^3)^4$

$$(x^3)^4 = x^{3 \cdot 4}$$
$$= x^{12}$$

Now Try:

5. Use power rule (a) for exponents to simplify.

a. $(4^5)^3$

____

b. $(7^2)^4$

____

c. $(x^5)^6$

____

Review these examples for Objective 4:

6. Use power rule (b) for exponents to simplify.

a. $(5xy)^3$

$$(5xy)^3 = 5^3x^3y^3$$
$$= 125x^3y^3$$

b. $6(rs)^4$

$$6(rs)^4 = 6(r^4s^4)$$
$$= 6r^4s^4$$

c. $4(3m^4n^5)^3$

$$4(3m^4n^5)^3 = 4\left[3^3(m^4)^3(n^5)^3\right]$$
$$= 4\left[3^3m^{12}n^{15}\right]$$
$$= 108m^{12}n^{15}$$

Now Try:

6. Use power rule (b) for exponents to simplify.

a. $(4ab)^3$

____

b. $7(pq)^3$

____

c. $8(5x^4y^3)^2$

____

Name: Date:
Instructor: Section:

d. $\left(-6^3\right)^5$

$$\begin{aligned}\left(-6^3\right)^5 &= \left(-1\cdot 6^3\right)^5\\ &= (-1)^5\left(6^3\right)^5\\ &= -1\cdot 6^{15}\\ &= -6^{15}\end{aligned}$$

d. $\left(-8^5\right)^7$

____

Review these examples for Objective 5:

7. Use power rule (c) for exponents to simplify.

a. $\left(\frac{3}{5}\right)^4$

$$\left(\frac{3}{5}\right)^4 = \frac{3^4}{5^4} = \frac{81}{625}$$

b. $\left(\frac{x}{y}\right)^5$ (where $y \neq 0$)

$$\left(\frac{x}{y}\right)^5 = \frac{x^5}{y^5}$$

c. $\left(\frac{1}{8}\right)^3$

$$\left(\frac{1}{8}\right)^3 = \frac{1^3}{8^3} = \frac{1}{512}$$

Now Try:

7. Use power rule (c) for exponents to simplify.

a. $\left(\frac{5}{6}\right)^3$

____

b. $\left(\frac{p}{q}\right)^6$

____

c. $\left(\frac{1}{4}\right)^5$

____

Review these examples for Objective 6:

8. Simplify each expression.

a. $\left(\frac{3}{4}\right)^3 \cdot 3^2$

$$\begin{aligned}\left(\frac{3}{4}\right)^3 \cdot 3^2 &= \frac{3^3}{4^3}\cdot\frac{3^2}{1}\\ &= \frac{3^3\cdot 3^2}{4^3\cdot 1}\\ &= \frac{3^{3+2}}{4^3}\\ &= \frac{3^5}{4^3}, \text{ or } \frac{243}{64}\end{aligned}$$

Now Try:

8. Simplify each expression.

a. $\left(\frac{5}{2}\right)^3 \cdot 5^2$

____

Name: Date:
Instructor: Section:

b. $(4x)^5(4x)^2$

$$(4x)^5(4x)^2 = (4x)^7$$
$$= 4^7x^7$$

c. $\left(5x^3y^4\right)^2\left(4x^2y\right)^3$

$$\left(5x^3y^4\right)^2\left(4x^2y\right)^3 = 5^2\left(x^3\right)^2\left(y^4\right)^2\cdot 4^3\left(x^2\right)^3y^3$$
$$= 5^2x^6y^8\cdot 4^3x^6y^3$$
$$= 5^2\cdot 4^3x^6x^6y^8y^3$$
$$= 25\cdot 64x^{12}y^{11}$$
$$= 1600x^{12}y^{11}$$

d. $\left(-x^5y\right)^4\left(-x^6y^5\right)^3$

$$\left(-x^5y\right)^4\left(-x^6y^5\right)^3$$
$$= \left(-1x^5y\right)^4\left(-1x^6y^5\right)^3$$
$$= (-1)^4\left(x^5\right)^4\left(y^4\right)\cdot(-1)^3\left(x^6\right)^3\left(y^5\right)^3$$
$$= (-1)^4\left(x^{20}\right)\left(y^4\right)\cdot(-1)^3\left(x^{18}\right)\left(y^{15}\right)$$
$$= (-1)^7x^{20+18}y^{4+15}$$
$$= -1x^{38}y^{19}$$
$$= -x^{38}y^{19}$$

Review these examples for Objective 7:

9. Find the area of the figure.

a.

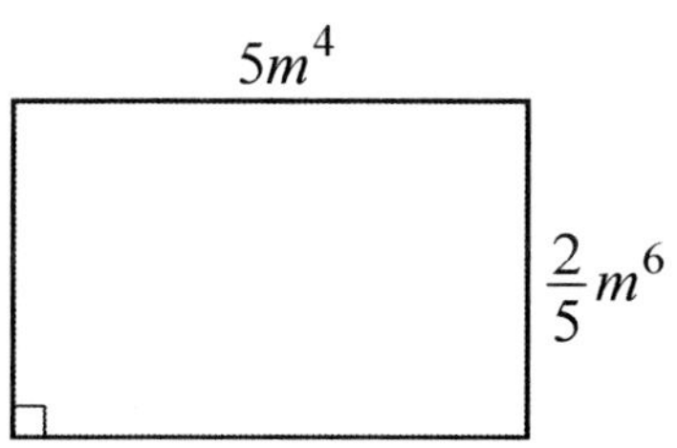

Use the formula for the area of a rectangle.

$$A = LW$$
$$A = \left(5m^4\right)\left(\frac{2}{5}m^6\right)$$
$$A = 5\cdot\frac{2}{5}\cdot m^{4+6}$$
$$A = 2m^{10}$$

b. $(6x)^6(6x)^5$

____

c. $\left(3x^3y^2\right)^3\left(4xy^3\right)^2$

____

d. $\left(-x^5y\right)^3\left(-x^6y^5\right)^2$

____

Now Try:

9. Find the area of the figure.

a.

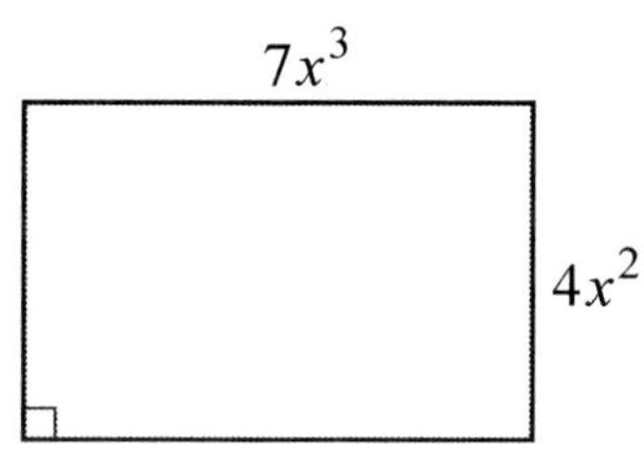

____

Name: Date:
Instructor: Section:

b.

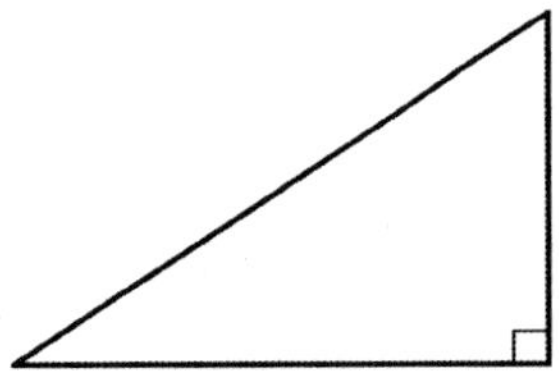

Use the formula for the area of a triangle.

$$A = \frac{1}{2}bh$$

$$A = \frac{1}{2}(3a^3)(4a^4)$$

$$A = \frac{1}{2}(3 \cdot 4 \cdot a^{3+4})$$

$$A = 6a^7$$

b.

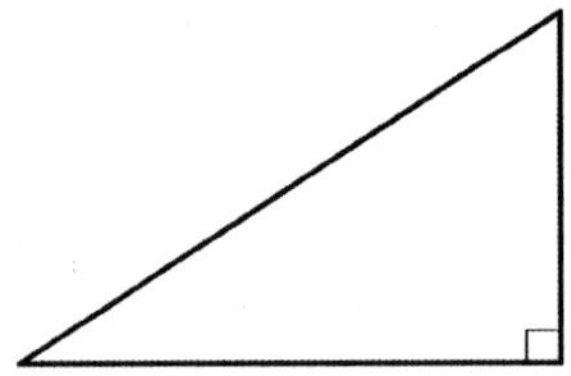

____

Objective 1 Use exponents.

For extra help, see Examples 1–2 on page 859 of your text and Section Lecture video for Section 13.2 and Exercise Solutions Clip 9, 13, and 15.

Write the expression in exponential form and evaluate, if possible.

1. $\left(\frac{1}{3}\right)\left(\frac{1}{3}\right)\left(\frac{1}{3}\right)\left(\frac{1}{3}\right)\left(\frac{1}{3}\right)$ **1.** __________

Evaluate each exponential expression. Name the base and the exponent.

2. $(-4)^4$ **2.** __________

base__________

exponent__________

3. -3^8 **3.** __________

base__________

exponent__________

Objective 2 Use the product rule for exponents.

For extra help, see Examples 3–4 on pages 860–861 of your text and Section Lecture video for Section 13.2 and Exercise Solutions Clip 21, 27, 31, and 33.

Use the product rule to simplify each expression, if possible. Write each answer in exponential form.

4. $7^4 \cdot 7^3$ **4.** __________

Name: Date:
Instructor: Section:

5. $\left(-2c^7\right)\left(-4c^8\right)$ 5. ________________

6. $\left(3k^7\right)\left(-8k^2\right)\left(-2k^9\right)$ 6. ________________

Objective 3 Use the rule $\left(a^m\right)^n = a^{mn}$.

For extra help, see Example 5 on page 861 of your text and Section Lecture video for Section 13.2 and Exercise Solutions Clip 37 and 39.

Simplify each expression. Write all answers in exponential form.

7. $\left(7^3\right)^4$ 7. ________________

8. $-\left(v^4\right)^9$ 8. ________________

9. $\left[(-3)^3\right]^7$ 9. ________________

Objective 4 Use the rule $(ab)^m = a^m b^m$.

For extra help, see Example 6 on page 862 of your text and Section Lecture video for Section 13.2 and Exercise Solutions Clip 47.

Simplify each expression.

10. $\left(5r^3t^2\right)^4$ 10. ________________

11. $\left(-0.2a^4b\right)^3$ 11. ________________

12. $\left(-2w^3z^7\right)^4$ 12. ________________

Name: Date:
Instructor: Section:

Objective 5 Use the rule $\left(\frac{a}{b}\right)^m = \frac{a^m}{b^m}$.

For extra help, see Example 7 on page 863 of your text and Section Lecture video for Section 13.2 and Exercise Solutions Clip 51.

Simplify each expression.

13. $\left(-\frac{2x}{5}\right)^3$ **13.** ______

14. $\left(\frac{xy}{z^2}\right)^4$ **14.** ______

15. $\left(\frac{-2a}{b^2}\right)^7$ **15.** ______

Objective 6 Use combinations of the rules for exponents.

For extra help, see Example 8 on page 863–864 of your text and Section Lecture video for Section 13.2 and Exercise Solutions Clip 63 and 75.

Simplify. Write all answers in exponential form.

16. $(-x^3)^2(-x^5)^4$ **16.** ______

17. $(2ab^2c)^5(ab)^4$ **17.** ______

18. $(5x^2y^3)^7(5xy^4)^4$ **18.** ______

Name: Date:
Instructor: Section:

Objective 7 Use the rules for exponents in a geometry problem.

For extra help, see Example 9 on page 864 of your text and Section Lecture video for Section 13.2.

Find a polynomial that represents the area of each figure.

19.

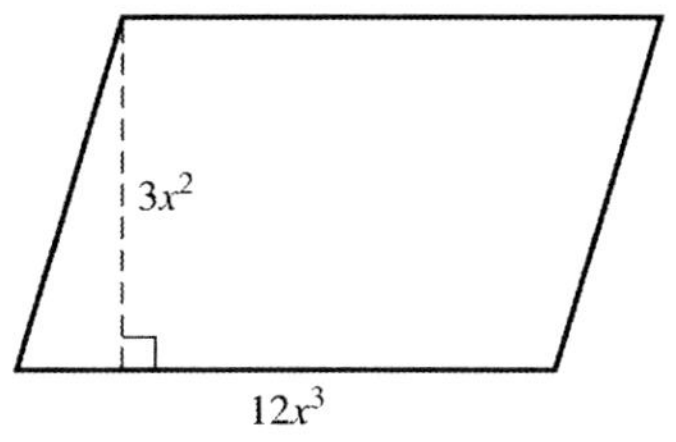

19. ________________

20.

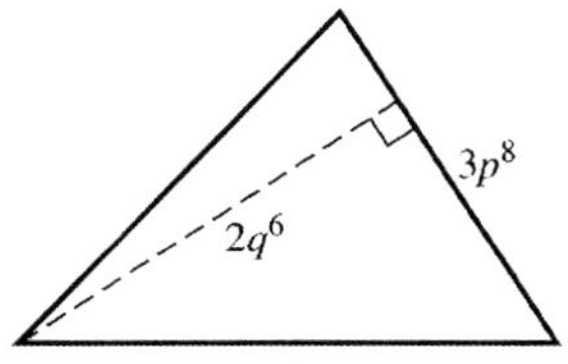

20. ________________

21.

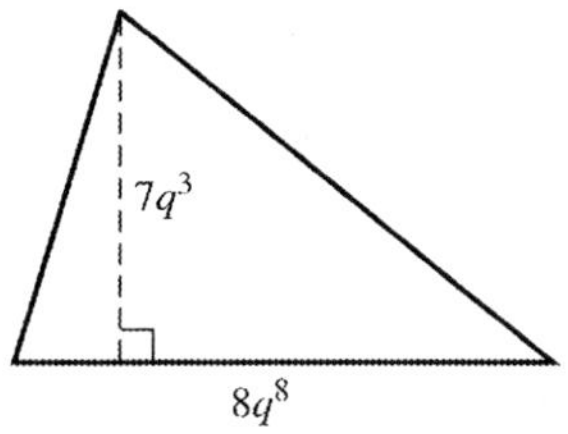

21. ________________

Name: Date:
Instructor: Section:

Chapter 13 EXPONENTS AND POLYNOMIALS

13.3 Multiplying Polynomials

Learning Objectives

1	Multiply a monomial and a polynomial.
2	Multiply two polynomials.
3	Multiply binomials by the FOIL method.

Key Terms

Use the vocabulary terms listed below to complete each statement in exercises 1–3.

FOIL **outer product** **inner product**

1. The ______________________________ of $(2y-5)(y+8)$ is $-5y$.

2. ____________________ is a shortcut method for finding the product of two binomials.

3. The ______________________________ of $(2y-5)(y+8)$ is $16y$.

Guided Examples

Review these examples for Objective 1:

1. Find each product.

a. $5x^2(7x+3)$

Use the distributive property.

$$5x^2(7x+3) = 5x^2(7x) + 5x^2(3)$$
$$= 35x^3 + 15x^2$$

b. $-9n^4(6n^4+5n^3+7n-1)$

Use the distributive property.

$$-9n^4(6n^4+5n^3+7n-1)$$
$$= -9n^4(6n^4) - 9n^4(5n^3) - 9n^4(7n) - 9n^4(-1)$$
$$= -54n^8 - 45n^7 - 63n^5 + 9n^4$$

Now Try:

1. Find each product.

a. $8x^3(4x+8)$

____

b. $-7m^5(5m^3-6m^2+4m-1)$

____

Name: Date:
Instructor: Section:

Review these examples for Objective 2:

2. Multiply $(x^2+6)(5x^3-4x^2+3x)$.

Multiply each term of the second polynomial by each term of the first.

$$(x^2+6)(5x^3-4x^2+3x)$$
$$=x^2(5x^3)+x^2(-4x^2)+x^2(3x)$$
$$+6(5x^3)+6(-4x^2)+6(3x)$$
$$=5x^5-4x^4+3x^3+30x^3-24x^2+18x$$
$$=5x^5-4x^4+33x^3-24x^2+18x$$

3. Multiply $(2x^3+7x^2+5x-1)(4x+6)$ vertically.

Write the polynomials vertically.

$$\begin{array}{r} 2x^3+7x^2+5x-1 \\ \underline{4x+6} \end{array}$$

Begin by multiplying each term in the top row by 6.

$$\begin{array}{rrrr} 2x^3 & +7x^2 & +5x & -1 \\ & & 4x & +6 \\ \hline 12x^3 & +42x^2 & +30x & -6 \end{array}$$

Now multiply each term in the top row by $4x$. Then add like terms.

$$\begin{array}{rrrrr} & 2x^3 & +7x^2 & +5x & -1 \\ & & & 4x & +6 \\ \hline & 12x^3 & +42x^2 & +30x & -6 \\ 8x^4 & +28x^3 & +20x^2 & -4x & \\ \hline 8x^4 & +40x^3 & +62x^2 & +26x & -6 \end{array}$$

The product is $8x^4+40x^3+62x^2+26x-6$.

Now Try:

2. Multiply
$(x^3+9)(4x^4-2x^2+x)$

____

3. Multiply
$(4x^3-3x^2+6x+5)(7x-3)$
vertically.

____

Review these examples for Objective 3:

4. Use the FOIL method to find the product $(x+7)(x-5)$.

Step 1 F Multiply the first terms: $x(x)=x^2$.

Step 2 O Find the outer product: $x(-5)=-5x$.

Step 3 I Find the inner product: $7(x)=7x$.
Add the outer and inner products mentally:
$$-5x+7x=2x$$

Now Try:

4. Use the FOIL method to find the product $(x+9)(x-6)$.

____

Name: Date:
Instructor: Section:

Step 4 L Multiply the last terms: $7(-5)=-35$.

The product $(x+7)(x-5)$ is $x^2+2x-35$.

5. Multiply $(7x-3)(4y+5)$.

First $7x(4y)=28xy$

Outer $7x(5)=35x$

Inner $-3(4y)=-12y$

Last $-3(5)=-15$

The product $(7x-3)(4y+5)$ is $28xy+35x-12y-15$.

6. Find each product.

a. $(3k+7m)(2k+9m)$

$$(3k+7m)(2k+9m)$$
$$=3k(2k)+3k(9m)+7m(2k)+7m(9m)$$
$$=6k^2+27km+14km+63m^2$$
$$=6k^2+41km+63m^2$$

b. $(8p-5q)(8p+q)$

$$(8p-5q)(8p+q)$$
$$=64p^2+8pq-40pq-5q^2$$
$$=64p^2-32pq-5q^2$$

c. $3x^2(x-5)(4x+7)$

$$3x^2(x-5)(4x+7)$$
$$=3x^2(4x^2-13x-35)$$
$$=12x^4-39x^3-105x^2$$

5. Multiply $(8y-7)(2x+9)$.

____

6. Find each product.

a. $(5k+8n)(3k+4n)$

____

b. $(9p+4q)(5p-q)$

____

c. $5x^3(x-6)(7x+3)$

____

Name: Date:
Instructor: Section:

Objective 1 Multiply a monomial and a polynomial.

For extra help, see Example 1 on page 867 of your text and Section Lecture video for Section 13.3 and Exercise Solutions Clip 13.

Find each product.

1. $7z(5z^3+2)$ **1.** ________

2. $2m(3+7m^2+3m^3)$ **2.** ________

3. $-3y^2(2y^3+3y^2-4y+11)$ **3.** ________

Objective 2 Multiply two polynomials.

For extra help, see Examples 2–3 on page 868 of your text and Section Lecture video for Section 13.3 and Exercise Solutions Clip 27, 31, and 35.

Find each product.

4. $(x+3)(x^2-3x+9)$ **4.** ________

5. $(2m^2+1)(3m^3+2m^2-4m)$ **5.** ________

6. $(3x^2+x)(2x^2+3x-4)$ **6.** ________

Name: Date:
Instructor: Section:

Objective 3 Multiply binomials by the FOIL method.

For extra help, see Examples 4–6 on page 870 of your text and Section Lecture video for Section 13.3 and Exercise Solutions Clip 41 and 52.

Find each product.

7. $(5a-b)(4a+3b)$ **7.** ________________

8. $(3+4a)(1+2a)$ **8.** ________________

9. $(2m+3n)(-3m+4n)$ **9.** ________________

Name: Date:
Instructor: Section:

Chapter 13 EXPONENTS AND POLYNOMIALS

13.4 Special Products

Learning Objectives
1. Square binomials.
2. Find the product of the sum and difference of two terms.
3. Find greater powers of binomials.

Key Terms

Use the vocabulary terms listed below to complete each statement in exercises 1–2.

conjugate **binomial**

1. A polynomial with two terms is called a ________________________.
2. The ____________________ of $a + b$ is $a - b$.

Guided Examples

Review these examples for Objective 1:

1. Find $(m+5)^2$.

$$(m+5)^2 = (m+5)(m+5)$$
$$= m^2 + 5m + 5m + 25$$
$$= m^2 + 10m + 25$$

2. Square each binomial.

a. $(7z-4)^2$

$$(7z-4)^2 = (7z)^2 - 2(7z)(4) + (-4)^2$$
$$= 7^2 z^2 - 56z + 16$$
$$= 49z^2 - 56z + 16$$

b. $(6x+3y)^2$

$$(6x+3y)^2 = (6x)^2 + 2(6x)(3y) + (3y)^2$$
$$= 36x^2 + 36xy + 9y^2$$

Now Try:

1. Find $(x+6)^2$.

____

2. Square each binomial.

a. $(8a-3b)^2$

____

b. $(2a+9k)^2$

____

Name: Date:
Instructor: Section:

c. $(5a-7x)^2$

$$(5a-7x)^2=(5a)^2-2(5a)(7x)+(7x)^2$$
$$=25a^2-70ax+49x^2$$

d. $\left(6n+\frac{1}{4}\right)^2$

$$\left(6n+\frac{1}{4}\right)^2=(6n)^2+2(6n)\left(\frac{1}{4}\right)+\left(\frac{1}{4}\right)^2$$
$$=36n^2+3n+\frac{1}{16}$$

e. $2x(5x-4)^2$

$$2x(5x-4)^2=2x\left(25x^2-40x+16\right)$$
$$=50x^3-80x^2+32x$$

c. $(3a-11b)^2$

____

d. $\left(3p+\frac{1}{6}\right)^2$

____

e. $5x(3x-7)^2$

____

Review these examples for Objective 2:

3. Find each product.

a. $(x+5)(x-5)$

Use the rule for the product of the sum and difference of two terms.

$$(x+5)(x-5)=x^2-5^2$$
$$=x^2-25$$

b. $\left(\frac{3}{4}-y\right)\left(\frac{3}{4}+y\right)$

$$\left(\frac{3}{4}-y\right)\left(\frac{3}{4}+y\right)=\left(\frac{3}{4}\right)^2-y^2$$
$$=\frac{9}{16}-y^2$$

c. $x(x+7)(x-7)$

Find the product of the sum and difference of two terms. Then use the distributive property.

$$x(x+7)(x-7)=x\left(x^2-49\right)$$
$$=x^3-49x$$

Now Try:

3. Find each product.

a. $(x+9)(x-9)$

____

b. $\left(\frac{5}{6}+a\right)\left(\frac{5}{6}-a\right)$

____

c. $x(x-12)(x+12)$

____

Name: Date:
Instructor: Section:

4. Find each product.

a. $(7m+9)(7m-9)$

Use the rule for the product of the sum and difference of two terms.

$$(7m+9)(7m-9) = (7m)^2 - 9^2$$
$$= 49m^2 - 81$$

b. $(6x+w)(6x-w)$

$$(6x+w)(6x-w) = (6x)^2 - w^2$$
$$= 36x^2 - w^2$$

c. $\left(z-\frac{6}{7}\right)\left(z+\frac{6}{7}\right)$

$$\left(z-\frac{6}{7}\right)\left(z+\frac{6}{7}\right) = z^2 - \left(\frac{6}{7}\right)^2$$
$$= z^2 - \frac{36}{49}$$

d. $3q\left(q^2+4\right)\left(q^2-4\right)$

First, multiply the conjugates.

$$3q\left(q^2+4\right)\left(q^2-4\right) = 3q\left(q^4-16\right)$$
$$= 3q^5 - 48q$$

4. Find each product.

a. $(8p+5)(8p-5)$

____

b. $(11x-y)(11x+y)$

____

c. $\left(x+\frac{3}{5}\right)\left(x-\frac{3}{5}\right)$

____

d. $4p\left(p^2+6\right)\left(p^2-6\right)$

____

Review these examples for Objective 3:

5. Find each product.

a. $(x+4)^3$

$$(x+4)^3$$
$$= (x+4)^2(x+4)$$
$$= \left(x^2+8x+16\right)(x+4)$$
$$= x^3 + 8x^2 + 16x + 4x^2 + 32x + 64$$
$$= x^3 + 12x^2 + 48x + 64$$

Now Try:

5. Find each product.

a. $(x+6)^3$

____

Name: Date:
Instructor: Section:

b. $(5y-4)^4$

$(5y-4)^4$
$=(5y-4)^2(5y-4)^2$
$=(25y^2-40y+16)(25y^2-40y+16)$
$=625y^4-1000y^3+400y^2$
$\quad -1000y^3+1600y^2-640y$
$\quad +400y^2-640y+256$
$=625y^4-2000y^3+2400y^2-1280y+256$

b. $(3x-5)^4$

____

c. $-5r(r+5)^3$

$-5r(r+5)^3$
$=-5r(r+5)(r+5)^2$
$=-5r(r+5)(r^2+10r+25)$
$=-5r(r^3+10r^2+25r+5r^2+50r+125)$
$=-5r(r^3+15r^2+75r+125)$
$=-5r^4-75r^3-375r^2-625r$

c. $-4x(x+3)^3$

____

Objective 1 Square binomials.

For extra help, see Examples 1–2 on pages 874–875 of your text and Section Lecture video for Section 13.4 and Exercise Solutions Clip 5 and 24.

Find each square by using the pattern for the square of a binomial.

1. $(7+x)^2$ **1.** ______________

2. $(2m-3p)^2$ **2.** ______________

3. $(4y-0.7)^2$ **3.** ______________

Name: Date:
Instructor: Section:

Objective 2 Find the product of the sum and difference of two terms.

For extra help, see Examples 3–4 on page 876 of your text and Section Lecture video for Section 13.4 and Exercise Solutions Clip 41 and 48.

Find each product by using the pattern for the sum and difference of two terms.

4. $(12+x)(12-x)$ 4. ______________

5. $(8k+5p)(8k-5p)$ 5. ______________

6. $\left(\frac{4}{7}t+2u\right)\left(\frac{4}{7}t-2u\right)$ 6. ______________

Objective 3 Find greater powers of binomials.

For extra help, see Example 5 on page 877 of your text and Section Lecture video for Section 13.4.

Find each product.

7. $(a-3)^3$ 7. ______________

8. $(j+3)^4$ 8. ______________

9. $(4s+3t)^4$ 9. ______________

Name: Date:
Instructor: Section:

Chapter 13 EXPONENTS AND POLYNOMIALS

13.5 Integer Exponents and the Quotient Rule

Learning Objectives	
1	Use 0 as an exponent.
2	Use negative numbers as exponents.
3	Use the quotient rule for exponents.
4	Use combinations of rules.

Key Terms

Use the vocabulary terms listed below to complete each statement in exercises 1–3.

exponent **base** **product rule for exponents**

power rule for exponents

1. The statement "If *m* and *n* are any integers, then $(a^m)^n = a^{mn}$" is an example of the ______________________________.

2. In the expression a^m, *a* is the ______________ and *m* is the ______________.

3. The statement "If *m* and *n* are any integers, then $a^m \cdot a^n = a^{m+n}$ is an example of the ______________________________.

Guided Examples

Review these examples for Objective 1:

1. Evaluate.

a. $75^0 = 1$

b. $(-75)^0 = 1$

c. $-75^0 = -(1)$ or -1

d. $x^0 = 1 \quad (x \neq 0)$

Now Try:

1. Evaluate.

a. 88^0

____

b. $(-88)^0$

____

c. -88^0

____

d. $a^0 \quad (a \neq 0)$

____

Name: Date:
Instructor: Section:

e. $9x^0 = 9(1)$, or $9 \quad (x \neq 0)$

f. $(9x)^0 = 1 \quad (x \neq 0)$

e. $88a^0 \quad (a \neq 0)$

____

f. $(88a)^0 \quad (a \neq 0)$

____

Review these examples for Objective 2:

2. Simplify by writing with positive exponents. Assume that all variables represent nonzero real numbers.

a. 5^{-2}

$$5^{-2} = \frac{1}{5^2}, \text{ or } \frac{1}{25}$$

b. 4^{-3}

$$4^{-3} = \frac{1}{4^3}, \text{ or } \frac{1}{64}$$

c. $\left(\frac{1}{3}\right)^{-3}$

$$\left(\frac{1}{3}\right)^{-3} = 3^3, \text{ or } 27$$

d. $\left(\frac{3}{5}\right)^{-4}$

$$\left(\frac{3}{5}\right)^{-4} = \left(\frac{5}{3}\right)^4 \quad \text{The reciprocal of } \frac{3}{5} \text{ is } \frac{5}{3}.$$

$$= \frac{5^4}{3^4}$$

$$= \frac{625}{81}$$

e. $\left(\frac{2}{3}\right)^{-5}$

$$\left(\frac{2}{3}\right)^{-5} = \left(\frac{3}{2}\right)^5$$

$$= \frac{3^5}{2^5}$$

$$= \frac{243}{32}$$

Now Try:

2. Simplify by writing with positive exponents. Assume that all variables represent nonzero real numbers.

a. 8^{-2}

____

b. 3^{-3}

____

c. $\left(\frac{1}{5}\right)^{-2}$

____

d. $\left(\frac{6}{7}\right)^{-2}$

____

e. $\left(\frac{3}{2}\right)^{-3}$

____

f. $5^{-1}-3^{-1}$

$$5^{-1}-3^{-1}=\frac{1}{5}-\frac{1}{3}$$
$$=\frac{3}{15}-\frac{5}{15}$$
$$=-\frac{2}{15}$$

g. $4q^{-3}$

$$4q^{-3}=\frac{4}{1}\cdot\frac{1}{q^3}$$
$$=\frac{4}{q^3}$$

h. $\frac{1}{x^{-6}}$

$$\frac{1}{x^{-6}}=\frac{1^{-6}}{x^{-6}}$$
$$=\left(\frac{1}{x}\right)^{-6}$$
$$=x^6$$

i. x^5y^{-8}

$$x^5y^{-8}=\frac{x^5}{1}\cdot\frac{1}{y^8}$$
$$=\frac{x^5}{y^8}$$

f. $4^{-1}-8^{-1}$ __________

g. $11p^{-4}$ __________

h. $\frac{1}{x^{-8}}$ __________

i. $a^{-5}b^4$ __________

3. Simplify. Assume that all variables represent nonzero real numbers.

a. $\frac{3^{-4}}{7^{-2}}=\frac{7^2}{3^4}$, or $\frac{49}{81}$

b. $\frac{a^{-6}}{b^{-1}}=\frac{b^1}{a^6}$, or $\frac{b}{a^6}$

c. $\frac{x^{-3}y}{4z^{-4}}=\frac{yz^4}{4x^3}$

3. Simplify. Assume that all variables represent nonzero real numbers.

a. $\frac{6^{-2}}{5^{-3}}$ __________

b. $\frac{x^{-7}}{y^{-1}}$ __________

c. $\frac{p^{-3}q}{4r^{-5}}$ __________

Name: Date:
Instructor: Section:

d. $\left(\frac{p}{3q}\right)^{-3}$

$$\left(\frac{p}{3q}\right)^{-3}=\left(\frac{3q}{p}\right)^{3}$$
$$=\frac{3^3q^3}{p^3}$$
$$=\frac{27q^3}{p^3}$$

d. $\left(\frac{a}{5b}\right)^{-4}$ __________

Review these examples for Objective 3:

4. Simplify. Assume that all variables represent nonzero real numbers.

a. $\frac{4^9}{4^6}=4^{9-6}=4^3=64$

b. $\frac{5^3}{5^7}=5^{3-7}=5^{-4}=\frac{1}{5^4}=\frac{1}{625}$

c. $\frac{3^{-4}}{3^{-9}}=3^{-4-(-9)}=3^5=243$

d. $\frac{p^6}{p^{-4}}=p^{6-(-4)}=p^{10}$

e. $\frac{4^3x^6}{4^5x^2}$

$$\frac{4^3x^6}{4^5x^2}=\frac{4^3}{4^5}\cdot\frac{x^6}{x^2}$$
$$=4^{3-5}\cdot x^{6-2}$$
$$=4^{-2}x^4$$
$$=\frac{x^4}{4^2},\quad \text{or}\quad \frac{x^4}{16}$$

Now Try:

4. Simplify. Assume that all variables represent nonzero real numbers.

a. $\frac{3^{18}}{3^{16}}$ __________

b. $\frac{2^6}{2^{10}}$ __________

c. $\frac{8^{-4}}{8^{-6}}$ __________

d. $\frac{z^6}{z^{-4}}$ __________

e. $\frac{5^4a^6}{5^6a^3}$ __________

Name: Date:
Instructor: Section:

f. $\dfrac{(x+7)^{-3}}{(x+7)^{-5}} \quad x \neq -7$

$$\frac{(x+7)^{-3}}{(x+7)^{-5}} = (x+7)^{-3-(-5)}$$
$$= (x+7)^{-3+5}$$
$$= (x+7)^2$$

g. $\dfrac{8x^{-4}y^3}{5^{-1}x^3y^{-4}}$

$$\frac{8x^{-4}y^3}{5^{-1}x^3y^{-4}} = \frac{8 \cdot 5y^3y^4}{x^3x^4}$$
$$= \frac{40y^7}{x^7}$$

f. $\dfrac{(a-b)^{-8}}{(a-b)^{-10}} \quad a \neq b$

____

g. $\dfrac{9a^{-5}b^3}{4^{-1}a^3b^{-4}}$

____

Review these examples for Objective 4:

5. Simplify each expression. Assume that all variables represent nonzero real numbers.

a. $\dfrac{(5^4)^2}{5^6}$

$$\frac{(5^4)^2}{5^6} = \frac{5^8}{5^6}$$
$$= 5^{8-6}$$
$$= 5^2$$
$$= 25$$

b. $(3a)^4(3a)^2$

$$(3a)^4(3a)^2 = (3a)^6$$
$$= 3^6a^6$$
$$= 729a^6$$

Now Try:

5. Simplify each expression. Assume that all variables represent nonzero real numbers.

a. $\dfrac{(6^3)^2}{6^5}$

____

b. $(5b)^3(5b)^2$

____

Name: Date:
Instructor: Section:

c. $\left(\frac{3x^4}{4}\right)^{-5}$

$$\left(\frac{3x^4}{4}\right)^{-5} = \left(\frac{4}{3x^4}\right)^5 = \frac{4^5}{3^5x^{20}} = \frac{1024}{243x^{20}}$$

c. $\left(\frac{2p^4}{3}\right)^{-5}$ ________________

d. $\left(\frac{5x^{-3}}{2^{-1}y^4}\right)^{-2}$

$$\left(\frac{5x^{-3}}{2^{-1}y^4}\right)^{-2} = \frac{5^{-2}x^6}{2^2y^{-8}} = \frac{x^6y^8}{5^2\cdot 2^2} = \frac{x^6y^8}{100}$$

d. $\left(\frac{4x^{-4}}{5^{-1}y^5}\right)^{-3}$ ________________

e. $\frac{(2m)^{-5}}{(3m)^{-6}}$

$$\frac{(2m)^{-5}}{(3m)^{-6}} = \frac{2^{-5}m^{-5}}{3^{-6}m^{-6}} = \frac{3^6m^6}{2^5m^5} = \frac{3^6m^{6-5}}{2^5} = \frac{3^6m}{2^5} = \frac{729m}{32}$$

e. $\frac{(7x)^{-2}}{(14x)^{-3}}$ ________________

Objective 1 Use 0 as an exponent.

For extra help, see Example 1 on page 881 of your text and Section Lecture video for Section 13.5 and Exercise Solutions Clip 9.

Evaluate each expression.

1. -12^0

1. ________________

Name: Date:
Instructor: Section:

2. $-15^0 - (-15)^0$ **2.** ________________

3. $\frac{0^8}{8^0}$ **3.** ________________

Objective 2 Use negative numbers as exponents.

For extra help, see Examples 2–3 on pages 882–884 of your text and Section Lecture video for Section 13.5 and Exercise Solutions Clip 23 and 27.

Evaluate or simplify each expression, and write it using only positive exponents. Assume that all variables represent nonzero real numbers.

4. $-2k^{-4}$ **4.** ________________

5. $(m^2n)^{-9}$ **5.** ________________

6. $\frac{2x^{-4}}{3y^{-7}}$ **6.** ________________

Objective 3 Use the quotient rule for exponents.

For extra help, see Example 4 on page 885 of your text and Section Lecture video for Section 13.5 and Exercise Solutions Clip 35, 37, and 49.

Use the quotient rule to simplify each expression, and write it using only positive exponents. Assume that all variables represent nonzero real numbers.

7. $\frac{4k^7m^{10}}{8k^3m^5}$ **7.** ________________

8. $\frac{a^4b^3}{a^{-2}b^{-3}}$ **8.** ________________

9. $\frac{3^{-1}m^{-4}p^6}{3^4m^{-1}p^{-2}}$ **9.** ________________

Name: Date:
Instructor: Section:

Objective 4 Use combinations of rules.

For extra help, see Example 5 on pages 886–887 of your text and Section Lecture video for Section 13.5 and Exercise Solutions Clip 59, 65, and 69.

Simplify each expression, and write it using only positive exponents. Assume that all variables represent nonzero real numbers.

10. $(9xy)^7(9xy)^{-8}$ **10.** ________________

11. $\dfrac{(a^{-1}b^{-2})^{-4}(ab^2)^6}{(a^3b)^{-2}}$ **11.** ________________

12. $\left(\dfrac{k^3t^4}{k^2t^{-1}}\right)^{-4}$ **12.** ________________

Name: Date:
Instructor: Section:

Chapter 13 EXPONENTS AND POLYNOMIALS

13.6 Dividing a Polynomial by a Monomial

Learning Objectives
1 Divide a polynomial by a monomial.

Key Terms

Use the vocabulary terms listed below to complete each statement in exercises 1–3.

quotient **dividend** **divisor**

1. In the division $\frac{5x^5-10x^3}{5x^2}=x^3-2x$, the expression $5x^5-10x^3$ is the ________________.

2. In the division $\frac{5x^5-10x^3}{5x^2}=x^3-2x$, the expression x^3-2x is the __________.

3. In the division $\frac{5x^5-10x^3}{5x^2}=x^3-2x$, the expression $5x^2$ is the ____________.

Guided Examples

Review these examples for Objective 1:

1. Divide $6x^4-18x^3$ by $6x^2$.

$$\frac{6x^4-18x^3}{6x^2}=\frac{6x^4}{6x^2}-\frac{18x^3}{6x^2}$$
$$=x^2-3x$$

Check Multiply. $6x^2(x^2-3x)=6x^4-18x^3$

2. Divide. $\frac{25a^6-15a^4+10a^2}{5a^3}$

Divide each term by $5a^3$.

$$\frac{25a^6-15a^4+10a^2}{5a^3}=\frac{25a^6}{5a^3}-\frac{15a^4}{5a^3}+\frac{10a^2}{5a^3}$$
$$=5a^3-3a+\frac{2}{a}$$

Check $5a^3\left(5a^3-3a+\frac{2}{a}\right)$
$$=5a^3(5a^3)+5a^3(-3a)+5a^3\left(\frac{2}{a}\right)$$
$$=25a^6-15a^4+10a^2$$

Now Try:

1. Divide $6x^4-18x^3$ by $6x^2$.

____

2. Divide. $\frac{27n^5-36n^4-18n^2}{9n^3}$

____

Name: Date:
Instructor: Section:

3. Divide $-12x^4 + 15x^5 - 5x$ by $-5x$.

Write the polynomial in descending powers before dividing.

$$\frac{15x^5 - 12x^4 - 5x}{-5x} = \frac{15x^5}{-5x} - \frac{12x^4}{-5x} - \frac{5x}{-5x}$$
$$= -3x^4 + \frac{12}{5}x^3 + 1$$

Check $-5x\left(-3x^4 + \frac{12}{5}x^3 + 1\right)$
$$= -5x\left(-3x^4\right) - 5x\left(\frac{12}{5}x^3\right) - 5x(1)$$
$$= 15x^5 - 12x^4 - 5x$$

3. Divide $-8z^5 + 7z^6 - 10z - 6$ by $2z^2$.

____

4. Divide $225x^5y^9 - 150x^3y^7 + 110x^2y^5 - 80xy^3 + 75y^2$ by $-25xy^2$.

$$\frac{225x^5y^9 - 150x^3y^7 + 110x^2y^5 - 80xy^3 + 75y^2}{-25xy^2}$$
$$= \frac{225x^5y^9}{-25xy^2} - \frac{150x^3y^7}{-25xy^2} + \frac{110x^2y^5}{-25xy^2} - \frac{80xy^3}{-25xy^2} + \frac{75y^2}{-25xy^2}$$
$$= -9x^4y^7 + 6x^2y^5 - \frac{22xy^3}{5} + \frac{16y}{5} - \frac{3}{x}$$

Check by multiplying the quotient by the divisor.

4. Divide $80a^5b^3 + 160a^4b^2 - 120a^2b$ by $-40a^2b$.

____

Objective 1 Divide a polynomial by a monomial.

For extra help, see Examples 1–4 on pages 892–893 of your text and Section Lecture video for Section 13.6 and Exercise Solutions Clip 11, 13, 17, 25, 27, and 31.

Perform each division.

1. $\dfrac{16a^5 - 24a^3}{8a^2}$

1. ____________

2. $\dfrac{12z^5 + 28z^4 - 8z^3 + 3z}{4z^3}$

2. ____________

3. $\dfrac{39m^4 - 12m^3 + 15}{-3m^2}$

3. ____________

Name: Date:
Instructor: Section:

Chapter 13 EXPONENTS AND POLYNOMIALS

13.7 Dividing a Polynomial by a Polynomial

Learning Objectives	
1	Divide a polynomial by a polynomial.
2	Apply division to a geometry problem.

Key Terms

Use the vocabulary terms listed below to complete each statement in exercises 1–3.

quotient **dividend** **divisor**

1. In the division $\frac{6x^2-9x-12}{2x-5}=3x+3+\frac{3}{2x-5}$, the expression $2x-5$ is the ____________.

2. In the division $\frac{6x^2-9x-12}{2x-5}=3x+3+\frac{3}{2x-5}$, the expression $3x+3+\frac{3}{2x-5}$ is the ____________.

3. In the division $\frac{6x^2-9x-12}{2x-5}=3x+3+\frac{3}{2x-5}$, the expression $6x^2-9x-12$ is the ____________.

Guided Examples

Review these examples for Objective 1:

1. Divide $8x+9x^3-7-9x^2$ by $3x-1$.

Write the dividend in descending powers as $9x^3-9x^2+8x-7$.

Step 1 $9x^3$ divided by $3x$ is $3x^2$.
$3x^2(3x-1)=9x^3-3x^2$

Step 2 Subtract. Bring down the next term.

Step 3 $-6x^2$ divided by $3x$ is $-2x$.
$-2x(3x-1)=-6x^2+2x$

Step 4 Subtract. Bring down the next term.

Step 5 $6x$ divided by $3x$ is 2.
$2(3x-1)=6x-2$

Now Try:

1. Divide $-12x^2+10x^3-3-8x$ by $5x-1$.

____

$$
\begin{array}{r}
3x^2 - 2x + 2 \\
3x-1\overline{)9x^3 - 9x^2 + 8x - 7} \\
\underline{9x^3 - 3x^2} \qquad\qquad \\
-6x^2 + 8x \qquad \\
\underline{-6x^2 + 2x} \qquad \\
6x - 7 \\
\underline{6x - 2} \\
-5
\end{array}
$$

$$\frac{9x^3 - 9x^2 + 8x - 7}{3x-1} = 3x^2 - 2x + 2 + \frac{-5}{3x-1}$$

Step 7 Multiply to check.

Check $(3x-1)\left(3x^2 - 2x + 2 + \frac{-5}{3x-1}\right)$

$= (3x-1)(3x^2) + (3x-1)(-2x)$

$+ (3x-1)(2) + (3x-1)\left(\frac{-5}{3x-1}\right)$

$= 9x^3 - 3x^2 - 6x^2 + 2x + 6x - 2 - 5$

$= 9x^3 - 9x^2 + 8x - 7$

2. Divide $x^3 - 64$ by $x - 4$.

Here the dividend is missing the x^2-term and the x-term. We use 0 as the coefficient for each missing term.

$$
\begin{array}{r}
x^2 + 4x + 16 \\
x-4\overline{)x^3 + 0x^2 + 0x - 64} \\
\underline{x^3 - 4x^2} \qquad\qquad \\
4x^2 + 0x \qquad \\
\underline{4x^2 - 16x} \qquad \\
16x - 64 \\
\underline{16x - 64} \\
0
\end{array}
$$

The remainder is 0. The quotient is $x^2 + 4x + 16$.

Check $(x-4)(x^2 + 4x + 16)$

$= x^3 + 4x^2 + 16x - 4x^2 - 16x - 64$

$= x^3 - 64$

2. Divide $x^3 - 1000$ by $x - 10$.

____

Name: Date:
Instructor: Section:

3. Divide $x^4 - 3x^3 + 7x^2 - 8x + 14$ by $x^2 + 2$.

Since $x^2 + 2$ is missing the x-term, we write it as $x^2 + 0x + 2$.

$$\begin{array}{r} x^2 - 3x + 5 \\ x^2 + 0x + 2 \overline{) x^4 - 3x^3 + 7x^2 - 8x + 14} \\ \underline{x^4 + 0x^3 + 2x^2 } \\ -3x^3 + 5x^2 - 8x \\ \underline{-3x^3 + 0x^2 - 6x } \\ 5x^2 - 2x + 14 \\ \underline{5x^2 + 0x + 10} \\ -2x + 4 \end{array}$$

The quotient is $x^2 - 3x + 5 + \dfrac{-2x + 4}{x^2 + 2}$

The check shows that the quotient multiplied by the divisor gives the original dividend.

3. Divide $3x^4 + 5x^3 - 7x^2 - 12x + 9$ by $x^2 - 4$.

____

4. Divide $5x^3 + 8x^2 + 12x - 1$ by $5x + 5$.

$$\begin{array}{r} x^2 + \frac{3}{5}x + \frac{9}{5} \\ 5x + 5 \overline{) 5x^3 + 8x^2 + 12x - 1} \\ \underline{5x^3 + 5x^2 } \\ 3x^2 + 12x \\ \underline{3x^2 + 3x } \\ 9x - 1 \\ \underline{9x + 9} \\ -10 \end{array}$$

The answer is $x^2 + \dfrac{3}{5}x + \dfrac{9}{5} - \dfrac{10}{5x + 5}$.

4. Divide $8x^3 - 7x^2 + 4x + 1$ by $8x - 8$.

____

Review this example for Objective 2:

5. The area of a rectangle is given by $12p^3 - 7p^2 + 5p - 1$ square units, and the width is $4p - 1$ units. What is the length of the rectangle?

For a rectangle, $A = LW$. Solving for L gives $L = \dfrac{A}{W}$. Divide the area, $12p^3 - 7p^2 + 5p - 1$ by the width $4p - 1$.

Now Try:

5. The area of a rectangle is given by $6r^3 - 5r^2 + 16r - 5$ square units, and the width is $3r - 1$ units. What is the length of the rectangle?

____

Name: Date:
Instructor: Section:

$$\begin{array}{r} 3p^2 - p + 1 \\ 4p-1 \overline{)12p^3 - 7p^2 + 5p - 1} \\ \underline{12p^3 - 3p^2} \qquad\qquad \\ -4p^2 + 5p \quad \\ \underline{-4p^2 + p} \quad \\ 4p - 1 \\ \underline{4p - 1} \\ 0 \end{array}$$

The length is $3p^2 - p + 1$ units.

Objective 1 Divide a polynomial by a polynomial.

For extra help, see Examples 1–4 on pages 897–899 of your text and Section Lecture video for Section 13.7 and Exercise Solutions Clip 5, 15, 17, 21, and 27.

Perform each division.

1. $\dfrac{-6x^2 + 23x - 20}{2x - 5}$

1. ________________

2. $\dfrac{6x^4 - 12x^3 + 13x^2 - 5x - 1}{2x^2 + 3}$

2. ________________

3. $\dfrac{2a^4 + 5a^2 + 3}{2a^2 + 3}$

3. ________________

Name: Date:
Instructor: Section:

Objective 2 Apply division to a geometry problem.

For extra help, see Example 5 on page 899 of your text and Section Lecture video for Section 13.7 and Exercise Solutions Clip 33.

Work each problem.

4. The area of a parallelogram is given by $4y^3 - 44y - 600$ square units, and the height is $y - 6$ units. What is the base of the parallelogram?

4. ________________

5. The area of a parallelogram is given by $3t^3 + 16t^2 - 32t - 64$ square units, and the base is $t^2 + 4t - 16$ units. What is the height of the parallelogram?

5. ________________

Name: Date:
Instructor: Section:

Chapter 13 EXPONENTS AND POLYNOMIALS

13.8 An Application of Exponents: Scientific Notation

Learning Objectives	
1	Express numbers in scientific notation.
2	Convert numbers in scientific notation to numbers without exponents.
3	Use scientific notation in calculations.

Key Terms

Use the vocabulary terms listed below to complete each statement in exercises 1−3.

scientific notation **quotient rule** **power rule**

1. A number written as $a \times 10^n$, where $1 \le |a| < 10$ and *n* is an integer, is written in ______________________________.

2. The statement "If *m* and *n* are any integers and $b \neq 0$, then $\left(\frac{a}{b}\right)^m = \frac{a^m}{b^m}$" is an example of the ______________________________.

3. The statement "If *m* and *n* are any integers and $b \neq 0$, then $\frac{a^m}{a^n} = a^{m-n}$" is an example of the ______________________________.

Guided Examples

Review these examples for Objective 1:

1. Write each number in scientific notation.

a. 54,000,000

Move the decimal point to follow the 5. Count the number of places the decimal point was moved: 7 places.

$54,000,000 = 5.4 \times 10^7$

b. 84,300,000,000

Move the decimal point 10 places to the left.

$84,300,000,000 = 8.43 \times 10^{10}$

c. 7.406

$7.406 = 7.406 \times 10^0$

Now Try:

1. Write each number in scientific notation.

a. 680,000,000

____

b. 47,710,000,000

____

c. 9.991

____

d. 0.00573

The first nonzero digit is 5. Count the places. Move the decimal point 3 places to the right.

$$0.00573 = 5.73 \times 10^{-3}$$

e. –0.0000617

Move the decimal point 5 places to the right.

$$-0.0000617 = -6.17 \times 10^{-5}$$

d. 0.0463

____

e. –0.000848

____

Review these examples for Objective 2:

2. Write each number without exponents.

a. 7.4×10^4

Move the decimal point 4 places to the right, and add three zeros.

$$7.4 \times 10^4 = 74{,}000$$

b. 3.57×10^6

Move the decimal point 6 places to the right, and add four zeros.

$$3.57 \times 10^6 = 3{,}570{,}000$$

c. -8.98×10^{-3}

Move the decimal point 3 places to the left.

$$-8.98 \times 10^{-3} = -0.00898$$

Now Try:

2. Write each number without exponents.

a. 8.35×10^5

____

b. 2.796×10^7

____

c. -1.64×10^{-4}

____

Review these examples for Objective 3:

3. Perform each calculation. Write answers in scientific notation and also without exponents.

a. $(8 \times 10^4)(7 \times 10^3)$

$$\begin{aligned}(8 \times 10^4)(7 \times 10^3) &= (8 \times 7)(10^4 \times 10^3)\\ &= 56 \times 10^7\\ &= (5.6 \times 10^1) \times 10^7\\ &= 5.6 \times 10^8\\ &= 560{,}000{,}000\end{aligned}$$

Now Try:

3. Perform each calculation. Write answers in scientific notation and also without exponents.

a. $(9 \times 10^5)(3 \times 10^2)$

____

Name: Date:
Instructor: Section:

b. $\dfrac{6\times10^{-4}}{3\times10^{2}}$

$$\frac{6\times10^{-4}}{3\times10^{2}}=\frac{6}{3}\times\frac{10^{-4}}{10^{2}}$$
$$=2\times10^{-6}$$
$$=0.000002$$

b. $\dfrac{39\times10^{-3}}{13\times10^{5}}$

____

4. A light-year is the distance that light travels in one year. The speed of light is about 3×10^{5} km per second. How many kilometers are in a light-year?

Convert km per second to km per year.

$$\frac{3\times10^{5}\text{km}}{1\text{ sec}}\cdot\frac{60\text{ sec}}{1\text{ min}}\cdot\frac{60\text{ min}}{1\text{ hr}}\cdot\frac{24\text{ hr}}{1\text{ day}}\cdot\frac{365\text{ day}}{1\text{ yr}}$$
$$=\left(3\times60^{2}\times24\times365\right)\times10^{5}$$
$$=94608000\times10^{5}$$
$$=\left(9.4608\times10^{7}\right)\times10^{5}$$
$$\approx9.46\times10^{12}\text{ km/yr}$$

Thus, there are 9.46×10^{12} km in a light-year.

4. Earth has a mass of 6×10^{24} kilograms and a volume of 1.1×10^{21} cubic meters. What is Earth's density in kilograms per cubic meter? Round to the nearest hundredth.

____

5. The Sahara desert covers approximately 3.5×10^{6} square miles. Its sand is, on average, 12 feet deep. Find the volume, in cubic feet, of sand in the Sahara. $\left(\text{Hint: }1\text{ mi}^{2}=5280^{2}\text{ ft}^{2}\right)$ Round your answer to two decimal places.

$$\left(3.5\times10^{6}\right)\left(5280^{2}\right)(12)$$
$$=97574400(12)\times10^{6}$$
$$=1170892800\times10^{6}$$
$$\approx1.17\times10^{15}\text{ ft}^{3}$$

The volume is 1.17×10^{15} cubic feet.

5. The Sahara desert covers approximately 3.5×10^{6} square miles. Its sand is, on average, 12 feet deep. The volume of a single grain of sand is approximately 1.3×10^{-9} cubic feet. About how many grains of sand are in the Sahara?

____

Objective 1 Express numbers in scientific notation.

For extra help, see Example 1 on page 903 of your text and Section Lecture video for Section 13.8 and Exercise Solutions Clip 13 and 17.

Write each number in scientific notation.

1. 23,651

1. ____________

Name: Date:
Instructor: Section:

2. $-429{,}600{,}000{,}000$ **2.** ________________

3. -0.0002208 **3.** ________________

Objective 2 Convert numbers in scientific notation to numbers without exponents.

For extra help, see Example 2 on page 904 of your text and Section Lecture video for Section 13.8 and Exercise Solutions Clip 21 and 29.

Write each number in scientific notation.

4. -2.45×10^{6} **4.** ________________

5. 6.4×10^{-3} **5.** ________________

6. -4.02×10^{0} **6.** ________________

Objective 3 Use scientific notation in calculations.

For extra help, see Examples 3–5 on pages 904–905 of your text and Section Lecture video for Section 13.8 and Exercise Solutions Clip 35, 49, and 61.

Perform the indicated operations, and write the answers in scientific notation.

7. $\left(2.3\times10^{4}\right)\times\left(1.1\times10^{-2}\right)$ **7.** ________________

8. $\dfrac{9.39\times10^{1}}{3\times10^{3}}$ **8.** ________________

Work the problem. Give answer in scientific notation.

9. There are about 6×10^{23} atoms in a mole of atoms. How many atoms are there in 8.1×10^{-5} mole? **9.** ________________

Name: Date:
Instructor: Section:

Chapter 14 FACTORING AND APPLICATIONS

14.1 Factors; The Greatest Common Factor

Learning Objectives

1. Find the greatest common factor of a list of numbers.
2. Find the greatest common factor of a list of variable terms.
3. Factor out the greatest common factor.
4. Factor by grouping.

Key Terms

Use the vocabulary terms listed below to complete each statement in exercises 1–4.

factor **factored form** **greatest common factor (GCF)**

factoring

1. The process of writing a polynomial as a product is called ________________.

2. An expression is in ________________________ when it is written as a product.

3. The ________________________ is the largest quantity that is a factor of each of a group of quantities.

4. An expression A is a ___________________ of an expression B if B can be divided by A with 0 remainder.

Guided Examples

Review these examples for Objective 1:

1. Find the greatest common factor for each list of numbers.

   a. 60, 45

   Write the prime factored form of each number.

   $60 = 2 \cdot 2 \cdot 3 \cdot 5$

   $45 = 3 \cdot 3 \cdot 5$

   Use each prime the least number of times it appears in all the factored forms.

   $\text{GCF} = 3^1 \cdot 5^1 = 15$

Now Try:

1. Find the greatest common factor for each list of numbers.

   a. 18, 36, 42

   ____

b. 90, 36, 108

Write the prime factored form of each number.

$90 = 2 \cdot 3 \cdot 3 \cdot 5$

$36 = 2 \cdot 2 \cdot 3 \cdot 3$

$108 = 2 \cdot 2 \cdot 3 \cdot 3 \cdot 3$

There is one factor of 2 and two factors of 3.

$\text{GCF} = 2^1 \cdot 3^2 = 18$

c. 17, 18, 24

Write the prime factored form of each number.

$17 = 17$

$18 = 2 \cdot 3 \cdot 3$

$24 = 2 \cdot 2 \cdot 2 \cdot 3$

There are no primes common to all three numbers, so the GCF is 1.

b. 32, 40, 72

____

c. 26, 27, 28

____

Review these examples for Objective 2:

2. Find the greatest common factor for each list of terms.

a. $28m^4$, $35m^6$, $49m^9$, $70m^5$

$28m^4 = 2 \cdot 2 \cdot 7 \cdot m^4$

$35m^6 = 5 \cdot 7 \cdot m^6$

$49m^9 = 7 \cdot 7 \cdot m^9$

$70m^5 = 2 \cdot 5 \cdot 7 \cdot m^5$

Then, $\text{GCF} = 7m^4$.

b. a^5b^3, a^7b^4, a^3b^5, b^9

$a^5b^3 = a^5 \cdot b^3$

$a^7b^4 = a^7 \cdot b^4$

$a^3b^5 = a^3 \cdot b^5$

$b^9 = b^9$

Then, $\text{GCF} = b^3$.

Now Try:

2. Find the greatest common factor for each list of terms.

a. $54x^5$, $48x^7$, $42x^9$, $30x^4$

____

b. p^5q^7, p^4q^3, p^8q^5, q^9

____

Name: Date:
Instructor: Section:

Review these examples for Objective 3:

3. Write in factored form by factoring out the greatest common factor.

a. $7a^2 + 14a$

GCF = $7a$

$$7a^2 + 14a = 7a(a) + 7a(2)$$
$$= 7a(a+2)$$

Check Multiply the factored form.

$$7a(a+2) = 7a(a) + 7a(2)$$
$$= 7a^2 + 14a$$

b. $12x^5 + 27x^4 - 15x^3$

GCF = $3x^3$

$$12x^5 + 27x^4 - 15x^3$$
$$= 3x^3(4x^2) + 3x^3(9x) + 3x^3(-5)$$
$$= 3x^3(4x^2 + 9x - 5)$$

Check Multiply the factored form.

$$3x^3(4x^2 + 9x - 5)$$
$$= 3x^3(4x^2) + 3x^3(9x) + 3x^3(-5)$$
$$= 12x^5 + 27x^4 - 15x^3$$

c. $y^7 + y^4$

GCF = y^4

$$y^7 + y^4 = y^4(y^3) + y^4(1)$$
$$= y^4(y^3 + 1)$$

Check mentally by distributing y^4 over each term inside the parentheses.

d. $40y^5z^3 - 56y^3z^5$

GCF = $8y^3z^3$

$$40y^5z^3 - 56y^3z^5 = 8y^3z^3(5y^2) - 8y^3z^3(7z^2)$$
$$= 8y^3z^3(5y^2 - 7z^2)$$

Check mentally by distributing $8y^3z^3$ over each term inside the parentheses.

Now Try:

3. Write in factored form by factoring out the greatest common factor.

a. $8x^5 + 24x$

____

b. $20y^4 - 12y^3 + 4y^2$

____

c. $n^8 + n^7$

____

d. $64x^5 - 40x^4 + 8x^3$

____

Name: Date:
Instructor: Section:

4. Write $-15x^5+35x^4-5x^2$ in factored form.

$-5x^2$ is a common factor.

$$-15x^5+35x^4-5x^2$$
$$=-5x^2(3x^3)-5x^2(-7x^2)-5x^2(1)$$
$$=-5x^2(3x^3-7x^2+1)$$

Check

$$-5x^2(3x^3-7x^2+1)$$
$$=-5x^2(3x^3)-5x^2(-7x^2)-5x^2(1)$$
$$=-15x^5+35x^4-5x^2$$

5. Write in factored form by factoring out the greatest common factor.

a. $x(x+9)+7(x+9)$

Factor out $x+9$.

$$x(x+9)+7(x+9)=(x+9)(x+7)$$

b. $a^2(a+6)-7(a+6)$

Factor out $a+6$.

$$a^2(a+6)-7(a+6)=(a+6)(a^2-7)$$

4. Write $-9p^2q^3-27p^4q^2+9pq$ in factored form.

____

5. Write in factored form by factoring out the greatest common factor.

a. $y(y+8)+4(y+8)$

____

b. $z^2(z+5)-11(z+5)$

____

Review these examples for Objective 4:

6. Factor by grouping.

a. $5x+15+bx+3b$

Group the first two terms and the last two terms.

$$5x+15+bx+3b$$
$$=(5x+15)+(bx+3b)$$
$$=5(x+3)+b(x+3)$$
$$=(x+3)(5+b)$$

Check Use the FOIL method.

$$(x+3)(5+b)=5x+bx+3(5)+3b$$
$$=5x+15+bx+3b$$

Now Try:

6. Factor by grouping.

a. $2x+xy+14+7y$

____

Name: Date:
Instructor: Section:

b. $7by+28y+b+4$

Group the terms, then factor each group.

$$7by+28y+b+4$$
$$=(7by+28y)+(b+4)$$
$$=7y(b+4)+1(b+4)$$
$$=(b+4)(7y+1)$$

Check Use the FOIL method.

$$(b+4)(7y+1)=7by+1b+4(7y)+4$$
$$=7by+28y+b+4$$

c. $3x^2-18x+5xy-30y$

$$3x^2-18x+5xy-30y$$
$$=(3x^2-18x)+(5xy-30y)$$
$$=3x(x-6)+5y(x-6)$$
$$=(x-6)(3x+5y)$$

Check by multiplying using the FOIL method.

d. $r^3+3r^2-5r-15$

$$r^3+3r^2-5r-15$$
$$=(r^3+3r^2)+(-5r-15)$$
$$=r^2(r+3)-5(r+3)$$
$$=(r+3)(r^2-5)$$

Check by multiplying using the FOIL method.

7. Factor by grouping.

a. $18x^2-20y+24x-15xy$

Group the terms, then factor each group.

$$18x^2-20y+24x-15xy$$
$$=2(9x^2-10y)+3x(8-5y)$$

This does not lead to a common factor, so we try rearranging the terms.

$$18x^2-20y+24x-15xy$$
$$=18x^2-15xy+24x-20y$$
$$=(18x^2-15xy)+(24x-20y)$$
$$=3x(6x-5y)+4(6x-5y)$$
$$=(6x-5y)(3x+4)$$

b. $36x+4tx+9+t$

____

c. $4x^2-28x+5xy-35y$

____

d. $x^3+7x^2-2x-14$

____

7. Factor by grouping.

a. $56x^2+32x-21xy-12y$

____

Check Use the FOIL method.

$$(6x-5y)(3x+4)$$
$$=18x^2+24x-15xy-20y$$
$$=18x^2-20y+24x-15xy$$

b. $5xy+36-4y-45x$

Rearrange the terms to get two groups that each have a common factor.

$$5xy+36-4y-45x$$
$$=(5xy-4y)+(-45x+36)$$
$$=y(5x-4)-9(5x-4)$$
$$=(5x-4)(y-9)$$

Check by multiplying using the FOIL method.

b. $9xy-18x-5y+10$

____

Objective 1 Find the greatest common factor of a list of numbers.

For extra help, see Example 1 on page 920–921 of your text and Section Lecture video for Section 14.1 and Exercise Solutions Clip 5 and 9.

Find the greatest common factor for each group of numbers.

1. 84, 280, 112 **1.** __________

2. 56, 21, 49 **2.** __________

3. 42, 48, 72 **3.** __________

Objective 2 Find the greatest common factor of a list of variable terms.

For extra help, see Example 2 on page 921 of your text and Section Lecture video for Section 14.1 and Exercise Solutions Clip 11 and 15.

Find the greatest common factor for each list of terms.

4. $12ab^3$, $18a^2b^4$, $26ab^2$, $32a^2b^2$ **4.** __________

5. $6k^2m^4n^5$, $8k^3m^7n^4$, $k^4m^8n^7$ **5.** __________

Name: Date:
Instructor: Section:

6. $9xy^4,\ 72x^4y^7,\ 27xy^2,\ 108x^2y^5$ 6. ______

Objective 3 Factor out the greatest common factor.

For extra help, see Examples 3–5 on pages 922–924 of your text and Section Lecture video for Section 14.1 and Exercise Solutions Clip 39 and 59.

Factor out the greatest common factor or a negative common factor if the coefficient of the term of greatest degree is negative.

7. $20x^2 + 40x^2y - 70xy^2$ 7. ______

8. $2a(x-2y)+9b(x-2y)$ 8. ______

9. $26x^8 - 13x^{12} + 52x^{10}$ 9. ______

Objective 4 Factor by grouping.

For extra help, see Examples 6–7 on pages 924–926 of your text and Section Lecture video for Section 14.1 and Exercise Solutions Clip 73 and 91.

Factor each polynomial by grouping.

10. $15 - 5x - 3y + xy$ 10. ______

11. $2x^2 - 14xy + xy - 7y^2$ 11. ______

12. $3r^3 - 2r^2s + 3s^2r - 2s^3$ 12. ______

Name: Date:
Instructor: Section:

Chapter 14 FACTORING AND APPLICATIONS

14.2 Factoring Trinomials

Learning Objectives
1. Factor trinomials with a coefficient of 1 for the second-degree term.
2. Factor such trinomials after factoring out the greatest common factor.

Key Terms

Use the vocabulary terms listed below to complete each statement in exercises 1–3.

prime polynomial **factoring** **greatest common factor**

1. ______________________ is the process of writing a polynomial as a product.

2. The ______________________ of a polynomial is the greatest term that is a factor of all the terms in the polynomial.

3. A ______________________ is a polynomial that cannot be factored using only integers.

Guided Examples

Review these examples for Objective 1:

1. Factor $m^2+8m+15$.

   Look for integers whose product is 15 and whose sum is 8. Only positive signs are needed.

Factors of 15	Sums of Factors
15, 1	$15+1=16$
5, 3	$5+3=8$

From the table, 5 and 3 are the required integers.

$m^2+8m+15$ factors as $(m+5)(m+3)$

Check Use the FOIL method.

$$(m+5)(m+3)=m^2+3m+5m+15$$
$$=m^2+8m+15$$

Now Try:

1. Factor $x^2+11x+24$.

____

Name: Date:
Instructor: Section:

2. Factor $x^2 - 11x + 28$.

Look for integers whose product is 28 and whose sum is –11. Since the numbers have a positive product and a negative sum, we consider only pairs of negative integers.

Factors of 28	Sums of Factors
$-28, -1$	$-28 + (-1) = -29$
$-14, -2$	$-14 + (-2) = -16$
$-7, -4$	$-7 + (-4) = -11$

The required integers are –7 and –4.

$x^2 - 11x + 28$ factors as $(x-7)(x-4)$

Check Use the FOIL method.

$$(x-7)(x-4) = x^2 - 4x - 7x + 28$$
$$= x^2 - 11x + 28$$

3. Factor $x^2 + 2x - 15$.

Look for integers whose product is –15 and whose sum is 2. To get a negative product, the pairs of integers must have different signs.

Factors of -15	Sums of Factors
$15, -1$	$15 + (-1) = 14$
$-15, 1$	$-15 + 1 = -14$
$5, -3$	$5 + (-3) = 2$

The required integers are 5 and –3.

$x^2 + 2x - 15$ factors as $(x+5)(x-3)$

Check Use the FOIL method.

$$(x+5)(x-3) = x^2 - 3x + 5x - 15$$
$$= x^2 + 2x - 15$$

2. Factor $y^2 - 12y + 35$.

____

3. Factor $p^2 + 6p - 27$.

____

Name: Date:
Instructor: Section:

4. Factor $x^2 - 10x - 39$.

Look for integers whose product is –39 and whose sum is –10. Because the constant term, –39, is negative, we need pairs of integers with different signs.

Factors of -39	Sums of Factors
$39, -1$	$39 + (-1) = 38$
$-39, 1$	$-39 + 1 = -38$
$3, -13$	$3 + (-13) = -10$

The required integers are –13 and 3.

$x^2 - 10x - 39$ factors as $(x - 13)(x + 3)$

Check Use the FOIL method.

$$(x-13)(x+3) = x^2 + 3x - 13x - 39$$
$$= x^2 - 10x - 39$$

4. Factor $a^2 - 15a - 34$.

____

5. Factor each trinomial.

a. $x^2 - 7x + 18$

Look for integers whose product is 18 and whose sum is –7. Since the numbers have a positive product and a negative sum, we consider only pairs of negative integers.

Factors of 18	Sums of Factors
$-18, -1$	$-18 + (-1) = -19$
$-9, -2$	$-9 + (-2) = -11$
$-6, -3$	$-6 + (-3) = -9$

None of the pairs of integers has a sum of –7.

$x^2 - 7x + 18$ cannot be factored.

It is a prime polynomial.

b. $k^2 - 5x + 13$

There are no pairs of integers whose product is 13 and whose sum is –5, so $k^2 - 5x + 13$ is a prime polynomial.

5. Factor each trinomial.

a. $m^2 - 7m + 5$

____

b. $x^2 - 3x + 10$

____

6. Factor $x^2 - 6xy - 7y^2$.

Here, the coefficient of *x* in the middle term is –6*y*, so we need to find two expressions whose product is $-7y^2$ and whose sum is –6*y*.

6. Factor $p^2 - 5pq - 14q^2$.

____

Name: Date:
Instructor: Section:

Factors of $-7y^2$	Sums of Factors
$7y, -y$	$7y+(-y)=6y$
$-7y, y$	$-7y+y=-6y$

$x^2-6xy-7y^2$ factors as $(x-7y)(x+y)$

Check Use the FOIL method.

$$(x-7y)(x+y)=x^2+xy-7xy+7y^2$$
$$=x^2-6xy-7y^2$$

Review this example for Objective 2:

7. Factor $5x^5-45x^4+90x^3$.

There is no second-degree term. Look for a common factor.

$$5x^5-45x^4+90x^3=5x^3(x^2-9x+18)$$

Now factor $x^2-9x+18$. The integers –3 and –6 have a product of 18 and a sum of –9.

$5x^5-45x^4+90x^3$ factors as $5x^3(x-6)(x-3)$

Check Use the FOIL method.

$$5x^3(x-6)(x-3)$$
$$=5x^3(x^2-3x-6x+18)$$
$$=5x^3(x^2-9x+18)$$
$$=5x^5-45x^4+90x^3$$

Now Try:

7. Factor $7x^6-49x^5+70x^4$.

____

Objective 1 Factor trinomials with a coefficient of 1 for the second-degree term.

For extra help, see Examples 1–6 on pages 931–933 of your text and Section Lecture video for Section 14.2 and Exercise Solutions Clip 27, 33, 35, 37, 39, and 47.

Factor completely. If a polynomial cannot be factored, write prime.

1. r^2+r+3 **1.** ____________

2. $x^2-11x+28$ **2.** ____________

3. $x^2-8x-33$ **3.** ____________

Name: Date:
Instructor: Section:

Objective 2 Factor such trinomials after factoring out the greatest common factor.

For extra help, see Example 7 on page 934 of your text and Section Lecture video for Section 14.2 and Exercise Solutions Clip 55.

Factor completely. If a polynomial cannot be factored, write **prime**.

4. $2n^4 - 16n^3 + 30n^2$ **4.** ________________

5. $2a^3b - 10a^2b^2 + 12ab^3$ **5.** ________________

6. $10k^6 + 70k^5 + 100k^4$ **6.** ________________

Name: Date:
Instructor: Section:

Chapter 14 FACTORING AND APPLICATIONS

14.3 Factoring Trinomials by Grouping

Learning Objectives
1 Factor trinomials by grouping when the coefficient of the second-degree term is not 1.

Key Terms

Use the vocabulary terms listed below to complete each statement in exercises 1–2.

coefficient **trinomial**

1. In the term $6x^2y$, 6 is the ______________________.

2. A polynomial with three terms is a ______________________.

Guided Examples

Review these examples for Objective 1:

1. Factor $3x^2+11x+10$.

Look for two positive integers whose product is $3\cdot 10=30$ and whose sum is 11.

The integers are 5 and 6, since $5\cdot 6=30$ and $5+6=11$.

$$\begin{aligned}3x^2+11x+10 &= 3x^2+5x+6x+10\\ &= (3x^2+5x)+(6x+10)\\ &= x(3x+5)+2(3x+5)\\ &= (3x+5)(x+2)\end{aligned}$$

Check Multiply $(3x+5)(x+2)$ to obtain $3x^2+11x+10$.

2. Factor each trinomial.

a. $8x^2-2x-1$

We must find two integers with a product of $8(-1)=-8$ and a sum of –2. The integers are –4 and 2. We write the middle term as $-4x+2x$.

Now Try:

1. Factor $5x^2+17x+6$.

____

2. Factor each trinomial.

a. $14x^2-3x-5$

____

$$\begin{aligned}8x^2-2x-1&=8x^2-4x+2x-1\\&=\left(8x^2-4x\right)+(2x-1)\\&=4x(2x-1)+1(2x-1)\\&=(2x-1)(4x+1)\end{aligned}$$

Check Multiply $(2x-1)(4x+1)$ to obtain $8x^2-2x-1$.

b. $15z^2+z-2$

Look for two integers whose product is $15(-2)=-30$ and whose sum is 1.

The integers are 6 and –5.

$$\begin{aligned}15z^2+z-2&=15z^2-5z+6z-2\\&=\left(15z^2-5z\right)+(6z-2)\\&=5z(3z-1)+2(3z-1)\\&=(3z-1)(5z+2)\end{aligned}$$

Check Multiply $(3z-1)(5z+2)$ to obtain $15z^2+z-2$.

c. $12r^2+5rs-2s^2$

Two integers whose product is $12(-2)=-24$ and whose sum is 5 are 8 and –3. Rewrite the trinomial with four terms.

$$\begin{aligned}12r^2+5rs-2s^2&=12r^2+8rs-3rs-2s^2\\&=\left(12r^2+8rs\right)+\left(-3rs-2s^2\right)\\&=4r(3r+2s)-s(3r+2s)\\&=(3r+2s)(4r-s)\end{aligned}$$

Check Multiply $(3r+2s)(4r-s)$ to obtain $12r^2+5rs-2s^2$.

3. Factor $100x^5+140x^4-15x^3$.

Factor out the greatest common factor, $5x^3$.

$$100x^5+140x^4-15x^3=5x^3\left(20x^2+28x-3\right)$$

To factor $20x^2+28x-3$, find two integers whose product is $20(-3)=-60$ and whose sum is 28. Factor 60 into prime factors.

b. $3m^2-m-14$

____

c. $10x^2+xy-3y^2$

____

3. Factor $30x^5+87x^4-63x^3$.

____

$60 = 2 \cdot 2 \cdot 3 \cdot 5$

Combine the prime factors in pairs using one positive factor and one negative factor to get –60. The factors of 30 and –2 have the correct sum, 28.

$$\begin{aligned}
&100x^5 + 140x^4 - 15x^3 \\
&= 5x^3(20x^2 + 28x - 3) \\
&= 5x^3(20x^2 + 30x - 2x - 3) \\
&= 5x^3[(20x^2 + 30x) + (-2x - 3)] \\
&= 5x^3[10x(2x + 3) - 1(2x + 3)] \\
&= 5x^3(2x + 3)(10x - 1)
\end{aligned}$$

Objective 1 Factor trinomials by grouping when the coefficient of the second-degree term is not 1.

For extra help, see Examples 1–3 on page 937–938 of your text and Section Lecture video for Section 14.3 and Exercise Solutions Clip 17, 31, 33, and 37.

Factor each trinomial by grouping.

1. $8b^2 + 18b + 9$ **1.** ________________

2. $7a^2b + 18ab + 8b$ **2.** ________________

3. $10c^2 - 29ct + 21t^2$ **3.** ________________

Name: Date:
Instructor: Section:

Chapter 14 FACTORING AND APPLICATIONS

14.4 Factoring Trinomials By Using the FOIL Method

Learning Objectives
1 Factor trinomials by using the FOIL method.

Key Terms

Use the vocabulary terms listed below to complete each statement in exercises 1–3.

FOIL **outer product** **inner product**

1. The ______________________________ of $(2y-5)(y+8)$ is $-5y$.

2. ________________ is a shortcut method for finding the product of two binomials.

3. The ______________________________ of $(2y-5)(y+8)$ is $16y$.

Guided Examples

Review these examples for Objective 1:

1. Factor $5x^2+14x+8$.

The product of the first terms of the binomial is $5x^2$. The possible factors are 5x and x, since we consider only positive factors. So we have

$$5x^2+14x+8=(5x+\_\_\_)(x+\_\_\_)$$

The product of the last terms is 8. The positive factors are $1\cdot 8$, $8\cdot 1$, $4\cdot 2$, or $2\cdot 4$. We want the middle term of 14.

Try $1\cdot 8$ in $(5x+\_\_\_)(x+\_\_\_)$.

$$(5x+1)(x+8)$$

gives middle term $40x+x=41x$. Incorrect.

Try $8\cdot 1$ in $(5x+\_\_\_)(x+\_\_\_)$.

$$(5x+8)(x+1)$$

gives middle term $5x+8x=13x$. Incorrect.

Try $4\cdot 2$ in $(5x+\_\_\_)(x+\_\_\_)$.

$$(5x+4)(x+2)$$

gives middle term $10x+4x=14x$. Correct.

Thus, $5x^2+14x+8=(5x+4)(x+2)$.

Check. Multiply $(5x+4)(x+2)$ to obtain $5x^2+14x+8$.

Now Try:

1. Factor $3x^2+17x+10$.

____

Name: Date:
Instructor: Section:

2. Factor $6x^2 + 13x + 7$.

The number 6 has several possible pairs of factors, but 7 has only 1 and 7, or –1 and –7. We choose positive factors since all coefficients in the trinomial are positive.

$(\_\_\_ + 7)(\_\_\_ + 1)$

The possible pairs of $6x^2$ are $6x$ and x, or $3x$ and $2x$.

$(3x + 7)(2x + 1)$

gives middle term $3x + 14x = 17x$. Incorrect.

$(2x + 7)(3x + 1)$

gives middle term $2x + 21x = 23x$. Incorrect.

$(6x + 7)(x + 1)$

gives middle term $6x + 7x = 13x$. Correct.

$6x^2 + 13x + 7$ factors as $(6x + 7)(x + 1)$.

Check. Multiply $(6x + 7)(x + 1)$ to obtain $6x^2 + 13x + 7$.

2. Factor $15x^2 + 26x + 7$.

____

3. Factor $10x^2 - 19x + 7$.

Since 7 has only 1 and 7 or –1 and –7 as factors, it is better to begin by factoring 7. We need two negative factors because the product of two negative factors is positive and their sum is negative, as required.
We try –1 and –7.

$(\_\_\_ - 1)(\_\_\_ - 7)$

The factors of $10x^2$ are $10x$ and x, or $5x$ and $2x$.

$(10x - 1)(x - 7)$

has middle term $-70x - x = -71x$. Incorrect.

$(5x - 1)(2x - 7)$

has middle term $-35x - 2x = -37x$. Incorrect.

$(2x - 1)(5x - 7)$

has middle term $-14x - 5x = -19x$. Correct.

Thus $10x^2 - 19x + 7$ factors as $(2x - 1)(5x - 7)$.

3. Factor $20x^2 - 13x + 2$.

____

Name: Date:
Instructor: Section:

4. Factor $6x^2 - x - 15$.

The integer 6 has several possible pairs of factors, as does –15. Since the constant term is negative, one positive factor and one negative factor of –15 are needed. Since the coefficient of the middle term is relatively small, it is wise to avoid large factors. We try 3*x* and 2*x* as factors of $6x^2$ and 5 and –3 as factors of –15.

$(3x+5)(2x-3)$
has middle term $-9x+10x=x$. Incorrect.

$(3x-5)(2x+3)$
has middle term $9x-10x=-x$. Correct.

$6x^2 - x - 15$ factors as $(3x-5)(2x+3)$.

5. Factor $18x^2 - 3xy - 28y^2$.

There are several factors of $18x^2$, including 18*x* and *x*, 9*x* and 2*x*, and 6*x* and 3*x*.
There are many possible pairs of factors of $-28y^2$, including
28*y* and –*y*, –28*y* and *y*, 14*y* and –2*y*, –14*y* and 2*y*, 7*y* and –4*y*, –7*y* and 4*y*.

Once again, since the coefficient of the middle term is relatively small, avoid the larger factors. Try the factors of 6*x* and 3*x*, and 4*y* and –7*y*.

$(6x+4y)(3x-7y)$ Incorrect

The first binomial has a common factor of 2.

$(6x-7y)(3x+4y)$
has middle term $24xy-21xy=3xy$. Incorrect.

Interchange the signs of the last two terms.

$(6x+7y)(3x-4y)$
has middle term $-24xy+21xy=-3xy$. Correct.

Thus, $18x^2 - 3xy - 28y^2$ factors as $(6x+7y)(3x-4y)$

4. Factor $8x^2 + 2x - 21$.

____

5. Factor $24x^2 - 2xy - 15y^2$.

____

Name: Date:
Instructor: Section:

6. Factor each trinomial.

a. $20x^4 + 148x^3 + 56x^2$

Factor out the greatest common factor, $4x^2$.

$$20x^4 + 148x^3 + 56x^2$$
$$= 4x^2(5x^2 + 37x + 14)$$

Now factor $5x^2 + 37x + 14$. Try $5x$ and x as factors of $5x^2$ and 2 and 7 as factors of 14.

$(5x+2)(x+7) = 5x^2 + 37x + 14$ Correct.

This leads to the completely factored form

$$20x^4 + 148x^3 + 56x^2 = 4x^2(5x+2)(x+7)$$

Check. $4x^2(5x+2)(x+7)$

$$= 4x^2(5x^2 + 37x + 14)$$
$$= 20x^4 + 148x^3 + 56x^2$$

b. $-105a^3 + 65a^2 - 10a$

The common factor is $-5a$. Then use trial and error.

$$-105a^3 + 65a^2 - 10a = -5a(21a^2 - 13a + 2)$$
$$= -5a(3a-1)(7a-2)$$

Check.

$$-5a(3a-1)(7a-2) = -5a(21a^2 - 13a + 2)$$
$$= -105a^3 + 65a^2 - 10a$$

6. Factor each trinomial.

a. $10y^3 + 22y^2 + 12y$

____

b. $-18a^3 + 66a^2 - 60a$

____

Objective 1 Factor trinomials by using the FOIL method.

For extra help, see Examples 1–6 on pages 941–944 of your text and Section Lecture video for Section 14.4 and Exercise Solutions Clip 7, 15, 23, 27, 35, 41, and 43.

Factor each trinomial completely.

1. $8q^2 + 10q + 3$ **1.** ______________

2. $3a^2 + 8ab + 4b^2$ **2.** ______________

3. $4c^2 + 14cd - 8d^2$ **3.** ______________

Name: Date:
Instructor: Section:

Chapter 14 FACTORING AND APPLICATIONS

14.5 Special Factoring Techniques

Learning Objectives
1 Factor a difference of squares.
2 Factor a perfect square trinomial.

Key Terms

Use the vocabulary terms listed below to complete each statement in exercises 1–2.

perfect square trinomial **difference**

1. A ____________________ is the result of a subtraction.

2. A ________________________ is a trinomial that can be factored as the square of a binomial.

Guided Examples

Review these examples for Objective 1:

1. Factor each binomial, if possible.

a. $x^2 - 64$

$x^2 - 64 = x^2 - 8^2 = (x+8)(x-8)$

b. $p^2 - q^2$

$p^2 - q^2 = (p+q)(p-q)$

c. $y^2 - 27$

Because 27 is not a square of an integer, this binomial is not a difference in squares. It is a prime polynomial.

d. $c^2 + 25$

Since $c^2 + 25$ is a sum of squares, it is a prime polynomial.

Now Try:

1. Factor each binomial, if possible.

a. $z^2 - 36$ ____________

b. $r^2 - s^2$ ____________

c. $y^2 - 110$ ____________

d. $x^2 + 100$ ____________

Name: Date:
Instructor: Section:

2. Factor each difference of squares.

a. $36x^2 - 25$

$$36x^2 - 25 = (6x)^2 - 5^2$$
$$= (6x+5)(6x-5)$$

b. $81y^2 - 49$

$$81y^2 - 49 = (9y)^2 - 7^2$$
$$= (9y+7)(9y-7)$$

c. $100a^2 - 9b^2$

$$100a^2 - 9b^2 = (10a)^2 - (3b)^2$$
$$= (10a+3b)(10a-3b)$$

2. Factor each difference of squares.

a. $4x^2 - 81$

____

b. $25t^2 - 49$

____

c. $64r^2 - 9s^2$

____

3. Factor completely.

a. $28y^2 - 175$

$$28y^2 - 175 = 7(4y^2 - 25)$$
$$= 7\left[(2y)^2 - 5^2\right]$$
$$= 7(2y+5)(2y-5)$$

b. $y^4 - 49$

$$y^4 - 49 = (y^2)^2 - 7^2$$
$$= (y^2+7)(y^2-7)$$

c. $p^4 - 81$

$$p^4 - 81 = (p^2)^2 - 9^2$$
$$= (p^2+9)(p^2-9)$$
$$= (p^2+9)(p+3)(p-3)$$

3. Factor completely.

a. $90x^2 - 490$

____

b. $x^4 - 64$

____

c. $p^4 - 256$

____

Review this example for Objective 2:

4. Factor $x^2 + 20x + 100$.

The terms x^2 and 100 are perfect squares.

$$x^2 + 20x + 100 = (x+10)^2$$

Check the middle term. $2(x)(10) = 20x$

The trinomial is a perfect square.

Now Try:

4. Factor $p^2 + 16p + 64$.

____

Name:
Instructor:

Date:
Section:

5. Factor each trinomial.

a. $x^2 - 26x + 169$

The first and last terms are perfect squares. Check to see if the middle term is twice the product of the first and last terms of the binomial $x - 13$.

$2 \cdot x \cdot (-13) = -26x$

Thus, $x^2 - 26x + 169$ is a perfect square trinomial.

$x^2 - 26x + 169$ factors as $(x - 13)^2$.

b. $49m^2 - 70m + 25$

$$49m^2 - 70m + 25$$
$$= (7m)^2 + 2(7m)(-5) + (-5)^2$$
$$= (7m - 5)^2$$

c. $36y^2 + 42y + 49$

The first and last terms are perfect squares.

$36y^2 = (6y)^2$ and $49 = 7^2$

Twice the product of the first and last terms of the binomial is $2 \cdot (6y)(7) = 84y$, which is not the middle term of $36y^2 + 42y + 49$.

It is a prime polynomial.

d. $128z^3 + 192z^2 + 72z$

$$128z^3 + 192z^2 + 72z$$
$$= 8z(16z^2 + 24z + 9)$$
$$= 8z\left[(4z)^2 + 2(4z)(3) + 3^2\right]$$
$$= 8z(4z + 3)^2$$

5. Factor each trinomial.

a. $x^2 - 24x + 144$

____

b. $64m^2 + 48m + 9$

____

c. $100y^2 - 70y + 49$

____

d. $20x^3 + 100x^2 + 125x$

____

Name: Date:
Instructor: Section:

Objective 1 Factor a difference of squares.

For extra help, see Examples 1–3 on pages 947–948 of your text and Section Lecture video for Section 14.5 and Exercise Solutions Clip 5, 9, 13, 15, 21, and 27.

Factor each binomial completely. If a binomial cannot be factored, write **prime**.

1. $x^2 - 49$ **1.** ________

2. $81x^4 - 16$ **2.** ________

3. $9x^2 + 16$ **3.** ________

Objective 2 Factor a perfect square trinomial.

For extra help, see Examples 4–5 on pages 949–950 of your text and Section Lecture video for Section 14.5 and Exercise Solutions Clip 31, 33, and 47.

Factor each trinomial completely. It may be necessary to factor out the greatest common factor first.

4. $z^2 - \frac{4}{3}z + \frac{4}{9}$ **4.** ________

5. $9j^2 + 12j + 4$ **5.** ________

6. $-12a^2 + 60ab - 75b^2$ **6.** ________

Name: Date:
Instructor: Section:

Chapter 14 FACTORING AND APPLICATIONS

14.6 Solving Quadratic Equations by Factoring

Learning Objectives

1. Solve quadratic equations by factoring.
2. Solve other equations by factoring.

Key Terms

Use the vocabulary terms listed below to complete each statement in exercises 1–2.

quadratic equation **standard form**

1. An equation written in the form $ax^2 + bx + c = 0$ is written in the ____________________ of a quadratic equation.

2. An equation that can written in the form $ax^2 + bx + c = 0$, with $a \neq 0$, is a ____________________.

Guided Examples

Review these examples for Objective 1:

1. Solve each equation.

a. $(x+9)(5x-6)=0$

By the zero-factor property, either $x+9=0$ or $5x-6=0$, or both.

$$x+9=0 \quad \text{or} \quad 5x-6=0$$
$$x=-9 \quad \text{or} \quad 5x=6$$
$$x=\frac{6}{5}$$

Check:
Let $x = -9$.

$$(x+9)(5x-6)=0$$
$$(-9+9)[5(-9)-6]\stackrel{?}{=}0$$
$$0(-51)\stackrel{?}{=}0$$
$$0=0 \quad \text{True}$$

Now Try:

1. Solve each equation.

a. $(x+12)(4x-7)=0$

____

Let $x = \frac{6}{5}$.

$$(x+9)(5x-6)=0$$

$$\left(\frac{6}{5}+9\right)\left[5\left(\frac{6}{5}\right)-6\right] \stackrel{?}{=} 0$$

$$\left(\frac{51}{5}\right)(6-6) \stackrel{?}{=} 0$$

$$0=0 \text{ True}$$

Both values check, so the solution set is $\left\{-9, \frac{6}{5}\right\}$.

b. $x(8x-11)=0$

Use the zero-factor property.

$$x=0 \quad \text{or} \quad 8x-11=0$$

$$8x=11$$

$$x=\frac{11}{8}$$

Check these solutions by substituting each in the original equation. The solution set is $\left\{0, \frac{11}{8}\right\}$.

b. $x(6x-11)=0$

____

2. Solve each equation.

a. $x^2-6x=-5$

First, write the equation in standard form.

$$x^2-6x=-5$$

$$x^2-6x+5=0$$

Factor and use the zero-factor property.

$$(x-1)(x-5)=0$$

$$x-1=0 \quad \text{or} \quad x-5=0$$

$$x=1 \quad \text{or} \quad x=5$$

Check these solutions by substituting each in the original equation. The solution set is $\{1, 5\}$.

b. $y^2=2y+15$

Write the equation in standard form, factor, and use the zero-factor property.

$$y^2=2y+15$$

$$y^2-2y-15=0$$

$$(y-5)(y+3)=0$$

2. Solve each equation.

a. $x^2-5x=24$

____

b. $x^2=x+72$

____

$$y-5=0 \quad \text{or} \quad y+3=0$$
$$y=5 \quad \text{or} \quad y=-3$$

Check by substituting in the original equation.
The solution set is {–3, 5}.

3. Solve $8p^2+30=46p$.

$$8p^2+30=46p$$
$$8p^2-46p+30=0 \quad \text{Standard form}$$
$$2(4p^2-23p+15)=0 \quad \text{Factor out 2.}$$
$$4p^2-23p+15=0 \quad \text{Divide each side by 2}$$
$$(4p-3)(p-5)=0 \quad \text{Factor.}$$
$$4p-3=0 \quad \text{or} \quad p-5=0 \quad \text{Zero-factor}$$
$$p=\frac{3}{4} \quad \text{or} \quad p=5 \quad \text{property}$$

The solution set is $\left\{\frac{3}{4}, 5\right\}$.

4. Solve each equation.

a. $64m^2-49=0$

$$64m^2-49=0$$
$$(8m+7)(8m-7)=0$$
$$8m+7=0 \quad \text{or} \quad 8m-7=0$$
$$8m=-7 \quad \text{or} \quad 8m=7$$
$$m=-\frac{7}{8} \quad \text{or} \quad m=\frac{7}{8}$$

The solution set is $\left\{-\frac{7}{8}, \frac{7}{8}\right\}$.

b. $y^2=9y$

$$y^2=9y$$
$$y^2-9y=0$$
$$y(y-9)=0$$
$$y=0 \quad \text{or} \quad y-9=0$$
$$y=9$$

The solution set is {0, 9}.

3. Solve $15p^2+36=57p$.

____

4. Solve each equation.

a. $100x^2-9=0$

____

b. $y^2=11y$

____

Name: Date:
Instructor: Section:

c. $k(3k+5)=2$

$$k(3k+5)=2$$
$$3k^2+5k=2$$
$$3k^2+5k-2=0$$
$$(3k-1)(k+2)=0$$
$$3k-1=0 \quad \text{or} \quad k+2=0$$
$$3k=1 \quad \text{or} \quad k=-2$$
$$k=\frac{1}{3}$$

The solution set is $\left\{-2, \frac{1}{3}\right\}$.

5. Solve each equation.

a. $z^2-26z+169=0$

$$z^2-26z+169=0$$
$$(z-13)(z-13)=0$$
$$z-13=0 \quad \text{or} \quad z-13=0$$

Because the two factors are identical, they both lead to the same solution, called a double solution.

$z=13$

Check in the original equation.

$$z^2-26z+169=0$$
$$13^2-26(13)+169 \stackrel{?}{=} 0$$
$$169-338+169 \stackrel{?}{=} 0$$
$$0=0 \quad \text{True}$$

The solution set is $\{13\}$.

b. $25t^2-70t=-49$

$$25t^2-70t=-49$$
$$25t^2-70t+49=0$$
$$(5t-7)^2=0$$
$$5t-7=0 \quad \text{or} \quad 5t-7=0$$
$$5t=7$$
$$t=\frac{7}{5}$$

The check verifies that the solution set is $\left\{\frac{7}{5}\right\}$.

c. $k(4k-23)=6$

____

5. Solve each equation.

a. $z^2-16z+64=0$

____

b. $16t^2+72t=-81$

____

Name: Date:
Instructor: Section:

Review these examples for Objective 2:

6. Solve $12z^3 - 3z = 0$.

$$12z^3 - 3z = 0$$
$$3z(4z^2 - 1) = 0$$
$$3z(2z+1)(2z-1) = 0$$

By an extension of the zero-factor property, we have

$3z = 0$ or $2z+1=0$ or $2z-1=0$

$z = 0$ or $z = -\frac{1}{2}$ or $z = \frac{1}{2}$

Check by substituting each value in the original equation. The solution set is $\left\{-\frac{1}{2}, 0, \frac{1}{2}\right\}$.

7. Solve $(3x-1)(x^2 - 13x + 36) = 0$.

$$(3x-1)(x^2 - 13x + 36) = 0$$
$$(3x-1)(x-4)(x-9) = 0$$

$3x - 1 = 0$ or $x - 4 = 0$ or $x - 9 = 0$

$x = \frac{1}{3}$ or $x = 4$ or $x = 9$

Check by substituting each value in the original equation. The solution set is $\left\{\frac{1}{3}, 4, 9\right\}$.

Now Try:

6. Solve $3r^3 = 75r$.

____

7. Solve $(5x-2)(x^2 - 11x + 18) = 0$.

____

Objective 1 Solve quadratic equations by factoring.

For extra help, see Examples 1–5 on pages 957–961 of your text and Section Lecture video for Section 14.6 and Exercise Solutions Clip 11, 25, 29, 45, and 49.

Solve each equation and check your solutions.

1. $2x^2 - 3x - 20 = 0$ **1.** ____________

2. $25x^2 = 20x$ **2.** ____________

Name: Date:
Instructor: Section:

3. $c(5c+17)=12$ **3.** ________________

Objective 2 Solve other equations by factoring.

For extra help, see Examples 6–7 on page 962 of your text and Section Lecture video for Section 14.6 and Exercise Solutions Clip 59.

Solve each equation and check your solutions.

4. $x^3+2x^2-8x=0$ **4.** ________________

5. $z^4+8z^3-9z^2=0$ **5.** ________________

6. $(y^2-5y+6)(y^2-36)=0$ **6.** ________________

Name: Date:
Instructor: Section:

Chapter 14 FACTORING AND APPLICATIONS

14.7 Applications of Quadratic Equations

Learning Objectives	
1	Solve problems involving geometric figures.
2	Solve problems involving consecutive integers.
3	Solve problems by applying the Pythagorean formula.
4	Solve problems by using given quadratic models.

Key Terms

Use the vocabulary terms listed below to complete each statement in exercises 1–2.

hypotenuse **legs**

1. In a right triangle, the sides that form the right angle are the _______________.

2. The longest side of a right triangle is the ___________________________.

Guided Examples

Review this example for Objective 1:

1. The length of a rectangle is three times its width. If the width was increased by 4 and the length remained the same, the resulting rectangle would have an area of 231 square inches. Find the dimensions of the original rectangle.

Step 1 Read the problem carefully. Find the dimensions of the original rectangle.

Step 2 Assign a variable.
Let x = width.
$3x$ = length
$x + 4$ = new width
$3x$ = length

Step 3 Write an equation. The area of the rectangle is given by $\text{Area} = \text{Length} \times \text{Width}$. Substitute 231 for area, $3x$ for length, and $x + 4$ for width.

$$231 = 3x(x+4)$$

Now Try:

1. Mr. Fixxall is building a box which will have a volume of 60 cubic meters. The height of the box will be 4 meters, and the length will be 2 meters more than the width. Find the width and length of the box.

____

Step 4 Solve.

$$231 = 3x(x+4)$$
$$231 = 3x^2 + 12x$$
$$3x^2 + 12x - 231 = 0$$
$$3(x^2 + 4x - 77) = 0$$
$$x^2 + 4x - 77 = 0$$
$$(x+11)(x-7) = 0$$
$$x + 11 = 0 \quad \text{or} \quad x - 7 = 0$$
$$x = -11 \quad \text{or} \quad x = 7$$

Step 5 State the answer. The solutions are –11 and 7. A rectangle cannot have a side of negative length, so we discard –11. The width is 7 inches. The length is $3(7) = 21$ inches.

Step 6 Check. The new width is 7 + 4 = 11. The new area of the rectangle is $11(21) = 231$ square inches.

Review these examples for Objective 2:

2. Find two consecutive integers such that the sum of the squares of the two integers is 3 more than the opposite (additive inverse) of the smaller integer.

Step 1 Read the problem. Note that the numbers are consecutive.

Step 2 Assign a variable.
Let x = the first number.
Then $x + 1$ = the second number.

Step 3 Write an equation. The sum of squares is three more than the opposite of the smaller integer.

$$x^2 + (x+1)^2 = 3 + (-x)$$

Step 4 Solve.

$$x^2 + x^2 + 2x + 1 = 3 - x$$
$$2x^2 + 3x - 2 = 0$$
$$(2x-1)(x+2) = 0$$
$$2x + 1 = 0 \quad \text{or} \quad x + 2 = 0$$
$$x = -\frac{1}{2} \quad \text{or} \quad x = -2$$

Now Try:

2. If the square of the sum of two consecutive integers is reduced by twice their product, the result is 25. Find the integers.

____

Step 5 State the answer. The solutions are $-\frac{1}{2}$ and –2. We discard $-\frac{1}{2}$ since it is not an integer. If $x = -2$, then $x + 1 = -2 + 1 = -1$. So, the numbers are –2 and –1.

Step 6 Check. $(-2)^2 + (-1)^2 = 3 + [-(-2)]$

$$4 + 1 = 3 + 2$$

$$5 = 5$$

3. Find two consecutive positive even integers whose product is 168.

Step 1 Read carefully. Note that the integers are positive consecutive even integers.

Step 2 Assign a variable.
Let x = the first integer.
Then $x + 2$ = the next even integer.

Step 3 Write an equation. The product is 168.
$x(x+2) = 168$

Step 4 Solve.

$$x^2 + 2x = 168$$

$$x^2 + 2x - 168 = 0$$

$$(x+14)(x-12) = 0$$

$$x + 14 = 0 \quad \text{or} \quad x - 12 = 0$$

$$x = -14 \quad \text{or} \quad x = 12$$

Step 5 State the answer. The solutions are –14 and 12. We discard –14 since the integers are required to be positive. If $x = 12$, then $x + 2 = 14$. So, the numbers are 12 and 14.

Step 6 Check. The numbers 12 and 14 are consecutive positive even integers, and their product is 168.

3. The product of two consecutive even positive integers is 10 more than seven times the larger. Find the integers.

____

Name: Date:
Instructor: Section:

Review this example for Objective 3:

4. Penny and Carla started biking from the same corner. Penny biked east and Carla biked south. When they were 26 miles apart, Carla had biked 14 miles further than Penny. Find the distance each biked.

Step 1 Read carefully. Find the two distances.

Step 2 Assign a variable.
Let x = Penny's distance.
Then $x + 14$ = Carla's distance.

Step 3 Write an equation. Substitute into the Pythagorean theorem.

$$a^2 + b^2 = c^2$$
$$x^2 + (x+14)^2 = 26^2$$

Step 4 Solve.

$$x^2 + x^2 + 28x + 196 = 676$$
$$2x^2 + 28x - 480 = 0$$
$$2(x^2 + 14x - 240) = 0$$
$$x^2 + 14x - 240 = 0$$
$$(x+24)(x-10) = 0$$
$$x + 24 = 0 \quad \text{or} \quad x - 10 = 0$$
$$x = -24 \quad \text{or} \quad x = 10$$

Step 5 State the answer. Since –24 cannot be a distance, 10 is the distance for Penny, and $10 + 14 = 24$ is the distance for Carla.

Step 6 Check. Since $10^2 + 24^2 = 26^2$ is true, the answer is correct.

Now Try:

4. A ladder is leaning against a building. The distance from the bottom of the ladder to the building is 8 feet less than the length of the ladder. How high up the side of the building is the top of the ladder if that distance is 4 feet less than the length of the ladder?

____

Review these examples for Objective 4:

5. Jeff threw a stone straight upward at 46 feet per second from a dock 6 feet above a lake. The height of the stone above the lake t seconds after it is thrown is given by $h = -16t^2 + 46t + 6$. How long will it take for the stone to reach a height of 39 feet?

Substitute 39 for h.

$$39 = -16t^2 + 46t + 6$$

Now Try:

5. A ball is dropped from the roof of a 19.6 meter high building. Its height h (in meters) t seconds later is given by the equation $h = -4.9t^2 + 19.6$. After how many second is the height 14.7 meters?

____

Solve for t.

$$16t^2 - 46t + 33 = 0$$

$$(8t - 11)(2t - 3) = 0$$

$$8t - 11 = 0 \quad \text{or} \quad 2t - 3 = 0$$

$$t = \frac{11}{8} \quad \text{or} \quad t = \frac{3}{2}$$

Since we have found two acceptable answers, the stone will be at height of 39 feet twice (once on its way up and once on its way down) —at $\frac{11}{8}$ sec or $\frac{3}{2}$ sec.

6. The unemployment rate in a certain community can be modeled by the equation $y = 0.0248x^2 - 0.4810x + 7.8543$, where y is the unemployment rate (percent) and x is the month ($x = 1$ represents January, $x = 2$ represents February, etc.) Use the model to find the unemployment rate in August. Round your answer to the nearest tenth.

Since $x = 1$ represents January, then $x = 8$ represents August. Substitute 8 for x in the model.

$$y = 0.0248(8)^2 - 0.4810(8) + 7.8543$$

$$y \approx 5.6 \quad \text{Rounding to the nearest tenth.}$$

The unemployment rate in August is 5.6%.

6. The total cost of a product can be modeled by the equation $C = 400 - 100x + x^2$ where x represents the number of items produced. How many items can be produced at a cost of \$1500?

____

Objective 1 Solve problems involving geometric figures.

For extra help, see Example 1 on pages 966–967 of your text and Section Lecture video for Section 14.7 and Exercise Solutions Clip 7.

Solve each problem. Check your answers to be sure they are reasonable.

1. A book is three times as long as it is wide. Find the length and width of the book in inches if its area is numerically 128 more than its perimeter.

1. width____________

length ____________

Name: Date:
Instructor: Section:

2. The area of a triangle is 42 square centimeters. The base is 2 centimeters less than twice the height. Find the base and height of the triangle.

2. base____________

height ____________

3. The volume of a box is 192 cubic feet. If the length of the box is 8 feet and the width is 2 feet more than the height, find the height and width of the box.

3. height ____________

width____________

Objective 2 Solve problems involving consecutive integers.

For extra help, see Examples 2–3 on pages 967–968 of your text and Section Lecture video for Section 14.7 and Exercise Solutions Clip 21.

Solve each problem.

4. Find all possible pairs of consecutive odd integers whose sum is equal to their product decreased by 47.

4. ____________

5. Find two consecutive positive even integers whose product is six more than three times its sum.

5. ____________

Name: Date:
Instructor: Section:

6. Find three consecutive positive odd integers such that four times the sum of all three equals 13 more than the product of the smaller two.

6. ________________

Objective 3 Solve problems by applying the Pythagorean formula.

For extra help, see Example 4 on page 969 of your text and Section Lecture video for Section 14.7 and Exercise Solutions Clip 27.

Solve each problem.

7. A field is in the shape of a right triangle. The shorter leg measures 45 meters. The hypotenuse measures 45 meters less than twice the longer the leg. Find the dimensions of the lot.

7. ________________

8. A train and a car leave a station at the same time, the train traveling due north and the car traveling west. When they are 100 miles apart, the train has traveled 20 miles farther than the car. Find the distance each has traveled.

8. car ________________

train ________________

9. Two ships left a dock at the same time. When they were 25 miles apart, the ship that sailed due south had gone 10 miles less than twice the distance traveled by the ship that sailed due west. Find the distance traveled by the ship that sailed due south.

9. ________________

Name: Date:
Instructor: Section:

Objective 4 Solve problems by using given quadratic models.

For extra help, see Examples 5–6 on pages 970–971 of your text and Section Lecture video for Section 14.7 and Exercise Solutions Clip 31.

Solve each problem.

10. If an object is propelled upward from a height of 16 feet with an initial velocity of 48 feet per second, its height h (in feet) t seconds later is given by the equation $h = -16t^2 + 48t + 16$.

(a) After how many seconds is the height 52 feet?

(b) After how many seconds is the height 48 feet?

10. a. ______________

b. ______________

11. A company determines that its daily revenue R (in dollars) for selling x items is modeled by the equation $R = x(150 - x)$. How many items must be sold for its revenue to be \$4400?

11. ______________

12. If a ball is batted at an angle of 35°, the distance that the ball travels is given approximately by $D = 0.029v^2 + 0.021v - 1$, where v is the bat speed in miles per hour and D is the distance traveled in feet. Find the distance a batted ball will travel if the ball is batted with a velocity of 90 miles per hour. Round your answer to the nearest whole number.

12. ______________

Name: Date:
Instructor: Section:

Chapter 15 RATIONAL EXPRESSIONS AND APPLICATIONS

15.1 The Fundamental Property of Rational Expressions

Learning Objectives

1	Find the numerical value of a rational expression.
2	Find the values of the variable for which a rational expression is undefined.
3	Write rational expressions in lowest terms.
4	Recognize equivalent forms of rational expressions.

Key Terms

Use the vocabulary terms listed below to complete each statement in exercises 1–2.

rational expression **lowest terms**

1. The quotient of two polynomials with denominator not 0 is called a ________________________.

2. A rational expression is written in ________________________ if the greatest common factor of its numerator and denominator is 1.

Guided Examples

Review these examples for Objective 1:

1. Find the numerical value of $\frac{4x+8}{3x-6}$ for each value of x.

a. $x=1$

$$\frac{4x+8}{3x-6}=\frac{4(1)+8}{3(1)-6}=\frac{12}{-3}=-4$$

b. $x=0$

$$\frac{4x+8}{3x-6}=\frac{4(0)+8}{3(0)-6}=\frac{8}{-6}, \text{ or } -\frac{4}{3}$$

c. $x=2$

$$\frac{4x+8}{3x-6}=\frac{4(2)+8}{3(2)-6}=\frac{16}{0} \quad \text{undefined}$$

d. $x=-2$

$$\frac{4x+8}{3x-6}=\frac{4(-2)+8}{3(-2)-6}=\frac{0}{-12}=0$$

Now Try:

1. Find the numerical value of $\frac{x-7}{x-5}$ for each value of x.

a. $x=1$ ____________

b. $x=0$ ____________

c. $x=5$ ____________

d. $x=7$ ____________

Name: Date:
Instructor: Section:

Review these examples for Objective 2:

2. Find any values of the variable for which each rational expression is undefined.

a. $\dfrac{3x+8}{5x+4}$

Step 1 Set the denominator equal to 0.

$$5x+4=0$$

Step 2 Solve.

$$5x=-4$$

$$x=-\frac{4}{5}$$

Step 3 The given expression is undefined for $-\frac{4}{5}$, so $x \neq -\frac{4}{5}$.

b. $\dfrac{3m^2}{m^2-4m-5}$

Set the denominator equal to 0.

$$m^2-4m-5=0$$

$$(m+1)(m-5)=0$$

$$m+1=0 \quad \text{or} \quad m-5=0$$

$$m=-1 \quad \text{or} \quad m=5$$

The given expression is undefined for –1 and 5, so $m \neq -1$, $m \neq 5$.

c. $\dfrac{6r}{r^2+49}$

This denominator will not equal 0 for any value of r. There are no values for which this expression is undefined.

Now Try:

2. Find any values of the variable for which each rational expression is undefined.

a. $\dfrac{y+6}{7y-1}$

____

b. $\dfrac{15m^2}{m^2-m-20}$

____

c. $\dfrac{12t^2}{t^2+100}$

____

Review these examples for Objective 3:

3. Write each expression in lowest terms.

a. $\dfrac{42}{90}$

Begin by factoring, and then group any factors common to the numerator and denominator.

$$\frac{42}{90}=\frac{2\cdot 3\cdot 7}{2\cdot 3\cdot 3\cdot 5}=\frac{(2\cdot 3)\cdot 7}{(2\cdot 3)\cdot 3\cdot 5}=\frac{7}{3\cdot 5}, \text{ or } \frac{7}{15}$$

Now Try:

3. Write each expression in lowest terms.

a. $\dfrac{96}{252}$

____

Name: Date:
Instructor: Section:

b. $\dfrac{15k^3}{3k^4}$

Write k^3 as $k \cdot k \cdot k$ and k^4 as $k \cdot k \cdot k \cdot k$.

$$\frac{15k^3}{3k^4} = \frac{3 \cdot 5 \cdot k \cdot k \cdot k}{3 \cdot k \cdot k \cdot k \cdot k} = \frac{5 \cdot (3 \cdot k \cdot k \cdot k)}{k \cdot (3 \cdot k \cdot k \cdot k)} = \frac{5}{k}$$

b. $\dfrac{12k^5}{4k^8}$

____

4. Write each rational expression in lowest terms.

a. $\dfrac{6x-18}{5x-15}$

$$\frac{6x-18}{5x-15} = \frac{6(x-3)}{5(x-3)} = \frac{6}{5}$$

b. $\dfrac{6y^2-96}{6y-24}$

$$\frac{6y^2-96}{6y-24} = \frac{6\left(y^2-16\right)}{6(y-4)} = \frac{6(y+4)(y-4)}{6(y-4)} = y+4$$

c. $\dfrac{m^2+5m-24}{3m^2-5m-12}$

$$\frac{m^2+5m-24}{3m^2-5m-12} = \frac{(m+8)(m-3)}{(3m+4)(m-3)} = \frac{m+8}{3m+4}$$

5. Write $\dfrac{3x-2y}{2y-3x}$ in lowest terms.

Factor –1 from the denominator.

$$\frac{3x-2y}{2y-3x} = \frac{3x-2y}{-1(-2y+3x)} = \frac{3x-2y}{-1(3x-2y)} = -1$$

4. Write each rational expression in lowest terms.

a. $\dfrac{7x-35}{9x-45}$

____

b. $\dfrac{4y^2-4y-24}{4y-12}$

____

c. $\dfrac{m^2-3m-54}{2m^2-15m-27}$

____

5. Write $\dfrac{4y-5x}{5x-4y}$ in lowest terms.

____

Name: Date:
Instructor: Section:

6. Write each rational expression in lowest terms.

a. $\dfrac{6-5m}{5m-6}$

Since $6-5m$ and $5m-6$ are opposites, this expression equals -1.

b. $\dfrac{25x^2-49}{21-15x}$

$$\frac{25x^2-49}{21-15x}=\frac{(5x+7)(5x-7)}{3(7-5x)}$$
$$=\frac{(5x+7)(5x-7)}{3(-1)(5x-7)}$$
$$=\frac{5x+7}{3(-1)}$$
$$=\frac{5x+7}{-3}, \text{ or } -\frac{5x+7}{3}$$

c. $\dfrac{6+y}{6-y}$

The rational expression is already in lowest terms.

6. Write each rational expression in lowest terms.

a. $\dfrac{11-8m}{8m-11}$

____

b. $\dfrac{100x^2-81}{36-40x}$

____

c. $\dfrac{13-k}{13+k}$

____

Review this example for Objective 4:

7. Write four equal forms of the following rational expression.

$$-\frac{4x+3}{x-8}$$

If we apply the negative sign to the numerator, we obtain the first two equivalent forms.

$$\frac{-(4x+3)}{x-8} \text{ and } \frac{-4x-3}{x-8}$$

If we apply the negative sign to the denominator, we obtain the last two equivalent forms.

$$\frac{4x+3}{-(x-8)} \text{ and } \frac{4x+3}{-x+8}$$

Now Try:

7. Write four equal forms of the following rational expression.

$$-\frac{10x-7}{4x-3}$$

____

Name: Date:
Instructor: Section:

Objective 1 Find the numerical value of a rational expression.

For extra help, see Example 1 on page 988 of your text and Section Lecture video for Section 15.1 and Exercise Solutions Clip 6.

Find the numerical value of each expression when (a) $x = 4$ *and (b)* $x = -1$.

1. $\dfrac{-3x+1}{2x+1}$

1. (a)______________

(b)______________

2. $\dfrac{2x^2-4}{x^2-2}$

2. (a)______________

(b)______________

3. $\dfrac{2x-5}{2-x-x^2}$

3. (a)______________

(b)______________

Objective 2 Find the values of the variable for which a rational expression is undefined.

For extra help, see Example 2 on page 989 of your text and Section Lecture video for Section 15.1 and Exercise Solutions Clip 17.

Find any value(s) of the variable for which each rational expression is undefined. Write answers with $\neq$.

4. $\dfrac{4x^2}{x+7}$

4. ______________

5. $\dfrac{2x^2}{x^2+4}$

5. ______________

6. $\dfrac{2y-5}{2y^2+4y-16}$

6. ______________

Name: Date:
Instructor: Section:

Objective 3 Write rational expressions in lowest terms.

For extra help, see Examples 3–6 on pages 990–993 of your text and Section Lecture video for Section 15.1 and Exercise Solutions Clip 25, 31, 47, and 49.

Write each rational expression in lowest terms. Assume that no values of any variable make any denominator zero.

7. $\dfrac{15ab^3c^9}{-24ab^2c^{10}}$

7. ______________

8. $\dfrac{16-x^2}{2x-8}$

8. ______________

9. $\dfrac{9x^2-9x-108}{2x-8}$

9. ______________

Objective 4 Recognize equivalent forms of rational expressions.

For extra help, see Example 7 on page 994 of your text and Section Lecture video for Section 15.1 and Exercise Solutions Clip 55.

Write four equivalent forms of the following rational expressions. Assume that no values of any variable make any denominator zero.

10. $-\dfrac{4x+5}{3-6x}$

10. ______________

11. $\dfrac{2p-1}{1-4p}$

11. ______________

12. $-\dfrac{2x-3}{x+2}$

12. ______________

Name: Date:
Instructor: Section:

Chapter 15 RATIONAL EXPRESSIONS AND APPLICATIONS

15.2 Multiplying and Dividing Rational Expressions

Learning Objectives	
1	Multiply rational expressions.
2	Find reciprocals.
3	Divide rational expressions.

Key Terms

Use the vocabulary terms listed below to complete each statement in exercises 1–3.

rational expression **reciprocal** **lowest terms**

1. The ______________________ of the expression $\frac{4x-5}{x+2}$ is $\frac{x+2}{4x-5}$.

2. A ______________________________ is the quotient of two polynomials with denominator not 0.

3. A rational expression is written in ________________________ when the numerator and denominator have no common terms.

Guided Examples

Review these examples for Objective 1:

1. Multiply. Write each answer in lowest terms.

a. $\frac{7}{12}\cdot\frac{3}{14}$

$$\frac{7}{12}\cdot\frac{3}{14}=\frac{7\cdot 3}{12\cdot 14}$$
$$=\frac{7\cdot 3}{2\cdot 2\cdot 3\cdot 2\cdot 7}$$
$$=\frac{1}{8}$$

b. $\frac{9}{x^2}\cdot\frac{x^3}{15}$

$$\frac{9}{x^2}\cdot\frac{x^3}{15}=\frac{9\cdot x^3}{x^2\cdot 15}$$
$$=\frac{3\cdot 3\cdot x\cdot x\cdot x}{3\cdot 5\cdot x\cdot x}$$
$$=\frac{3x}{5}$$

Now Try:

1. Multiply. Write each answer in lowest terms.

a. $\frac{5}{9}\cdot\frac{12}{25}$

____

b. $\frac{8}{x^3}\cdot\frac{x^2}{6}$

____

Name: Date:
Instructor: Section:

2. Multiply. Write the answer in lowest terms.

$$\frac{3x+2y}{5x}\cdot\frac{x^2}{(3x+2y)^2}$$

Multiply numerators, multiply denominators, factor, and identify the common factors.

$$\frac{3x+2y}{5x}\cdot\frac{x^2}{(3x+2y)^2}=\frac{(3x+2y)x^2}{5x(3x+2y)^2}$$

$$=\frac{(3x+2y)\cdot x\cdot x}{5x(3x+2y)(3x+2y)}$$

$$=\frac{x}{5(3x+2y)}$$

2. Multiply. Write the answer in lowest terms.

$$\frac{r-s}{6s}\cdot\frac{s^3}{(r-s)^2}$$

____

3. Multiply. Write the answer in lowest terms.

$$\frac{x^2+10x+9}{x^2-5x}\cdot\frac{x^2+3x-40}{x^2+9x+8}$$

$$\frac{x^2+10x+9}{x^2-5x}\cdot\frac{x^2+3x-40}{x^2+9x+8}$$

$$=\frac{(x^2+10x+9)(x^2+3x-40)}{(x^2-5x)(x^2+9x+8)}$$

$$=\frac{(x+9)(x+1)(x+8)(x-5)}{x(x-5)(x+1)(x+8)}$$

$$=\frac{x+9}{x}$$

The quotients $\frac{x+1}{x+1}$, $\frac{x+8}{x+8}$, and $\frac{x-5}{x-5}$ are all equal to 1, justifying the final product $\frac{x+9}{x}$.

3. Multiply. Write the answer in lowest terms.

$$\frac{7x^2-7}{x^2+x-2}\cdot\frac{5x+10}{x^2+x}$$

____

Review these examples for Objective 2:

4. Find the reciprocal of each rational expression.

a. $\frac{7x^3}{8y}$

$\frac{7x^3}{8y}$ has reciprocal $\frac{8y}{7x^3}$.

b. $\frac{m^2-16}{m^2-2m-15}$

$\frac{m^2-16}{m^2-2m-15}$ has reciprocal $\frac{m^2-2m-15}{m^2-16}$.

Now Try:

4. Find the reciprocal of each rational expression.

a. $\frac{11a^3}{10b}$

____

b. $\frac{w^2-6w}{w^2+6w-7}$

____

Name: Date:
Instructor: Section:

Review these examples for Objective 3:

5. Divide. Write each answer in lowest terms.

a. $\frac{7}{9} \div \frac{4}{27}$

Multiply the first expression by the reciprocal of the second.

$$\frac{7}{9} \div \frac{4}{27} = \frac{7}{9} \cdot \frac{27}{4} = \frac{7 \cdot 27}{9 \cdot 4} = \frac{7 \cdot 3 \cdot 9}{9 \cdot 4} = \frac{21}{4}$$

b. $\frac{y}{y-5} \div \frac{7y}{y+4}$

$$\frac{y}{y-5} \div \frac{7y}{y+4} = \frac{y}{y-5} \cdot \frac{y+4}{7y} = \frac{y(y+4)}{(y-5)(7y)} = \frac{y+4}{7(y-5)}$$

6. Divide. Write the answer in lowest terms.

$$\frac{(5m)^2}{(3p)^3} \div \frac{10m^4}{24p^2}$$

Multiply by the reciprocal.

$$\frac{(5m)^2}{(3p)^3} \div \frac{10m^4}{24p^2} = \frac{(5m)^2}{(3p)^3} \cdot \frac{24p^2}{10m^4} = \frac{25m^2}{27p^3} \cdot \frac{24p^2}{10m^4} = \frac{25 \cdot 24m^2p^2}{27 \cdot 10m^4p^3} = \frac{5 \cdot 5 \cdot 3 \cdot 2 \cdot 4 \cdot m^2 \cdot p^2}{3 \cdot 9 \cdot 2 \cdot 5 \cdot m^2 \cdot m^2 \cdot p^2 \cdot p} = \frac{20}{9m^2p}$$

Now Try:

5. Divide. Write each answer in lowest terms.

a. $\frac{9}{10} \div \frac{3}{25}$

____

b. $\frac{y}{y+2} \div \frac{6y}{y-2}$

____

6. Divide. Write the answer in lowest terms.

$$\frac{(7c)^2}{(5d)^3} \div \frac{49c^2}{10d^3}$$

____

Name: Date:
Instructor: Section:

7. Divide. Write the answer in lowest terms.

$$\frac{m^2-16}{(m-5)(m-4)} \div \frac{(m+4)(m-5)}{-6m}$$

Multiply by the reciprocal.

$$\frac{m^2-16}{(m-5)(m-4)} \div \frac{(m+4)(m-5)}{-6m}$$

$$= \frac{m^2-16}{(m-5)(m-4)} \cdot \frac{-6m}{(m+4)(m-5)}$$

$$= \frac{-6m(m^2-16)}{(m-5)(m-4)(m+4)(m-5)}$$

$$= \frac{-6m(m+4)(m-4)}{(m-5)(m-4)(m+4)(m-5)}$$

$$= \frac{-6m}{(m-5)^2}, \text{ or } -\frac{6m}{(m-5)^2}$$

8. Divide. Write the answer in lowest terms.

$$\frac{x^2-25}{x^2-9} \div \frac{3x^2-15x}{3-x}$$

Multiply by the reciprocal.

$$\frac{x^2-25}{x^2-9} \div \frac{3x^2-15x}{3-x}$$

$$= \frac{x^2-25}{x^2-9} \cdot \frac{3-x}{3x^2-15x}$$

$$= \frac{(x^2-25)(3-x)}{(x^2-9)(3x^2-15x)}$$

$$= \frac{(x+5)(x-5)(3-x)}{(x+3)(x-3)(3x)(x-5)}$$

$$= \frac{-1(x+5)}{3x(x+3)}, \text{ or } \frac{-x-5}{3x(x+3)}$$

7. Divide. Write the answer in lowest terms.

$$\frac{x^2-49}{(x-7)(x-3)} \div \frac{(x+7)(x-3)}{8x}$$

____

8. Divide. Write the answer in lowest terms.

$$\frac{m^2-64}{m^2-81} \div \frac{5m^2+40m}{9-m}$$

____

Name: Date:
Instructor: Section:

Objective 1 Multiply rational expressions.

For extra help, see Examples 1–3 on pages 997–998 of your text and Section Lecture video for Section 15.2 and Exercise Solutions Clip 5, 7, and 37.

Multiply. Write each answer in lowest terms.

1. $\frac{8m^4n^3}{3} \cdot \frac{5}{4mn^2}$

1. ________________

2. $\frac{m^2-16}{m-3} \cdot \frac{9-m^2}{4-m}$

2. ________________

3. $\frac{3x+12}{6x-30} \cdot \frac{x^2-x-20}{x^2-16}$

3. ________________

Objective 2 Find reciprocals.

For extra help, see Example 4 on page 998 of your text and Section Lecture video for Section 15.2 and Exercise Solutions Clip 13 and 15.

Write the reciprocal of each rational expression.

4. $\frac{7}{2y}$

4. ________________

5. $4r^2+2r+3$

5. ________________

6. $\frac{3x+4y}{5x-2y}$

6. ________________

Name: Date:
Instructor: Section:

Objective 3 Divide rational expressions.

For extra help, see Examples 5–8 on pages 999–1000 of your text and Section Lecture video for Section 15.2 and Exercise Solutions Clip 17, 19, 21, 33, and 35.

Divide. Write each answer in lowest terms.

7. $\frac{b-7}{16} \div \frac{7-b}{8}$

7. ________________

8. $\frac{m^2+2mn+n^2}{m^2+m} \div \frac{m^2-n^2}{m^2-1}$

8. ________________

9. $\frac{27-3k^2}{3k^2+8k-3} \div \frac{k^2-6k+9}{6k^2-19k+3}$

9. ________________

Name: Date:
Instructor: Section:

Chapter 15 RATIONAL EXPRESSIONS AND APPLICATIONS

15.3 Least Common Denominators

Learning Objectives
1 Find the least common denominator for a list of fractions.
2 Write equivalent rational expressions.

Key Terms

Use the vocabulary terms listed below to complete each statement in exercises 1–2.

least common denominator **equivalent expressions**

1. $\frac{24x-8}{9x^2-1}$ and $\frac{8}{3x+1}$ are ______________________________.

2. The simplest expression that is divisible by all denominators is called the ______________________________.

Guided Examples

Review these examples for Objective 1:

1. Find the LCD for each pair of fractions.

a. $\frac{1}{35}, \frac{11}{45}$

Step 1 Factor each denominator into prime factors.

$35 = 5 \cdot 7$

$45 = 3 \cdot 3 \cdot 5 = 3^2 \cdot 5$

Step 2 List each denominator the greatest number of times it appears as a factor in any of the denominators.

The factor 3 appears two times, and the factors 5 and 7 each appear once.

Step 3 Multiply to get the LCD.

$$\begin{aligned} \text{LCD} &= 3 \cdot 3 \cdot 5 \cdot 7 \\ &= 3^2 \cdot 5 \cdot 7 \\ &= 315 \end{aligned}$$

Now Try:

1. Find the LCD for each pair of fractions.

a. $\frac{5}{18}, \frac{13}{24}$

____

b. $\frac{1}{15x}, \frac{9}{20x}$

Step 1

$15x = 3 \cdot 5 \cdot x$

$20x = 2 \cdot 2 \cdot 5 \cdot x = 2^2 \cdot 5 \cdot x$

Step 2

Here 2 appears twice, and 3, 5, and x each appear once.

Step 3

$\text{LCD} = 2 \cdot 2 \cdot 3 \cdot 5 \cdot x$

$= 2^2 \cdot 3 \cdot 5 \cdot x$

$= 60x$

2. Find the LCD for $\frac{9}{28s^3}$ and $\frac{5}{42s^2}$.

Step 1

$28s^3 = 2 \cdot 2 \cdot 7 \cdot s^3$

$42s^2 = 2 \cdot 3 \cdot 7 \cdot s^2$

Step 2

Here s appears three times, 2 appears twice, and 3 and 7 each appear once.

Step 3

$\text{LCD} = 2^2 \cdot 3 \cdot 7 \cdot s^3 = 84s^3$

3. Find the LCD for the fractions in each list.

a. $\frac{5}{7b}, \frac{8}{b^2 - 5b}$

$7b = 7 \cdot b$

$b^2 - 5b = b(b-5)$

$\text{LCD} = 7 \cdot b(b-5) = 7b(b-5)$

b. $\frac{6}{c^2 - 5c - 6}, \frac{10}{c^2 + 3c - 54}, \frac{4}{c^2 - 12c + 36}$

$c^2 - 5c - 6 = (c-6)(c+1)$

$c^2 + 3c - 54 = (c-6)(c+9)$

$c^2 - 12c + 36 = (c-6)^2$

Use each factor the greatest number of times it appears as a factor.

$\text{LCD} = (c+1)(c+9)(c-6)^2$

b. $\frac{7}{8x}, \frac{11}{12x}$

____

2. Find the LCD for $\frac{15}{40a^2}$ and $\frac{13}{24a^4}$.

____

3. Find the LCD for the fractions in each list.

a. $\frac{7}{9w}, \frac{13}{w^2 - 2w}$

____

b. $\frac{12}{b^2 - 16}, \frac{6}{b^2 - 3b - 4}, \frac{9}{b^2 - 8b + 16}$

____

Name: Date:
Instructor: Section:

c. $\frac{1}{a-8}, \frac{7}{8-a}$

$-(a-8)=-a+8=8-a$

Therefore either $8-a$ or $a-8$ can be used as the LCD.

Review these examples for Objective 2:

4. Write each rational expression as an equivalent expression with the indicated denominator.

a. $\frac{7}{9}=\frac{?}{45}$

Step 1 Factor both denominators.

$\frac{7}{9}=\frac{?}{5\cdot 9}$

Step 2 A factor of 5 is missing.

Step 3 Multiply $\frac{7}{9}$ by $\frac{5}{5}$.

$\frac{7}{9}=\frac{7}{9}\cdot\frac{5}{5}=\frac{35}{45}$

b. $\frac{8k}{27}=\frac{?}{81k}$

Step 1 Factor both denominators.

$\frac{8k}{27}=\frac{?}{3\cdot 27k}$

Step 2 Factors of 3 and *k* are missing.

Step 3 Multiply $\frac{8k}{27}$ by $\frac{3k}{3k}$.

$\frac{8k}{27}=\frac{8k}{27}\cdot\frac{3k}{3k}=\frac{24k^2}{81k}$

5. Write each rational expression as an equivalent expression with the indicated denominator.

a. $\frac{9}{5x+2}=\frac{?}{15x+6}$

Factor the denominator on the right.

$\frac{9}{5x+2}=\frac{?}{3(5x+2)}$

The missing factor is 3, so multiply by $\frac{3}{3}$.

$\frac{9}{5x+2}\cdot\frac{3}{3}=\frac{27}{15x+6}$

c. $\frac{13}{p-14}, \frac{12}{14-p}$

____

Now Try:

4. Write each rational expression as an equivalent expression with the indicated denominator.

a. $\frac{13}{6}=\frac{?}{30}$

____

b. $\frac{9n}{22}=\frac{?}{66n}$

____

5. Write each rational expression as an equivalent expression with the indicated denominator.

a. $\frac{19}{6c-5}=\frac{?}{24c-20}$

____

b. $\dfrac{7}{q^2+6q}=\dfrac{?}{q^3+q^2-30q}$

Factor the denominator in each rational expression.

$$\frac{7}{q(q+6)}=\frac{?}{q(q+6)(q-5)}$$

The missing factor is $(q-5)$, so multiply by $\dfrac{q-5}{q-5}$.

$$\frac{7}{q(q+6)}=\frac{7}{q(q+6)}\cdot\frac{q-5}{q-5}$$
$$=\frac{7(q-5)}{q^3-q^2-30q}$$
$$=\frac{7q-35}{q^3-q^2-30q}$$

b. $\dfrac{3}{z^2-7z}=\dfrac{?}{z^3-5z^2-14z}$

____

Objective 1 Find the least common denominator for a list of fractions.

For extra help, see Examples 1–3 on pages 1003–1004 of your text and Section Lecture video for Section 15.3 and Exercise Solutions Clip 11, 19, and 23.

Find the least common denominator for each list of rational expressions.

1. $\dfrac{13}{36b^4},\ \dfrac{17}{27b^2}$

1. ____________

2. $\dfrac{-7}{a^2-2a},\ \dfrac{3a}{2a^2+a-10}$

2. ____________

3. $\dfrac{8}{w^3-9w},\ \dfrac{4w}{w^2+w-6}$

3. ____________

Name: Date:
Instructor: Section:

Objective 2 Write equivalent rational expressions.

For extra help, see Examples 4–5 on pages 1005–1006 of your text and Section Lecture video for Section 15.3 and Exercise Solutions Clip 29, 39, and 41.

Rewrite each rational expression with the indicated denominator. Give the numerator of the new fraction.

4. $\dfrac{5a}{8a-3}=\dfrac{?}{6-16a}$ **4.** ____________

5. $\dfrac{3}{5r-10}=\dfrac{?}{50r^2-100r}$ **5.** ____________

6. $\dfrac{3}{k^2+3k}=\dfrac{?}{k^3+10k^2+21k}$ **6.** ____________

Name: Date:
Instructor: Section:

Chapter 15 RATIONAL EXPRESSIONS AND APPLICATIONS

15.4 Adding and Subtracting Rational Expressions

Learning Objectives

1	Add rational expressions having the same denominator.
2	Add rational expressions having different denominators.
3	Subtract rational expressions.

Key Terms

Use the vocabulary terms listed below to complete each statement in exercises 1–2.

least common multiple **greatest common factor**

1. The ______________________ of $2m^2-5m-3$ and $2m-6$ is $m-3$.

2. The ______________________ of $2m^2-5m-3$ and $2m-6$ is $2(m-3)(2m+1)$.

Guided Examples

Review these examples for Objective 1:

1. Add. Write each answer in lowest terms.

a. $\dfrac{1}{8}+\dfrac{3}{8}$

The denominators are the same, so add the numerators and keep the common denominator.

$$\frac{1}{8}+\frac{3}{8}=\frac{1+3}{8}=\frac{4}{8}=\frac{4\cdot 1}{4\cdot 2}=\frac{1}{2}$$

b. $\dfrac{4x}{x+2}+\dfrac{8}{x+2}$

$$\frac{4x}{x+2}+\frac{8}{x+2}=\frac{4x+8}{x+2}=\frac{4(x+2)}{x+2}=4$$

Now Try:

1. Add. Write each answer in lowest terms.

a. $\dfrac{9}{20}+\dfrac{7}{20}$

____

b. $\dfrac{2x^2}{x+4}+\dfrac{8x}{x+4}$

____

Name: Date:
Instructor: Section:

Review these examples for Objective 2:

2. Add. Write each answer in lowest terms.

a. $\frac{5}{18}+\frac{7}{24}$

Step 1 Find the LCD.

$$18=2\cdot 3\cdot 3=2\cdot 3^2$$

$$24=2\cdot 2\cdot 2\cdot 3=2^3\cdot 3$$

$$\text{LCD}=2^3\cdot 3^2=72$$

Step 2 Now write each expression as an equivalent expression with the LCD as the denominator.

$$\frac{5}{18}+\frac{7}{24}=\frac{5(4)}{18(4)}+\frac{7(3)}{24(3)}$$

$$=\frac{20}{72}+\frac{21}{72}$$

Step 3 Add the numerators.

Step 4 Write in lowest terms, if necessary.

$$=\frac{20+21}{72}$$

$$=\frac{41}{72}$$

b. $\frac{5}{6z}+\frac{7}{9z}$

Step 1 Find the LCD.

$$6z=2\cdot 3\cdot z$$

$$9z=3\cdot 3\cdot z=3^2\cdot z$$

$$\text{LCD}=2\cdot 3^2\cdot z=18z$$

Step 2 Now write each expression as an equivalent expression with the LCD as the denominator.

$$\frac{5}{6z}+\frac{7}{9z}=\frac{5(3)}{6z(3)}+\frac{7(2)}{9z(2)}$$

$$=\frac{15}{18z}+\frac{14}{18z}$$

Step 3 Add the numerators.

Step 4 Write in lowest terms, if necessary.

$$=\frac{15+14}{18z}$$

$$=\frac{29}{18z}$$

Now Try:

2. Add. Write each answer in lowest terms.

a. $\frac{9}{35}+\frac{8}{45}$

____

b. $\frac{4}{9y}+\frac{2}{7y}$

____

Name: Date:
Instructor: Section:

3. Add. Write the answer in lowest terms.

$$\frac{10x}{x^2-9}+\frac{-5}{x+3}$$

Step 1 Find the LCD.

$x^2-9=(x+3)(x-3)$

$x+3$ is prime

$\text{LCD}=(x+3)(x-3)$

Step 2 Now write each expression as an equivalent expression with the LCD as the denominator.

$$\frac{10x}{x^2-9}+\frac{-5}{x+3}$$

$$=\frac{10x}{(x+3)(x-3)}+\frac{-5(x-3)}{(x+3)(x-3)}$$

$$=\frac{10x}{(x+3)(x-3)}+\frac{-5x+15}{(x+3)(x-3)}$$

Step 3 Add the numerators.

$$=\frac{10x-5x+15}{(x+3)(x-3)}$$

$$=\frac{5x+15}{(x+3)(x-3)}$$

Step 4 Write in lowest terms, if necessary.

$$=\frac{5(x+3)}{(x+3)(x-3)}$$

$$=\frac{5}{x-3}$$

3. Add. Write the answer in lowest terms.

$$\frac{8x}{x^2-100}+\frac{-4}{x+10}$$

____

4. Add. Write the answer in lowest terms.

$$\frac{3x}{x^2-x-20}+\frac{5}{x^2-2x-15}$$

The LCD is $(x+3)(x+4)(x-5)$.

$$=\frac{3x}{(x+4)(x-5)}+\frac{5}{(x+3)(x-5)}$$

$$=\frac{3x(x+3)}{(x+3)(x+4)(x-5)}+\frac{5(x+4)}{(x+3)(x+4)(x-5)}$$

$$=\frac{3x(x+3)+5(x+4)}{(x+3)(x+4)(x-5)}$$

$$=\frac{3x^2+9x+5x+20}{(x+3)(x+4)(x-5)}$$

$$=\frac{3x^2+14x+20}{(x+3)(x+4)(x-5)}$$

4. Add. Write the answer in lowest terms.

$$\frac{7x}{x^2-2x-8}+\frac{4}{x^2-3x-4}$$

____

Name: Date:
Instructor: Section:

5. Add. Write the answer in lowest terms.

$$\frac{y}{y-5}+\frac{6}{5-y}$$

The denominators are opposites.

$$\frac{y}{y-5}+\frac{6}{5-y}=\frac{y}{y-5}+\frac{6(-1)}{(5-y)(-1)}$$
$$=\frac{y}{y-5}+\frac{-6}{-5+y}$$
$$=\frac{y}{y-5}+\frac{-6}{y-5}$$
$$=\frac{y-6}{y-5}$$

5. Add. Write the answer in lowest terms.

$$\frac{x}{4x-3y}+\frac{y}{3y-4x}$$

____

Review these examples for Objective 3:

6. Subtract. Write the answer in lowest terms.

$$\frac{7x}{x-2}-\frac{x+5}{x-2}$$

$$\frac{7x}{x-2}-\frac{x+5}{x-2}=\frac{7x-(x+5)}{x-2}$$
$$=\frac{7x-x-5}{x-2}$$
$$=\frac{6x-5}{x-2}$$

7. Subtract. Write the answer in lowest terms.

$$\frac{7}{x-5}-\frac{6}{x}$$

The LCD is $x(x-5)$.

$$\frac{7}{x-5}-\frac{6}{x}=\frac{7x}{x(x-5)}-\frac{6(x-5)}{x(x-5)}$$
$$=\frac{7x-6(x-5)}{x(x-5)}$$
$$=\frac{7x-6x+30}{x(x-5)}$$
$$=\frac{x+30}{x(x-5)}$$

Now Try:

6. Subtract. Write the answer in lowest terms.

$$\frac{9x}{x-3}-\frac{2x+7}{x-3}$$

____

7. Subtract. Write the answer in lowest terms.

$$\frac{12}{x-9}-\frac{7}{x}$$

____

Name: Date:
Instructor: Section:

8. Subtract. Write the answer in lowest terms.

$$\frac{5x}{x-7}-\frac{x-42}{7-x}$$

The denominators are opposites.

$$\frac{5x}{x-7}-\frac{x-42}{7-x}=\frac{5x}{x-7}-\frac{(x-42)(-1)}{(7-x)(-1)}$$
$$=\frac{5x}{x-7}-\frac{-x+42}{x-7}$$
$$=\frac{5x+x-42}{x-7}$$
$$=\frac{6x-42}{x-7}$$
$$=\frac{6(x-7)}{x-7}$$
$$=6$$

9. Subtract. Write the answer in lowest terms.

$$\frac{7x}{x^2-4x+4}-\frac{2}{x^2-4}$$
$$=\frac{7x}{(x-2)^2}-\frac{2}{(x+2)(x-2)}$$

The LCD is $(x+2)(x-2)^2$.

$$=\frac{7x}{(x-2)^2}-\frac{2}{(x+2)(x-2)}$$
$$=\frac{7x(x+2)}{(x+2)(x-2)^2}-\frac{2(x-2)}{(x+2)(x-2)^2}$$
$$=\frac{7x(x+2)-2(x-2)}{(x+2)(x-2)^2}$$
$$=\frac{7x^2+14x-2x+4}{(x+2)(x-2)^2}$$
$$=\frac{7x^2+12x+4}{(x+2)(x-2)^2}$$

10. Subtract. Write the answer in lowest terms.

$$\frac{2q}{q^2-2q-8}-\frac{7}{3q^2-14q+8}$$

The LCD is $(q+2)(3q-2)(q-4)$.

8. Subtract. Write the answer in lowest terms.

$$\frac{4x}{x-9}-\frac{3x-63}{9-x}$$

____

9. Subtract. Write the answer in lowest terms.

$$\frac{8x}{x^2-10x+25}-\frac{3}{x^2-25}$$

____

10. Subtract. Write the answer in lowest terms.

$$\frac{5}{p^2-6p+5}-\frac{4}{p^2-1}$$

____

$$= \frac{2q}{(q+2)(q-4)} - \frac{7}{(3q-2)(q-4)}$$

$$= \frac{2q(3q-2)}{(q+2)(3q-2)(q-4)} - \frac{7(q+2)}{(q+2)(3q-2)(q-4)}$$

$$= \frac{6q^2 - 4q - 7q - 14}{(q+2)(3q-2)(q-4)}$$

$$= \frac{6q^2 - 11q - 14}{(q+2)(3q-2)(q-4)}$$

Objective 1 Add rational expressions having the same denominator.

For extra help, see Example 1 on page 1009 of your text and Section Lecture video for Section 15.4 and Exercise Solutions Clip 9 and 17.

Add. Write each answer in lowest terms.

1. $\frac{5}{3w^2} + \frac{7}{3w^2}$

1. ________________

2. $\frac{b}{b^2 - 4} + \frac{2}{b^2 - 4}$

2. ________________

3. $\frac{2x+3}{x^2 + 3x - 10} + \frac{2-x}{x^2 + 3x - 10}$

3. ________________

Objective 2 Add rational expressions having different denominators.

For extra help, see Examples 2–5 on pages 1010–1011 of your text and Section Lecture video for Section 15.4 and Exercise Solutions Clip 23, 39, 47, and 65.

Add. Write each answer in lowest terms.

4. $\frac{7}{x-5} + \frac{4}{x+5}$

4. ________________

Name: Date:
Instructor: Section:

5. $\dfrac{3z}{z^2-4}+\dfrac{4z-3}{z^2-4z+4}$

5. ________________

6. $\dfrac{4z}{z^2+6z+8}+\dfrac{2z-1}{z^2+5z+6}$

6. ________________

Objective 3 Subtract rational expressions.

For extra help, see Examples 6–10 on pages 1012–1014 of your text and Section Lecture video for Section 15.4 and Exercise Solutions Clip 15 and 59.

Subtract. Write each answer in lowest terms.

7. $\dfrac{z+2}{z-2}-\dfrac{z-2}{z+2}$

7. ________________

8. $\dfrac{-4}{x^2-4}-\dfrac{3}{4-2x}$

8. ________________

9. $\dfrac{m}{m^2-4}-\dfrac{1-m}{m^2+4m+4}$

9. ________________

Name: Date:
Instructor: Section:

Chapter 15 RATIONAL EXPRESSIONS AND APPLICATIONS

15.5 Complex Fractions

Learning Objectives

1. Simplify a complex fraction by writing it as a division problem (Method 1).
2. Simplify a complex fraction by multiplying numerator and denominator by the least common denominator (Method 2).

Key Terms

Use the vocabulary terms listed below to complete each statement in exercises 1–2.

complex fraction **LCD**

1. A ______________________ is a rational expression with one or more fractions in the numerator, denominator, or both.

2. To simplify a complex fraction, multiply the numerator and denominator by the ______________________ of all the fractions within the complex fraction.

Guided Examples

Review these examples for Objective 1:

1. Simplify each complex fraction.

a. $\dfrac{\frac{5}{6}+\frac{7}{12}}{\frac{1}{5}+\frac{8}{15}}$

Step 1 Write the numerator as a single fraction.

$$\frac{5}{6}+\frac{7}{12}=\frac{5(2)}{6(2)}+\frac{7}{12}=\frac{10}{12}+\frac{7}{12}=\frac{17}{12}$$

Do the same with each denominator.

$$\frac{1}{5}+\frac{8}{15}=\frac{1(3)}{5(3)}+\frac{8}{15}=\frac{3}{15}+\frac{8}{15}=\frac{11}{15}$$

Step 2 Write the equivalent complex fraction as a division problem.

$$\frac{\frac{17}{12}}{\frac{11}{15}}=\frac{17}{12}\div\frac{11}{15}$$

Step 3 Divide by multiplying by the reciprocal.

$$\frac{17}{12}\div\frac{11}{15}=\frac{17}{12}\cdot\frac{15}{11}=\frac{17\cdot5\cdot3}{3\cdot4\cdot11}=\frac{85}{44}$$

Now Try:

1. Simplify each complex fraction.

a. $\dfrac{\frac{1}{6}+\frac{2}{9}}{\frac{3}{10}+\frac{4}{15}}$

____

Name: Date:
Instructor: Section:

b. $\dfrac{8+\dfrac{4}{x}}{\dfrac{x}{6}+\dfrac{1}{12}}$

Step 1 Write the numerator as a single fraction.

$$8+\frac{4}{x}=\frac{8}{1}+\frac{4}{x}=\frac{8x}{x}+\frac{4}{x}=\frac{8x+4}{x}$$

Do the same with each denominator.

$$\frac{x}{6}+\frac{1}{12}=\frac{x(2)}{6(2)}+\frac{1}{12}=\frac{2x}{12}+\frac{1}{12}=\frac{2x+1}{12}$$

Step 2 Write the equivalent complex fraction as a division problem.

$$\frac{\dfrac{8x+4}{x}}{\dfrac{2x+1}{12}}=\frac{8x+4}{x}\div\frac{2x+1}{12}$$

Step 3 Divide by multiplying by the reciprocal.

$$\frac{8x+4}{x}\div\frac{2x+1}{12}=\frac{8x+4}{x}\cdot\frac{12}{2x+1}$$

$$=\frac{4(2x+1)}{x}\cdot\frac{12}{2x+1}$$

$$=\frac{48}{x}$$

b. $\dfrac{9+\dfrac{3}{x}}{\dfrac{x}{10}+\dfrac{1}{30}}$

____

2. Simplify the complex fraction.

$$\frac{\dfrac{rs^2}{t^3}}{\dfrac{s^3}{r^3t}}$$

Use the definition of division and then the fundamental property.

$$=\frac{rs^2}{t^3}\div\frac{s^3}{r^3t}$$

$$=\frac{rs^2}{t^3}\cdot\frac{r^3t}{s^3}$$

$$=\frac{r^4}{st^2}$$

2. Simplify the complex fraction.

$$\frac{\dfrac{a^3b^2}{c}}{\dfrac{a^5b}{c^3}}$$

____

Name: Date:
Instructor: Section:

3. Simplify the complex fraction.

$$\frac{\frac{30}{x+4}-6}{\frac{4}{x+4}+1}=\frac{\frac{30}{x+4}-\frac{6(x+4)}{x+4}}{\frac{4}{x+4}+\frac{1(x+4)}{x+4}}$$

$$=\frac{\frac{30-6(x+4)}{x+4}}{\frac{4+1(x+4)}{x+4}}$$

$$=\frac{\frac{30-6x-24}{x+4}}{\frac{4+x+4}{x+4}}$$

$$=\frac{\frac{6-6x}{x+4}}{\frac{x+8}{x+4}}$$

$$=\frac{6-6x}{x+4}\cdot\frac{x+4}{x+8}$$

$$=\frac{6-6x}{x+8}$$

3. Simplify the complex fraction.

$$\frac{\frac{20}{x-5}-9}{\frac{5}{x-5}+2}$$

____

Review these examples for Objective 2:

4. Simplify each complex fraction.

a. $\dfrac{\frac{3}{10}+\frac{7}{20}}{\frac{1}{5}+\frac{2}{15}}$

Step 1 Find the LCD for all the denominators. The LCD for 5, 15, 10, and 20 is 60.

Step 2 Multiply the numerator and denominator of the complex fraction by the LCD.

$$\frac{\frac{3}{10}+\frac{7}{20}}{\frac{1}{5}+\frac{2}{15}}=\frac{60\left(\frac{3}{10}+\frac{7}{20}\right)}{60\left(\frac{1}{5}+\frac{2}{15}\right)}$$

$$=\frac{60\left(\frac{3}{10}\right)+60\left(\frac{7}{20}\right)}{60\left(\frac{1}{5}\right)+60\left(\frac{2}{15}\right)}$$

$$=\frac{18+21}{12+8}$$

$$=\frac{39}{20}$$

Now Try:

4. Simplify each complex fraction.

a. $\dfrac{\frac{1}{3}+\frac{4}{9}}{\frac{5}{8}+\frac{3}{4}}$

____

b. $\dfrac{12+\dfrac{4}{x}}{\dfrac{x}{5}+\dfrac{1}{15}}$

Step 1 Find the LCD for all the denominators.

The LCD for x, 5, and 15 is $15x$.

Step 2 Multiply the numerator and denominator of the complex fraction by the LCD.

$$\frac{12+\frac{4}{x}}{\frac{x}{5}+\frac{1}{15}}=\frac{15x\left(12+\frac{4}{x}\right)}{15x\left(\frac{x}{5}+\frac{1}{15}\right)}$$

$$=\frac{180x+60}{3x^2+x}$$

$$=\frac{60(3x+1)}{x(3x+1)}$$

$$=\frac{60}{x}$$

b. $\dfrac{4+\dfrac{2}{x}}{\dfrac{x}{3}+\dfrac{1}{6}}$

____

5. Simplify the complex fraction.

$$\frac{\frac{9}{7n}-\frac{3}{n^2}}{\frac{8}{3n}+\frac{5}{6n^2}}$$

The LCD is $42n^2$.

$$=\frac{42n^2\left(\frac{9}{7n}-\frac{3}{n^2}\right)}{42n^2\left(\frac{8}{3n}+\frac{5}{6n^2}\right)}$$

$$=\frac{42n^2\left(\frac{9}{7n}\right)-42n^2\left(\frac{3}{n^2}\right)}{42n^2\left(\frac{8}{3n}\right)+42n^2\left(\frac{5}{6n^2}\right)}$$

$$=\frac{54n-126}{112n+35}, \text{ or } \frac{18(3n-7)}{7(16n+5)}$$

5. Simplify the complex fraction.

$$\frac{\frac{2}{9n}-\frac{2}{5n^2}}{\frac{4}{5n}+\frac{2}{3n^2}}$$

____

Name: Date:
Instructor: Section:

6. Simplify each complex fraction. Use either method.

a. $\dfrac{\dfrac{6}{y}+\dfrac{5}{y+5}}{\dfrac{8}{y}-\dfrac{2}{y+5}}$

Method 2. Multiply by the LCD $y(y+5)$.

$$=\frac{\left(\frac{6}{y}+\frac{5}{y+5}\right)y(y+5)}{\left(\frac{8}{y}-\frac{2}{y+5}\right)y(y+5)}$$

$$=\frac{\left(\frac{6}{y}\right)y(y+5)+\left(\frac{5}{y+5}\right)y(y+5)}{\left(\frac{8}{y}\right)y(y+5)-\left(\frac{2}{y+5}\right)y(y+5)}$$

$$=\frac{6(y+5)+5y}{8(y+5)-2y}$$

$$=\frac{6y+30+5y}{8y+40-2y}$$

$$=\frac{11y+30}{6y+40}$$

b. $\dfrac{2-\dfrac{5}{x}-\dfrac{7}{x^2}}{1-\dfrac{3}{x}-\dfrac{4}{x^2}}$

Method 2. The LCD is x^2.

$$=\frac{\left(2-\frac{5}{x}-\frac{7}{x^2}\right)x^2}{\left(1-\frac{3}{x}-\frac{4}{x^2}\right)x^2}$$

$$=\frac{2x^2-5x-7}{x^2-3x-4}$$

$$=\frac{(2x-7)(x+1)}{(x-4)(x+1)}$$

$$=\frac{2x-7}{x-4}$$

6. Simplify each complex fraction. Use either method.

a. $\dfrac{\dfrac{2}{x}+\dfrac{3}{x+3}}{\dfrac{5}{x}-\dfrac{4}{x+3}}$

____

b. $\dfrac{1-\dfrac{5}{x}+\dfrac{4}{x^2}}{1-\dfrac{7}{x}+\dfrac{12}{x^2}}$

____

Name: Date:
Instructor: Section:

c. $\dfrac{\dfrac{x+5}{x-4}}{\dfrac{x^2-25}{x^2-16}}$

Method 1.

$$=\frac{x+5}{x-4}\div\frac{x^2-25}{x^2-16}$$

$$=\frac{x+5}{x-4}\cdot\frac{x^2-16}{x^2-25}$$

$$=\frac{(x+5)(x+4)(x-4)}{(x-4)(x+5)(x-5)}$$

$$=\frac{x+4}{x-5}$$

c. $\dfrac{\dfrac{3x+5}{x-2}}{\dfrac{9x^2-25}{x^2-4}}$

____

Objective 1 Simplify a complex fraction by writing it as a division problem (Method 1).

For extra help, see Examples 1–3 on pages 1020–1021 of your text and Section Lecture video for Section 15.5 and Exercise Solutions Clip 9.

Simplify each complex fraction by writing it as a division problem.

1. $\dfrac{\dfrac{49m^3}{18n^5}}{\dfrac{21m}{27n^2}}$

1. _______________

2. $\dfrac{\dfrac{p}{2}-\dfrac{1}{3}}{\dfrac{p}{3}+\dfrac{1}{6}}$

2. _______________

3. $\dfrac{3+\dfrac{4}{s}}{2s+\dfrac{2}{3}}$

3. _______________

Name: Date:
Instructor: Section:

Objective 2 Simplify a complex fraction by multiplying numerator and denominator by the least common denominator (Method 2).

For extra help, see Examples 4–6 on pages 1022–1024 of your text and Section Lecture video for Section 15.5 and Exercise Solutions Clip 4, 15, 19, and 23.

Simplify each complex fraction by multiplying numerator and denominator by the least common denominator.

4. $$\frac{\frac{9}{x^2}-1}{\frac{3}{x}-1}$$

4. ______________

5. $$\frac{\frac{x-2}{x+2}}{\frac{x}{x-2}}$$

5. ______________

6. $$\frac{\frac{6}{k+1}-\frac{5}{k-3}}{\frac{3}{k-3}+\frac{2}{k+2}}$$

6. ______________

Name: Date:
Instructor: Section:

Chapter 15 RATIONAL EXPRESSIONS AND APPLICATIONS

15.6 Solving Equations with Rational Expressions

Learning Objectives

1. Distinguish between operations with rational expressions and equations with terms that are rational expressions.
2. Solve equations with rational expressions.
3. Solve a formula for a specified variable.

Key Terms

Use the vocabulary terms listed below to complete each statement in exercises 1–2.

proposed solution **extraneous solution**

1. A solution that is not an actual solution of a given equation is called a(n) ______________________.

2. A value of the variable that appears to be a solution after both sides of an equation with rational expressions are multiplied by a variable expression is called a(n) ____________________ .

Guided Examples

Review these examples for Objective 1:

1. Identify each of the following as an expression or an equation. Then simplify the expression or solve the equation.

a. $\frac{8}{9}x-\frac{5}{6}x$

This is a difference of two terms. It represents an expression since there is no equality symbol. The LCD is 18.

$$\frac{8}{9}x-\frac{5}{6}x=\frac{2\cdot 8}{2\cdot 9}x-\frac{3\cdot 5}{3\cdot 6}x$$
$$=\frac{16}{18}x-\frac{15}{18}x$$
$$=\frac{1}{18}x$$

Now Try:

1. Identify each of the following as an expression or an equation. Then simplify the expression or solve the equation.

a. $\frac{4}{5}x-\frac{3}{10}x$

____

Name: Date:
Instructor: Section:

b. $\frac{8}{9}x-\frac{5}{6}x=\frac{2}{3}$

Because there is an equality symbol, this is an equation to be solved. The LCD is 18.

$$18\left(\frac{8}{9}x-\frac{5}{6}x\right)=18\left(\frac{2}{3}\right)$$
$$18\left(\frac{8}{9}x\right)-18\left(\frac{5}{6}x\right)=18\left(\frac{2}{3}\right)$$
$$16x-15x=12$$
$$x=12$$

Check $\frac{8}{9}x-\frac{5}{6}x=\frac{2}{3}$

$$\frac{8}{9}(12)-\frac{5}{6}(12)\stackrel{?}{=}\frac{2}{3}$$
$$\frac{32}{3}-10\stackrel{?}{=}\frac{2}{3}$$
$$\frac{2}{3}=\frac{2}{3}\quad\text{True}$$

The solution set is $\{12\}$.

b. $\frac{4}{5}x-\frac{3}{10}x=7$

____

Review this example for Objective 2:

2. Solve, and check the solution.

$$\frac{x}{6}+\frac{x}{5}=8+\frac{x}{2}$$

Multiply by the LCD, 30, to clear fractions.

$$30\left(\frac{x}{6}+\frac{x}{5}\right)=30\left(8+\frac{x}{2}\right)$$
$$30\left(\frac{x}{6}\right)+30\left(\frac{x}{5}\right)=30(8)+30\left(\frac{x}{2}\right)$$
$$5x+6x=240+15x$$
$$11x=240+15x$$
$$-4x=240$$
$$x=-60$$

Check $\frac{x}{6}+\frac{x}{5}=8+\frac{x}{2}$

$$\frac{-60}{6}+\frac{-60}{5}\stackrel{?}{=}8+\frac{-60}{2}$$
$$-10-12\stackrel{?}{=}8-30$$
$$-22=-22\quad\text{True}$$

The solution set is $\{-60\}$.

Now Try:

2. Solve, and check the solution.

$$\frac{x}{3}+\frac{x}{9}=5+\frac{x}{6}$$

____

Name: Date:
Instructor: Section:

3. Solve, and check the solution.

$$\frac{m}{4}-\frac{m-2}{5}=1$$

Multiply by the LCD, 20.

$$20\left(\frac{m}{4}-\frac{m-2}{5}\right)=20\cdot 1$$
$$20\left(\frac{m}{4}\right)-20\left(\frac{m-2}{5}\right)=20$$
$$5m-4(m-2)=20$$
$$5m-4m-4(-2)=20$$
$$m+8=20$$
$$m=12$$

Check that $\{12\}$ is the solution set by replacing m with 12 in the original equation.

4. Solve, and check the proposed solution.

$$\frac{x}{x-3}-\frac{3}{x-3}=3$$

Note that x cannot equal 3, since 3 causes both denominators to equal 0.
Multiply by the LCD, $x-3$.

$$(x-3)\left(\frac{x}{x-3}-\frac{3}{x-3}\right)=(x-3)(3)$$
$$(x-3)\left(\frac{x}{x-3}\right)-(x-3)\left(\frac{3}{x-3}\right)=(x-3)(3)$$
$$x-3=3x-9$$
$$-2x-3=-9$$
$$-2x=-6$$
$$x=3$$

Check $\frac{x}{x-3}-\frac{3}{x-3}=3$

$$\frac{3}{3-3}-\frac{3}{3-3}\stackrel{?}{=}3$$
$$\frac{3}{0}-\frac{3}{0}\stackrel{?}{=}3$$

Division by 0 is undefined.

Thus, the proposed solution 3 must be rejected, and the solution set is $\varnothing$.

3. Solve, and check the solution.

$$\frac{n}{8}-\frac{n-2}{12}=\frac{1}{2}$$

____

4. Solve, and check the proposed solution.

$$\frac{8x}{x+1}+\frac{8}{x+1}=4$$

____

Name: Date:
Instructor: Section:

5. Solve, and check the proposed solution.

$$\frac{3}{x^2-4}=\frac{5}{x^2-2x}$$

Step 1 Factor the denominators to find the LCD, $x(x+2)(x-2)$. Multiply by the LCD.

$$\frac{3}{(x+2)(x-2)}=\frac{5}{x(x-2)}$$

$$x(x+2)(x-2)\frac{3}{(x+2)(x-2)}$$
$$=x(x+2)(x-2)\frac{5}{x(x-2)}$$

Step 2 Solve for x.

$$3x=5(x+2)$$
$$3x=5x+10$$
$$-2x=10$$
$$x=-5$$

Step 3 Check the proposed solution.

$$\frac{3}{x^2-4}=\frac{5}{x^2-2x}$$
$$\frac{3}{(-5)^2-4}\stackrel{?}{=}\frac{5}{(-5)^2-2(-5)}$$
$$\frac{3}{25-4}\stackrel{?}{=}\frac{5}{25+10}$$
$$\frac{1}{7}=\frac{1}{7}\quad\text{True}$$

The solution set is $\{-5\}$.

6. Solve, and check the proposed solution.

$$\frac{2x}{x^2-25}+\frac{5}{x-5}=\frac{3}{x+5}$$

$x\neq 5,-5$ or a denominator is 0.
Factor the denominator.

$$\frac{2x}{(x+5)(x-5)}+\frac{5}{x-5}=\frac{3}{x+5}$$

The LCD is $(x+5)(x-5)$.

5. Solve, and check the proposed solution.

$$\frac{4}{x^2-3x}=\frac{6}{x^2-9}$$

____

6. Solve, and check the proposed solution.

$$\frac{8r}{9r^2-1}=\frac{4}{3r+1}+\frac{4}{3r-1}$$

____

$$(x+5)(x-5)\left(\frac{2x}{(x+5)(x-5)}+\frac{5}{x-5}\right)$$
$$=(x+5)(x-5)\frac{3}{x+5}$$
$$(x+5)(x-5)\frac{2x}{(x+5)(x-5)}+(x+5)(x-5)\frac{5}{x-5}$$
$$=(x+5)(x-5)\frac{3}{x+5}$$
$$2x+5(x+5)=3(x-5)$$
$$2x+5x+25=3x-15$$
$$7x+25=3x-15$$
$$4x=-40$$
$$x=-10$$

A check will verify that {–10} is the solution set.

7. Solve, and check the proposed solution(s).

$$\frac{6}{x^2-1}=1-\frac{3}{x+1}$$

$x \neq 1, -1$ or a denominator is 0.
Factor the denominator.

$$\frac{6}{(x+1)(x-1)}=1-\frac{3}{x+1}$$

The LCD is $(x+1)(x-1)$.

$$(x+1)(x-1)\frac{6}{(x+1)(x-1)}$$
$$=(x+1)(x-1)\left(1-\frac{3}{x+1}\right)$$
$$(x+1)(x-1)\frac{6}{(x+1)(x-1)}$$
$$=(x+1)(x-1)-(x+1)(x-1)\frac{3}{x+1}$$
$$6=(x+1)(x-1)-3(x-1)$$
$$6=x^2-1-3x+3$$
$$0=x^2-3x-4$$
$$0=(x+1)(x-4)$$
$$x+1=0 \quad \text{or} \quad x-4=0$$
$$x=-1 \quad \text{or} \quad x=4$$

Since –1 makes the original denominator equal 0, the proposed solution –1 is an extraneous value.
A check shows that {4} is the solution set.

7. Solve, and check the proposed solution(s).

$$\frac{x}{x+1}+\frac{4}{x}=\frac{4}{x^2+x}$$

____

Name: Date:
Instructor: Section:

8. Solve, and check the proposed solution(s).

$$\frac{3}{b^2+6b+8}+\frac{1}{3b+12}=\frac{1}{6b+12}$$

Factor each denominator.

$$\frac{3}{(b+4)(b+2)}+\frac{1}{3(b+4)}=\frac{1}{6(b+2)}$$

The LCD is $6(b+4)(b+2)$.

$$6(b+4)(b+2)\left(\frac{3}{(b+4)(b+2)}+\frac{1}{3(b+4)}\right)$$
$$=6(b+4)(b+2)\frac{1}{6(b+2)}$$

$$6(b+4)(b+2)\left(\frac{3}{(b+4)(b+2)}\right)$$
$$+6(b+4)(b+2)\left(\frac{1}{3(b+4)}\right)$$
$$=6(b+4)(b+2)\frac{1}{6(b+2)}$$

$$18+2(b+2)=b+4$$
$$18+2b+4=b+4$$
$$22+2b=b+4$$
$$22+b=4$$
$$b=-18$$

The check shows that $\{-18\}$ is the solution set.

8. Solve, and check the proposed solution(s).

$$\frac{4}{y+2}-\frac{3}{y+3}=\frac{8}{y^2+5y+6}$$

____

Review these examples for Objective 3:

9. Solve each formula for the specified variable.

a. $r=\dfrac{s+w}{v}$ for w

Isolate w. Multiply by v.

$$r=\frac{s+w}{v}$$
$$rv=s+w$$
$$rv-s=w$$

Check $\quad r=\dfrac{s+w}{v}$

$$r\stackrel{?}{=}\frac{s+rv-s}{v}$$
$$r\stackrel{?}{=}\frac{rv}{v}$$
$$r=r \quad \text{True}$$

Now Try:

9. Solve each formula for the specified variable.

a. $a=\dfrac{b-c}{q}$ for b

____

Name: Date:
Instructor: Section:

b. $S = \dfrac{a_1}{1-r}$ for r

Isolate r. Multiply by $1-r$.

$$S = \frac{a_1}{1-r}$$
$$S(1-r) = a_1$$
$$S - rS = a_1$$
$$-rS = a_1 - S$$
$$r = \frac{a_1 - S}{-S}, \text{ or } \frac{S - a_1}{S}$$

b. $w = \dfrac{x}{y+z}$ for y

____

10. Solve the formula $\dfrac{1}{p} + \dfrac{1}{q} = \dfrac{1}{r}$ for p.

Isolate p, the specified variable. Multiply by the LCD, pqr.

$$pqr\left(\frac{1}{p} + \frac{1}{q}\right) = pqr\left(\frac{1}{r}\right)$$
$$pqr\left(\frac{1}{p}\right) + pqr\left(\frac{1}{q}\right) = pqr\left(\frac{1}{r}\right)$$
$$qr + pr = pq$$
$$qr = pq - pr$$
$$qr = p(q - r)$$
$$\frac{qr}{q-r} = p$$

10. Solve the formula $\dfrac{1}{x} = \dfrac{1}{y} + \dfrac{1}{z}$ for x.

____

Objective 1 Distinguish between operations with rational expressions and equations with terms that are rational expressions.

For extra help, see Example 1 on page 1027 of your text and Section Lecture video for Section 15.6 and Exercise Solutions Clip 3.

Identify each of the following as an expression or an equation. Then simplify the expression or solve the equation.

1. $\dfrac{3x}{5} - \dfrac{4x}{3} = \dfrac{22}{15}$

1. ______________

Name: Date:
Instructor: Section:

2. $\frac{4x}{5}-\frac{5x}{10}$ **2.** ________________

3. $\frac{2x}{5}+\frac{7x}{3}$ **3.** ________________

Objective 2 Solve equations with rational expressions.

For extra help, see Examples 2–8 on pages 1028–1032 of your text and Section Lecture video for Section 15.6 and Exercise Solutions Clip 35, 37, 41, 47, 49, 51, and 53.

Solve each equation and check your solutions.

4. $\frac{4}{n+2}-\frac{2}{n}=\frac{1}{6}$ **4.** ________________

5. $\frac{x}{3x+16}=\frac{4}{x}$ **5.** ________________

6. $\frac{-16}{n^2-8n+12}=\frac{3}{n-2}+\frac{n}{n-6}$ **6.** ________________

Name: Date:
Instructor: Section:

Objective 3 Solve a formula for a specified variable.

For extra help, see Examples 9–10 on pages 1033–1034 of your text and Section Lecture video for Section 15.6 and Exercise Solutions Clip 65.

Solve each formula for the specified variable.

7. $\frac{1}{f}=\frac{1}{d_0}+\frac{1}{d_1}$ for f **7.** ________________

8. $m=\frac{y_2-y_1}{x_2-x_1}$ for y_1 **8.** ________________

9. $A=\frac{2pf}{b(q+1)}$ for q **9.** ________________

Name: Date:
Instructor: Section:

Chapter 15 RATIONAL EXPRESSIONS AND APPLICATIONS

15.7 Applications of Rational Expressions

Learning Objectives
1 Solve problems about numbers.
2 Solve problems about distance, rate, and time.
3 Solve problems about work.

Key Terms

Use the vocabulary terms listed below to complete each statement in exercises 1–3.

reciprocal **numerator** **denominator**

1. In the fraction $\frac{x+5}{x-2}$, $x + 5$ is the ______________________.

2. In the fraction $\frac{x+5}{x-2}$, $x -2$ is the ______________________.

3. The fraction $\frac{x+5}{x-2}$ is the ____________________ of the fraction $\frac{x-2}{x+5}$.

Guided Examples

Review this example for Objective 1:

1. If a certain number is added to the numerator and twice that number is subtracted from the denominator of the fraction $\frac{3}{5}$, the result is equal to 5. Find the number.

Step 1 Read the problem carefully. We are trying to find a number.

Step 2 Assign a variable.
Let x = the number.

Step 3 Write an equation. The fraction $\frac{3+x}{5-2x}$ represents adding the number to the numerator and twice that number is subtracted from the denominator of the fraction $\frac{3}{5}$. The result is equal to 5.

$$\frac{3+x}{5-2x} = 5$$

Now Try:

1. If the same number is added to the numerator and denominator of the fraction $\frac{5}{9}$, the value of the resulting fraction is $\frac{2}{3}$. Find the number.

____

Step 4 Solve. Multiply by the LCD, $5-2x$.

$$(5-2x)\frac{3+x}{5-2x}=(5-2x)5$$
$$3+x=25-10x$$
$$11x=22$$
$$x=2$$

Step 5 State the answer. The number is 2.

Step 6 Check the solution in the original problem. If 2 is added to the numerator, and twice 2 is subtracted from the denominator of $\frac{3}{5}$, the result is $\frac{3+2}{5-2(2)}=\frac{5}{1}$, or 5, as required.

Review these examples for Objective 2:

2. Solve each problem using a form of the distance formula.

a. Find the distance sound travels in 8 sec, if the speed of sound is 1088 ft/sec.

We must find the distance, given rate and time, using $d=rt$ (or $rt=d$).

1088	×	8	=	8704 ft
↑		↑	↑	↑
Rate	×	Time	=	Distance

b. Find the time to travel 400 miles at 81.5 mph.

We must find the time, given the rate and distance using $t=\frac{d}{r}$ (or $\frac{d}{r}=t$).

$$\begin{matrix}\text{Distance}\rightarrow \\ \text{Rate}\rightarrow\end{matrix}\frac{400}{81.5}=4.9 \text{ hr (rounded)} \leftarrow \text{Time}$$

c. Find the speed running 200 m in 19.5 sec.

We must find the rate, given the distance and time using $r=\frac{d}{t}$ (or $\frac{d}{t}=r$).

$$\begin{matrix}\text{Distance}\rightarrow \\ \text{Time}\rightarrow\end{matrix}\frac{200}{19.5}=10.3 \text{ m/sec (rounded)} \leftarrow \text{Rate}$$

Now Try:

2. Solve each problem using a form of the distance formula.

a. Find the distance 255 mph for 3 hrs.

____

b. Find the time to travel 675 miles at 165 mph.

____

c. Find the speed to travel 900 miles in 3.7 hrs.

____

Name: Date:
Instructor: Section:

3. Carl traveled to his destination at an average speed of 70 miles per hour. Coming home, his average speed was 50 miles per hour and the trip took 2 hours longer. How far did he travel each way?

Step 1 Read the problem carefully. We must find the distance.

Step 2 Assign a variable.
Let x = the distance for the trip.

Step 3 Write an equation. We use $t = \frac{d}{r}$ for each trip. Then, the difference, in time, between the two trips was 2 hours.

$$\frac{x}{50} - \frac{x}{70} = 2$$

Step 4 Solve. The LCD is 350.

$$350\left(\frac{x}{50} - \frac{x}{70}\right) = 350(2)$$
$$7x - 5x = 700$$
$$2x = 700$$
$$x = 350 \text{ miles}$$

Step 5 State the answer. The trip was 350 miles.

Step 6 Check.
$\frac{350}{50} = 7$ hr and $\frac{350}{70} = 5$ hr
The difference in time is 2 hours.

4. A boat goes 6 miles per hour in still water. It takes as long to go 40 miles upstream as 80 miles downstream. Find the speed of the current.

Step 1 Read the problem carefully. Find the speed of the current.

Step 2 Assign a variable.
Let x = the speed of the current.
The rate of traveling upstream is $6 - x$.
The rate of traveling downstream is $6 + x$.

	d	r	t
Upstream	40	$6-x$	$\frac{40}{6-x}$
Downstream	80	$6+x$	$\frac{80}{6+x}$

The times are equal.

3. Wendy drove 250 miles to her aunt's house at a speed that was 10 miles per hour faster than her speed on her return trip. If it took Wendy $\frac{5}{6}$ hour longer on the return trip, what was her speed on the return trip?

____

4. The Cuyahoga River has a current of 2 miles per hour. Ali can paddle 10 miles downstream in the time it takes her to paddle 2 miles upstream. How fast can Ali paddle?

____

Step 3 Write an equation.

$$\frac{40}{6-x}=\frac{80}{6+x}$$

Step 4 Solve. The LCD is $(6+x)(6-x)$.

$$(6+x)(6-x)\frac{40}{6-x}=(6+x)(6-x)\frac{80}{6+x}$$
$$40(6+x)=80(6-x)$$
$$240+40x=480-80x$$
$$240+120x=480$$
$$120x=240$$
$$x=2$$

Step 5 State the answer. The speed of the current is 2 miles per hour.

Step 6 Check.

Upstream: $\frac{40}{6-2}=10$ hr

Downstream: $\frac{80}{6+2}=10$ hr

The time upstream is the same as the time downstream, as required.

Review this example for Objective 3:

5. One pipe can fill a swimming pool in 8 hours and another pipe can fill the pool in 12 hours. How long will it take to fill the pool if both pipes are open?

Step 1 Read the problem carefully. Find the time working together.

Step 2 Assign a variable.
Let x = the number of hours working together.

The rate for the first pipe is $\frac{1}{8}$.

The rate for the second pipe is $\frac{1}{12}$.

Step 3 Write an equation. The sum of the fractional part for each pipe multiplied by the time working together is the whole job.

$$\frac{1}{8}x+\frac{1}{12}x=1$$

Now Try:

5. Chuck can weed the garden in $\frac{1}{2}$ hour, but David takes 2 hours. How long does it take them to weed the garden if they work together?

____

Step 4 Solve. The LCD is 48.

$$48\left(\frac{1}{8}x+\frac{1}{12}x\right)=48(1)$$

$$48\left(\frac{1}{8}x\right)+48\left(\frac{1}{12}x\right)=48$$

$$6x+4x=48$$

$$10x=48$$

$$x=\frac{48}{10}=\frac{24}{5}=4\frac{4}{5}$$

Step 5 State the answer. Working together, it takes $4\frac{4}{5}$ hours to fill the pool.

Step 6 Check. Substitute $\frac{24}{5}$ for x in the equation from Step 3.

$$\frac{1}{8}x+\frac{1}{12}x=1$$

$$\frac{1}{8}\left(\frac{24}{5}\right)+\frac{1}{12}\left(\frac{24}{5}\right)=1$$

$$\frac{3}{5}+\frac{2}{5}=1 \quad \text{True}$$

Objective 1 Solve problems about numbers.

For extra help, see Example 1 on page 1041 of your text and Section Lecture video for Section 15.7 and Exercise Solutions Clip 8.

Solve each problem. Check your answers to be sure they are reasonable.

1. If three times a number is subtracted from twice its reciprocal, the result is –1. Find the number.

1. ________________

2. If two times a number is added to one-half of its reciprocal, the result is $\frac{13}{6}$. Find the number.

2. ________________

3. The denominator of a fraction is 1 less than twice the numerator. If the numerator and the denominator are each increased by 3, the resulting fraction simplifies to $\frac{3}{4}$. Find the original fraction. 3. ________________

Objective 2 Solve problems about distance, rate, and time.

For extra help, see Examples 2–4 on pages 1042–1045 of your text and Section Lecture video for Section 15.7 and Exercise Solutions Clip 21.

Solve each problem.

4. A boat travels 15 miles per hour in still water. The boat travels 20 miles downstream in the same time it takes the boat to travel 10 miles upstream. How fast is the current? 4. ________________

5. A ship goes 120 miles downriver in $2\frac{2}{3}$ hours less than it takes to go the same distance upriver. If the speed of the current is 6 miles per hour, find the speed of the ship. 5. ________________

6. On Saturday, Pablo jogged 6 miles. On Monday, jogging at the same speed, it took him 30 minutes longer to cover 10 miles. How fast did Pablo jog? 6. ________________

Name: Date:
Instructor: Section:

Objective 3 Solve problems about work.

For extra help, see Example 5 on page 1046 of your text and Section Lecture video for Section 15.7 and Exercise Solutions Clip 31 and 34.

Solve each problem.

7. Kelly can clean the house in 6 hours, but it takes Linda 4 hours. How long would it take them to clean the house if they worked together?

7. ________________

8. Michael can type twice as fast as Sharon. Together they can type a certain job in 2 hours. How long would it take Michael to type the entire job by himself?

8. ________________

9. A swimming pool can be filled by an inlet pipe in 18 hours and emptied by an outlet pipe in 24 hours. How long will it take to fill the empty pool if the outlet pipe is accidentally left open at the same time as the inlet pipe is opened?

9. ________________

Name: Date:
Instructor: Section:

Chapter 15 RATIONAL EXPRESSIONS AND APPLICATIONS

15.8 Variation

Learning Objectives
1 Solve direct variation problems.
2 Solve inverse variation problems.

Key Terms

Use the vocabulary terms listed below to complete each statement in exercises 1–3.

direct variation **constant of variation** **inverse variation**

1. In the equation $y = kx$, the number k is called the ______________________.

2. If two positive quantities x and y are in ______________________ and the constant of variation is positive, then as x increases, y also increases.

3. If two positive quantities x and y are in ______________________ and the constant of variation is positive, then as x increases, y decreases.

Guided Examples

Review this example for Objective 1:

1. If w varies directly as v, and $w = 24$ when $v = 20$, find w when $v = 25$.

Since w varies directly as v, there is a constant k such that $w = kv$. We let $w = 24$, $v = 20$, and solve for k.

$$w = kv$$
$$24 = k \cdot 20$$
$$\frac{6}{5} = k$$

Since $w = kv$ and $k = \frac{6}{5}$, we have $w = \frac{6}{5}v$.

Now we can find the value of w when $v = 25$.

$$w = \frac{6}{5} \cdot 25 = 30$$

Thus, $w = 30$ when $v = 25$.

Now Try:

1. If a varies directly as b, and $a = 61.5$ when $b = 82$, find a when $b = 224$.

____

Name: Date:
Instructor: Section:

Review this example for Objective 2:

2. If g varies inversely as f, and $g = 6$ when $f = 12$, find g when $f = 18$.

Since g varies inversely as f, there is a constant k such that $g = \frac{k}{f}$. We know that $g = 6$ when $f = 12$, so we can find k.

$$g = \frac{k}{f}$$
$$6 = \frac{k}{12}$$
$$72 = k$$

Since $g = \frac{72}{f}$, we let $f = 18$ and solve for g.

$$g = \frac{72}{f} = \frac{72}{18} = 4$$

Therefore, when $f = 18$, $g = 4$.

3. For a specified distance, time varies inversely with speed. If Ramona walks a certain distance on a treadmill in 40 minutes at 4.2 miles per hour, how long will it take her to walk the same distance at 3.5 miles per hour?

Let t = time and s = speed.
Since t varies inversely as s, there is a constant k such that $t = \frac{k}{s}$. Recall that $40 \text{ min} = \frac{40}{60} \text{ hr}$.

$$t = \frac{k}{s}$$
$$\frac{40}{60} = \frac{k}{4.2}$$
$$2.8 = k$$

Now use $t = \frac{k}{s}$ to find the value of t when $s = 3.5$.

$$t = \frac{2.8}{3.5} = \frac{4}{5}$$

It takes $\frac{4}{5}$ hr, or 48 min to walk the same distance.

Now Try:

2. If y varies inversely as x, and $y = 10$ when $x = 2$, find y when $x = 4$.

____

3. The force required to compress a spring varies directly as the change in the length of the spring. If a force of 25 pounds is required to compress a spring 4 inches, how much force is required to compress the spring 8 inches?

____

Name: Date:
Instructor: Section:

Objective 1 Solve direct variation problems.

For extra help, see Example 1 on page 1052 of your text and Section Lecture video for Section 15.8 and Exercise Solutions Clip 15, 16, and 18.

Solve each problem involving direct variation.

1. If y varies directly as x, and $x = 14$ when $y = 42$, find y when $x = 4$. **1.** ________________

2. If c varies directly as d, and $c = 100$ when $d = 5$, find c when $d = 3$. **2.** ________________

3. If h varies directly as m, and $h = 9$ when $m = 6$, find h when $m = 10$. **3.** ________________

Objective 2 Solve inverse variation problems.

For extra help, see Examples 2–3 on page 1053 of your text and Section Lecture video for Section 15.8 and Exercise Solutions Clip 19, 20, 22, 29, and 30.

Solve each problem involving indirect variation.

4. If y varies inversely as x, and $y = 20$ when $x = 4$, find y when $x = 12$. **4.** ________________

5. If n varies inversely as m, and $n = 10.5$ when $m = 1.2$, find n when $m = 5.6$. **5.** ________________

Solve the problem

6. The length of a violin string varies inversely with the frequency of its vibrations. A 10-inch violin string vibrates at a frequency of 512 cycles per second. Find the frequency of an 8-inch string. **6.** ________________

Name: Date:
Instructor: Section:

Chapter 16 ROOTS AND RADICALS

16.1 Evaluating Roots

Learning Objectives

1. Find square roots.
2. Decide whether a given root is rational, irrational, or not a real number.
3. Find decimal approximations for irrational square roots.
4. Use the Pythagorean theorem.
5. Find cube, fourth, and other roots.

Key Terms

Use the vocabulary terms listed below to complete each statement in exercises 1–9.

square root	**principal square root**	**radicand**
radical	**radical expression**	**perfect square**
irrational number	**cube root**	**index (order)**

1. The number or expression inside a radical sign is called the ________________.
2. A number with a rational square root is called a ________________________.
3. In a radical of the form $\sqrt[n]{a}$, the number n is the ________________________.
4. The number b is a ________________________ of a if $b^2 = a$.
5. The expression $\sqrt[n]{a}$ is called a ________________________________.
6. The positive square root of a number is its ________________________.
7. A real number that is not rational is called an ________________________.
8. A ________________________ is a radical sign and the number or expression in it.
9. The number b is a ________________________ of a if $b^3 = a$.

Guided Examples

Review this example for Objective 1:

1. Find the square roots of 64.

   What number multiplied by itself equals 64?

   $8^2 = 64$ and $(-8)^2 = 64$.

   Thus, 64 has two square roots: 8 and –8.

Now Try:

1. Find the square roots of 81.

   ____

Name: Date:
Instructor: Section:

2. Find each square root.

a. $\sqrt{121}$

$11^2 = 121$, so $\sqrt{121} = 11$.

b. $-\sqrt{576}$

This is the negative square root of 576.
Since $\sqrt{576} = 24$, then $-\sqrt{576} = -24$.

c. $\sqrt{\frac{16}{25}}$

$\sqrt{\frac{16}{25}} = \frac{4}{5}$

d. $\sqrt{0.36}$

$\sqrt{0.36} = 0.6$

3. Find the square of each radical expression.

a. $\sqrt{17}$

The square of $\sqrt{17}$ is $(\sqrt{17})^2 = 17$.

b. $-\sqrt{31}$

$(-\sqrt{31})^2 = 31$

c. $\sqrt{w^2+3}$

$\left(\sqrt{w^2+3}\right)^2 = w^2+3$

2. Find each square root.

a. $\sqrt{169}$ __________

b. $\sqrt{1681}$ __________

c. $\sqrt{\frac{9}{49}}$ __________

d. $\sqrt{0.64}$ __________

3. Find the square of each radical expression.

a. $\sqrt{19}$ __________

b. $-\sqrt{37}$ __________

c. $\sqrt{n^2+5}$ __________

Review these examples for Objective 2:

4. Tell whether each square root is rational, irrational, or not a real number.

a. $\sqrt{5}$

Because 5 is not a perfect square, $\sqrt{5}$ is irrational.

b. $\sqrt{81}$

81 is a perfect square, 9^2, so $\sqrt{81} = 9$ is a rational number.

Now Try:

4. Tell whether each square root is rational, irrational, or not a real number.

a. $\sqrt{11}$ __________

b. $\sqrt{100}$ __________

Name: Date:
Instructor: Section:

c. $\sqrt{-16}$

There is no real number whose square is –16. Therefore, $\sqrt{-16}$ is not a real number.

c. $\sqrt{-9}$

____

Review these examples for Objective 3:

5. Find a decimal approximation for reach square root. Round answers to the nearest thousandth.

a. $\sqrt{12}$

Use the square root key on a calculator.
$\sqrt{12} \approx 3.464$

b. $\sqrt{42}$

$\sqrt{42} \approx 6.481$

c. $-\sqrt{596}$

$-\sqrt{596} \approx -24.413$

Now Try:

5. Find a decimal approximation for reach square root. Round answers to the nearest thousandth.

a. $\sqrt{14}$

____

b. $\sqrt{51}$

____

c. $-\sqrt{678}$

____

Review these examples for Objective 4:

6. Find the length of the unknown side of each right triangle with sides a, b, and c, where c is the hypotenuse.

a. $a = 5,\ b = 12$

Use the Pythagorean theorem.

$$a^2 + b^2 = c^2$$
$$5^2 + 12^2 = c^2$$
$$25 + 144 = c^2$$
$$169 = c^2$$

Since the length of a side of a triangle must be a positive number, find the positive square root of 169 to get c.

$$c = \sqrt{169} = 13$$

Now Try:

6. Find the length of the unknown side of each right triangle with sides a, b, and c, where c is the hypotenuse.

a. $a = 8,\ b = 15$

____

Name: Date:
Instructor: Section:

b. $b = 17,\ c = 18$

Use the Pythagorean theorem.

$$a^2 + b^2 = c^2$$
$$a^2 + 17^2 = 18^2$$
$$a^2 + 289 = 324$$
$$a^2 = 35$$

Use a calculator to find the positive square root of 35 to approximate a.

$$a = \sqrt{35} \approx 5.916$$

b. $b = 12,\ c = 14$

____

7. A ladder 25 feet long leans against a wall. The foot of the ladder is 7 feet from the base of the wall. How high up the wall does the top of the ladder rest?

Step 1 Read the problem again.

Step 2 Assign a variable. Let a represent the height of the top of the ladder when measured straight down to the ground.

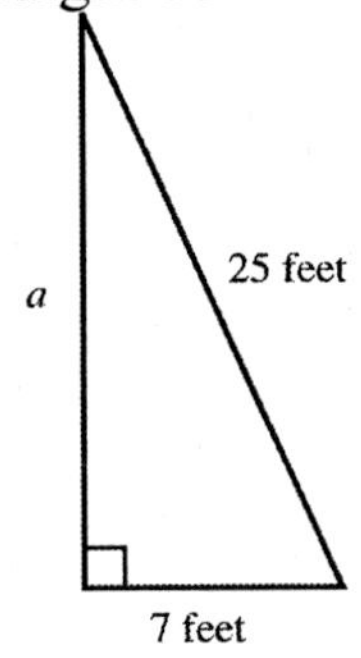

Step 3 Write an equation using the Pythagorean theorem.

$$a^2 + b^2 = c^2$$
$$a^2 + 7^2 = 25^2$$

Step 4 Solve.

$$a^2 + 49 = 625$$
$$a^2 = 576$$
$$a = 24$$

Choose the positive square root of 576 since a represents a length.

Step 5 State the answer. The top of the ladder rests 24 ft up the wall.

7. Susan started to drive due south at the same time John started to drive due west. John drove 21 miles in the same time that Susan drove 28 miles. How far apart were they at that time?

____

Name: Date:
Instructor: Section:

Step 6 Check. From the figure, we have the following.

$24^2 + 7^2 \stackrel{?}{=} 25^2$

$576 + 49 = 625$

The check confirms that the top of the ladder rests 24 ft up the wall.

Review these examples for Objective 5:

8. Find each cube root.

a. $\sqrt[3]{64}$

Because $4^3 = 64$, $\sqrt[3]{64} = 4$.

b. $\sqrt[3]{-64}$

$\sqrt[3]{-64} = -4$, because $(-4)^3 = -64$.

c. $\sqrt[3]{729}$

$\sqrt[3]{729} = 9$, because $9^3 = 729$.

9. Find each root.

a. $\sqrt[4]{81}$

$\sqrt[4]{81} = 3$, because 3 is positive and $3^4 = 81$.

b. $-\sqrt[4]{81}$

From part (a), $\sqrt[4]{81} = 3$, so the negative root is $-\sqrt[4]{81} = -3$.

c. $\sqrt[4]{-81}$

For a real number fourth root, the radicand must be nonnegative. There is no real number that equals $\sqrt[4]{-81}$.

d. $-\sqrt[5]{1024}$

$-\sqrt[5]{1024} = -4$

e. $\sqrt[5]{-1024}$

$\sqrt[5]{-1024} = -4$, because $(-4)^5 = -1024$.

Now Try:

8. Find each cube root.

a. $\sqrt[3]{125}$ __________

b. $\sqrt[3]{-125}$ __________

c. $\sqrt[3]{343}$ __________

9. Find each root.

a. $\sqrt[4]{625}$ __________

b. $-\sqrt[4]{625}$ __________

c. $\sqrt[4]{-625}$ __________

d. $-\sqrt[5]{3125}$ __________

e. $\sqrt[5]{-3125}$ __________

Name: Date:
Instructor: Section:

Objective 1 Find square roots.

For extra help, see Examples 1–3 on pages 1070–1071 of your text and Section Lecture video for Section 16.1 and Exercise Solutions Clip 7, 25, 29, and 43.

Find all square roots of each number.

1. 625 **1.** ________________

2. $\frac{121}{196}$ **2.** ________________

Find the square root

3. $\sqrt{\frac{900}{49}}$ **3.** ________________

Objective 2 Decide whether a given root is rational, irrational, or not a real number.

For extra help, see Example 4 on page 1072 of your text and Section Lecture video for Section 16.1.

Tell whether each square root is rational, irrational, *or* not a real number.

4. $\sqrt{72}$ **4.** ________________

5. $\sqrt{-36}$ **5.** ________________

6. $\sqrt{6400}$ **6.** ________________

Objective 3 Find decimal approximations for irrational square roots.

For extra help, see Example 5 on page 1073 of your text and Section Lecture video for Section 16.1 and Exercise Solutions Clip 51, 53, 57, and 59.

Use a calculator to find a decimal approximation for each square root. Round answers to the nearest thousandth.

7. $\sqrt{32}$ **7.** ________________

8. $-\sqrt{131}$ **8.** ________________

9. $\sqrt{210}$ **9.** ________________

Name: Date:
Instructor: Section:

Objective 4 Use the Pythagorean theorem.

For extra help, see Examples 6–7 on pages 1073–1074 of your text and Section Lecture video for Section 16.1 and Exercise Solutions Clip 65 and 73.

Find the length of the unknown side of each right triangle with sides a, b, and c, where c is the hypotenuse. If necessary, round your answer to the nearest thousandth.

10. $a = 5,\ b = 9$ **10.** ______________

11. $c = 15,\ a = 12$ **11.** ______________

Use the Pythagorean formula to solve the problem. If necessary, round your answer to the nearest thousandth.

12. A plane flies due east for 35 miles and then due south until it is 37 miles from its starting point. How far south did the plane fly? **12.** ______________

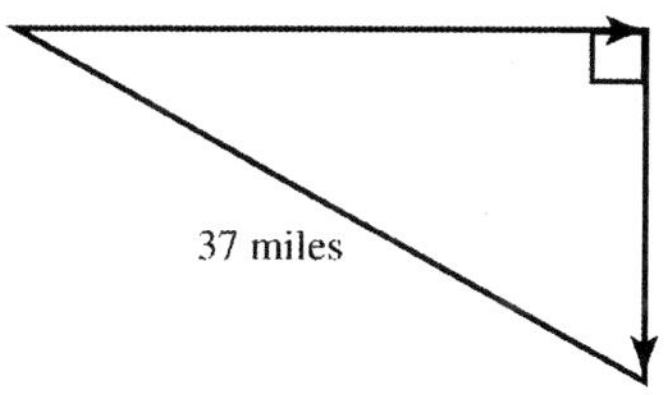

Objective 5 Find cube, fourth, and other roots.

For extra help, see Examples 8–9 on pages 1075–1076 of your text and Section Lecture video for Section 16.1 and Exercise Solutions Clip 85, 89, 96, and 103.

Find each root.

13. $\sqrt[3]{-64}$ **13.** ______________

14. $\sqrt[4]{256}$ **14.** ______________

15. $\sqrt[7]{-1}$ **15.** ______________

Name: Date:
Instructor: Section:

Chapter 16 ROOTS AND RADICALS

16.2 Multiplying, Dividing, and Simplifying Radicals

Learning Objectives
1. Multiply square root radicals.
2. Simplify radicals by using the product rule.
3. Simplify radicals by using the quotient rule.
4. Simplify radicals involving variables.
5. Simplify other roots.

Key Terms

Use the vocabulary terms listed below to complete each statement in exercises 1–3.

perfect cube **radical** **radicand**

1. A root of a number is called a ____________________.

2. A number with a rational cube root is called a ____________________.

3. The ____________________ is the number or expression inside a radical sign.

Guided Examples

Review these examples for Objective 2:

1. Use the product rule for radicals to find each product.

   a. $\sqrt{10}\cdot\sqrt{7}$

   $\sqrt{10}\cdot\sqrt{7}=\sqrt{10\cdot 7}=\sqrt{70}$

   b. $\sqrt{3}\cdot\sqrt{13}$

   $\sqrt{3}\cdot\sqrt{13}=\sqrt{39}$

   c. $\sqrt{17}\cdot\sqrt{b}\quad(b\geq 0)$

   $\sqrt{17}\cdot\sqrt{b}=\sqrt{17b}$

Now Try:

1. Use the product rule for radicals to find each product.

   a. $\sqrt{13}\cdot\sqrt{7}$

   ____

   b. $\sqrt{6}\cdot\sqrt{7}$

   ____

   c. $\sqrt{15}\cdot\sqrt{c}\quad(c\geq 0)$

   ____

Review these examples for Objective 2:

2. Simplify each radical.

   a. $\sqrt{40}$

   $$\begin{aligned}\sqrt{40}&=\sqrt{4\cdot 10}\\&=\sqrt{4}\cdot\sqrt{10}\\&=2\sqrt{10}\end{aligned}$$

Now Try:

2. Simplify each radical.

   a. $\sqrt{12}$

   ____

Name: Date:
Instructor: Section:

b. $\sqrt{75}$

$$\sqrt{75} = \sqrt{25 \cdot 2}$$
$$= \sqrt{25} \cdot \sqrt{2}$$
$$= 5\sqrt{2}$$

c. $\sqrt{48}$

$$\sqrt{48} = \sqrt{16 \cdot 3} = \sqrt{16} \cdot \sqrt{3} = 4\sqrt{3}$$

d. $\sqrt{26}$

Since 26 has no perfect square factors, except 1, $\sqrt{26}$ cannot be simplified.

b. $\sqrt{98}$

____

c. $\sqrt{80}$

____

d. $\sqrt{33}$

____

3. Find each product and simplify.

a. $\sqrt{16} \cdot \sqrt{20}$

$$\sqrt{16} \cdot \sqrt{20} = 4\sqrt{20}$$
$$= 4\sqrt{4 \cdot 5}$$
$$= 4\sqrt{4} \cdot \sqrt{5}$$
$$= 4 \cdot 2 \cdot \sqrt{5}$$
$$= 8\sqrt{5}$$

b. $\sqrt{27} \cdot \sqrt{50}$

$$\sqrt{27} \cdot \sqrt{50} = \sqrt{27 \cdot 50}$$
$$= \sqrt{9 \cdot 3 \cdot 25 \cdot 2}$$
$$= \sqrt{9} \cdot \sqrt{25} \cdot \sqrt{3 \cdot 2}$$
$$= 3 \cdot 5 \cdot \sqrt{6}$$
$$= 15\sqrt{6}$$

c. $7\sqrt{10} \cdot 4\sqrt{5}$

$$7\sqrt{10} \cdot 4\sqrt{5} = 7 \cdot 4\sqrt{10 \cdot 5}$$
$$= 28\sqrt{50}$$
$$= 28\sqrt{25 \cdot 2}$$
$$= 28\sqrt{25} \cdot \sqrt{2}$$
$$= 28 \cdot 5\sqrt{2}$$
$$= 140\sqrt{2}$$

3. Find each product and simplify.

a. $\sqrt{25} \cdot \sqrt{18}$

____

b. $\sqrt{6} \cdot \sqrt{12}$

____

c. $4\sqrt{5} \cdot 3\sqrt{10}$

____

Name: Date:
Instructor: Section:

Review these examples for Objective 3:

4. Use the quotient rule to simplify each radical.

a. $\sqrt{\frac{121}{4}}$

$$\sqrt{\frac{121}{4}} = \frac{\sqrt{121}}{\sqrt{4}} = \frac{11}{2}$$

b. $\frac{\sqrt{75}}{\sqrt{3}}$

$$\frac{\sqrt{75}}{\sqrt{3}} = \sqrt{\frac{75}{3}} = \sqrt{25} = 5$$

c. $\sqrt{\frac{5}{16}}$

$$\sqrt{\frac{5}{16}} = \frac{\sqrt{5}}{\sqrt{16}} = \frac{\sqrt{5}}{4}$$

5. Simplify.

$\frac{32\sqrt{30}}{8\sqrt{5}}$

$$\frac{32\sqrt{30}}{8\sqrt{5}} = \frac{32}{8} \cdot \frac{\sqrt{30}}{\sqrt{5}}$$
$$= \frac{32}{8} \cdot \sqrt{\frac{30}{5}}$$
$$= 4\sqrt{6}$$

6. Simplify.

$\sqrt{\frac{5}{7}} \cdot \sqrt{\frac{2}{7}}$

$$\sqrt{\frac{5}{7}} \cdot \sqrt{\frac{2}{7}} = \sqrt{\frac{5}{7} \cdot \frac{2}{7}}$$
$$= \sqrt{\frac{10}{49}}$$
$$= \frac{\sqrt{10}}{\sqrt{49}}, \quad \text{or} \quad \frac{\sqrt{10}}{7}$$

Now Try:

4. Use the quotient rule to simplify each radical.

a. $\sqrt{\frac{169}{9}}$ ____________

b. $\frac{\sqrt{405}}{\sqrt{5}}$ ____________

c. $\sqrt{\frac{7}{64}}$ ____________

5. Simplify.

$\frac{9\sqrt{30}}{3\sqrt{3}}$ ____________

6. Simplify.

$\sqrt{\frac{7}{6}} \cdot \sqrt{\frac{1}{6}}$ ____________

Name: Date:
Instructor: Section:

Review these examples for Objective 4:

7. Simplify each radical. Assume that all variables represent nonnegative real numbers.

a. $\sqrt{x^6}$

$\sqrt{x^6} = x^3$, since $(x^3)^2 = x^6$

b. $\sqrt{16m^{10}}$

$\sqrt{16m^{10}} = \sqrt{16} \cdot \sqrt{m^{10}} = 4m^5$

c. $\sqrt{27q^{12}}$

$$\begin{aligned}\sqrt{27q^{12}} &= \sqrt{9 \cdot 3 \cdot q^{12}} \\ &= \sqrt{9} \cdot \sqrt{3} \cdot \sqrt{q^{12}} \\ &= 3 \cdot \sqrt{3} \cdot q^6 \\ &= 3q^6\sqrt{3}\end{aligned}$$

d. $\sqrt{r^7}$

$$\begin{aligned}\sqrt{r^7} &= \sqrt{r^6 \cdot r} \\ &= \sqrt{r^6}\sqrt{r} \\ &= r^3\sqrt{r}\end{aligned}$$

e. $\sqrt{\dfrac{7}{y^2}}$, $y \neq 0$

$$\sqrt{\frac{7}{y^2}} = \frac{\sqrt{7}}{\sqrt{y^2}} = \frac{\sqrt{7}}{y}$$

Now Try:

7. Simplify each radical. Assume that all variables represent nonnegative real numbers.

a. $\sqrt{x^{20}}$ __________

b. $\sqrt{49x^8}$ __________

c. $\sqrt{72x^4}$ __________

d. $\sqrt{r^{19}}$ __________

e. $\sqrt{\dfrac{11}{y^4}}$ __________

Review these examples for Objective 5:

8. Simplify each radical.

a. $\sqrt[3]{40}$

$\sqrt[3]{40} = \sqrt[3]{8 \cdot 5} = \sqrt[3]{8} \cdot \sqrt[3]{5} = 2\sqrt[3]{5}$

b. $\sqrt[4]{1250}$

$\sqrt[4]{1250} = \sqrt[4]{625 \cdot 2} = \sqrt[4]{625} \cdot \sqrt[4]{2} = 5\sqrt[4]{2}$

Now Try:

8. Simplify each radical.

a. $\sqrt[3]{54}$ __________

b. $\sqrt[4]{162}$ __________

Name: Date:
Instructor: Section:

c. $\sqrt[3]{\frac{64}{27}}$

$\sqrt[3]{\frac{64}{27}} = \frac{\sqrt[3]{64}}{\sqrt[3]{27}} = \frac{4}{3}$

c. $\sqrt[3]{\frac{8}{343}}$ ____________

9. Simplify each radical.

a. $\sqrt[4]{m^8}$

$\sqrt[4]{m^8} = m^2$

b. $\sqrt[3]{64x^9}$

$\sqrt[3]{64x^9} = \sqrt[3]{64} \cdot \sqrt[3]{x^9} = 4x^3$

c. $\sqrt[3]{40a^5}$

$\sqrt[3]{40a^5} = \sqrt[3]{8a^3 \cdot 5a^2}$

$= \sqrt[3]{8a^3} \cdot \sqrt[3]{5a^2}$

$= 2a\sqrt[3]{5a^2}$

d. $\sqrt[3]{\frac{y^6}{1000}}$

$\sqrt[3]{\frac{y^6}{1000}} = \frac{\sqrt[3]{y^6}}{\sqrt[3]{1000}} = \frac{y^2}{10}$

9. Simplify each radical.

a. $\sqrt[4]{m^{12}}$ ____________

b. $\sqrt[3]{125x^6}$ ____________

c. $\sqrt[4]{80a^5}$ ____________

d. $\sqrt[3]{\frac{x^{21}}{216}}$ ____________

Objective 1 Multiply square root radicals.

For extra help, see Example 1 on page 1081 of your text and Section Lecture video for Section 16.2 and Exercise Solutions Clip 5, 9, and 11.

Use the product rule for radicals to find each product.

1. $\sqrt{13} \cdot \sqrt{5}$ 1. ____________

2. $\sqrt{11} \cdot \sqrt{2}$ 2. ____________

3. $\sqrt{3x} \cdot \sqrt{7},\ x > 0$ 3. ____________

Name: Date:
Instructor: Section:

Objective 2 Simplify radicals by using the product rule.

For extra help, see Examples 2–3 on pages 1081–1083 of your text and Section Lecture video for Section 16.2 and Exercise Solutions Clip 17, 25, 29, and 47.

Simplify each radical.

4. $\sqrt{405}$ **4.** ________________

Find each product and simplify.

5. $\sqrt{11} \cdot \sqrt{33}$ **5.** ________________

6. $\sqrt{18} \cdot \sqrt{24}$ **6.** ________________

Objective 3 Simplify radicals by using the quotient rule.

For extra help, see Examples 4–6 on pages 1083–1084 of your text and Section Lecture video for Section 16.2 and Exercise Solutions Clip 53, 61, and 63.

Use the quotient rule and product rule, as necessary to simplify each expression.

7. $\sqrt{\frac{25}{81}}$ **7.** ________________

8. $\frac{\sqrt{24}}{2\sqrt{6}}$ **8.** ________________

9. $\sqrt{\frac{2}{125}} \cdot \sqrt{\frac{2}{5}}$ **9.** ________________

Name: Date:
Instructor: Section:

Objective 4 Simplify radicals involving variables.

For extra help, see Example 7 on page 1085 of your text and Section Lecture video for Section 16.2 and Exercise Solutions Clip 65 and 81.

Simplify each radical. Assume that all variables represent positive real numbers.

10. $\sqrt{p^2q^6}$ **10.** ________________

11. $\sqrt{32x^4y^5}$ **11.** ________________

12. $\sqrt{\frac{81}{25x^6}}$ **12.** ________________

Objective 5 Simplify other roots.

For extra help, see Examples 8–9 on page 1086 of your text and Section Lecture video for Section 16.2 and Exercise Solutions Clip 93, 101, 105, 107, and 113.

Simplify each expression.

13. $\sqrt[5]{-64}$ **13.** ________________

14. $\sqrt[3]{\frac{1728}{1000}}$ **14.** ________________

15. $\sqrt[4]{\frac{625}{256}}$ **15.** ________________

Name: Date:
Instructor: Section:

Chapter 16 ROOTS AND RADICALS

16.3 Adding and Subtracting Radicals

Learning Objectives

1. Add and subtract radicals.
2. Simplify radical sums and differences.
3. Simplify more complicated radical expressions.

Key Terms

Use the vocabulary terms listed below to complete each statement in exercises 1–3.

like radicals **index** **unlike radicals**

1. In the expression, $\sqrt[4]{x^2}$, the "4" is called the ______________________________.

2. The expressions $2\sqrt{2}$ and $6\sqrt[3]{2}$ are ______________________________.

3. The expressions $2\sqrt{2}$ and $7\sqrt{2}$ are ______________________________.

Guided Examples

Review these examples for Objective 1:

1. Add or subtract, as indicated.

a. $7\sqrt{11}+6\sqrt{11}$

$7\sqrt{11}+6\sqrt{11}=(7+6)\sqrt{11}=13\sqrt{11}$

b. $9\sqrt{13}-12\sqrt{13}$

$9\sqrt{13}-12\sqrt{13}=(9-12)\sqrt{13}=-3\sqrt{13}$

c. $\sqrt{15}+3\sqrt{15}$

$$\begin{aligned}\sqrt{15}+3\sqrt{15}&=1\sqrt{15}+3\sqrt{15}\\&=(1+3)\sqrt{15}\\&=4\sqrt{15}\end{aligned}$$

d. $\sqrt{19}+\sqrt{19}$

$$\begin{aligned}\sqrt{19}+\sqrt{19}&=1\sqrt{19}+1\sqrt{19}\\&=(1+1)\sqrt{19}\\&=2\sqrt{19}\end{aligned}$$

Now Try:

1. Add or subtract, as indicated.

a. $8\sqrt{21}+2\sqrt{21}$

____

b. $7\sqrt{17}-13\sqrt{17}$

____

c. $\sqrt{14}+5\sqrt{14}$

____

d. $\sqrt{30}+\sqrt{30}$

____

Name: Date:
Instructor: Section:

e. $\sqrt{5}+\sqrt{10}$

$\sqrt{5}+\sqrt{10}$ cannot be combined. They are unlike radicals.

e. $\sqrt{10}+\sqrt{15}$

____

Review these examples for Objective 2:

2. Add or subtract, as indicated.

a. $4\sqrt{5}+\sqrt{20}$

$$\begin{aligned}4\sqrt{5}+\sqrt{20}&=4\sqrt{5}+\sqrt{4\cdot 5}\\&=4\sqrt{5}+\sqrt{4}\cdot\sqrt{5}\\&=4\sqrt{5}+2\sqrt{5}\\&=6\sqrt{5}\end{aligned}$$

b. $\sqrt{40}-\sqrt{50}$

$$\begin{aligned}\sqrt{40}-\sqrt{50}&=\sqrt{4\cdot 10}-\sqrt{25\cdot 2}\\&=\sqrt{4}\cdot\sqrt{10}-\sqrt{25}\cdot\sqrt{2}\\&=2\sqrt{10}-5\sqrt{2}\end{aligned}$$

c. $2\sqrt{28}+8\sqrt{63}$

$$\begin{aligned}2\sqrt{28}+8\sqrt{63}&=2(\sqrt{4}\cdot\sqrt{7})+8(\sqrt{9}\cdot\sqrt{7})\\&=2(2\sqrt{7})+8(3\sqrt{7})\\&=4\sqrt{7}+24\sqrt{7}\\&=28\sqrt{7}\end{aligned}$$

Now Try:

2. Add or subtract, as indicated.

a. $3\sqrt{6}+\sqrt{150}$

____

b. $\sqrt{72}-\sqrt{50}$

____

c. $3\sqrt{24}+7\sqrt{54}$

____

Review these examples for Objective 3:

3. Simplify each radical expression. Assume that all variables represent nonnegative real numbers.

a. $\sqrt{3}\cdot\sqrt{6}+5\sqrt{2}$

$$\begin{aligned}\sqrt{3}\cdot\sqrt{6}+5\sqrt{2}&=\sqrt{3\cdot 6}+5\sqrt{2}\\&=\sqrt{18}+5\sqrt{2}\\&=\sqrt{9}\cdot\sqrt{2}+5\sqrt{2}\\&=3\sqrt{2}+5\sqrt{2}\\&=8\sqrt{2}\end{aligned}$$

Now Try:

3. Simplify each radical expression. Assume that all variables represent nonnegative real numbers.

a. $\sqrt{3}\cdot\sqrt{15}+7\sqrt{5}$

____

Name: Date:
Instructor: Section:

b. $\sqrt{75k}+\sqrt{108k}$

$$\begin{aligned}\sqrt{75k}+\sqrt{108k} &= \sqrt{25\cdot 3k}+\sqrt{36\cdot 3k}\\ &= \sqrt{25}\cdot\sqrt{3k}+\sqrt{36}\cdot\sqrt{3k}\\ &= 5\sqrt{3k}+6\sqrt{3k}\\ &= 11\sqrt{3k}\end{aligned}$$

c. $7x\sqrt{20}+4\sqrt{5x^2}$

$$\begin{aligned}7x\sqrt{20}+4\sqrt{5x^2} &= 7x\sqrt{4\cdot 5}+4\sqrt{5\cdot x^2}\\ &= 7x\sqrt{4}\cdot\sqrt{5}+4\sqrt{5}\cdot\sqrt{x^2}\\ &= 7x\cdot 2\sqrt{5}+4x\sqrt{5}\\ &= 14x\sqrt{5}+4x\sqrt{5}\\ &= 18x\sqrt{5}\end{aligned}$$

d. $8\sqrt[3]{81x^3}-\sqrt[3]{375x^3}$

$$\begin{aligned}&8\sqrt[3]{81x^3}-\sqrt[3]{375x^3}\\ &= 8\sqrt[3]{(27x^3)(3)}-\sqrt[3]{(125x^3)(3)}\\ &= 8(3x)\sqrt[3]{3}-5x\sqrt[3]{3}\\ &= 24x\sqrt[3]{3}-5x\sqrt[3]{3}\\ &= 19x\sqrt[3]{3}\end{aligned}$$

b. $\sqrt{28k}+\sqrt{63k}$ ____________

c. $6x\sqrt{75}+2\sqrt{3x^2}$ ____________

d. $7\sqrt[3]{16m^3}-\sqrt[3]{54m^3}$ ____________

Objective 1 Add and subtract radicals.

For extra help, see Example 1 on page 1091 of your text and Section Lecture video for Section 16.3 and Exercise Solutions Clip 7, 9, and 11.

Add or subtract wherever possible.

1. $5\sqrt{2}+\sqrt{3}$ **1.** ____________

2. $6\sqrt{3}-2\sqrt{3}+4\sqrt{3}$ **2.** ____________

3. $3\sqrt{5}-9\sqrt{5}+\sqrt{5}$ **3.** ____________

Name: Date:
Instructor: Section:

Objective 2 Simplify radical sums and differences.

For extra help, see Example 2 on pages 1091–1092 of your text and Section Lecture video for Section 16.3 and Exercise Solutions Clip 19, 21, and 25.

Simplify and add or subtract wherever possible.

4. $4\sqrt{128}+2\sqrt{32}$ **4.** ________________

5. $7\sqrt{162}-9\sqrt{32}$ **5.** ________________

6. $5\sqrt{32}-8\sqrt{18}+2\sqrt{20}$ **6.** ________________

Objective 3 Simplify more complicated radical expressions.

For extra help, see Example 3 on page 1092 of your text and Section Lecture video for Section 16.3 and Exercise Solutions Clip 37, 39, 45, 47, and 49.

Perform the indicated operations. Assume that all variables represent nonnegative real numbers.

7. $\sqrt{5}\cdot\sqrt{7}+3\sqrt{35}$ **7.** ________________

8. $3\sqrt{125x}-\sqrt{80x}+2\sqrt{45x}$ **8.** ________________

9. $11\sqrt{5w}\cdot\sqrt{30w}-8w\sqrt{24}$ **9.** ________________

Name: Date:
Instructor: Section:

Chapter 16 ROOTS AND RADICALS

16.4 Rationalizing the Denominator

Learning Objectives

1	Rationalize denominators with square roots.
2	Write radicals in simplified form.
3	Rationalize denominators with cube roots.

Key Terms

Use the vocabulary terms listed below to complete each statement in exercises 1–3.

rationalizing the denominator **product rule** **quotient rule**

1. The ________________________________ states that $\sqrt{a} \cdot \sqrt{b} = \sqrt{ab}$.

2. The process of ____ is changing the denominator of a fraction from a radical to an expression not involving a radical.

3. The ________________________________ states that $\sqrt{\frac{a}{b}} = \frac{\sqrt{a}}{\sqrt{b}}$.

Guided Examples

Review these examples for Objective 1:

1. Rationalize each denominator.

a. $\frac{10}{\sqrt{5}}$

$\frac{10}{\sqrt{5}} = \frac{10 \cdot \sqrt{5}}{\sqrt{5} \cdot \sqrt{5}}$ Multiply by $\frac{\sqrt{5}}{\sqrt{5}} = 1$.

$= \frac{10\sqrt{5}}{5}$

$= 2\sqrt{5}$ Write in lowest terms.

b. $\frac{18}{\sqrt{27}}$

$\frac{18}{\sqrt{27}} = \frac{18}{3\sqrt{3}}$

$= \frac{18 \cdot \sqrt{3}}{3\sqrt{3} \cdot \sqrt{3}}$ Multiply by $\frac{\sqrt{3}}{\sqrt{3}} = 1$.

$= \frac{18\sqrt{3}}{3 \cdot 3}$

$= 2\sqrt{3}$

Now Try:

1. Rationalize each denominator.

a. $\frac{14}{\sqrt{7}}$

____

b. $\frac{5}{\sqrt{75}}$

____

Name: Date:
Instructor: Section:

Review these examples for Objective 2:

2. Simplify.

$$\sqrt{\frac{75}{7}}$$

$$\sqrt{\frac{75}{7}} = \frac{\sqrt{75}\cdot\sqrt{7}}{\sqrt{7}\cdot\sqrt{7}}$$

$$= \frac{\sqrt{25\cdot3}\cdot\sqrt{7}}{7}$$

$$= \frac{\sqrt{25}\cdot\sqrt{3}\cdot\sqrt{7}}{7}$$

$$= \frac{5\sqrt{3}\cdot\sqrt{7}}{7}$$

$$= \frac{5\sqrt{21}}{7}$$

3. Simplify.

$$\sqrt{\frac{7}{15}}\cdot\sqrt{\frac{1}{10}}$$

$$\sqrt{\frac{7}{15}}\cdot\sqrt{\frac{1}{10}} = \sqrt{\frac{7}{15}\cdot\frac{1}{10}}$$

$$= \sqrt{\frac{7}{150}}$$

$$= \frac{\sqrt{7}}{\sqrt{150}}$$

$$= \frac{\sqrt{7}}{\sqrt{25}\cdot\sqrt{6}}$$

$$= \frac{\sqrt{7}}{5\sqrt{6}}$$

$$= \frac{\sqrt{7}\cdot\sqrt{6}}{5\sqrt{6}\cdot\sqrt{6}}$$

$$= \frac{\sqrt{42}}{5\cdot6}$$

$$= \frac{\sqrt{42}}{30}$$

Now Try:

2. Simplify.

$$\sqrt{\frac{7}{12}}$$

____

3. Simplify.

$$\sqrt{\frac{1}{3}}\cdot\sqrt{\frac{2}{15}}$$

____

Name: Date:
Instructor: Section:

4. Simplify. Assume that all variables represent positive real numbers.

a. $\dfrac{\sqrt{9m}}{\sqrt{n}}$

$$\frac{\sqrt{9m}}{\sqrt{n}}=\frac{\sqrt{9m}\cdot\sqrt{n}}{\sqrt{n}\cdot\sqrt{n}}=\frac{\sqrt{9mn}}{n}=\frac{\sqrt{9}\cdot\sqrt{mn}}{n}=\frac{3\sqrt{mn}}{n}$$

b. $\sqrt{\dfrac{5a^2b}{7}}$

$$\sqrt{\frac{5a^2b}{7}}=\frac{\sqrt{5a^2b}}{\sqrt{7}}=\frac{\sqrt{5a^2b}\cdot\sqrt{7}}{\sqrt{7}\cdot\sqrt{7}}=\frac{\sqrt{35a^2b}}{7}=\frac{\sqrt{a^2}\cdot\sqrt{35b}}{7}=\frac{a\sqrt{35b}}{7}$$

4. Simplify. Assume that all variables represent positive real numbers.

a. $\dfrac{\sqrt{25a}}{\sqrt{b}}$

____

b. $\sqrt{\dfrac{11m^2n}{5}}$

____

Review these examples for Objective 3:

5. Rationalize each denominator.

a. $\sqrt[3]{\dfrac{2}{5}}$

$$\sqrt[3]{\frac{2}{5}}=\frac{\sqrt[3]{2}\cdot\sqrt[3]{5\cdot5}}{\sqrt[3]{5}\cdot\sqrt[3]{5\cdot5}}=\frac{\sqrt[3]{2\cdot5\cdot5}}{\sqrt[3]{5\cdot5\cdot5}}=\frac{\sqrt[3]{50}}{5}$$

b. $\dfrac{\sqrt[3]{7}}{\sqrt[3]{25}}$

$$\frac{\sqrt[3]{7}}{\sqrt[3]{25}}=\frac{\sqrt[3]{7}\cdot\sqrt[3]{5}}{\sqrt[3]{25}\cdot\sqrt[3]{5}}=\frac{\sqrt[3]{35}}{\sqrt[3]{125}}=\frac{\sqrt[3]{35}}{5}$$

Now Try:

5. Rationalize each denominator.

a. $\sqrt[3]{\dfrac{5}{3}}$

____

b. $\dfrac{\sqrt[3]{11}}{\sqrt[3]{9}}$

____

Name: Date:
Instructor: Section:

c. $\dfrac{\sqrt[3]{13}}{\sqrt[3]{5x^2}}$

$$\frac{\sqrt[3]{13}}{\sqrt[3]{5x^2}} = \frac{\sqrt[3]{13}\cdot\sqrt[3]{5\cdot 5\cdot x}}{\sqrt[3]{5x^2}\cdot\sqrt[3]{5\cdot 5\cdot x}} = \frac{\sqrt[3]{325x}}{5x}$$

c. $\dfrac{\sqrt[3]{5}}{\sqrt[3]{4x}}$ ____________

Objective 1 Rationalize denominators with square roots.

For extra help, see Example 1 on page 1095 of your text and Section Lecture video for Section 16.4 and Exercise Solutions Clip 17 and 23.

Rationalize each denominator.

1. $\dfrac{15}{\sqrt{10}}$ **1.** ____________

2. $\dfrac{6}{\sqrt{28}}$ **2.** ____________

3. $\dfrac{3\sqrt{5}}{\sqrt{125}}$ **3.** ____________

Objective 2 Write radicals in simplified form.

For extra help, see Examples 2–4 on pages 1096–1097 of your text and Section Lecture video for Section 16.4 and Exercise Solutions Clip 13, 21, 29, and 31.

Perform the indicated operations and write all answers in simplest form. Rationalize all denominators. Assume that all variables represent positive real numbers.

4. $\sqrt{\dfrac{3}{2}}\cdot\sqrt{\dfrac{5}{6}}$ **4.** ____________

Name: Date:
Instructor: Section:

5. $\dfrac{\sqrt{k^2m^4}}{\sqrt{k^5}}$ **5.** ________________

6. $\sqrt{\dfrac{5a^2b^3}{6}}$ **6.** ________________

Objective 3 Rationalize denominators with cube roots.

For extra help, see Example 5 on page 1098 of your text and Section Lecture video for Section 16.4 and Exercise Solutions Clip 48 and 49.

Rationalize each denominator. Assume that all variables in the denominator represent nonzero real numbers.

7. $\dfrac{\sqrt[3]{6}}{\sqrt[3]{9}}$ **7.** ________________

8. $\sqrt[3]{\dfrac{8t}{125u}}$ **8.** ________________

9. $\sqrt[3]{\dfrac{5}{49x}}$ **9.** ________________

Name: Date:
Instructor: Section:

Chapter 16 ROOTS AND RADICALS

16.5 More Simplifying and Operations with Radicals

Learning Objectives

1. Simplify products of radical expressions.
2. Use conjugates to rationalize denominators of radical expressions.
3. Write radical expressions with quotients in lowest terms.

Key Terms

Use the vocabulary terms listed below to complete each statement in exercises 1–2.

conjugate **rationalize the denominator**

1. To ______________________________ of $\frac{3}{\sqrt{5}}$, multiply both the numerator and the denominator by $\sqrt{5}$.

2. The ______________________________ of $a + b$ is $a - b$.

Guided Examples

Review these examples for Objective 1:

1. Find each product, and simplify.

a. $\sqrt{7}\left(\sqrt{27}-\sqrt{48}\right)$

$$\begin{aligned}\sqrt{7}\left(\sqrt{27}-\sqrt{48}\right) &= \sqrt{7}\left(3\sqrt{3}-4\sqrt{3}\right)\\ &= \sqrt{7}\left(-\sqrt{3}\right)\\ &= -\sqrt{7\cdot 3}\\ &= -\sqrt{21}\end{aligned}$$

b. $\left(\sqrt{7}+5\sqrt{2}\right)\left(\sqrt{7}-8\sqrt{2}\right)$

Use the FOIL method to multiply.

$$\begin{aligned}&\left(\sqrt{7}+5\sqrt{2}\right)\left(\sqrt{7}-8\sqrt{2}\right)\\ &=\sqrt{7}\left(\sqrt{7}\right)+\sqrt{7}\left(-8\sqrt{2}\right)+5\sqrt{2}\left(\sqrt{7}\right)+5\sqrt{2}\left(-8\sqrt{2}\right)\\ &=7-8\sqrt{14}+5\sqrt{14}-40\cdot 2\\ &=7-3\sqrt{14}-80\\ &=-73-3\sqrt{14}\end{aligned}$$

Now Try:

1. Find each product, and simplify.

a. $\sqrt{2}\left(\sqrt{45}-\sqrt{20}\right)$

____

b. $\left(\sqrt{6}+3\sqrt{5}\right)\left(\sqrt{6}-5\sqrt{5}\right)$

____

Name: Date:
Instructor: Section:

c. $\left(\sqrt{6}+\sqrt{15}\right)\left(\sqrt{6}-\sqrt{5}\right)$

$$\begin{aligned}&\left(\sqrt{6}+\sqrt{15}\right)\left(\sqrt{6}-\sqrt{5}\right)\\&=\sqrt{6}\left(\sqrt{6}\right)+\sqrt{6}\left(-\sqrt{5}\right)+\sqrt{15}\left(\sqrt{6}\right)+\sqrt{15}\left(-\sqrt{5}\right)\\&=6-\sqrt{30}+\sqrt{90}-\sqrt{75}\\&=6-\sqrt{30}+\sqrt{9}\cdot\sqrt{10}-\sqrt{25}\cdot\sqrt{3}\\&=6-\sqrt{30}+3\sqrt{10}-5\sqrt{3}\end{aligned}$$

c. $\left(\sqrt{7}+\sqrt{27}\right)\left(\sqrt{7}-\sqrt{3}\right)$

____

2. Find each product.

a. $\left(\sqrt{15}-6\right)^2$

$$\begin{aligned}\left(\sqrt{15}-6\right)^2&=\left(\sqrt{15}\right)^2-2\left(\sqrt{15}\right)(6)+6^2\\&=15-12\sqrt{15}+36\\&=51-12\sqrt{15}\end{aligned}$$

b. $\left(3\sqrt{5}+7\right)^2$

$$\begin{aligned}\left(3\sqrt{5}+7\right)^2&=\left(3\sqrt{5}\right)^2+2\left(3\sqrt{5}\right)(7)+7^2\\&=45+42\sqrt{5}+49\\&=94+42\sqrt{5}\end{aligned}$$

c. $\left(7-\sqrt{a}\right)^2$ Assume that $a\neq 0$.

$$\begin{aligned}\left(7-\sqrt{a}\right)^2&=7^2-2(7)\sqrt{a}+\left(\sqrt{a}\right)^2\\&=49-14\sqrt{a}+a\end{aligned}$$

2. Find each product.

a. $\left(\sqrt{11}-5\right)^2$

____

b. $\left(5\sqrt{2}+8\right)^2$

____

c. $\left(12-\sqrt{m}\right)^2$
Assume that $m\neq 0$.

____

3. Find each product.

a. $\left(7+\sqrt{5}\right)\left(7-\sqrt{5}\right)$

Recall $(a+b)(a-b)=a^2-b^2$.

$$\begin{aligned}\left(7+\sqrt{5}\right)\left(7-\sqrt{5}\right)&=7^2-\left(\sqrt{5}\right)^2\\&=49-5\\&=44\end{aligned}$$

b. $\left(\sqrt{a}-\sqrt{10}\right)\left(\sqrt{a}+\sqrt{10}\right)$ Assume that $a\neq 0$.

$$\begin{aligned}\left(\sqrt{a}-\sqrt{10}\right)\left(\sqrt{a}+\sqrt{10}\right)&=\left(\sqrt{a}\right)^2-\left(\sqrt{10}\right)^2\\&=a-10\end{aligned}$$

3. Find each product.

a. $\left(11+\sqrt{4}\right)\left(11-\sqrt{4}\right)$

____

b. $\left(\sqrt{b}-\sqrt{20}\right)\left(\sqrt{b}+\sqrt{20}\right)$
Assume that $b\neq 0$.

____

Name: Date:
Instructor: Section:

Review these examples for Objective 2:

4. Simplify by rationalizing each denominator.

a. $\dfrac{7}{4+\sqrt{7}}$

$$\frac{7}{4+\sqrt{7}}=\frac{7(4-\sqrt{7})}{(4+\sqrt{7})(4-\sqrt{7})}$$
$$=\frac{7(4-\sqrt{7})}{4^2-(\sqrt{7})^2}$$
$$=\frac{7(4-\sqrt{7})}{16-7}, \text{ or } \frac{7(4-\sqrt{7})}{9}$$

b. $\dfrac{8+\sqrt{3}}{\sqrt{3}-4}$

$$\frac{8+\sqrt{3}}{\sqrt{3}-4}=\frac{(8+\sqrt{3})(\sqrt{3}+4)}{(\sqrt{3}-4)(\sqrt{3}+4)}$$
$$=\frac{8\sqrt{3}+32+3+4\sqrt{3}}{3-16}$$
$$=\frac{12\sqrt{3}+35}{-13}$$
$$=\frac{-12\sqrt{3}-35}{13}$$

c. $\dfrac{5}{2-\sqrt{z}}$ Assume that $z\neq 0$.

$$\frac{5}{2-\sqrt{z}}=\frac{5(2+\sqrt{z})}{(2-\sqrt{z})(2+\sqrt{z})}$$
$$=\frac{5(2+\sqrt{z})}{4-z}$$

Now Try:

4. Simplify by rationalizing each denominator.

a. $\dfrac{10}{7+\sqrt{10}}$ __________

b. $\dfrac{5+\sqrt{6}}{\sqrt{6}-3}$ __________

c. $\dfrac{7}{5-\sqrt{b}}$

Assume that $b\neq 0$. __________

Review this example for Objective 3:

5. Write $\dfrac{5\sqrt{2}+10}{35}$ in lowest terms.

$$\frac{5\sqrt{2}+10}{35}=\frac{5(\sqrt{2}+2)}{5(7)}$$
$$=1\cdot\frac{\sqrt{2}+2}{7} \quad \text{Divide out the common factor; } \tfrac{5}{5}=1$$
$$=\frac{\sqrt{2}+2}{7}$$

Now Try:

5. Write $\dfrac{4\sqrt{5}+20}{36}$ in lowest terms. __________

Name: Date:
Instructor: Section:

Objective 1 Simplify products of radical expressions.

For extra help, see Examples 1–3 on pages 1101–1103 of your text and Section Lecture video for Section 16.5 and Exercise Solutions Clip 7, 11, 13, 17, 19, and 23.

Find each product, and simplify.

1. $\sqrt{7}\left(2\sqrt{8}-9\sqrt{7}\right)$ **1.** ________________

2. $\left(4\sqrt{5}+\sqrt{3}\right)\left(\sqrt{2}-\sqrt{7}\right)$ **2.** ________________

3. $\left(2\sqrt{3}-5\sqrt{2}\right)\left(2\sqrt{3}+5\sqrt{2}\right)$ **3.** ________________

Objective 2 Use conjugates to rationalize denominators of radical expressions.

For extra help, see Example 4 on pages 1103–1104 of your text and Section Lecture video for Section 16.5.

Rationalize each denominator. Write quotients in lowest terms.

4. $\dfrac{\sqrt{2}}{\sqrt{5}-2}$ **4.** ________________

5. $\dfrac{\sqrt{6}+2}{\sqrt{2}-4}$ **5.** ________________

Name: Date:
Instructor: Section:

6. $\dfrac{\sqrt{5}-2}{\sqrt{3}+2}$

6. ________________

Objective 3 Write radical expressions with quotients in lowest terms.

For extra help, see Example 5 on page 1104 of your text and Section Lecture video for Section 16.5 and Exercise Solutions Clip 55 and 59.

Write each quotient in lowest terms.

7. $\dfrac{3+\sqrt{27}}{9}$

7. ________________

8. $\dfrac{12+6\sqrt{6}}{8}$

8. ________________

9. $\dfrac{135\sqrt{3}+25}{5}$

9. ________________

Name: Date:
Instructor: Section:

Chapter 16 ROOTS AND RADICALS

16.6 Solving Equations with Radicals

Learning Objectives

1. Solve radical equations having square root radicals.
2. Identify equations with no solutions.
3. Solve equations by squaring a binomial.
4. Solve problems using formulas that involve radicals.

Key Terms

Use the vocabulary terms listed below to complete each statement in exercises 1–2.

radical equation **extraneous solution**

1. An ______________________ is a potential solution to an equation that does not satisfy the equation.

2. An equation with a variable in the radicand is a ______________________.

Guided Examples

Review this example for Objective 1:

1. Solve $\sqrt{p+2}=5$.

$$\left(\sqrt{p+2}\right)^2=5^2$$
$$p+2=25$$
$$p=23$$

Check $\sqrt{p+2}=5$
$$\sqrt{23+2}\stackrel{?}{=}5$$
$$\sqrt{25}\stackrel{?}{=}5$$
$$5=5 \quad \text{True}$$

The solution set is $\{23\}$.

2. Solve $4\sqrt{x}=\sqrt{x+30}$.

$$\left(4\sqrt{x}\right)^2=\left(\sqrt{x+30}\right)^2$$
$$4^2\left(\sqrt{x}\right)^2=\left(\sqrt{x+30}\right)^2$$
$$16x=x+30$$
$$15x=30$$
$$x=2$$

Now Try:

1. Solve $\sqrt{p+5}=4$.

____

2. Solve $5\sqrt{x}=\sqrt{x+72}$.

____

Check $4\sqrt{x} = \sqrt{x+30}$

$4\sqrt{2} \stackrel{?}{=} \sqrt{2+30}$

$4\sqrt{2} \stackrel{?}{=} \sqrt{32}$

$4\sqrt{2} = 4\sqrt{2}$ True

The solution set is $\{2\}$.

Review this example for Objective 2:

3. Solve $\sqrt{x} = -25$.

$$(\sqrt{x})^2 = (-25)^2$$
$$x = 625$$

Check $\sqrt{x} = -25$

$\sqrt{625} \stackrel{?}{=} -25$

$25 = -25$ False

Because the statement $25 = -25$ is false, the number 625 is not a solution. It is an extraneous solution and must be rejected. There is no solution. The solution set is $\varnothing$.

4. Solve $x = \sqrt{x^2 + 7x + 21}$.

Step 1 The radical is already isolated on the right side of the equation.

Step 2 Square each side.

$$x^2 = \left(\sqrt{x^2 + 7x + 21}\right)^2$$
$$x^2 = x^2 + 7x + 21$$

Step 3 $0 = 7x + 21$

Step 4 This step is not needed.

Step 5 $-21 = 7x$

$-3 = x$

Step 6 Check

$$x = \sqrt{x^2 + 7x + 21}$$
$$-3 \stackrel{?}{=} \sqrt{(-3)^2 + 7(-3) + 21}$$
$$-3 \stackrel{?}{=} \sqrt{9 - 21 + 21}$$
$$-3 = 3 \quad \text{False}$$

Since substituting –3 for x leads to a false result, the equation has no solution, and the solution set is $\varnothing$.

Now Try:

3. Solve $\sqrt{x} = -10$.

____

4. Solve $x = \sqrt{x^2 + 8x + 40}$.

____

Name: Date:
Instructor: Section:

Review this example for Objective 3:

5. Solve $\sqrt{2x-4}=2-x$.

$$\left(\sqrt{2x-4}\right)^2=(2-x)^2$$
$$2x-4=4-4x+x^2$$
$$0=x^2-6x+8$$
$$0=(x-2)(x-4)$$
$$x-2=0 \quad \text{or} \quad x-4=0$$
$$x=2 \quad \text{or} \quad x=4$$

Check Let $x=2$.

$$\sqrt{2x-4}=2-x$$
$$\sqrt{2(2)-4}\overset{?}{=}2-2$$
$$\sqrt{4-4}\overset{?}{=}0$$
$$0=0 \quad \text{True}$$

Let $x=4$.

$$\sqrt{2x-4}=2-x$$
$$\sqrt{2(4)-4}\overset{?}{=}2-4$$
$$\sqrt{8-4}\overset{?}{=}-2$$
$$\sqrt{4}\overset{?}{=}-2$$
$$2=-2 \quad \text{False}$$

Only 2 is a valid solution. (4 is extraneous.)
The solution set is $\{2\}$.

6. Solve $\sqrt{9x}+2=4x+1$.

$$\sqrt{9x}+2=4x+1$$
$$\sqrt{9x}=4x-1$$
$$\left(\sqrt{9x}\right)^2=(4x-1)^2$$
$$9x=16x^2-8x+1$$
$$0=16x^2-17x+1$$
$$0=(16x-1)(x-1)$$
$$16x-1=0 \quad \text{or} \quad x-1=0$$
$$x=\frac{1}{16} \quad \text{or} \quad x=1$$

Check Let $x=\frac{1}{16}$

$$\sqrt{9x}+2=4x+1$$
$$\sqrt{9\left(\frac{1}{16}\right)}+2\overset{?}{=}4\left(\frac{1}{16}\right)+1$$
$$\frac{3}{4}+2\overset{?}{=}\frac{1}{4}+1$$
$$\frac{11}{4}=\frac{5}{4} \quad \text{False}$$

Let $x=1$.

$$\sqrt{9x}+2=4x+1$$
$$\sqrt{9(1)}+2\overset{?}{=}4(1)+1$$
$$3+2\overset{?}{=}4+1$$
$$5=5 \quad \text{True}$$

Only 1 is a valid solution. ($\frac{1}{16}$ is extraneous.)

The solution set is $\{1\}$.

Now Try:

5. Solve $\sqrt{3x-5}=x-3$.

____

6. Solve $\sqrt{x+2}+4=x+6$.

____

Name: Date:
Instructor: Section:

7. Solve $\sqrt{40+x}=4+\sqrt{x}$.

$$\left(\sqrt{40+x}\right)^2=\left(4+\sqrt{x}\right)^2$$
$$40+x=16+8\sqrt{x}+x$$
$$24=8\sqrt{x}$$
$$3=\sqrt{x}$$
$$9=x$$

Check $\sqrt{40+x}=4+\sqrt{x}$
$$\sqrt{40+9}\stackrel{?}{=}4+\sqrt{9}$$
$$\sqrt{49}\stackrel{?}{=}4+3$$
$$7=7 \quad \text{True}$$

The solution set is $\{9\}$.

7. Solve $\sqrt{p+4}-\sqrt{p-1}=1$.

____

Review this example for Objective 4:

8. The sides of a triangle have lengths 84 ft, 68 ft, and 40 ft. Find the area of the triangle.

First find the semiperimeter, s.
$$s=\frac{1}{2}(a+b+c)$$
$$s=\frac{1}{2}(84+68+40)=96$$

Now use Heron's formula to find the area.
$$A=\sqrt{s(s-a)(s-b)(s-c)}$$
$$A=\sqrt{96(96-84)(96-68)(96-40)}$$
$$A=\sqrt{96(12)(28)(56)}$$
$$A=\sqrt{1{,}806{,}336}$$
$$A=1344 \text{ ft}^2$$

The area is 1344 ft^2.

Now Try:

8. The sides of a triangle have lengths 90 m, 153 m, and 189 m. Find the area of the triangle.

____

Objective 1 Solve radical equations having square root radicals.

For extra help, see Examples 1–2 on pages 1111–1112 of your text and Section Lecture video for Section 16.6 and Exercise Solutions Clip 11, 19, and 23.

Solve each equation.

1. $\sqrt{3x+1}=3$ **1.** _______________

Name: Date:
Instructor: Section:

2. $\sqrt{2+4k}=3\sqrt{k}$ **2.** ________________

3. $\sqrt{4x+3}=\sqrt{3x+5}$ **3.** ________________

Objective 2 Identify equations with no solutions.

For extra help, see Examples 3–4 on pages 1112–1113 of your text and Section Lecture video for Section 16.6 and Exercise Solutions Clip 9 and 25.

Solve each equation.

4. $\sqrt{x+2}+7=0$ **4.** ________________

5. $\sqrt{2m+3}=3\sqrt{m+5}$ **5.** ________________

6. $r=\sqrt{r^2-6r+12}$ **6.** ________________

Objective 3 Solve equations by squaring a binomial.

For extra help, see Examples 5–7 on pages 1114–1115 of your text and Section Lecture video for Section 16.6 and Exercise Solutions Clip 31, 35, 37, 39, and 43.

Solve each equation.

7. $\sqrt{b-4}=b-6$ **7.** ________________

Name: Date:
Instructor: Section:

8. $3\sqrt{p+6}=p+6$ **8.** ________________

9. $q-1=\sqrt{q^2-4q+7}$ **9.** ________________

Objective 4 Solve problems using formulas that involve radicals.

For extra help, see Example 8 on page 1116 of your text and Section Lecture video for Section 16.6.

Use Heron's formula $\mathcal{A}=\sqrt{s(s-a)(s-b)(s-c)}$ *to find the area of the triangle with the given sides. Round answers to the nearest whole number.*

10. $a = 10$ in., $b = 10$ in., $c = 12$ in. **10.** ________________

11. $a = 42.3$ cm, $b = 29.8$ cm, $c = 33.7$ cm **11.** ________________

12. $a = 23$ m, $b = 19$ m, $c = 12$ m **12.** ________________

Name: Date:
Instructor: Section:

Chapter 17 QUADRATIC EQUATIONS

17.1 Solving Quadratic Equations by the Square Root Property

Learning Objectives

1. Review the zero-factor property.
2. Solve equations of the form $x^2 = k$, where $k > 0$.
3. Solve equations of the form $(ax+b)^2 = k$, where $k > 0$.
4. Use formulas involving squared variables.

Key Terms

Use the vocabulary terms listed below to complete each statement in exercises 1–2.

quadratic equation **zero-factor property**

1. An equation that can be written in the form $ax^2 + bx + c = 0$ is a ____________________.

2. The ____________________ states that if a product equals 0, then at least one of the factors of the product also equals zero.

Guided Examples

Review these examples for Objective 1:

1. Solve each equation by the zero-factor property.

a. $x^2 + 5x + 4 = 0$

$$x^2 + 5x + 4 = 0$$
$$(x+4)(x+1) = 0$$
$$x+4 = 0 \quad \text{or} \quad x+1 = 0$$
$$x = -4 \quad \text{or} \quad x = -1$$

The solution set is $\{-4, -1\}$.

b. $x^2 = 64$

$$x^2 = 64$$
$$x^2 - 64 = 0$$
$$(x+8)(x-8) = 0$$
$$x+8 = 0 \quad \text{or} \quad x-8 = 0$$
$$x = -8 \quad \text{or} \quad x = 8$$

The solution set is $\{-8, 8\}$.

Now Try:

1. Solve each equation by the zero-factor property.

a. $x^2 + 8x + 7 = 0$

____

b. $x^2 = 100$

____

Name: Date:
Instructor: Section:

Review these examples for Objective 2:

2. Solve each equation. Write radicals in simplified form.

a. $x^2 = 36$

By the square root property, if $x^2 = 36$, then
$x = \sqrt{36} = 6$ or $x = -\sqrt{36} = -6$
The solution set is $\{-6, 6\}$, or $\{\pm 6\}$.

b. $z^2 = 13$

The solutions are $z = \sqrt{13}$ or $z = -\sqrt{13}$
The solution set is $\{-\sqrt{13}, \sqrt{13}\}$, or $\{\pm\sqrt{13}\}$.

c. $7m^2 - 17 = 67$

$$7m^2 - 17 = 67$$
$$7m^2 = 84$$
$$m^2 = 12$$
$$m = \sqrt{12} \quad \text{or} \quad m = -\sqrt{12}$$
$$m = 2\sqrt{3} \quad \text{or} \quad m = -2\sqrt{3}$$

The solution set is $\{-2\sqrt{3}, 2\sqrt{3}\}$, or $\{\pm 2\sqrt{3}\}$.

d. $p^2 = -64$

Because –64 is a negative number and because the square of a real number cannot be negative, there is no real number solution of this equation. The solution set is $\varnothing$.

Now Try:

2. Solve each equation. Write radicals in simplified form.

a. $x^2 = 81$

____

b. $z^2 = 23$

____

c. $3m^2 - 15 = 210$

____

d. $n^2 = -25$

____

Review these examples for Objective 3:

3. Solve each equation.

a. $(x-4)^2 = 64$

$$x - 4 = \sqrt{64} \quad \text{or} \quad x - 4 = -\sqrt{64}$$
$$x - 4 = 8 \quad \text{or} \quad x - 4 = -8$$
$$x = 12 \quad \text{or} \quad x = -4$$

Check

$(x-4)^2 = 64$	$(x-4)^2 = 64$
$(12-4)^2 \stackrel{?}{=} 64$	$(-4-4)^2 \stackrel{?}{=} 64$
$8^2 \stackrel{?}{=} 64$	$(-8)^2 \stackrel{?}{=} 64$
$64 = 64$ True	$64 = 64$ True

The solution set is $\{-4, 12\}$.

Now Try:

3. Solve each equation.

a. $(x-5)^2 = 121$

____

Name:
Instructor:

Date:
Section:

b. $(x+2)^2 = 7$

$x+2=\sqrt{7}$ or $x+2=-\sqrt{7}$

$x=-2+\sqrt{7}$ or $x=-2-\sqrt{7}$

Check

$(-2+\sqrt{7}+2)^2=(\sqrt{7})^2=7$

$(-2-\sqrt{7}+2)^2=(-\sqrt{7})^2=7$

The solution set is $\{-2+\sqrt{7},-2-\sqrt{7}\}$, or $\{-2\pm\sqrt{7}\}$.

b. $(x+3)^2 = 13$

____

4. Solve $(5r-3)^2 = 12$.

$5r-3=\sqrt{12}$ or $5r-3=-\sqrt{12}$

$5r-3=2\sqrt{3}$ or $5r-3=-2\sqrt{3}$

$5r=3+2\sqrt{3}$ or $5r=3-2\sqrt{3}$

$r=\dfrac{3+2\sqrt{3}}{5}$ or $r=\dfrac{3-2\sqrt{3}}{5}$

Check

$$(5r-3)^2=12$$

$$\left[5\cdot\frac{3+2\sqrt{3}}{5}-3\right]^2 \stackrel{?}{=} 12$$

$$(3+2\sqrt{3}-3)^2 \stackrel{?}{=} 12$$

$$(2\sqrt{3})^2 \stackrel{?}{=} 12$$

$$12=12 \quad \text{True}$$

The check of the other solution is similar. The solution set is

The solution set is $\left\{\dfrac{3+2\sqrt{3}}{5}, \dfrac{3-2\sqrt{3}}{5}\right\}$.

4. Solve $(7r-3)^2 = 32$.

____

5. Solve $(x+2)^2 = -16$.

Because the square root of -16 is not a real number, there is no real number solution for this equation. The solution set is $\varnothing$.

5. Solve $(x-7)^2 = -49$.

____

Name: Date:
Instructor: Section:

Review this example for Objective 4:

6. We can approximate the weight of a bass, in pounds, given its length L and its girth (distance around) g, where both are measured in inches, using the following formula.

$$w = \frac{L^2 g}{1200}$$

Approximate the length of a bass weighing 2.40 lb and having girth 9 in.

Start with the formula and substitute 2.40 for w and 9 for g.

$$w = \frac{L^2 g}{1200}$$

$$2.40 = \frac{L^2 \cdot 9}{1200}$$

$$2880 = 9L^2$$

$$L^2 = 320$$

$$L = \sqrt{320} \text{ or } L = -\sqrt{320}$$

The calculator shows that $\sqrt{320} \approx 17.89$, so the length of the bass is almost 18 in. (We discard the solution $-\sqrt{320} \approx -17.89$, since L represents length.)

Now Try:

6. We can approximate the weight of a bass, in pounds, given its length L and its girth (distance around) g, where both are measured in inches, using the following formula.

$$w = \frac{L^2 g}{1200}$$

Approximate the length of a bass weighing 2.50 lb and having girth 10.5 in.

____

Objective 1 Review the zero-factor property.

For extra help, see Example 1 on page 1132 of your text and Section Lecture video for Section 17.1.

Solve each equation by using the zero-factor property.

1. $x^2 + 6x + 8 = 0$ **1.** __________

2. $x^2 = 121$ **2.** __________

3. $x^2 + 2x - 35 = 0$ **3.** __________

Name: Date:
Instructor: Section:

Objective 2 Solve equations of the form $x^2 = k$, where $k > 0$.

For extra help, see Example 2 on page 1133 of your text and Section Lecture video for Section 17.1 and Exercise Solutions Clip 13 and 15.

Solve each equation by using the square root property. Express all radicals in simplest form.

4. $r^2 = 900$ **4.** ________

5. $s^2 - 98 = 0$ **5** ________

6. $p^2 = -144$ **6.** ________

Objective 3 Solve equations of the form $(ax+b)^2 = k$, where $k > 0$.

For extra help, see Examples 3–5 on pages 1133–1134 of your text and Section Lecture video for Section 17.1 and Exercise Solutions Clip 35, 39, and 47.

Solve each equation by using the square root property. Express all radicals in simplest form.

7. $(y+2)^2 = 16$ **7.** ________

8. $(7p-4)^2 = 289$ **8.** ________

9. $(10m-5)^2 - 9 = 0$ **9.** ________

Name: Date:
Instructor: Section:

Objective 4 Use formulas involving squared variables.

For extra help, see Example 6 on page 1135 of your text and Section Lecture video for Section 17.1.

Solve each problem.

10. The formula $A = P(1+r)^2$ gives the amount A that P dollars invested at an annual rate of interest r will grow to in two years. Mary invests \$1500, and after two years, she has \$1653.75. What is the interest rate?

10. ________________

11. The volume of a cylinder is given by the formula $V = \pi r^2 h$, where V = the volume, r = the radius of the base of the cylinder, and h = the height of the cylinder. If the volume of a can is 20π in.^{3}, and its height is 5 inches, find the radius of the can.

11. ________________

12. One leg of a right triangle has length 5 cm, and the hypotenuse has length 10 cm. Find the length of the other leg. (Hint: Use the Pythagorean theorem.)

12. ________________

Name: Date:
Instructor: Section:

Chapter 17 QUADRATIC EQUATIONS

17.2 Solving Quadratic Equations by Completing the Square

Learning Objectives

1	Solve quadratic equations by completing the square when the coefficient of the second-degree term is 1.
2	Solve quadratic equations by completing the square when the coefficient of the second-degree term is not 1.
3	Simplify the terms of an equation before solving.
4	Solve applied problems that require quadratic equations.

Key Terms

Use the vocabulary terms listed below to complete each statement in exercises 1–3.

completing the square **perfect square trinomial** **square root property**

1. A ______________________________ can be written in the form $x^2 + 2kx + k^2$ or $x^2 - 2kx + k^2$

2. The ______________________________ says that, if k is positive and $a^2 = k$, then $a = \pm\sqrt{k}$.

3. Use the process called ________________________ in order to rewrite an equation so it can be solved using the square root property.

Guided Examples

Review these examples for Objective 1:

1. Complete each trinomial so that it is a perfect square. Then factor the trinomial.

a. $x^2 + 24x +$ ______

The perfect square trinomial will have the form $x^2 + 2kx + k^2$. Thus, the middle term 24x, must equal 2kx.

$$24x = 2kx$$
$$12 = k$$

Therefore, $k = 12$ and $k^2 = 12^2 = 144$. The perfect square trinomial is $x^2 + 24x + 144$, which factors as $(x + 12)^2$.

Now Try:

1. Complete each trinomial so that it is a perfect square. Then factor the trinomial.

a. $x^2 + 10x +$ ______

____

b. $x^2 - 36x +$ ______

The perfect square trinomial will have the form $x^2 - 2kx + k^2$. Thus, the middle term –36*x*, must equal –2*kx*.

$$-36x = -2kx$$
$$18 = k$$

Therefore, $k = 18$ and $k^2 = 18^2 = 324$. The required perfect square trinomial is $x^2 - 36x + 324$, which factors as $(x-18)^2$.

b. $x^2 - 22x +$ ______

____

2. Solve $x^2 + 8x + 3 = 0$.

$$x^2 + 8x + 3 = 0$$
$$x^2 + 8x = -3$$

The expression on the left must be written as a perfect square trinomial in the form $x^2 + 2kx + k^2$.

$$x^2 + 8x + \_\_\_\_\_$$

Here, $2kx = 8x$, so $k = 4$ and $k^2 = 16$. The required perfect square trinomial is $x^2 + 8x + 16$ which factors as $(x+4)^2$.

Therefore, if we add 16 to each side of $x^2 + 8x = -3$, the equation will have a perfect square trinomial on the left side, as needed.

$$x^2 + 8x + 16 = -3 + 16$$
$$(x+4)^2 = 13$$

Use the square root property.

$$x + 4 = \sqrt{13} \quad \text{or} \quad x + 4 = -\sqrt{13}$$
$$x = -4 + \sqrt{13} \quad \text{or} \quad x = -4 - \sqrt{13}$$

Check by substituting $-4 + \sqrt{13}$ and then $-4 - \sqrt{13}$ for *x* in the original equation. The solution set is $\{-4 + \sqrt{13}, -4 - \sqrt{13}\}$.

2. Solve $x^2 + 10x - 7 = 0$.

____

Name: Date:
Instructor: Section:

3. Complete the square to solve $x^2 - 18x = 11$.

To complete the square on $x^2 - 18x$, take half the coefficient of *x* and square it.

$\frac{1}{2}(-18) = -9$ and $(-9)^2 = 81$

Add the result, 81, to each side of the equation.

$$x^2 - 18x = 11$$
$$x^2 - 18x + 81 = 11 + 81$$
$$(x-9)^2 = 92$$

$x - 9 = \sqrt{92}$ or $x - 9 = -\sqrt{92}$

$x - 9 = 2\sqrt{23}$ $\quad$ $x - 9 = -2\sqrt{23}$

$x = 9 + 2\sqrt{23}$ or $x = 9 - 2\sqrt{23}$

A check indicates the solution set is $\{9 + 2\sqrt{23},\ 9 - 2\sqrt{23}\}$

3. Complete the square to solve $x^2 + 16x = 9$.

____

Review this example for Objective 2:

4. Solve $9x^2 + 18x = 7$.

Step 1 Before completing the square, the coefficient of x^2 must be 1, not 9. Divide each side by 9.

$$x^2 + 2x = \frac{7}{9}$$

Step 2 The equation is already in the correct form, with the variable terms on one side and the constant on the other.

Step 3 Complete the square. Take half the coefficient of *x*, and square it.

$\frac{1}{2}(2) = 1$ and $1^2 = 1$

$$x^2 + 2x + 1 = \frac{7}{9} + 1$$
$$(x+1)^2 = \frac{16}{9}$$

Now Try:

4. Solve $16x^2 - 64x = 55$.

____

Step 4 Solve the equation by using the square root property.

$$x+1=\sqrt{\frac{16}{9}} \quad \text{or} \quad x+1=-\sqrt{\frac{16}{9}}$$

$$x+1=\frac{4}{3} \qquad x+1=-\frac{4}{3}$$

$$x=\frac{1}{3} \quad \text{or} \quad x=-\frac{7}{3}$$

The two solutions $-\frac{7}{3}$ and $\frac{1}{3}$ check, so the solution set is $\left\{-\frac{7}{3}, \frac{1}{3}\right\}$.

5. Solve $5x^2-3x-8=0$.

Step 1 Transform so that 1 is the coefficient of the x^2-term.

$$x^2-\frac{3}{5}x-\frac{8}{5}=0$$

Step 2 Add $\frac{8}{5}$ to each side to get the variable terms on the left and the constant on the right.

$$x^2-\frac{3}{5}x=\frac{8}{5}$$

Step 3 To complete the square, take half the coefficient of x and square it.

$$\left[\frac{1}{2}\left(-\frac{3}{5}\right)\right]^2=\left(-\frac{3}{10}\right)^2=\frac{9}{100}$$

Add the result, $\frac{9}{100}$, to each side of the equation.

$$x^2-\frac{3}{5}x+\frac{9}{100}=\frac{8}{5}+\frac{9}{100}$$

$$\left(x-\frac{3}{10}\right)^2=\frac{169}{100}$$

Step 4 Solve by using the square root property.

$$x-\frac{3}{10}=\sqrt{\frac{169}{100}} \quad \text{or} \quad x-\frac{3}{10}=-\sqrt{\frac{169}{100}}$$

$$x=\frac{3}{10}+\frac{13}{10} \qquad x=\frac{3}{10}-\frac{13}{10}$$

$$x=\frac{16}{10}=\frac{8}{5} \quad \text{or} \quad x=-\frac{10}{10}=-1$$

A check confirms that the solution set is $\left\{-1, \frac{8}{5}\right\}$.

5. Solve $6x^2-19x+14=0$.

____

Name: Date:
Instructor: Section:

6. Solve $5p^2 - 20p + 21 = 0$.

$$5p^2 - 20p + 21 = 0$$
$$p^2 - 4p + \frac{21}{5} = 0$$
$$p^2 - 4p = -\frac{21}{5}$$

The coefficient of p is –4. Take half of –4, square the result, and add it to each side.

$$p^2 - 4p + 4 = -\frac{21}{5} + 4$$
$$(p-2)^2 = -\frac{1}{5}$$

We cannot use the square root property to solve this equation, because the square root of $-\frac{1}{5}$ is not a real number. This equation has no real number solution. The solution set is $\varnothing$.

6. Solve $6p^2 - 36p + 55 = 0$.

____

Review this example for Objective 3:

7. Solve $(x-6)(x+2) = 7$.

$$(x-6)(x+2) = 7$$
$$x^2 - 4x - 12 = 7 \quad \text{FOIL}$$
$$x^2 - 4x = 19 \quad \text{Add 12.}$$
$$x^2 - 4x + 4 = 19 + 4 \quad \text{Add } \left[\tfrac{1}{2}(-4)\right]^2 = 4$$
$$(x-2)^2 = 23$$
$$x - 2 = \sqrt{23} \quad \text{or} \quad x - 2 = -\sqrt{23}$$
$$x = 2 + \sqrt{23} \quad \text{or} \quad x = 2 - \sqrt{23}$$

The solution set is $\{2+\sqrt{23},\ 2-\sqrt{23}\}$.

Now Try:

7. Solve $(x+8)(x-2) = 5$.

____

Review this example for Objective 4:

8. If James throws an object upward from ground level with an initial velocity of 80 feet per second, its height s (in feet) after t seconds is given by the formula $s = -16t^2 + 80t$. After how many seconds will the object reach a height of 64 feet?

Since s represents the height, we substitute 64 for s in the formula, and then solve for t using completing the square.

Now Try:

8. A certain projectile is located at a distance of $d = 3t^2 - 6t + 1$ feet from its starting point after t seconds. How many seconds will it take the projectile to travel 10 feet?

____

$$64 = -16t^2 + 80t$$

$$-4 = t^2 - 5t \qquad \text{Divide by } -16.$$

$$t^2 - 5t = -4$$

$$t^2 - 5t + \frac{25}{4} = -4 + \frac{25}{4} \qquad \text{Add } \left[\frac{1}{2}(-5)\right]^2 = \frac{25}{4}.$$

$$\left(t - \frac{5}{2}\right)^2 = \frac{9}{4}$$

$$t - \frac{5}{2} = \frac{3}{2} \quad \text{or} \quad t - \frac{5}{2} = -\frac{3}{2}$$

$$t = 4 \quad \text{or} \quad t = 1$$

The ball reaches a height of 64 twice, once on the way up and again on the way down. It takes 1 second to reach 64 ft on the way up and then after 4 sec, the object reaches 64 ft on the way down.

Objective 1 Solve quadratic equations by completing the square when the coefficient of the second-degree term is 1.

For extra help, see Examples 1–3 on pages 1139–1140 of your text and Section Lecture video for Section 17.2 and Exercise Solutions Clip 9 and 13.

Solve each equation by completing the square.

1. $r^2 + 8r = -4$

1. ____________________

2. $x^2 - 4x = 2$

2. ____________________

3. $x^2 + 2x = 63$

3. ____________________

Name: Date:
Instructor: Section:

Objective 2 Solve quadratic equations by completing the square when the coefficient of the second-degree term is not 1.

For extra help, see Examples 4–6 on pages 1141–1140 of your text and Section Lecture video for Section 17.2 and Exercise Solutions Clip 25 and 27.

Solve each equation by completing the square.

4. $6x^2 - x = 15$ **4.** ________________

5. $3x^2 - 2x + 4 = 0$ **5.** ________________

6. $3t^2 + t - 2 = 0$ **6.** ________________

Objective 3 Simplify the terms of an equation before solving.

For extra help, see Example 7 on page 1143 of your text and Section Lecture video for Section 17.2 and Exercise Solutions Clip 29.

Simplify each of the following equations and then solve by completing the square.

7. $6y^2 + 3y = 4y^2 + y - 5$ **7.** ________________

8. $(b-1)(b+7) = 9$ **8.** ________________

9. $(s+3)(s+1) = 1$ **9.** ________________

Name: Date:
Instructor: Section:

Objective 4 Solve applied problems that require quadratic equations.

For extra help, see Example 8 on page 1144 of your text and Section Lecture video for Section 17.2 and Exercise Solutions Clip 37.

Solve.

10. A rule for estimating the number of board feet of lumber that can be cut from a log depends on the diameter and length of the log. To find the diameter *d* (in inches) needed to get *x* board feet of lumber from an 8-foot log, use the formula, $\left(\frac{d-4}{4}\right)^2 = x$. Find the diameter needed to get 20 board feet of lumber.

10. ____________

11. The commodities market is very unstable; money can be made or lost quickly on investments in soybeans, wheat, pork bellies, and so on. Suppose that an investor kept track of his total profit, P (in thousands of dollars), at time t (in months), after he began investing, and found that his profit was given by the formula $P = 4t^2 - 24t + 32$. Find the times at which he broke even on his investment.

11. ____________

12. George and Albert have found that the profit (in dollars) from their cigar shop is given by the formula $P = -10x^2 + 100x + 300$, where x is the number of units of cigars sold daily. How many units should be sold for a profit of $460?

12. ____________

Name: Date:
Instructor: Section:

Chapter 17 QUADRATIC EQUATIONS

17.3 Solving Quadratic Equations by the Quadratic Formula

Learning Objectives

1	Identify the values of *a*, *b*, and *c* in a quadratic equation.
2	Use the quadratic formula to solve quadratic equations.
3	Solve quadratic equations with only one solution.
4	Solve quadratic equations with fractions.

Key Terms

Use the vocabulary terms listed below to complete each statement in exercises 1–3.

quadratic formula **standard form** **constant**

1. A quadratic equation written in the form $ax^2 + bx + c = 0,\ a \neq 0$ is written in ______________________________.

2. A symbol that represents a value that doesn't change is a ______________.

3. The formula $x = \dfrac{-b \pm \sqrt{b^2 - 4ac}}{2a}$ is called the ____________________.

Guided Examples

Review these examples for Objective 1:

1. Identify the values of the variables *a*, *b*, and *c* in each quadratic equation $ax^2 + bx + c = 0$.

a. $7x^2 - 6x + 4 = 0$

Here $a = 7$, $b = -6$, and $c = 4$.

b. $-8x^2 + 3 = 5x$

Write the equation in standard form.

$-8x^2 - 5x + 3 = 0$

Here, $a = -8$, $b = -5$, and $c = 3$.

c. $9x^2 - 32 = 0$

The *x*-term is missing.

$9x^2 + 0x - 32 = 0$

Then, $a = 9$, $b = 0$, and $c = -32$.

Now Try:

1. Identify the values of the variables *a*, *b*, and *c* in each quadratic equation $ax^2 + bx + c = 0$.

a. $12x^2 - 10x + 7 = 0$

____

b. $-4x^2 + 8 = 110x$

____

c. $15x^2 - 8 = 0$

____

d. $-5x^2 = -x$

Write in standard form.

$-5x^2 + x = 0$

The constant term is missing, so $a = -5$, $b = 1$, and $c = 0$.

d. $-8x^2 = 3x$

____

e. $(3x-5)(x+6) = -29$

$(3x-5)(x+6) = -29$

$3x^2 + 13x - 30 = -29$

$3x^2 + 13x - 1 = 0$

Here, $a = 3$, $b = 13$, and $c = -1$.

e. $(5x-4)(6x+7) = 11$

____

Review this example for Objective 2:

2. Solve $2x^2 + 3x - 20 = 0$.

Here, $a = 2$, $b = 3$, and $c = -20$. Substitute into the quadratic formula.

$$x = \frac{-b \pm \sqrt{b^2 - 4ac}}{2a}$$

$$x = \frac{-3 \pm \sqrt{3^2 - 4(2)(-20)}}{2(2)}$$

$$x = \frac{-3 \pm \sqrt{9+160}}{4}$$

$$x = \frac{-3 \pm \sqrt{169}}{4}$$

$$x = \frac{-3 \pm 13}{4}$$

Find the two solutions by first using the plus symbol, and then using the minus symbol.

$$x = \frac{-3+13}{4} = \frac{10}{4} = \frac{5}{2} \text{ or}$$

$$x = \frac{-3-13}{4} = \frac{-16}{4} = -4$$

Check each solution in the original equation. The solution set is $\left\{-4, \frac{5}{2}\right\}$.

Now Try:

2. Solve $3x^2 - x - 14 = 0$.

____

Name: Date:
Instructor: Section:

3. Solve $x^2 = 4x - 1$.

Write the equation in standard form as $x^2 - 4x + 1 = 0$.

$$x = \frac{-b \pm \sqrt{b^2 - 4ac}}{2a}$$

$$x = \frac{-(-4) \pm \sqrt{(-4)^2 - 4(1)(1)}}{2(1)}$$

$$x = \frac{4 \pm \sqrt{16-4}}{2}$$

$$x = \frac{4 \pm \sqrt{12}}{2}$$

$$x = \frac{4 \pm 2\sqrt{3}}{2}$$

$$x = \frac{2\left(2 \pm \sqrt{3}\right)}{2}$$

$$x = 2 \pm \sqrt{3}$$

The solution set is $\left\{2+\sqrt{3},\ 2-\sqrt{3}\right\}$.

3. Solve $x^2 = 6x - 4$.

____

Review this example for Objective 3:

4. Solve $9x^2 + 49 = 42x$.

Write the equation in standard form.

$$9x^2 - 42x + 49 = 0$$

Here, $a = 9$, $b = -42$, and $c = 49$.

$$x = \frac{-(-42) \pm \sqrt{(-42)^2 - 4(9)(49)}}{2(9)} = \frac{42}{18} = \frac{7}{3}$$

In this case, $b^2 - 4ac = 0$, and the trinomial $9x^2 - 42x + 49$ is a perfect square. There is one distinct solution in the solution set $\left\{\frac{7}{3}\right\}$.

Now Try:

4. Solve $4x^2 - 44x + 121 = 0$.

____

Review this example for Objective 4:

5. Solve $\frac{1}{5}t^2 = \frac{1}{5}t - \frac{3}{10}$.

Clear fractions. Multiply by the LCD, 10.

$$\frac{1}{5}t^2 = \frac{1}{5}t - \frac{3}{10}$$

$$10\left(\frac{1}{5}t^2\right) = 10\left(\frac{1}{5}t - \frac{3}{10}\right)$$

Now Try:

5. Solve $\frac{5}{7}x^2 = \frac{9}{7}x - \frac{12}{7}$.

____

Name: Date:
Instructor: Section:

$$10\left(\frac{1}{5}t^2\right)=10\left(\frac{1}{5}t\right)-10\left(\frac{3}{10}\right)$$

$$2t^2=2t-3$$

$$2t^2-2t+3=0$$

Identify $a = 2$, $b = -2$, and $c = 3$.

$$t=\frac{-(-2)\pm\sqrt{(-2)^2-4(2)(3)}}{2(2)}$$

$$t=\frac{2\pm\sqrt{4-24}}{4}$$

$$t=\frac{2\pm\sqrt{-20}}{4}$$

Because $\sqrt{-20}$ does not represent a real number, the solution set is $\varnothing$.

Objective 1 Identify the values of *a*, *b*, and *c* in a quadratic equation.

For extra help, see Example 1 on page 1147 of your text and Section Lecture video for Section 17.3.

Write each equation in standard form, if necessary, and then identify the values of a, b, and c. Do not actually solve the equation.

1. $10x^2=-4x$ **1.** ______________

2. $4p=-4p^2+7$ **2.** ______________

3. $(z+1)(z+2)=-7$ **3.** ______________

Objective 2 Use the quadratic formula to solve quadratic equations.

For extra help, see Examples 2–3 on page 1149 of your text and Section Lecture video for Section 17.3 and Exercise Solutions Clip 9, 15, 21, and 29.

Use the quadratic formula to solve each equation. Write all radicals in simplified form, and write all answers in lowest terms.

4. $y^2=13-12y$ **4.** ______________

Name: Date:
Instructor: Section:

5. $-7r^2 = 5r + 3$ **5.** ________________

6. $5k^2 + 4k - 2 = 0$ **6.** ________________

Objective 3 Solve quadratic equations with only one solution.

For extra help, see Example 4 on page 1150 of your text and Section Lecture video for Section 17.3.

Use the quadratic formula to solve each equation. Write all radicals in simplified form, and write all answers in lowest terms.

7. $m^2 + 4 = 4m$ **7.** ________________

8. $4q^2 + 12q + 9 = 0$ **8.** ________________

9. $49a^2 - 126a = -81$ **9.** ________________

Name: Date:
Instructor: Section:

Objective 4 Solve quadratic equations with fractions.

For extra help, see Example 5 on page 1150 of your text and Section Lecture video for Section 17.3 and Exercise Solutions Clip 37.

Use the quadratic formula to solve each equation. Write all radicals in simplified form, and write all answers in lowest terms.

10. $\frac{1}{6}y^2 + \frac{1}{2}y = \frac{2}{3}$ **10.** ________________

11. $\frac{1}{4}t^2 - \frac{1}{3}t + \frac{5}{12} = 0$ **11.** ________________

12. $-\frac{1}{4}x^2 + 4 = \frac{1}{2}x$ **12.** ________________

Name: Date:
Instructor: Section:

Chapter 17 QUADRATIC EQUATIONS

17.4 Graphing Quadratic Equations

Learning Objectives
1. Graph quadratic equations.
2. Find the vertex of a parabola.

Key Terms

Use the vocabulary terms listed below to complete each statement in exercises 1–4.

parabola **vertex** **axis** **line of symmetry**

1. If a graph is folded on its ________________________________, the two sides coincide.

2. The ________________________________ of a parabola that opens upward or downward is the lowest or highest point on the graph.

3. The ________________________________ of a parabola that opens upward or downward is a vertical line through the vertex.

4. The graph of the quadratic equation $y = ax^2 + bx + c$ is called a __________________.

Guided Examples

Review these examples for Objective 1:

1. Graph $y = -x^2$.

Select several values for x. Then find the corresponding y-values. Then plot the points on the graph and join them with a smooth curve.

x	y
2	−4
1	−1
0	0
−1	−1
−2	−4

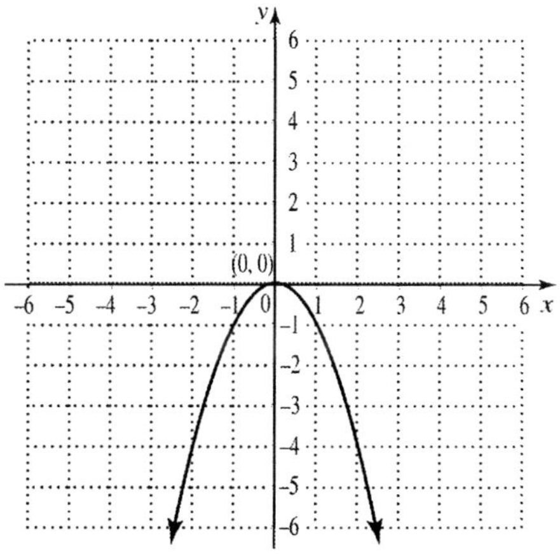

Now Try:

1. Graph $y = \frac{1}{3}x^2$.

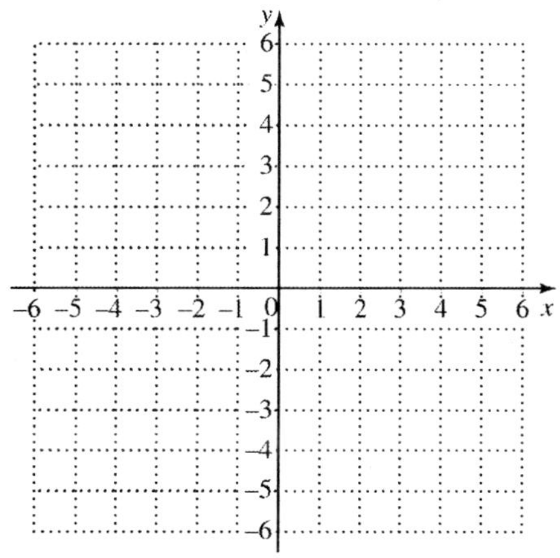

Name:
Instructor:

Date:
Section:

2. Graph $y = x^2 - 3$.

Find several ordered pairs. Let $x = 0$ to find the y-intercept.

$$y = x^2 - 3 = 0^2 - 3 = -3$$

This gives the ordered pair (0, –3). Select several values for x and find the corresponding values for y. Plot the ordered pairs and join them with a smooth curve.

x	y
2	1
1	−2
0	−3
−1	−2
−2	1

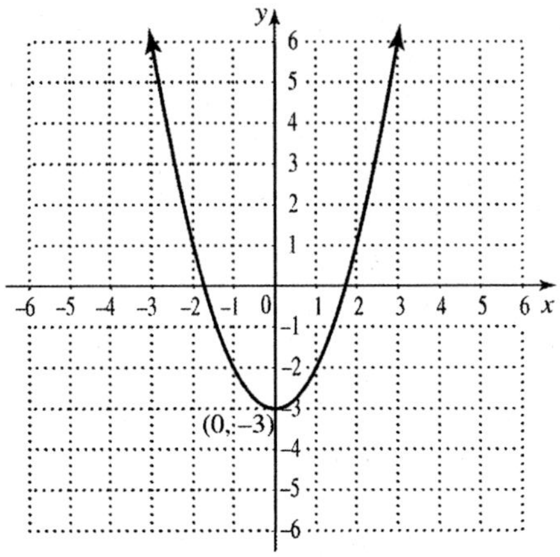

2. Graph $y = 9 - x^2$.

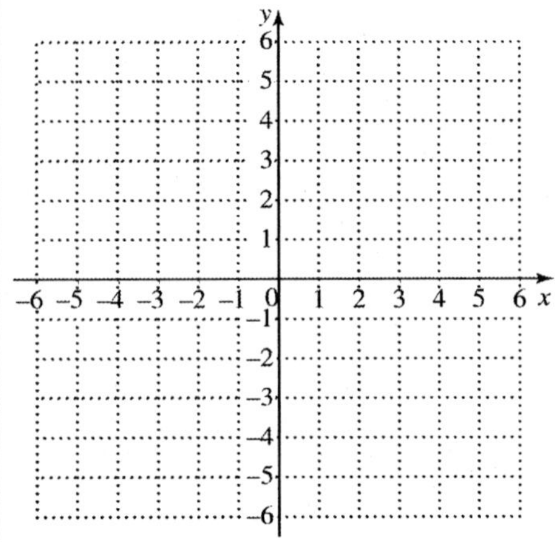

Review these examples for Objective 2:

3. Graph $y = x^2 + x - 2$.

Find the x-intercepts of the parabola. Let $y = 0$.

$$0 = x^2 + x - 2$$

$$0 = (x+2)(x-1)$$

$$x + 2 = 0 \quad \text{or} \quad x - 1 = 0$$

$$x = -2 \quad \text{or} \quad x = 1$$

There are two intercepts, (–2, 0) and (1, 0). Since the x-value of the vertex is halfway between the x-values of the x-intercepts, it is half their sum.

$$x = \frac{1}{2}(-2+1) = -\frac{1}{2}$$

We find the corresponding y-value.

$$y = \left(-\frac{1}{2}\right)^2 + \left(-\frac{1}{2}\right) - 2 = -\frac{9}{4}$$

The vertex is $\left(-\frac{1}{2}, -\frac{9}{4}\right)$. The axis of symmetry is the vertical line $x = -\frac{1}{2}$.

Now Try:

3. Graph $y = x^2 - 2x - 3$.

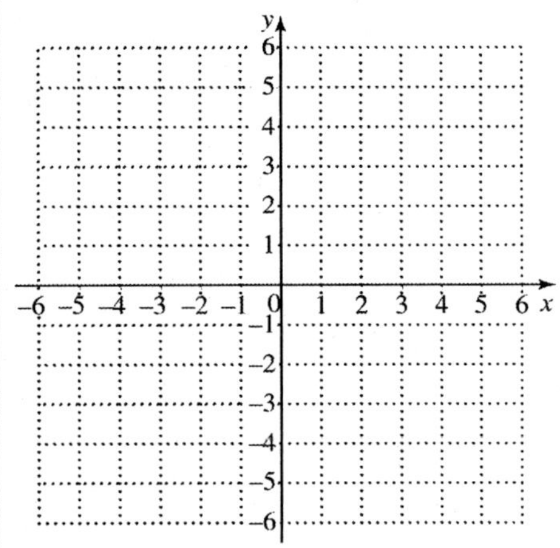

To find the y-intercept, we substitute 0 for x in the equation.

$$y = 0^2 + 0 - 2 = -2$$

We plot the three intercepts and the vertex, and find additional ordered pairs as needed.

x	y
-2	0
-1	-2
$-\frac{1}{2}$	$-\frac{9}{4}$
0	-2
1	0

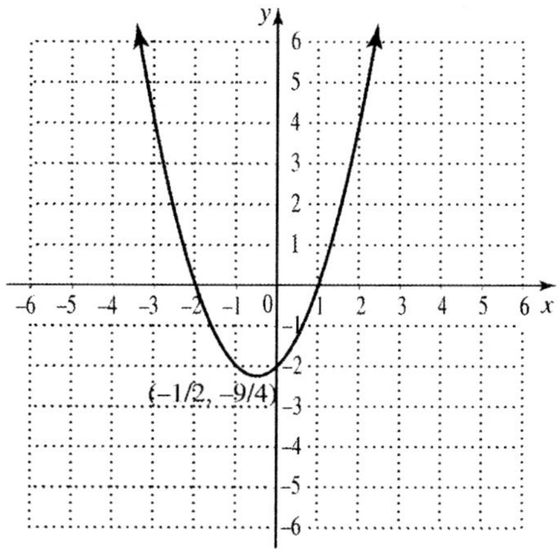

4. Graph $y = -x^2 + 4x - 1$.

Here $a = -1$ and $b = 4$, so we substitute to find the x-value of the vertex.

$$x = -\frac{b}{2a} = -\frac{4}{2(-1)} = 2$$

We find the y-value from the original equation.

$$y = -(2)^2 + 4(2) - 1$$
$$y = 3$$

The vertex is (2, 3). The axis is the line $x = 2$. Now we find the intercepts. Let $x = 0$ in the equation.

$$y = -0^2 + 4(0) - 1 = -1$$

The y-intercept is (0, –1). Let $y = 0$ to get the x-intercepts. We use the quadratic formula to solve for x.

$$x = \frac{-4 \pm \sqrt{(4)^2 - 4(-1)(-1)}}{2(-1)}$$
$$x = \frac{-4 \pm \sqrt{12}}{-2}$$
$$x = \frac{-4 \pm 2\sqrt{3}}{-2}$$
$$x = 2 \pm \sqrt{3}$$

Using a calculator, we find that the x-intercepts are (0.3, 0) and (3.7, 0).

4. Graph $y = x^2 + 8x + 14$.

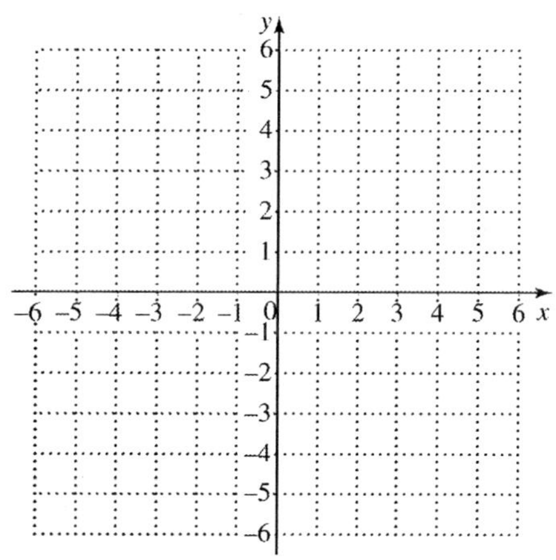

Name: Date:
Instructor: Section:

We plot the intercepts, vertex, and other ordered pairs, and join these points with a smooth curve.

x	y
0	-1
0.3	0
1	2
2	3
3	2
3.7	0

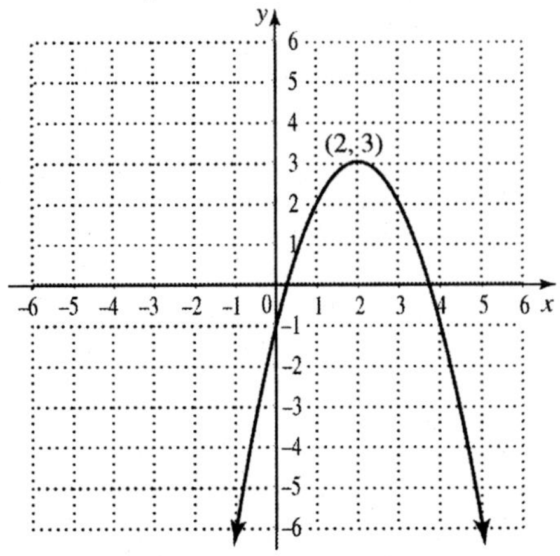

Objective 1 Graph quadratic equations.

For extra help, see Examples 1–2 on pages 1155–1156 of your text and Section Lecture video for Section 17.4 and Exercise Solutions Clip 9, 13, and 15.

Objective 2 Find the vertex of a parabola.

For extra help, see Examples 3–4 on pages 1157–1158 of your text and Section Lecture video for Section 17.4 and Exercise Solutions Clip 9, 13, 15, and 19.

Graph each equation. Give the coordinates of the vertex in each case.

1. $y = -x^2 - 1$

1.

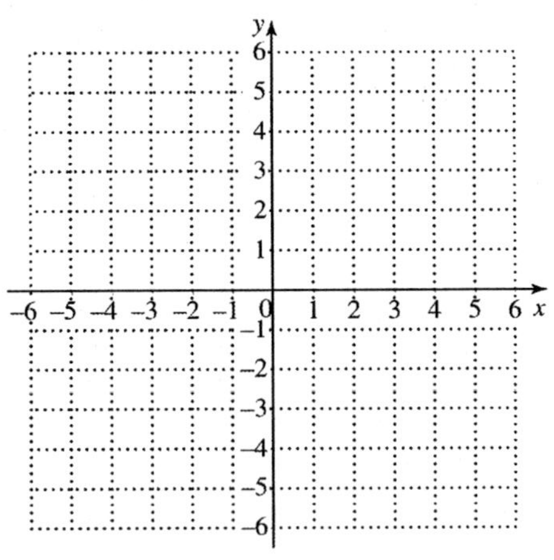

vertex: ____________________

Name: Date:
Instructor: Section:

2. $y=(x-2)^2$

2.

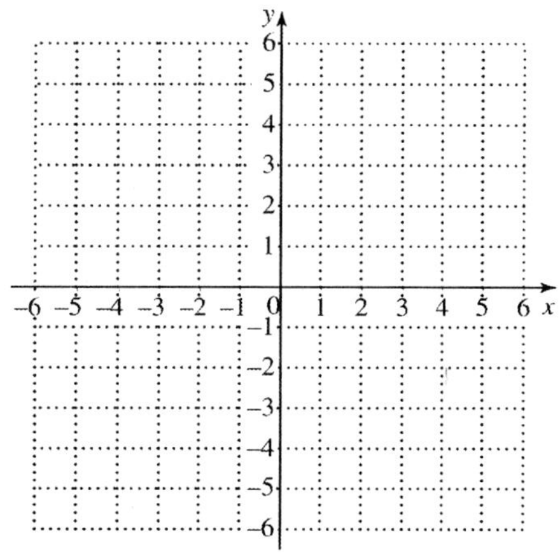

vertex: ______________________

3. $y=2x^2+4x$

3.

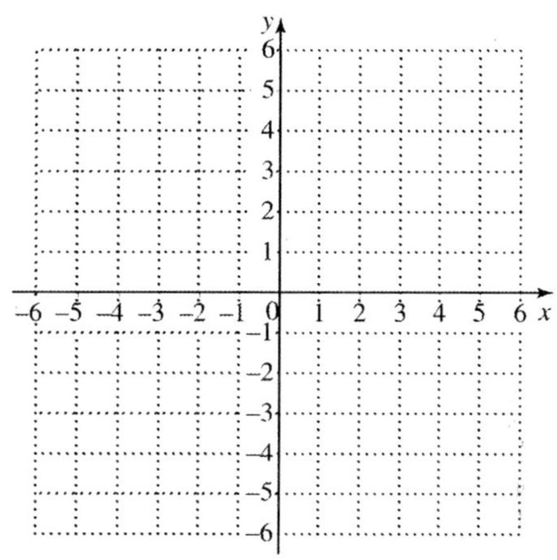

vertex: ______________________

4. $y=-x^2-3x+1$

4.

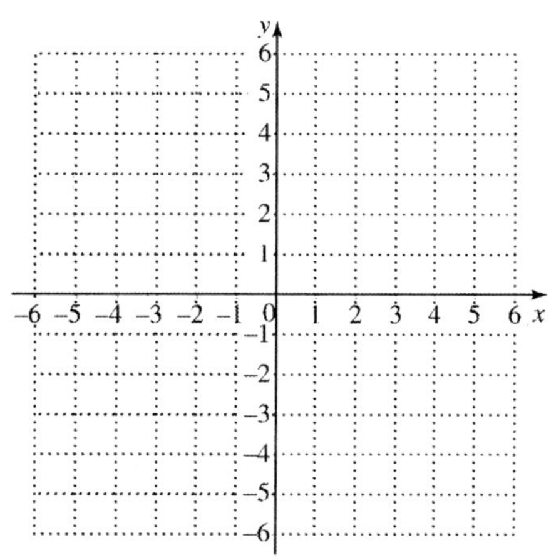

vertex: ______________________

Name: Date:
Instructor: Section:

5. $y=-\frac{1}{2}x^2+2x$

5.

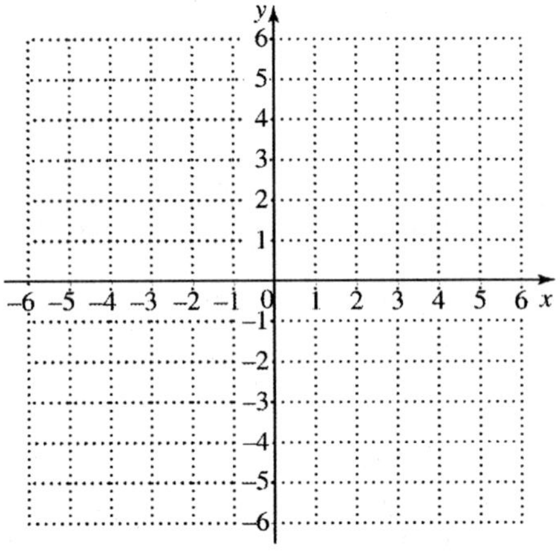

vertex: ______________________

Name: Date:
Instructor: Section:

Chapter 17 QUADRATIC EQUATIONS

17.5 Introduction to Functions

Learning Objectives

1 Understand the definition of a relation.
2 Understand the definition of a function.
3 Determine whether an equation defines a function.
4 Use function notation.
5 Apply the function concept in an application.

Key Terms

Use the vocabulary terms listed below to complete each statement in exercises 1–5.

components **relation** **domain** **range** **function**

1. Any set of ordered pairs is called a ________________________.

2. The set of all second components in the ordered pairs of a relation is the ____________________________ of the relation.

3. A ________________________ is a set of ordered pairs in which each first component corresponds to exactly one second component.

4. In an ordered pair (x, y), x and y are the ________________________.

5. The set of all first components in the ordered pairs of a relation is the _______________ of the relation.

Guided Examples

Review these examples for Objective 1:

1. Identify the domain and range of each relation.

 a. {(1, 0), (2, 7), (5, 9), (6, 2)}

 This relation has
 domain: {1, 2, 5, 6}
 and
 range: {0, 7, 9, 2}

 b. {(5, 7), (5, –10), (5, 8), (5, 4)}

 This relation has
 domain: {5}
 and
 range: {7, –10, 8, 4}

Now Try:

1. Identify the domain and range of each relation.

 a. {(6, 7), (8, 9), (10, 11), (12, 13)}

 ____

 b. {(1, 11), (1, 12), (1, 13), (1, 14)}

 ____

Name: Date:
Instructor: Section:

Review these examples for Objective 2:

2. Determine whether each relation is a function.

a. $\{(-5, 10), (-4, 8), (0, 0), (4, 8), (5, 10)\}$

Each first component appears once and only once. The relation is a function.

b. $\{(7, 4), (-7, 3), (7, 2)\}$

The first component 7 appears in two ordered pairs and corresponds to two different second components. Therefore, this relation is not a function.

Now Try:

2. Determine whether each relation is a function.

a. $\{(0, 1), (4, 2), (5, 9), (6, 15)\}$

____

b. $\{(10, 0), (10, 5), (10, 20)\}$

____

Review these examples for Objective 3:

3. Determine whether each relation represented by a graph or an equation is a function.

a.

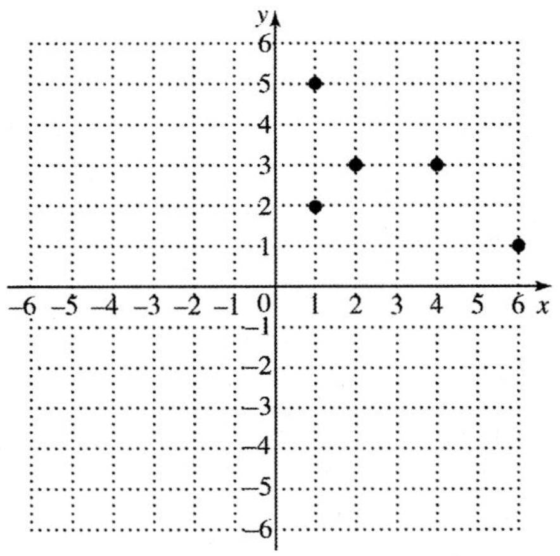

Because there are two ordered pairs with first component 1, this is not the graph of a function.

b. $y = 4x - 2$

This linear equation is in the form $y = mx + b$. Since the graph of this equation is a line that is not vertical, the equation defines a function.

c. $x = 3$

The graph of $x = 3$ is a vertical line, so the equation does not define a function.

Now Try:

3. Determine whether each relation represented by a graph or an equation is a function.

a.

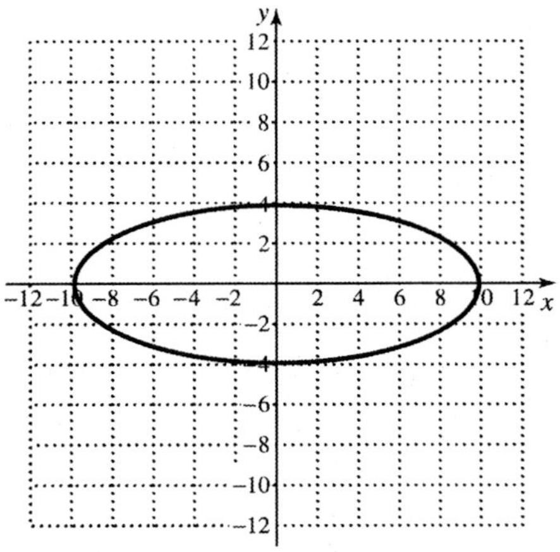

____

b. $y = -3x + 5$

____

c. $x = -6$

____

Name: Date:
Instructor: Section:

d.

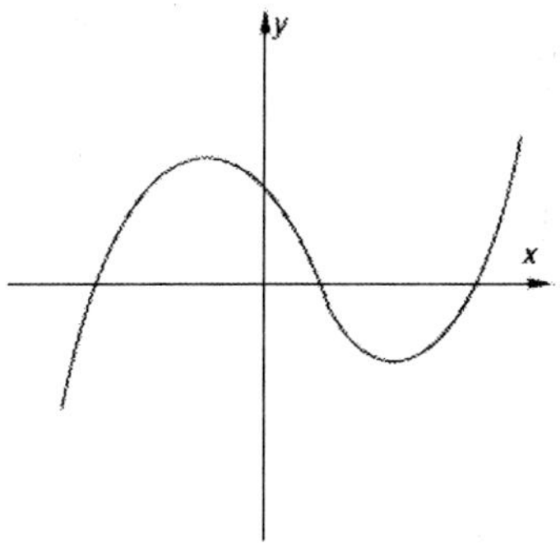

Use the vertical line test. Any vertical line intersects the graph just once, so this is the graph of a function.

d.

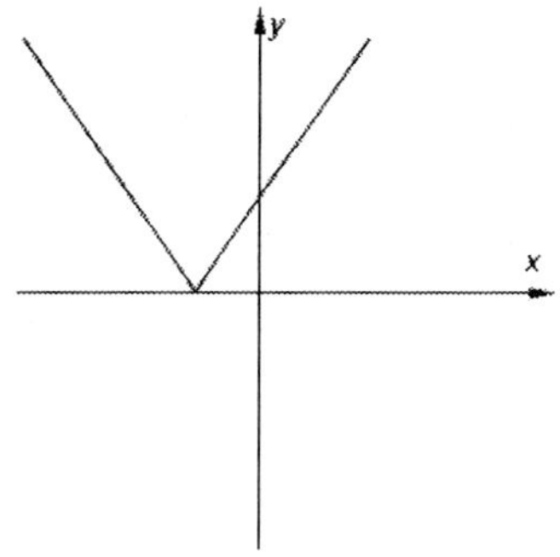

____

Review these examples for Objective 4:

4. For the function $f(x) = x^2 - 5$, find each function value.

a. $f(3)$

Substitute 3 for x.

$$f(x) = x^2 - 5$$
$$f(3) = 3^2 - 5$$
$$f(3) = 9 - 5$$
$$f(3) = 4$$

b. $f(0)$

$$f(0) = 0^2 - 5$$
$$f(0) = 0 - 5$$
$$f(0) = -5$$

c. $f(-4)$

$$f(-4) = (-4)^2 - 5$$
$$f(-4) = 16 - 5$$
$$f(-4) = 11$$

Now Try:

4. For the function $f(x) = 5x - 10$, find each function value.

a. $f(3)$

____

b. $f(0)$

____

c. $f(-1)$

____

Name: Date:
Instructor: Section:

Review these examples for Objective 5:

5.

Year	Profit (millions of dollars)
2006	34
2008	40
2010	46
2012	48

a. Write the information in the graph as a set of ordered pairs. Does this set define a function?

{(2006, 34), (2008, 40), (2010, 46), (2012, 48)}
Yes, this set defines a function.

b. Suppose that p is the name given to this relation. Give the domain and range of p.

The domain of the function is the set of years.
Domain: {2006, 2008, 2010, 2012}
The range of the function is the set of profits, in millions.
Range: {34, 40, 46, 48}

c. Find $p(2008)$ and $p(2012)$.

We refer to the table or ordered pairs from (a).
$p(2008) = 40$ million and $p(2012) = 48$ million

d. For what value of x does $p(x) = 46$ (million)?

We use the table or the ordered pairs from (a).
$p(2010) = 46$ million

Now Try:

5.

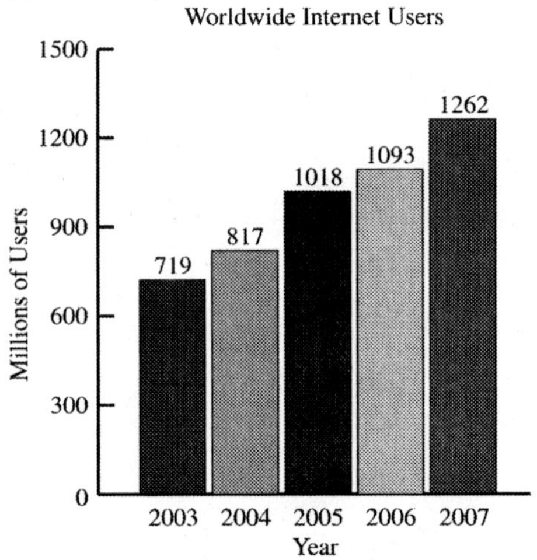

Source: www. internetworldstats.com-January, 2008

a. Write the information in the graph as a set of ordered pairs. Does this set define a function?

____

b. Suppose that w is the name given to this relation. Give the domain and range of w.

____

____

c. Find $w(2003)$ and $w(2006)$.

____

d. For what value of x does $w(x) = 1018$ (million)?

____

Name: Date:
Instructor: Section:

Objective 1 Understand the definition of a relation.

For extra help, see Example 1 on page 1161 of your text and Section Lecture video for Section 17.5.

Identify the domain and range of each relation.

1. $\{(2,7),(5,-4),(-3,-1),(0,-8),(5,2)\}$

1. ________________

domain:__________

range: ___________

2. $\{(3,5),(3,8),(3,-4),(3,1),(3,0)\}$

2. ________________

domain:__________

range: ___________

3. $\{(-3,5),(-2,5),(-1,0),(0,-5),(1,5)\}$

3. ________________

domain:__________

range: ___________

Objective 2 Understand the definition of a function.

For extra help, see Example 2 on page 1162 of your text and Section Lecture video for Section 17.5 and Exercise Solutions Clip 9.

Determine whether each relation is a function.

4. $\{(1,3),(5,7),(11,9),(8,-2),(6,-7),(-4,-3)\}$

4. ________________

5. $\{(-1.2,4),(1.8,-2.5),(3.7,-3.8),(3.7,3.8)\}$

5. ________________

6. $\{(-3,5),(-2,5),(-1,0),(0,-5),(1,5)\}$

6. ________________

Name: Date:
Instructor: Section:

Objective 3 Determine whether an equation defines a function.

For extra help, see Example 3 on page 1163 of your text and Section Lecture video for Section 17.5 and Exercise Solutions Clip 13 and 15.

Use the vertical line test to determine whether each relation graphed is a function.

7.

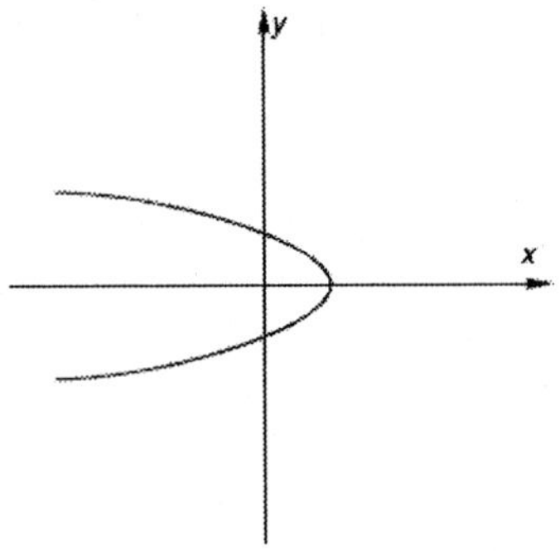

7. ______________

8.

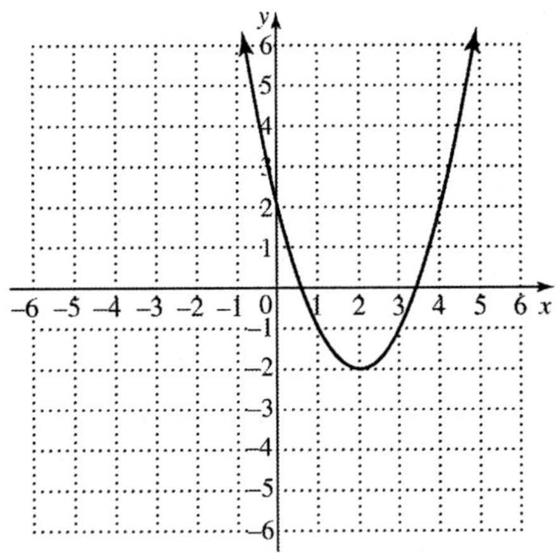

8. ______________

Decide whether the equation defines y as a function of x.

9. $y = 7$

9. ______________

Objective 4 Use function notation.

For extra help, see Example 4 on page 1164 of your text and Section Lecture video for Section 17.5 and Exercise Solutions Clip 25 and 27.

For each function f, find (a) $f(-2)$, (b) $f(0)$, *and* (c) $f(4)$.

10. $f(x) = 3x - 7$

10. a. ______________

b. ______________

c. ______________

Name: Date:
Instructor: Section:

11. $f(x) = x^2 + 2$

11. a.________________

b.________________

c.________________

12. $f(x) = 9$

12. a.________________

b.________________

c.________________

Objective 5 Apply the function concept in an application.

For extra help, see Example 5 on page 1165 of your text and Section Lecture video for Section 17.5 and Exercise Solutions Clip 33.

The yearly revenue for a small business is shown in the table below. Use the table to answer problems 13–15.

Year	Revenue (thousands of dollars)
2008	596
2009	625
2010	872
2011	795
2012	625

13. Write the information in the table as a set of ordered pairs. Does this set define a function?

13. ________________

14. Suppose that r is the name given to this relation. Give the domain and range of r.

14. domain ____________

range____________

15. Find $r(2009)$ and $r(2011)$.

15. ________________

Name: Date:
Instructor: Section:

Chapter R WHOLE NUMBERS REVIEW

R.1 Adding Whole Numbers

Learning Objectives

1	Add two or three single-digit numbers.
2	Add more than two numbers.
3	Add when regrouping is not required.
4	Add with regrouping.
5	Use addition to solve application problems.
6	Check the sum in addition.

Key Terms

Use the vocabulary terms listed below to complete each statement in exercises 1–6.

addition **addends** **sum (total)** **commutative property of addition**

associative property of addition **regrouping**

1. By the________________________________, changing the order of the addends in an addition problem does not change the sum.

2. In addition, the numbers being added are called the ________________.

3. The process of finding the total is called ________________.

4. By the________________________________, changing the grouping of the addends in an addition problem does not change the sum.

5. If the sum of the digits in any column is greater than 9, use the process called ____________________.

6. The answer to an addition problem is called the ________________.

Guided Examples

Review these examples for Objective 1:

1. Add, and then change the order of numbers to write another addition problem.

a. $4 + 5$

$4 + 5 = 9$ and $5 + 4 = 9$

b. $3 + 7$

$3 + 7 = 10$ and $7 + 3 = 10$

Now Try:

1. Add, and then change the order of numbers to write another addition problem.

a. $7 + 2$

____

b. $5 + 9$

____

c. $5 + 5$

$5 + 5 = 10$

c. $3 + 3$

____

Review this example for Objective 2:

2. Add 7, 2, 5, 3, and 8.

$$\begin{array}{r} 7 \\ 2 \\ 5 \\ 3 \\ +8 \\ \hline 25 \end{array} \quad 7+2=9,\quad 9+5=14,\quad 14+3=17,\quad 17+8=25$$

Now Try:

2. Add 9, 7, 2, 5, and 6.

____

Review this example for Objective 3:

3. Add 302, 23, 151, and 21.

First line up the numbers in columns, with the ones column at the right.

$$\begin{array}{r} 302 \\ 23 \\ 151 \\ +\ 21 \\ \hline \end{array}$$

Start at the right and add the ones digits. Add the tens digits next, and finally the hundreds digits.

$$\begin{array}{r} 302 \\ 23 \\ 151 \\ +\ 21 \\ \hline 497 \end{array}$$

Now Try:

3. Add 610, 42, 105, and 31.

____

Review these examples for Objective 4:

4. Add 38 and 27.

Add ones.

$$\begin{array}{r} 38 \\ +\ 27 \\ \hline \end{array}$$

8 ones and 7 ones = 15 ones

Regroup 15 ones as 1 ten and 5 ones. Write 5 ones in the ones column and write 1 ten in the tens column.

$$\begin{array}{r} {}^{1} \\ 38 \\ +\ 27 \\ \hline 5 \end{array}$$

Add the digits in the tens column, including the

Now Try:

4. Add 46 and 19.

____

regrouped 1.

$$\begin{array}{r} \scriptstyle 1 \\ 38 \\ +\ 27 \\ \hline 65 \end{array}$$ 1 ten + 3 tens + 2 tens = 6 tens

5. Add 3197 + 420 + 638 + 67.

Step 1 Add the digits in the ones column.

$$\begin{array}{r} \scriptstyle 2\ \ \\ 3197 \\ 420 \\ 638 \\ +\ \ 67 \\ \hline 2 \end{array}$$

Write 2 tens in the tens column.

Write 2 ones in the ones column.

Notice that 22 ones are regrouped as 2 tens and 2 ones.

Step 2 Add the digits in the tens column, including the regrouped 2.

$$\begin{array}{r} \scriptstyle 2\,2\ \ \\ 3197 \\ 420 \\ 638 \\ +\ \ 67 \\ \hline 22 \end{array}$$

Write 2 hundreds in the hundreds column.

Write 2 tens in the tens column.

Step 3 Add the hundreds column, including the regrouped 2.

$$\begin{array}{r} \scriptstyle 1\,2\,2\ \ \\ 3197 \\ 420 \\ 638 \\ +\ \ 67 \\ \hline 322 \end{array}$$

Write 1 thousand in the thousands column.

Write 3 hundreds in the hundreds column.

Step 4 Add the thousands column, including the regrouped 1.

$$\begin{array}{r} \scriptstyle 1\,2\,2\ \ \\ 3197 \\ 420 \\ 638 \\ +\ \ 67 \\ \hline 4322 \end{array}$$ Sum of the thousands column is 4.

Finally, 3197 + 420 + 638 + 67 = 4322.

5. Add 6835 + 97 + 246 + 4001.

____

Name: Date:
Instructor: Section:

Using the map below, find the shortest distance between the following cities. All distances are in miles.

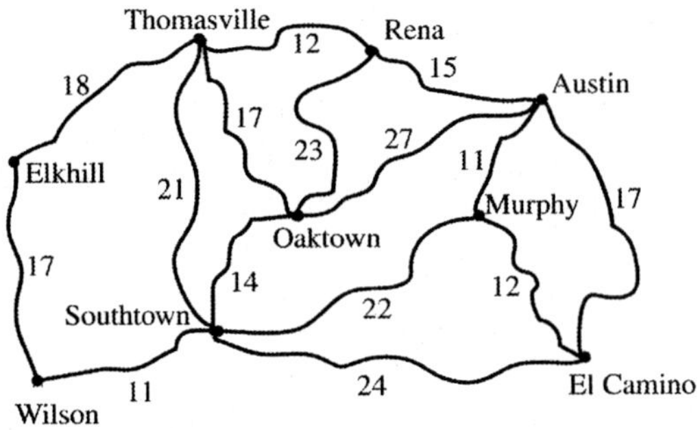

Review these examples for Objective 5:

6. Oaktown and El Camino

One way from Oaktown to El Camino is through Austin. Add the mileage numbers along this route.

$$\begin{array}{rl} 27 & \text{Oaktown to Austin} \\ +\ 17 & \text{Austin to El Camino} \\ \hline 44 & \end{array}$$

44 miles from Oaktown to El Camino going through Austin.

Another way from Oaktown to El Camino is through Southtown. Add the mileage numbers along this route.

$$\begin{array}{rl} 14 & \text{Oaktown to Southtown} \\ +\ 24 & \text{Southtown to El Camino} \\ \hline 38 & \end{array}$$

38 miles from Oaktown to El Camino going through Southtown.

The shortest route from Oaktown to El Camino is 38 miles through Southtown.

7. Murphy and Thomasville

Add the mileage from Murphy to Southtown to Thomasville.

$$\begin{array}{rl} 11 & \text{Murphy to Austin} \\ 15 & \text{Austin to Rena} \\ +\ 12 & \text{Rena to Thomasville} \\ \hline 38 & \end{array}$$

38 miles from Murphy to Thomasville.

Now Try:

6. Southtown and Austin

____

7. El Camino and Thomasville

____

Name: Date:
Instructor: Section:

Review this example for Objective 6:

8. Check the sum. If the sum is incorrect, find the correct sum.

$$\begin{array}{r} 73 \\ 9815 \\ 390 \\ +\ 7002 \\ \hline 16{,}270 \end{array}$$

To check, add up.

$$\begin{array}{r} \underline{17{,}280} \\ 73 \\ 9815 \\ 390 \\ +\ 7002 \\ \hline 16{,}270 \end{array}$$

Re-add to find the sum is 17,280.

Now Try:

8. Check the sum. If the sum is incorrect, find the correct sum.

$$\begin{array}{r} 723 \\ 681 \\ 29 \\ 412 \\ +\ 103 \\ \hline 1947 \end{array}$$

____

Objective 1 Add two or three single-digit numbers.

For extra help, see Example 1 on page 1180 of your text and Section Lecture video for Section R.1.

Add.

1. $3 + 9$ **1.** ____________

2. $8 + 7$ **2.** ____________

Objective 2 Add more than two numbers.

For extra help, see Example 2 on page 1180 of your text and Section Lecture video for Section R.1.

Add.

3.

$$\begin{array}{r} 3 \\ 4 \\ 9 \\ 2 \\ 7 \\ +\ 6 \\ \hline \end{array}$$

3. ____________

Name: Date:
Instructor: Section:

4.

$$\begin{array}{r} 6 \\ 4 \\ 9 \\ 8 \\ 4 \\ +\,3 \\ \hline \end{array}$$

4. ________________

Objective 3 Add when regrouping is not required.

For extra help, see Example 3 on page 1181 of your text and Section Lecture video for Section R.1.

Add.

5.

$$\begin{array}{r} 421 \\ 14 \\ 532 \\ +\ 20 \\ \hline \end{array}$$

5. ________________

6.

$$\begin{array}{r} 158 \\ 340 \\ +\ 401 \\ \hline \end{array}$$

6. ________________

Objective 4 Add with regrouping.

For extra help, see Examples 4–5 on pages 1181–1182 of your text and Section Lecture video for Section R.1.

Add.

7.

$$\begin{array}{r} 563 \\ +\,478 \\ \hline \end{array}$$

7. ________________

8.

$$\begin{array}{r} 7439 \\ +\,8376 \\ \hline \end{array}$$

8. ________________

9.

$$\begin{array}{r} 7033 \\ 809 \\ 2532 \\ +\quad 41 \\ \hline \end{array}$$

9. ________________

Name: Date:
Instructor: Section:

Objective 5 Use addition to solve application problems.

For extra help, see Examples 6–7 on page 1183 of your text and Section Lecture video for Section R.1.

Using the map below, find the shortest distance between the following cities. All distances are in miles.

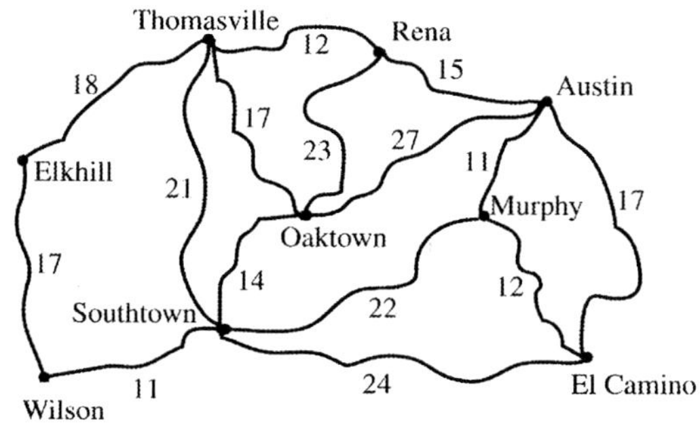

10. Murphy and Oaktown **10.** ________________

11. Wilson and Austin **11.** ________________

Objective 6 Check the sum in addition.

For extra help, see Example 8 on page 1184 of your text and Section Lecture video for Section R.1.

Check the following additions. If an answer is incorrect, give the correct answer.

12.
$$\begin{array}{r} 67 \\ 48 \\ +\ 83 \\ \hline 198 \end{array}$$
12. ________________

13.
$$\begin{array}{r} 3028 \\ 335 \\ 2914 \\ 688 \\ +\ 1647 \\ \hline 8482 \end{array}$$
13. ________________

Name: Date:
Instructor: Section:

Chapter R WHOLE NUMBERS REVIEW

R.2 Subtracting Whole Numbers

Learning Objectives	
1	Change addition problems to subtraction problems or the reverse.
2	Identify the minuend, subtrahend, and difference.
3	Subtract when no regrouping is needed.
4	Use addition to check subtraction answers.
5	Subtract with regrouping.
6	Use subtraction to solve application problems.

Key Terms

Use the vocabulary terms listed below to complete each statement in exercises 1–4.

minuend **subtrahend** **difference** **regrouping**

1. The number from which another number is being subtracted is called the ____________________.

2. In order to subtract 29 from 76, use a process called ____________________ from the tens place.

3. The ____________________ is the number being subtracted.

4. The answer to a subtraction problem is called the ____________________.

Guided Examples

Review these examples for Objective 1:

1. Change each addition problem to a subtraction problem.

a. $4 + 2 = 6$

Two subtraction problems are possible:
$6 - 2 = 4$ or $6 - 4 = 2$

b. $5 + 9 = 14$

$14 - 5 = 9$ or $14 - 9 = 5$

2. Change each subtraction problem to an addition problem.

a. $9 - 4 = 5$

Two addition problems are possible:
$9 = 4 + 5$ or $9 = 5 + 4$

b. $390 - 150 = 240$

$390 = 150 + 240$ or $390 = 240 + 150$

Now Try:

1. Change each addition problem to a subtraction problem.

a. $8 + 5 = 13$

____

b. $9 + 3 = 12$

____

2. Change each subtraction problem to an addition problem.

a. $9 - 7 = 2$

____

b. $320 - 190 = 130$

____

Name: Date:
Instructor: Section:

Review this example for Objective 2:

3. Identify the minuend, subtrahend, and difference in the following subtraction problem.

94 – 13 = 81

The number 94 is the minuend.
The number 13 is the subtrahend.
The number 81 is the difference.

Now Try:

3. Identify the minuend, subtrahend, and difference in the following subtraction problem.

87 – 12 = 75

Minuend ____________

Subtrahend ____________

Difference ____________

Review these examples for Objective 3:

4. Subtract.

a.
$$\begin{array}{r} 63 \\ -\ 31 \\ \hline \end{array}$$

3 ones – 1 one = 2 ones
6 tens – 3 tens = 3 tens

$$\begin{array}{r} 63 \\ -\ 31 \\ \hline 32 \end{array}$$

b.
$$\begin{array}{r} 463 \\ -\ 143 \\ \hline \end{array}$$

3 ones – 3 ones = 0 ones
6 tens – 4 tens = 2 tens
4 hundreds – 1 hundred = 3 hundreds

$$\begin{array}{r} 463 \\ -\ 143 \\ \hline 320 \end{array}$$

c.
$$\begin{array}{r} 9546 \\ -\ 305 \\ \hline \end{array}$$

6 ones – 5 ones = 1 ones
4 tens – 0 tens = 4 tens
5 hundreds – 3 hundreds = 2 hundreds
9 thousands – 0 thousand = 9 thousand

$$\begin{array}{r} 9546 \\ -\ 305 \\ \hline 9241 \end{array}$$

Now Try:

4. Subtract.

a.
$$\begin{array}{r} 85 \\ -\ 42 \\ \hline \end{array}$$

____

b.
$$\begin{array}{r} 538 \\ -\ 427 \\ \hline \end{array}$$

____

c.
$$\begin{array}{r} 6574 \\ -\ 364 \\ \hline \end{array}$$

____

Name: Date:
Instructor: Section:

Review these examples for Objective 4:

5. Use addition to check each answer. If the answer is incorrect, find the correct answer.

a.
$$\begin{array}{r} 58 \\ -\ 37 \\ \hline 21 \end{array}$$

Rewrite as an addition problem.
$$\begin{array}{r} 37 \\ +\ 21 \\ \hline 58 \end{array}$$

Because 37 + 21 = 58, the subtraction was done correctly.

b. 83 – 32 = 52

Rewrite as an addition problem.

32 + 52 = 83

But 32 + 52 = 84, not 83, so the subtraction was done incorrectly. Reworking the subtraction,

83 – 32 = 51.

c.
$$\begin{array}{r} 785 \\ -\ 162 \\ \hline 623 \end{array}$$

Since 162 + 623 = 785, the answer checks.

Now Try:

5. Use addition to check each answer. If the answer is incorrect, find the correct answer.

a.
$$\begin{array}{r} 56 \\ -\ 25 \\ \hline 31 \end{array}$$

____

b.
$$\begin{array}{r} 736 \\ -\ 514 \\ \hline 223 \end{array}$$

____

c.
$$\begin{array}{r} 9614 \\ -\ 8312 \\ \hline 1312 \end{array}$$

____

Review these examples for Objective 5:

6. Subtract 29 from 46.

Write the problem vertically. In the ones column 6 is less than 9, so we must regroup 1 ten as 10 ones.

$$\begin{array}{r} \scriptstyle 3\ 16 \\ \not{4}\ \not{6} \\ -\ 2\ 9 \\ \hline \end{array}$$

Now subtract 9 from 16 ones in the ones column. Then subtract 2 tens from 3 tens in the tens column.

$$\begin{array}{r} \scriptstyle 3\ 16 \\ \not{4}\ \not{6} \\ -\ 2\ 9 \\ \hline 1\ 7 \end{array}$$

Finally, 46 – 29 = 17. Check by adding 29 and 17; you should get 46.

Now Try:

6. Subtract 29 from 68.

____

Name: Date:
Instructor: Section:

7. Subtract by regrouping when necessary.

$$\begin{array}{r} 645 \\ -\ 449 \\ \hline \end{array}$$

Regroup 1 ten as 10 ones.

$$\begin{array}{rccc} & & 3 & 15 \\ & 6 & \cancel{4} & \cancel{5} \\ - & 4 & 4 & 9 \\ \hline & & & 6 \end{array}$$

Need to regroup further because 3 is less than 4 in the tens column.

$$\begin{array}{rccc} & 5 & 13 & 15 \\ & \cancel{6} & \cancel{4} & \cancel{5} \\ - & 4 & 4 & 9 \\ \hline & 1 & 9 & 6 \end{array}$$

The difference is 196.

7. Subtract by regrouping when necessary.

$$\begin{array}{r} 726 \\ -\ 238 \\ \hline \end{array}$$

____

8. Subtract.

$$\begin{array}{r} 5203 \\ -\ 2857 \\ \hline \end{array}$$

There are no tens that can be regrouped into ones. So you must first regroup 1 hundred as 10 tens.

$$\begin{array}{rcccc} & & 1 & 10 & \\ & 5 & \cancel{2} & \cancel{0} & 3 \\ - & 2 & 8 & 5 & 7 \\ \hline \end{array}$$

Now we may regroup from the tens position.

$$\begin{array}{rcccc} & & & 9 & \\ & & 1 & \cancel{10} & 13 \\ & 5 & \cancel{2} & \cancel{0} & \cancel{3} \\ - & 2 & 8 & 5 & 7 \\ \hline & & & & 6 \end{array}$$

Complete the process.

$$\begin{array}{rcccc} & & 11 & 9 & \\ & 4 & \cancel{1} & \cancel{10} & 13 \\ & \cancel{5} & \cancel{2} & \cancel{0} & \cancel{3} \\ - & 2 & 8 & 5 & 7 \\ \hline & 2 & 3 & 4 & 6 \end{array}$$

The difference is 2346.

8. Subtract.

$$\begin{array}{r} 4203 \\ -\ 1415 \\ \hline \end{array}$$

____

9. Subtract.

$$\begin{array}{r} 6000 \\ -\ 3888 \\ \hline \end{array}$$

There are no tens or hundreds that can be regrouped into ones. So you must first regroup 1 thousand as 10 hundreds, 1 hundred as 10 tens, and 1 ten as 10 ones.

9. Subtract.

$$\begin{array}{r} 8000 \\ -\ 2997 \\ \hline \end{array}$$

____

$$\begin{array}{r} \overset{5}{\cancel{6}}\overset{\overset{9}{\cancel{10}}}{\cancel{0}}\overset{\overset{9}{\cancel{10}}}{\cancel{0}}\overset{10}{\cancel{0}} \\ -\ 3\,8\,8\,8 \\ \hline 2\,1\,1\,2 \end{array}$$

The difference is 2112.

10. Use addition to check each answer.

a.
$$\begin{array}{r} 823 \\ -\ 585 \\ \hline 238 \end{array}$$

Check.
$$\begin{array}{r} 585 \\ +\ 238 \\ \hline 823 \end{array}$$

Correct.

b.
$$\begin{array}{r} 561 \\ -\ 273 \\ \hline 308 \end{array}$$

Check.
$$\begin{array}{r} 273 \\ +\ 308 \\ \hline 581 \end{array}$$

Incorrect. Rework the original problem to get the correct answer.

$$\begin{array}{r} \overset{4}{\cancel{5}}\overset{15}{\cancel{6}}\overset{11}{\cancel{1}} \\ -\ 2\,7\,3 \\ \hline 2\,8\,8 \end{array}$$

10. Use addition to check each answer.

a.
$$\begin{array}{r} 738 \\ -\ 249 \\ \hline 489 \end{array}$$

____

b.
$$\begin{array}{r} 614 \\ -\ 236 \\ \hline 488 \end{array}$$

____

Review this example for Objective 6:

11. Sally Tanner had $22,143 withheld from her paycheck last year for income tax. She actually owes only $16,959 in tax. What refund should she receive?

To find out the refund, subtract $16,959 from $22,143.

$$\begin{array}{r} \$\overset{1}{\cancel{2}}\overset{11}{\cancel{2}},\overset{10}{\cancel{1}}\overset{13}{\cancel{4}}\overset{13}{\cancel{3}} \\ -\ \$1\,6,\,9\,5\,9 \\ \hline \$5\,1\,8\,4 \end{array}$$

The refund should be $5184.

Now Try:

11. An airplane is carrying 234 passengers. When it lands in Atlanta, 139 passengers get off the plane. How many passengers are then left on the plane?

____

Name: Date:
Instructor: Section:

Objective 1 Change addition problems to subtraction problems or the reverse.

For extra help, see Examples 1–2 on pages 1187–1188 of your text and Section Lecture video for Section R.2.

Write two subtraction problems for each addition problem.

1. 149 + 38 = 187 **1.** ________________

2. 478 + 239 = 717 **2.** ________________

Write an addition problem for each subtraction problem.

3. 1211 – 426 = 785 **3.** ________________

Objective 2 Identify the minuend, subtrahend, and difference.

For extra help, see page 1188 of your text and Section Lecture video for Section R.2.

Identify the minuend, subtrahend, and difference in each of the following subtraction problems.

4. 98 – 36 = 62

4.
Minuend ____________
Subtrahend __________
Difference ___________

5. 35 – 9 = 24

5.
Minuend ____________
Subtrahend __________
Difference ___________

Objective 3 Subtract when no regrouping is needed.

For extra help, see Example 3 on page 1188 of your text and Section Lecture video for Section R.2.

Subtract.

6. $\begin{array}{r} 5573 \\ -\ 422 \\ \hline \end{array}$ **6.** ________________

7. $\begin{array}{r} 8539 \\ -\ 2527 \\ \hline \end{array}$ **7.** ________________

Name: Date:
Instructor: Section:

Objective 4 Use addition to check subtraction answers.

For extra help, see Example 4 on page 1189 of your text and Section Lecture video for Section R.2.

Check the following subtractions. If an answer is not correct, give the correct answer.

8. $\begin{array}{r} 31{,}146 \\ -\ 7312 \\ \hline 23{,}834 \end{array}$ **8.** ______________

9. $\begin{array}{r} 192 \\ -\ 39 \\ \hline 167 \end{array}$ **9.** ______________

10. $\begin{array}{r} 4847 \\ -\ 3768 \\ \hline 1121 \end{array}$ **10.** ______________

Objective 5 Subtract with regrouping.

For extra help, see Examples 5–9 on pages 1189–1192 of your text and Section Lecture video for Section R.2.

Subtract.

11. $\begin{array}{r} 927 \\ -\ 729 \\ \hline \end{array}$ **11.** ______________

12. $\begin{array}{r} 33{,}728 \\ -\ \ \ 7829 \\ \hline \end{array}$ **12.** ______________

13. $\begin{array}{r} 7000 \\ -\ 297 \\ \hline \end{array}$ **13.** ______________

Objective 6 Use subtraction to solve application problems.

For extra help, see Example 10 on page 1192 of your text and Section Lecture video for Section R.2.

Solve each application problem.

14. On Sunday, 7342 people went to a football game, while on Monday, 9138 people went. How many more people went on Monday? **14.** ______________

Name: Date:
Instructor: Section:

15. The Conrads now pay $439 per month for rent. If they rent a larger apartment, the payment will be $702 per month. How much extra will they pay each month?

15. ________________

16. One bid for painting a house was $2134. A second bid was $1954. How much would be saved using the second bid?

16. ________________

Name: Date:
Instructor: Section:

Chapter R WHOLE NUMBERS REVIEW

R.3 Multiplying Whole Numbers

Learning Objectives	
1	Identify the parts of a multiplication problem.
2	Do chain multiplications.
3	Multiply by single-digit numbers.
4	Multiply quickly by numbers ending in zeros.
5	Multiply by numbers having more than one digit.
6	Use multiplication to solve application problems.

Key Terms

Use the vocabulary terms listed below to complete each statement in exercises 1–7.

factors **product** **commutative property of multiplication**

associative property of multiplication **chain multiplication problem**

multiple **partial products**

1. By the________________________________, changing the order of the factors in a multiplication problem does not change the product.

2. In multiplication, the numbers being multiplied are called the _______________.

3. By the________________________________, changing the grouping of the factors in a multiplication problem does not change the product.

4. The product of two whole number factors is called a _______________ of either factor.

5. The answer to a multiplication problem is called the _________________.

6. The products found when multiplying numbers having two or more digits are called _________________.

7. A multiplication problem with more than two factors is a
_________________________________.

Guided Examples

Review these examples for Objective 1:

1. Multiply.

a. 3×7

$3\times7=21$

Now Try:

1. Multiply.

a. 7×6

____

b. $9 \cdot 0$

$9 \cdot 0 = 0$

c. $(5)(9)$

$(5)(9) = 45$

b. $0 \cdot 5$ __________

c. $(2)(8)$ __________

Review this example for Objective 2:

2. Multiply $4 \times 3 \times 2$.

Parentheses show what to do first.

$(4 \times 3) \times 2$

$12 \times 2 = 24$

Also,

$4 \times (3 \times 2)$

$4 \times \quad 6 = 24$

Either grouping results in the same product.

Now Try:

2. Multiply $5 \times 4 \times 2$. __________

Review these examples for Objective 3:

3. Multiply.

a. $\begin{array}{r} 64 \\ \times \ \ 3 \\ \hline \end{array}$

Start by multiplying in the ones column.

$4 \times 3 = 12$ ones

Write 1 ten in the tens column.

Write 2 ones in the ones column.

$\begin{array}{r} {}^{1} \\ 64 \\ \times \ \ 3 \\ \hline 2 \end{array}$

Next, multiply 3 times 6 tens.

3×6 tens $= 18$ tens

Add the 1 ten that was written at the top of the tens column.

18 tens + 1 ten = 19 tens

$\begin{array}{r} {}^{1} \\ 64 \\ \times \ \ 3 \\ \hline 192 \end{array}$

b. $\begin{array}{r} 645 \\ \times \ \ \ 4 \\ \hline \end{array}$

$5 \times 4 = 20$ ones; write 0 ones; write 2 tens in the tens column.

Now Try:

3. Multiply.

a. $\begin{array}{r} 43 \\ \times \ \ 6 \\ \hline \end{array}$ __________

b. $\begin{array}{r} 567 \\ \times \ \ \ 7 \\ \hline \end{array}$ __________

Name: Date:
Instructor: Section:

$$\begin{array}{r} \overset{2}{6}\,4\,5 \\ \times \quad 4 \\ \hline 0 \end{array}$$

$4\times4=16$ tens; add the 2 regrouped tens to get 18 tens; write 8 tens; write 1 hundred in the hundreds column.

$$\begin{array}{r} {\scriptstyle 1\,2} \\ 6\,4\,5 \\ \times \quad 4 \\ \hline 8\,0 \end{array}$$

$4\times6=24$ hundreds; add the 1 regrouped hundred to get 25 hundreds.

$$\begin{array}{r} {\scriptstyle 1\,2} \\ 6\,4\,5 \\ \times \quad 4 \\ \hline 2\,5\,8\,0 \end{array}$$

Review these examples for Objective 4:

4. Multiply.

a. 67×10

$67\times10=670$ Attach 0.

b. 39×100

$39\times100=3900$ Attach 00.

c. 74×1000

$74\times1000=74{,}000$ Attach 000.

5. Multiply.

a. 67×4000

Multiply 67 by 4, and then attach three zeros.

$$\begin{array}{r} 6\,7 \\ \times \;\; 4 \\ \hline 2\,6\,8 \end{array}$$

$67\times4000=268{,}000$

b. 170×80

Multiply 17 by 8, and then attach two zeros.

$$\begin{array}{r} 1\,7 \\ \times \;\; 8 \\ \hline 1\,3\,6 \end{array}$$

$170\times80=13{,}600$

Now Try:

4. Multiply.

a. 46×10

____

b. 708×100

____

c. 519×1000

____

5. Multiply.

a. 89×200

____

b. 140×60

____

Name: Date:
Instructor: Section:

Review these examples for Objective 5:

6. Multiply.

a. (53)(36)

First write in vertical format. Then start by multiplying 53 by 6.

$$\begin{array}{r} \scriptstyle 1 \\ 53 \\ \times\ \ 6 \\ \hline 318 \end{array} \quad \leftarrow 53\times 6 = 318$$

Now multiply 53 by 30.

$$\begin{array}{r} 53 \\ \times\ 30 \\ \hline 1590 \end{array} \quad \leftarrow 53\times 30 = 1590$$

Add the results.

$$\begin{array}{rl} 53 & \\ \times\ 36 & \\ \hline 318 & \leftarrow 53\times 6 \\ 1590 & \leftarrow 53\times 30 \\ \hline 1908 & \end{array}$$

Both 318 and 1590 are called partial products. As a common practice and to save time, the 0 in 1590 is usually not written.

$$\begin{array}{r} 53 \\ \times\ 36 \\ \hline 318 \\ 159 \\ \hline 1908 \end{array}$$

b.
$$\begin{array}{r} 231 \\ \times\ 322 \\ \hline \end{array}$$

Be sure to align numbers in columns of partial products

$$\begin{array}{r} 231 \\ \times\ 322 \\ \hline 462 \\ 462 \\ 693 \\ \hline 74,382 \end{array}$$

c.
$$\begin{array}{r} 647 \\ \times\ 54 \\ \hline \end{array}$$

First, multiply by 4.

Now Try:

6. Multiply.

a. (45)(64) ________

b.
$$\begin{array}{r} 132 \\ \times\ 213 \\ \hline \end{array}$$
____

c.
$$\begin{array}{r} 165 \\ \times\ 53 \\ \hline \end{array}$$
____

Name: Date:
Instructor: Section:

$$\begin{array}{r} {}^{1\,2} \\ 647 \\ \times\ 54 \\ \hline 2588 \end{array}$$

Now multiply by 5, being careful to line up the tens. Finally, add the partial products.

$$\begin{array}{r} {}^{2\,3} \\ {}^{1\,2} \\ 647 \\ \times\ 54 \\ \hline 2588 \\ 3235 \\ \hline 34,938 \end{array}$$

7. Multiply.

a. $\begin{array}{r} 146 \\ \times\ 205 \\ \hline \end{array}$

Align numbers in columns of partial products

$$\begin{array}{r} 146 \\ \times\ 205 \\ \hline 730 \\ 000 \\ 292 \\ \hline 29,930 \end{array}$$

b. $\begin{array}{r} 1062 \\ \times\ 3004 \\ \hline \end{array}$

Be careful to line up the zeros.

$$\begin{array}{r} 1062 \\ \times\ 3004 \\ \hline 4248 \\ 0000 \\ 0000 \\ 3186 \\ \hline 3,190,248 \end{array} \quad \text{or} \quad \begin{array}{r} 1062 \\ \times\ 3004 \\ \hline 4248 \\ 318600 \\ \hline 3,190,248 \end{array}$$

7. Multiply.

a. $\begin{array}{r} 843 \\ \times\ 106 \\ \hline \end{array}$ __________

b. $\begin{array}{r} 4625 \\ \times\ 7003 \\ \hline \end{array}$ __________

Review this example for Objective 6:

8. Find the total cost of 85 tickets at $29 each.

Multiply 85 by 29.

$$\begin{array}{r} 85 \\ \times\ 29 \\ \hline 765 \\ 170 \\ \hline 2465 \end{array}$$

The total cost of the tickets is $2465.

Now Try:

8. Find the total cost of 316 pair of gloves at $17 per pair.

____

Name: Date:
Instructor: Section:

Objective 1 Identify the parts of a multiplication problem.

For extra help, see Example 1 on page 1195 of your text and Section Lecture video for Section R.3.

Identify the factors and the product in each multiplication problem.

1. $108 = 9 \times 12$

1.
Factors ______________

Product ______________

Multiply.

2. $4 \cdot 6$

2. ______________

3. 8×9

3. ______________

Objective 2 Do chain multiplications.

For extra help, see Example 2 on page 1196 of your text and Section Lecture video for Section R.3.

Multiply.

4. $4 \times 4 \times 2$

4. ______________

5. $3 \times 4 \times 7$

5. ______________

6. $(6)(4)(8)$

6. ______________

Objective 3 Multiply by single-digit numbers.

For extra help, see Example 3 on pages 1196–1197 of your text and Section Lecture video for Section R.3.

Multiply.

7. $\begin{array}{r} 54 \\ \times\ 4 \\ \hline \end{array}$

7. ______________

8. $\begin{array}{r} 163 \\ \times\ \ 5 \\ \hline \end{array}$

8. ______________

Name: Date:
Instructor: Section:

9. $\begin{array}{r} 30{,}009 \\ \times \quad 6 \\ \hline \end{array}$

9. ______________

Objective 4 Multiply quickly by numbers ending in zeros.

For extra help, see Examples 4–5 on pages 1197–1198 of your text and Section Lecture video for Section R.3.

Multiply.

10. $(852)(30)$

10. ______________

11. 500×40

11. ______________

12. 8234×2000

12. ______________

Objective 5 Multiply by numbers having more than one digit.

For extra help, see Examples 6–7 on pages 1198–1199 of your text and Section Lecture video for Section R.3.

Multiply.

13. $\begin{array}{r} 92 \\ \times \ 48 \\ \hline \end{array}$

13. ______________

14. $\begin{array}{r} 644 \\ \times \ 19 \\ \hline \end{array}$

14. ______________

Name: Date:
Instructor: Section:

15. $\begin{array}{r} 4031 \\ \times \;\; 408 \\ \hline \end{array}$ **15.** ________________

Objective 6 Use multiplication to solve application problems.

For extra help, see Example 8 on page 1200 of your text and Section Lecture video for Section R.3.

Solve the following application problems.

16. On a recent trip the Jensen family drove 45 miles per hour on the average. They drove 22 hours altogether. How many miles did they drive altogether? **16.** ________________

Find the total cost of the following items.

17. 512 boxes of chalk at $19 per box **17.** ________________

18. 178 baseball caps at $9 per cap **18.** ________________

Name: Date:
Instructor: Section:

Chapter R WHOLE NUMBERS REVIEW

R.4 Dividing Whole Numbers

Learning Objectives	
1	Write division problems in three ways.
2	Identify the parts of a division problem.
3	Divide 0 by a number.
4	Recognize that division by 0 is undefined.
5	Divide a number by itself.
6	Use short division.
7	Use multiplication to check quotients.
8	Use tests for divisibility.

Key Terms

Use the vocabulary terms listed below to complete each statement in exercises 1–5.

dividend **divisor** **quotient** **short division** **remainder**

1. The number left over when two numbers do not divide exactly is the ________________.

2. The number being divided by another number in a division problem is the ________________.

3. The answer to a division problem is called the ________________.

4. In the problem $639 \div 9$, 9 is called the ________________.

5. ________________ is a method of dividing a number by a one-digit divisor.

Guided Examples

Review these examples for Objective 1:

1. Write each division problem using two other symbols.

a. $24 \div 8 = 3$

$8\overline{)24}$ with quotient 3 or $\frac{24}{8} = 3$

b. $\frac{27}{3} = 9$

$27 \div 3 = 9$ or $3\overline{)27}$ with quotient 9

Now Try:

1. Write each division problem using two other symbols.

a. $28 \div 4 = 7$

____

b. $\frac{72}{9} = 8$

____

Name: Date:
Instructor: Section:

c. $6\overline{)30}$ with quotient 5

$30 \div 6 = 5$ or $\frac{30}{6} = 5$

c. $8\overline{)32}$ with quotient 4

____

Review these examples for Objective 2:

2. Identify the dividend, divisor and quotient.

a. $45 \div 5 = 9$

$45 \div 5 = 9$ ← Quotient (45: Dividend; 5: Divisor)

b. $\frac{90}{30} = 3$

Dividend ↓ $\frac{90}{30} = 3$ ← Quotient; ↑ Divisor

c. $8\overline{)48}$ with quotient 6

6 ← Quotient
$8\overline{)48}$ ← Dividend
↑ Divisor

Now Try:

2. Identify the dividend, divisor and quotient.

a. $63 \div 9 = 7$

____

b. $\frac{56}{8} = 7$

____

c. $6\overline{)54}$ with quotient 9

____

Review these examples for Objective 3:

3. Divide.

a. $0 \div 17$

$0 \div 17 = 0$

b. $0 \div 9674$

$0 \div 9674 = 0$

c. $\frac{0}{668}$

$\frac{0}{668} = 0$

d. $139\overline{)0}$

$139\overline{)0}$ with quotient 0

Now Try:

3. Divide.

a. $0 \div 6$

____

b. $0 \div 1283$

____

c. $\frac{0}{25}$

____

d. $67\overline{)0}$

____

Name: Date:
Instructor: Section:

4. Change each division problem to a multiplication problem.

a. $\frac{24}{3} = 8$

$\frac{24}{3} = 8$ becomes $3 \cdot 8 = 24$ or $8 \cdot 3 = 24$

b. $7\overline{)42}$ with quotient 6

$7\overline{)42}$ (quotient 6) becomes $6 \cdot 7 = 42$ or $7 \cdot 6 = 42$

c. $63 \div 7 = 9$

$63 \div 7 = 9$ becomes $7 \cdot 9 = 63$ or $9 \cdot 7 = 63$

4. Change each division problem to a multiplication problem.

a. $\frac{36}{4} = 9$ __________

b. $3\overline{)18}$ with quotient 6 __________

c. $40 \div 8 = 5$ __________

Review these examples for Objective 4:

5. Find the quotient whenever possible.

a. $\frac{13}{0}$

$\frac{13}{0}$ is undefined.

b. $0\overline{)21}$

$0\overline{)21}$ is undefined.

c. $27 \div 0$

$27 \div 0$ is undefined.

Now Try:

5. Find the quotient whenever possible.

a. $\frac{9}{0}$ __________

b. $0\overline{)21}$ __________

c. $110 \div 0$ __________

Review these examples for Objective 5:

6. Divide.

a. $18 \div 18$

$18 \div 18 = 1$

b. $42\overline{)42}$

$42\overline{)42}$ with quotient 1

c. $\frac{95}{95}$

$\frac{95}{95} = 1$

Now Try:

6. Divide.

a. $7 \div 7$ __________

b. $19\overline{)19}$ __________

c. $\frac{89}{89}$ __________

Name: Date:
Instructor: Section:

Review these examples for Objective 6:

7. Divide: $4\overline{)84}$.

First, divide 8 by 4.

$$4\overline{)84}^{\;2} \quad \frac{8}{4}=2$$

Next, divide 4 by 4.

$$4\overline{)84}^{\;21} \quad \frac{4}{4}=1$$

8. Divide 167 by 3.

$$3\overline{)167}$$

Because 1 cannot be divided by 3, divide 16 by 3. Notice that the 5 is placed over the 6 in 16.

$$\begin{array}{r} 5 \\ 3\overline{)16^17} \end{array} \quad \frac{16}{3}=5 \text{ with 1 left over}$$

Next divide 17 by 3. The final number left over is the remainder.

$$\begin{array}{r} 5\;5 \text{ R2} \\ 3\overline{)16^17} \end{array} \quad \frac{17}{3}=5 \text{ with 2 left over}$$

9. Divide 1943 by 8.

Divide 19 by 8.

$$\begin{array}{r} 2 \\ 8\overline{)19^343} \end{array} \quad \frac{19}{8}=2 \text{ with 3 left over}$$

Divide 34 by 8.

$$\begin{array}{r} 2\;4 \\ 8\overline{)19^34^23} \end{array} \quad \frac{34}{8}=4 \text{ with 2 left over}$$

Divide 23 by 8.

$$\begin{array}{r} 2\;4\;2 \text{ R7} \\ 8\overline{)19^34^23} \end{array} \quad \frac{23}{8}=2 \text{ with 7 left over}$$

The quotient is 242 R7.

Now Try:

7. Divide: $3\overline{)69}$.

____

8. Divide 149 by 2.

____

9. Divide 2354 by 9.

____

Review these examples for Objective 7:

10. Check each answer.

a. $\begin{array}{r} 91 \text{ R3} \\ 6\overline{)549} \end{array}$

To check, multiply the divisor and the quotient, then add the remainder to get the dividend.

Now Try:

10. Check each answer.

a. $\begin{array}{r} 28 \text{ R1} \\ 3\overline{)85} \end{array}$

____

Name: Date:
Instructor: Section:

$(6 \times 91) + 3$

$546 \quad + 3 = 549$

549 matches the original dividend, so the division was done correctly.

b. $8\overline{)2315}$ = 288 R6

To check, multiply the divisor and the quotient, then add the remainder to get the dividend.

$(8 \times 288) + 6$

$2304 \quad + 6 = 2310$

The answer does not check. Rework the original problem to get the correct answer, 289 R 3.

b. $6\overline{)469}$ = 79 R4

____

Review these examples for Objective 8:

11. Are the following numbers divisible by 2?

a. 864

Because the number ends in 4, which is an even number, the number 864 is divisible by 2.

b. 4287

4287 is not divisible by 2.
Ends in 7, not 0, 2, 4, 6, or 8

12. Are the following numbers divisible by 3?

a. 537

Add the digits.

$5 + 3 + 7 = 15$

Because 15 is divisible by 3, the number 537 is also divisible by 3.

b. 156,304

Add the digits.

$1 + 5 + 6 + 3 + 0 + 4 = 19$

Because 19 is not divisible by 3, the number 156,304 is not divisible by 3.

13. Are the following numbers divisible by 5?

a. 43,760

43,760 ends in 0 and is divisible by 5.

b. 7565

7565 ends in 5 and is divisible by 5.

Now Try:

11. Are the following numbers divisible by 2?

a. 568 _______________

b. 873 _______________

12. Are the following numbers divisible by 3?

a. 843 _______________

b. 386,452 _______________

13. Are the following numbers divisible by 5?

a. 290 _______________

b. 970,405 _______________

Name: Date:
Instructor: Section:

c. 954

954 ends in 4 and is not divisible by 5.

14. Are the following numbers divisible by 10?

a. 950 and 8790

950 and 8790 both end in 0 and are divisible by 10.

b. 366 and 27,853

366 and 27,853 do not end in 0 and are not divisible by 10.

c. 7356

____

14. Are the following numbers divisible by 10?

a. 360

____

b. 565

____

Objective 1 Write division problems in three ways.

For extra help, see Example 1 on page 1203 of your text and Section Lecture video for Section R.4.

Write each division problem using two other symbols.

1. $15 \div 3 = 5$

1. ______________

2. $\frac{50}{25} = 2$

2. ______________

Objective 2 Identify the parts of a division problem.

For extra help, see Example 2 on pages 1203–1204 of your text and Section Lecture video for Section R.4.

Identify the dividend, divisor, and quotient.

3. $63 \div 7 = 9$

3.
Dividend ______________

Divisor ______________

Quotient ______________

4. $5\overline{)30}$ with quotient 6

4.
Dividend ______________

Divisor ______________

Quotient ______________

Name: Date:
Instructor: Section:

5. $\frac{44}{11} = 4$

5.
Dividend ____________

Divisor ____________

Quotient ____________

Objective 3 Divide 0 by a number.

For extra help, see Examples 3–4 on page 1204 of your text and Section Lecture video for Section R.4.

Divide.

6. $12\overline{)0}$ **6.** ____________

7. $\frac{0}{6}$ **7.** ____________

8. $0 \div 15$ **8.** ____________

Objective 4 Recognize that division by 0 is undefined.

For extra help, see Example 5 on page 1205 of your text and Section Lecture video for Section R.4.

Divide. If the division is not possible, write "**undefined**."

9. $0\overline{)72}$ **9.** ____________

10. $\frac{7}{0}$ **10.** ____________

11. $9 \div 0$ **11.** ____________

Objective 5 Divide a number by itself.

For extra help, see Example 6 on page 1206 of your text and Section Lecture video for Section R.4.

Divide.

12. $77 \div 77$ **12.** ____________

13. $\frac{421}{421}$ **13.** ____________

Name: Date:
Instructor: Section:

Objective 6 Use short division.

For extra help, see Examples 7–9 on pages 1206–1207 of your text and Section Lecture video for Section R.4.

Divide by using short division.

14. $724 \div 5$ **14.** ______________

15. $\dfrac{651}{9}$ **15.** ______________

16. $8\overline{)1135}$ **16.** ______________

Objective 7 Use multiplication to check quotients.

For extra help, see Example 10 on pages 1207–1208 of your text and Section Lecture video for Section R.4.

Use multiplication to check each answer. If an answer is incorrect, find the correct answer.

17. $\begin{array}{r} 1522 \text{ R4} \\ 6\overline{)9137} \end{array}$ **17.** ______________

18. $3852 \div 4 = 963$ **18.** ______________

19. $\dfrac{18{,}150}{3} = 650$ **19.** ______________

Objective 8 Use tests for divisibility.

For extra help, see Examples 11–14 on pages 1209–1210 of your text and Section Lecture video for Section R.4.

Determine if the following numbers are divisible by 2, 3, 5, or 10 Write **yes** *or* **no**.

20. 897 **20. 2:** ______________

3: ______________

5: ______________

10: ______________

Name: Date:
Instructor: Section:

21. 908

21. 2: ______________

3: ______________

5: ______________

10: ______________

22. 6205

22. 2: ______________

3: ______________

5: ______________

10: ______________

Name: Date:
Instructor: Section:

Chapter R WHOLE NUMBERS REVIEW

R.5 Long Division

Learning Objectives
1 Use long division.
2 Divide by multiples of 10.
3 Use multiplication to check quotients.

Key Terms

Use the vocabulary terms listed below to complete each statement in exercises 1–5.

long division **dividend** **divisor** **quotient** **remainder**

1. In the problem $751 \div 23 = 32$ R 15, 751 is called the ____________________.

2. In the problem $751 \div 23 = 32$ R 15, 15 is called the ____________________.

3. In the problem $751 \div 23 = 32$ R 15, 23 is called the ____________________.

4. In the problem $751 \div 23 = 32$ R 15, 32 is called the ____________________.

5. ____________________ is a method of dividing a number by a divisor with more than one digit.

Guided Examples

Review these examples for Objective 1:

1. Divide. $62\overline{)5146}$

Because 62 is closer to 60 than 70, use the first digit of the divisor as a trial divisor.
Try to divide the first digit of the dividend by 6. Since 5 cannot be divided by 6, use the first two digits, 51.

$$\frac{51}{6} = 8 \text{ R3}$$

$$\begin{array}{r} 8 \\ 62\overline{)5146} \end{array}$$

Multiply 8 and 62 to get 496; next, subtract 496 from 514.

$$\begin{array}{r} 8 \\ 62\overline{)5146} \\ \underline{496} \\ 18 \end{array}$$

Now Try:

1. Divide. $34\overline{)2482}$

____

Bring down the 6 on the right.

$$62\overline{)5146}$$ quotient 8

$$\begin{array}{r} 8 \\ 62\overline{)5146} \\ \underline{496} \\ 186 \end{array}$$

Use the trial divisor 6. $\frac{18}{6}=3$

$$\begin{array}{r} 83 \\ 62\overline{)5146} \\ \underline{496} \\ 186 \\ \underline{186} \\ 0 \end{array}$$

Check: $62\times 83=5146$

2. Divide. $78\overline{)4139}$

Use 8 as a trial divisor, since 78 is closer to 80 than 70.

$\frac{41}{8}=5$ with 1 left over

$$\begin{array}{r} 5 \\ 78\overline{)4139} \\ \underline{390} \\ 23 \end{array}$$

$\leftarrow 5\times 78=390$

$\leftarrow 413-390=23$ (smaller than 78, the divisor)

Bring down the 9.

$$\begin{array}{r} 5 \\ 78\overline{)4139} \\ \underline{390} \\ 239 \end{array}$$

$\frac{23}{8}=2$ with 7 left over

$$\begin{array}{r} 52 \\ 78\overline{)4139} \\ \underline{390} \\ 239 \\ \underline{156} \\ 83 \end{array}$$

$\leftarrow 2\times 78=156$

$\leftarrow$ Greater than 78

The remainder 83 is greater than the divisor 78, so 3 should be used instead of 2.

2. Divide. $59\overline{)2297}$

____

$$\begin{array}{r} 53 \text{ R5} \\ 78\overline{)4139} \\ \underline{390} \\ 239 \\ \underline{234} \\ 5 \end{array} \quad \leftarrow 3\times 78 = 234$$

3. Divide. $27\overline{)8340}$

$$\begin{array}{r} 3 \\ 27\overline{)8340} \\ \underline{81} \\ 2 \end{array} \quad \begin{array}{l} \leftarrow 3\times 27 = 81 \\ \leftarrow 83 - 81 = 2 \end{array}$$

Bring down the 4.

$$\begin{array}{r} 3 \\ 27\overline{)8340} \\ \underline{81} \\ 24 \end{array}$$

Since 24 cannot be divided by 27, write 0 in the quotient as a placeholder.

$$\begin{array}{r} 30 \\ 27\overline{)8340} \\ \underline{81} \\ 24 \end{array}$$

Bring down the final digit, the 0.

$$\begin{array}{r} 30 \\ 27\overline{)8340} \\ \underline{81} \\ 240 \end{array}$$

Complete the problem.

$$\begin{array}{r} 308 \text{ R24} \\ 27\overline{)8340} \\ \underline{81} \\ 240 \\ \underline{216} \\ 24 \end{array}$$

The quotient is 308 R24.

3. Divide. $26\overline{)5454}$

____

Review these examples for Objective 2:

4. Divide:

a. $40 \div 10$

Drop one 0 in divisor and dividend.
$40 \div 10 = 4$

Now Try:

4. Divide:

a. $50 \div 10$

____

b. $6500 \div 100$

Drop two zeros in divisor and dividend.
$6500 \div 100 = 65$

c. $865{,}000 \div 1000$

Drop three zeros in divisor and dividend.
$865{,}000 \div 1000 = 865$

5. Divide:

a. $30\overline{)11{,}250}$

Drop one zero from the divisor and the dividend.

$$
\begin{array}{r}
375 \\
3\overline{)1125} \\
\underline{9} \\
22 \\
\underline{21} \\
15 \\
\underline{15} \\
0
\end{array}
$$

Since $1125 \div 3$ is 375, then $11{,}250 \div 30$ is 375.

b. $5400\overline{)37{,}800}$

Drop two zeros from the divisor and the dividend.

$$
\begin{array}{r}
7 \\
54\overline{)378} \\
\underline{378} \\
0
\end{array}
$$

Since $378 \div 54$ is 7, then $37{,}800 \div 5400$ is 7.

b. $4700 \div 100$

____

c. $606{,}000 \div 1000$

____

5. Divide:

a. $60\overline{)8100}$

____

b. $160\overline{)320{,}960}$

____

Review this example for Objective 3:

6. Check the quotient:

$$
\begin{array}{r}
34 \text{ R}120 \\
415\overline{)15{,}890}
\end{array}
$$

Multiply the quotient and the divisor. Then add the remainder.

$$
\begin{array}{r}
4\,1\,5 \\
\times \quad 3\,4 \\
\hline
1\,6\,6\,0 \\
1\,2\,4\,5 \\
\hline
1\,4\,1\,1\,0 \\
+ \quad 1\,2\,0 \\
\hline
1\,4{,}2\,3\,0
\end{array}
$$

The answer does not check. Rework the original problem to get 38 R120.

Now Try:

6. Check the quotient:

$$
\begin{array}{r}
36 \text{ R}62 \\
532\overline{)19{,}746}
\end{array}
$$

____

Name: Date:
Instructor: Section:

Objective 1 Use long division.

For extra help, see Examples 1–3 on pages 1213–1215 of your text and Section Lecture video for Section R.5.

Divide using long division. Check each answer.

1. $42\overline{)3234}$ **1.** ________________

2. $94\overline{)29,047}$ **2.** ________________

3. $657\overline{)429,700}$ **3.** ________________

Objective 2 Divide by multiples of 10.

For extra help, see Examples 4–5 on pages 1215–1216 of your text and Section Lecture video for Section R.5.

Divide.

4. $80\overline{)560}$ **4.** ________________

5. $800\overline{)10,400}$ **5.** ________________

6. $750\overline{)25,500}$ **6.** ________________

Name: Date:
Instructor: Section:

Objective 3 Use multiplication to check division answers.

For extra help, see Example 6 on page 1216 of your text and Section Lecture video for Section R.5.

Check each answer. If an answer is incorrect, give the correct answer.

7. $89\overline{)5790}$ with quotient 65 R5 **7.** ________________

8. $103\overline{)4658}$ with quotient 44 R22 **8.** ________________

9. $428\overline{)196,883}$ with quotient 400 R30 **9.** ________________

Chapter 1 INTRODUCTION TO ALGEBRA: INTEGERS

1.1 Place Value

Key Terms

1. digits
2. whole numbers
3. place value system

Now Try

1. 0, 5, 59, 350
2. hundred-thousands and hundreds

3a. nine million, seventy-five thousand, eight hundred sixty-two

3b. fifty-four trillion, eight hundred billion, five hundred forty-three million, seven hundred thousand, one hundred

4a. 753,006   4b. 11,000,010,025

Objective 1

1. 2
3. 1, 365

Objective 2

5. ten-millions

Objective 3

7. fifty-nine billion, five hundred four million, eight hundred six thousand, eight hundred seventy-three

9. 987, 000,330

1.2 Introduction to Integers

Key Terms

1. absolute value
2. number line
3. integers

Now Try

1a. –150 feet   1b. +68°F or 68°F   2. 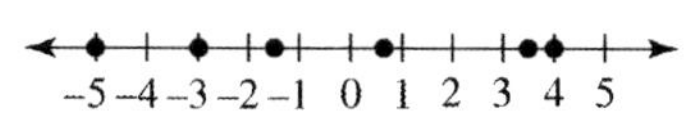

3a. 0 > –8   3b. –3 > –9   3c. 11 > –4

4a. 12   4b. 13   4c. 0

Objective 1

1. –46
3. –1,200,000

Objective 2

5. 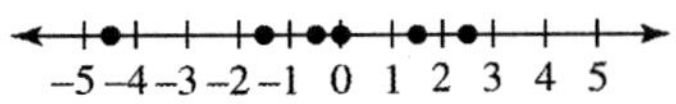

Objective 3

7. >

9. <

Objective 4

11. 21

1.3 Adding Integers

Key Terms

1. commutative property of addition
2. addends
3. associative property of addition
4. sum
5. addition property of 0

Now Try

1. –5

2a. –36

2b. 40

3a. –16

3b. 21

4. –3°

5a. $63+254=317$

5b. $-77+(-44)=-121$

6a. $-20+(19+(-19))=-20+0=-20$

6b. $8+2+(-17)=(8+2)+(-17)=10+(-17)=-7$

Objective 1

1. –3

3. 18 points

Objective 2

5. $8+(-15+(-5))=-12$

1.4 Subtracting Integers

Key Terms

1. opposite
2. additive inverse

Now Try

1a. $-25,\ 25+(-25)=0$

1b. $39,\ -39+39=0$

1c. $0,\ 0+0=0$

2a. $13+(-15)=-2$

2b. $-10+9=-1$

2c. $12+11=23$

2d. $-8+(-16)=-24$

3. –38

Objective 1

1. $11;\ -11+11=0$

3. $5;\ -5+5=0$

Objective 2

5. –14

Objective 3

7. 0

9. –22

1.5 Problem Solving: Rounding and Estimating

Key Terms

1. front end rounding
2. rounding
3. estimate

Now Try

1a. –$\underline{3}$2; –30

1b. \$$\underline{6}$42; \$600

1c. –7$\underline{6}$,542; –77,000

2. 700

3. 95,000

4a. –4740

4b. 57,000

5a. –80,000

5b. 246,000,000

6a. –800

6b. 1,000,000

7. \$2200; \$2412

Objective 1

1. 3

3. 4

Objective 2

5. 810,000

Objective 3

7. 200,000

9. 30°; 35°

1.6 Multiplying Integers

Key Terms

1. commutative property of multiplication
2. factors
3. associative property of multiplication
4. multiplication property of 0
5. product
6. distributive property
7. multiplication property of 1

Now Try

1a. $80 \cdot 6$ or 80(6) or (80)(6). The factors are 80 and 6. The product is 480.

1b. $8 \cdot 15$ or 8(15) or (8)(15). The factors are 8 and 15. The product is 120.

2a. –42 2b. 60 2c. –54

3a. –70 3b. –64

4a. 0; multiplication property of 0

4b. –92; multiplication property of 1

5a. $63 = 63$; commutative property of multiplication

5b. $-196 = -196$; associative property of multiplication

6a. $7(4) + 7(8)$; both results are 84

6b. $-9(-5) + (-9)(3)$; both results are 18

7. –1800; –1785

Objective 1

1. $-3 \cdot 6$; $(-3)(6)$
3. $4 \cdot 5$; $(4)(5)$

Objective 2

5. –62

Objective 3

7. 0
9. $3 \cdot ((-2) \cdot (-9)) = 54$

Objective 4

11. \$1300; \$1315

1.7 Dividing Integers

Key Terms

1. factors
2. quotient
3. product

Now Try

1a. –5
1b. 4
1c. –3

2a. 1 Any nonzero number divided by itself is 1.

2b. 93 Any number divided by 1 is the number.

2c. 0 Zero divided by any nonzero number is 0.

2d. undefined Division by 0 is undefined.

3a. 10

3b. –4

3c. –1

4. Est: $300; Exact: $309

5a. 6 packages with 2 hot dogs left without rolls, or 7 packages with 6 extra rolls

5b. 9 trips, 8 elevator trips would leave 6 people walking to the top

Objective 1

1. –7
3. –10

Objective 2

5. 1. Any nonzero number divided by itself is 1.

Objective 3

7. 2
9. –21

Objective 4

11. $23,000; $22,600

Objective 5

13. $96 \div 7 = 13$ R5. Every truck has at least 13 construction workers. The remainder of 5 means that 5 trucks each carry a 14th worker.

15. $1500 \div 468 = 3\,\text{R}\,96$. This means that each student receives three tickets, and there are 96 seats that are available for assignment.

1.8 Exponents and Order of Operations

Key Terms

1. order of operations
2. exponent

Now Try

1a. 5^4; 625; 5 to the fourth power

1b. 6^2; 36; 6 squared, or 6 to the second power

1c. 23^1; 23; 23 to the first power

2a. 49

2b. –1000

2c. 81

2d. –432

3a. –26

3b. 24

4. 13

5a. –24

5b. –42

6a. –17

6b. –53

7. –9

Objective 1

1. 12^3; twelve cubed
3. 7^7; seven to the seventh power

Objective 2

5. 144

Objective 3

7. 17
9. –16

Objective 4

11. 16

Chapter 2 UNDERSTANDING VARIABLES AND SOLVING EQUATIONS

2.1 Introduction to Variables

Key Terms

1. expression
2. variable
3. evaluate the expression
4. constant
5. coefficient

Now Try

1. Variable: c; constant: 5; $c + 5$

2a. 29; Order 29 lunches.

2b. 50; Order 50 lunches

3. 32 yards

4. 128

5a. 96

5b.

$10+21$ is 31	$7\cdot 6$ is 42
$-12+9$ is -3	$-9\cdot 5$ is -45
$93+0$ is 93	$31\cdot 0$ is 0

6. $\frac{0}{b}=0$

7a. $z\cdot z\cdot z\cdot z$

7b. $14\cdot r\cdot s\cdot s$

7c. $-13\cdot m\cdot m\cdot m\cdot n\cdot n\cdot n\cdot n\cdot n$

8a. 81

8b. 144

8c. –240

Objective 1

1. –7: constant; h: variable

3. 9: coefficient; k: variable; 1: constant

Objective 2

5. (a) \$10,040 (b) \$12,620

Objective 3

7. Multiplication is distributive over addition.

9. Zero added to any number equals that number.

Objective 4

11. $u\cdot u\cdot u\cdot v\cdot v\cdot w\cdot w$

2.2 Simplifying Expressions

Key Terms

1. like terms
2. term
3. simplify an expression
4. variable term
5. constant term

Now Try

1a. –8*a* and *a*; –8 and 1

1b. $21r^2s$ and $-5r^2s$; 21 and –5

1c. 26 and –2; they are constants.

2a. $20x$

2b. $-39z^2$

3a. $11pq+q$

3b. $7c-9$

4a. $108b$

4b. $-36x$

4c. $40y^2$

5a. $30x+12$

5b. $-54x-36$

5c. $28x-112$

6. $6x-6$

Objective 1

1. like terms: 2*x*, –3*x*; coefficients: 2, –3

3. $-19c^3z^4$

Objective 2

5. $-6s-5st-16t+8$

Objective 3

7. $-5t-20$

9. $20q-3$

2.3 Solving Equations Using Addition

Key Terms

1. solution
2. addition property of equality
3. equation
4. solve an equation
5. check the solution

Now Try

1. 29

2a. $c = 16$

2b. $x = -2$

3a. $x = -11$

3b. $b = 6$

Objective 1

1. 12

3. 7

Objective 2

5. $n = -66$

Objective 3

7. $a = 3$

9. $n = -44$

2.4 Solving Equations Using Division

Key Terms

1. division property of equality
2. addition property of equality

Now Try

1a. $s = 6$

1b. $w = -16$

2a. $y = 4$

2b. $h = 0$

3. $y = -13$

Objective 1

1. $m = -7$

3. $h = 28$

Objective 2

5. $m = 33$

Objective 3

7. $b = -19$

9. $v = -4$

2.5 Solving Equations with Several Steps

Key Terms

1. addition property of equality
2. distributive property
3. division property of equality

Now Try

1. $n = 7$
2. $x = -10$
3. $m = 0$
4. $m = -5$

Objective 1

1. $m = 0$
3. $q = -3$

Objective 2

5. $t = 6$

Chapter 3 SOLVING APPLICATION PROBLEMS

3.1 Problem Solving: Perimeter

Key Terms

1. triangle
2. perimeter
3. formula
4. square
5. rectangle
6. parallelogram

Now Try

1. 60 in.
2. 125 cm
3. 184 ft
4. 18 in.
5. 74 m
6. 70 mm
7. 306 in.

Objective 1

1. 28 ft
3. 19 mi

Objective 2

5. 13 ft

Objective 3

7. 22 in.
9. 255 cm

3.2 Problem Solving: Area

Key Terms

1. square
2. parallelogram
3. rectangle
4. area

Now Try

1a. 253 m^2
1b. 189 cm^2
2. 5 ft
3. 169 $in.^2$
4. 5 mi
5a. 171 cm^2
5b. 425 m^2
6. 9 ft
7. 1688 m; 169,984 m^2

Objective 1

1. 10 cm^2
3. 76 cm

Objective 2

5. 2025 cm^2

Objective 3

7. 208 m^2
9. 6 m

Objective 4

11. 8 ft

3.3 Solving Application Problems with One Unknown Quantity

Key Terms

1. sum; increased by
2. product; double
3. quotient; per
4. difference; less than

Now Try

1a. $x + 7$ or $7 + x$
1b. $43 + x$ or $x + 43$
1c. $x + 63$ or $63 + x$
1d. $-20 + x$ or $x + (-20)$
1e. $x + 75$ or $75 + x$
1f. $x - 44$
1g. $89 - x$
1h. $x - 89$
1i. $x - 16$
1j. $x - 36$
1k. $48 - x$
2a. $57x$

2b. $43x$ | 2c. $2x$ | 2d. $\frac{-13}{x}$

2e. $\frac{x}{19}$ | 2f. $6x - 81$ | 3. $7x + 21 = 56$; $x = 5$

4. 3 pounds | 5. 55 paper plates | 6. \$14.00

Objective 1

1. $-6x$
3. $1 + 3x$ or $3x + 1$

Objective 2

5. $3 + 7x = 31$; $x = 4$

Objective 3

7. 12 bananas
9. 7 celery sticks

3.4 Solving Application Problems with Two Unknown Quantities

Key Terms

1. added to; more than
2. times; triple
3. divided by; half
4. subtracted from; minus

Now Try

1. cat: \$55; dog: \$165
2. 38 m, 60 m
3. length: 31 in.; width: 12 in.

Objective 1

1. Greer: 8135 votes; Jones: 8651 votes
3. length: 36 yd; width: 12 yd

Chapter 4 RATIONAL NUMBERS: POSITIVE AND NEGATIVE FRACTIONS

4.1 Introduction to Signed Fractions

Key Terms

1. equivalent fractions
2. fraction
3. improper fraction
4. numerator
5. proper fraction
6. denominator

Now Try

1. shaded: $\frac{3}{7}$; unshaded: $\frac{4}{7}$

2. $\frac{9}{8}$

3a. Numerator: 5; denominator: 17; 17 equal parts in the whole

3b. Numerator: 6; denominator: 23; 23 equal parts in the whole

4a. $\frac{7}{8}, \frac{9}{11}, \frac{12}{29}, \frac{1}{5}$

4b. $\frac{22}{7}, \frac{13}{13}, \frac{17}{4}$

5. 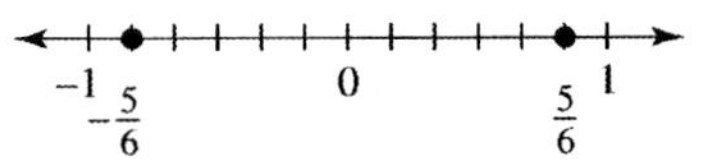

6. $\frac{9}{10}, \frac{9}{10}$

7a. $-\frac{24}{30}$

7b. $\frac{5}{6}$

8a. 1

8b. –6

8c. 13

Objective 1

1. $\frac{3}{5}; \frac{2}{5}$

3. $\frac{3}{5}; \frac{7}{5}$

Objective 2

5. 12; 5

Objective 3

7. 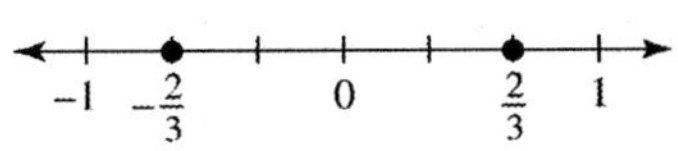

9. 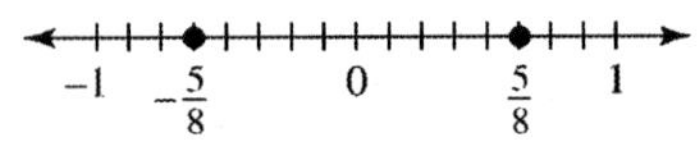

Objective 4

11. $\frac{8}{9}$

Objective 5

13. $-\frac{32}{48}$

15. $-\frac{1}{5}$

4.2 Writing Fractions in Lowest Terms

Key Terms

1. composite number
2. prime factorization
3. prime number
4. lowest terms

Now Try

1a. yes

1b. no; 9

2a. $\frac{5}{9}$

2b. $\frac{7}{8}$

2c. $-\frac{5}{13}$

2d. $\frac{11}{12}$

3. Prime: 29 and 31; composite: 8 and 21; neither: 1

4a. $2 \cdot 2 \cdot 2 \cdot 2 \cdot 2 \cdot 3$

4b. $2 \cdot 2 \cdot 2 \cdot 7$

5a. $2 \cdot 2 \cdot 2 \cdot 19$

5b. $2 \cdot 2 \cdot 2 \cdot 3 \cdot 7$

6a. $\frac{1}{3}$

6b. $\frac{3}{10}$

6c. $\frac{4}{5}$

7a. $\frac{5}{6x}$

7b. $\frac{1}{3}$

7c. $\frac{b^2}{4a}$

7d. $\frac{9p^2q}{r^3}$

Objective 1

1. yes

3. no; 4

Objective 2

5. $\frac{5}{7}$

Objective 3

7. $3 \cdot 5 \cdot 7$

9. $2 \cdot 2 \cdot 2 \cdot 2 \cdot 2 \cdot 2 \cdot 5$

Objective 4

11. $\frac{3 \cdot 3 \cdot 7}{3 \cdot 5 \cdot 7}; \frac{3}{5}$

Objective 5

13. $\frac{3r}{s^2}$

15. $\frac{4}{3x}$

4.3 Multiplying and Dividing Signed Fractions

Key Terms

1. reciprocals
2. indicator words
3. of
4. each

Now Try

1a. $\frac{40}{143}$

1b. $-\frac{45}{88}$

2a. $-\frac{2}{3}$

2b. $\frac{3}{20}$

3. 15

4a. $\frac{14}{15}$

4b. $\frac{2c}{15}$

5a. $\frac{1}{12}$

5b. -80

5c. 4

5d. undefined

5e. 0

6a. $\frac{5a}{b}$

6b. $\frac{10}{9b^2}$

7a. $\frac{21}{22}\text{ m}^2$

7b. 72 servings

Objective 1

1. $-\frac{1}{7}$

3. 45

Objective 2

5. $\frac{5}{2m}$

Objective 3

7. $-\frac{1}{24}$

9. $\frac{6}{7}$

Objective 4

11. $6x$

Objective 5

13. \$600

4.4 Adding and Subtracting Signed Fractions

Key Terms

1. unlike fractions
2. like fractions
3. least common denominator

Now Try

1a. $\frac{2}{3}$

1b. $\frac{1}{2}$

1c. $-\frac{3}{5}$

1d. $\frac{4}{a^2}$

2a. 27

2b. 30

3a. 100

3b. 90

4a. $\frac{17}{18}$

4b. $-\frac{1}{6}$

4c. $\frac{41}{72}$

4d. $\frac{61}{7}$

5a. $\frac{10+3x}{15}$

5b. $\frac{5x-64}{8x}$

Objective 1

1. $-\frac{2}{3}$

3. $\frac{4}{y^2}$

Objective 2

5. 63

Objective 3

7. $\frac{3}{10}$

9. $\frac{8}{15}$

Objective 4

11. $\frac{1+2x}{6}$

4.5 Problem Solving: Mixed Numbers and Estimating

Key Terms

1. mixed number
2. improper fraction

Now Try

1. 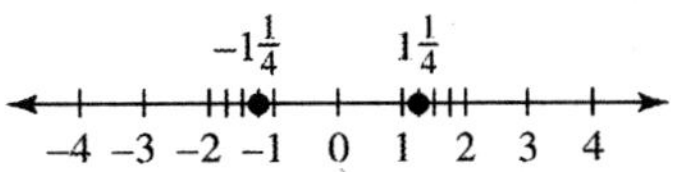

2. $\frac{35}{6}$

3a. $3\frac{3}{8}$

3b. $4\frac{1}{9}$

4a. 7

4b. 4

5a. 20; $17\frac{1}{2}$

5b. 36; 35

6a. $\frac{4}{7}$; $\frac{2}{3}$

6b. $\frac{7}{11}$; $\frac{5}{8}$

7a. 6; $6\frac{5}{9}$

7b. 3; $3\frac{7}{12}$

7c. 5; $4\frac{6}{7}$

8a. 6 oz; $7\frac{25}{48}$ oz

8b. 10 ft; $10\frac{3}{40}$ ft

Objective 1

1. 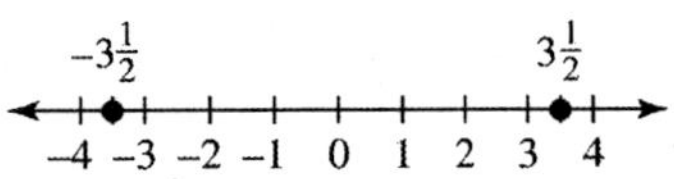

3. 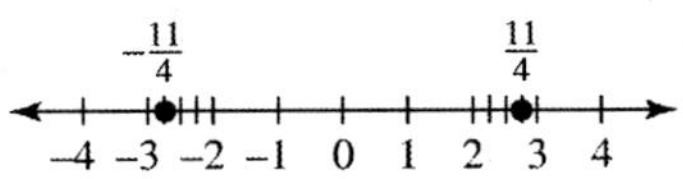

Objective 2

5. $-\frac{16}{9}$

Objective 3

7. 4; $4\frac{4}{21}$

9. $2\frac{1}{2}$; $2\frac{2}{9}$

Objective 4

11. 4; $3\frac{5}{9}$

Objective 5

13. 12 m^2; $12\frac{1}{2}$ m^2

4.6 Exponents, Order of Operations, and Complex Fractions

Key Terms

1. order of operations
2. exponent
3. complex fraction

Now Try

1a. $-\frac{1}{64}$

1b. $\frac{5}{28}$

2a. $\frac{11}{30}$

2b. $-\frac{77}{81}$

3a. $\frac{20}{21}$

3b. $\frac{1}{25}$

Objective 1

1. $-\frac{1}{9}$

3. $\frac{1}{36}$

Objective 2

5. $\frac{3}{4}$

Objective 3

7. –8

9. $\frac{25}{36}$

4.7 Problem Solving: Equations Containing Fractions

Key Terms

1. division property of equality
2. multiplication property of equality

Now Try

1a. 21

1b. –30

1c. $\frac{2}{5}$

2a. 30

2b. –21

3. 80 in. or 6 ft 8 in.

Objective 1

1. $b=-36$

3. $h=\frac{3}{16}$

Objective 2

5. $r=-4$

Objective 3

7. 20 years old

9. 30 years old

4.8 Geometry Applications: Area and Volume

Key Terms

1. area
2. volume

Now Try

1a. $\frac{77}{128}$ in.^{2}

1b. 232.56 cm^2

2. 534 m^2

3a. 48 m^3

3b. $62\frac{1}{2}$ in.^{3}

4. 25.8 ft^3

Objective 1

1. $P = 66$ m; $A = 200$ m^2

3. 1940 yd^2

Objective 2

5. $166\frac{3}{8}$ ft^3

Objective 3

7. 196 ft^3

9. $139\frac{1}{2}$ m^3

Chapter 5 RATIONAL NUMBERS: POSITIVE AND NEGATIVE DECIMALS

5.1 Reading and Writing Decimals

Key Terms

1. decimals
2. place value
3. decimal point

Now Try

1a. 0.6

1b. 0.05

1c. 0.37

1d. 0.007

1e. 0.049

1f. 0.518

2a. 8 hundreds; 6 tens; 2 ones; 9 tenths; 3 hundredths

2b. 0 ones; 0 tenths; 0 hundredths; 7 thousandths; 6 ten-thousandths; 9 hundred-thousandths

3a. nine tenths

3b. fifty-three hundredths

3c. seven hundredths

3d. five hundred two thousandths

3e. four hundred sixty-nine ten-thousandths

4a. four and seven tenths

4b. eighteen and nine thousandths

4c. eight-two ten-thousandths

4d. fifty-seven and nine hundred six thousandths

5a. $\frac{27}{100}$

5b. $\frac{303}{1000}$

5c. $3\frac{2636}{10,000}$

6a. $\frac{1}{5}$

6b. $\frac{7}{20}$

6c. $6\frac{1}{25}$

6d. $562\frac{101}{2500}$

Objective 1

1. $\frac{8}{10}$; 0.8; eight tenths

3. $\frac{58}{100}$; 0.58; fifty-eight hundredths

Objective 2

5. 7; 1

Objective 3

7. eight hundredths

9. ninety seven and eight thousandths

Objective 4

11. $3\frac{3}{5}$

5.2 Rounding Decimal Numbers

Key Terms

1. decimal places
2. rounding

Now Try

1. 43.803
2a. 0.80
2b. 7.78
2c. 22.040
2d. 0.6
3a. $2.08
3b. $425.10
4a. $38
4b. $307
4c. $881
4d. $6860
4e. $1

Objective 1

1. up

Objective 2

3. 489.8
5. 989.990

Objective 3

7. $11,840

5.3 Adding and Subtracting Signed Decimal Numbers

Key Terms

1. front end rounding
2. estimating

Now Try

1a. 15.630
1b. 21.709
2a. 36.844
2b. 1013.931
3a. 11.612
3b. 10.57
4a. 4.654
4b. 3.26
4c. 0.501
5a. –38.4
5b. –10.586
6a. –11.74
6b. 3.31
6c. 16.54
7a. 14; 13.847
7b. $50; $41.55
7c. –11; –11.097

Objective 1

1. 92.49
3. 72.453

Objective 2

5. 178

Objective 3

7. 680; 676.60
9. –15; –14.562

5.4 Multiplying Signed Decimal Numbers

Key Terms

1. factor
2. decimal places
3. product

Now Try

1. –10.793
2. 0.00186
3. 100; 112.8229

Objective 1

1. –90.71
3. 0.0037

Objective 2

5. 600; 756.6478

5.5 Dividing Signed Decimal Numbers

Key Terms

1. dividend
2. repeating decimal
3. quotient
4. divisor

Now Try

1a. –1.413

1b. 15.03

2. 20.178

3. 16.791

4a. 20,410

4b. 1.78

5. 40; 37.8

6a. 12.18

6b. 22.62

6c. –0.233

Objective 1

1. 6.96
3. 2.359

Objective 2

5. 128.25

Objective 3

7. reasonable
9. reasonable

Objective 4

11. 53.24

5.6 Fractions and Decimals

Key Terms

1. equivalent
2. numerator
3. denominator
4. mixed number

Now Try

1a. 0.0625
1b. 4.375
2. 0.417
3a. >
3b. =
3c. <
3d. =
4a. 0.3, 0.3057, 0.307
4b. $\frac{1}{5}, \frac{2}{9}$, 0.23, $\frac{3}{13}$

Objective 1

1. 0.125
3. 19.708

Objective 2

5. $\frac{3}{11}$, 0.29, $\frac{1}{3}$

5.7 Problem Solving with Statistics: Mean, Median, and Mode

Key Terms

1. mode
2. weighted mean
3. mean
4. median

Now Try

1. 272
2. 52.7
3. 40.2
4. 2.5
5. 18
6. 0.06

7a. 3
7b. 13 and 16
7c. no mode

Objective 1

1. 60.3
3. 273.1

Objective 2

5. 4.7

Objective 3

7. 232
9. 239.5

Objective 4

11. 24, 35, 39

5.8 Geometry Applications: Pythagorean Theorem and Square Roots

Key Terms

1. right triangle
2. hypotenuse
3. legs

Now Try

1a. 4.47
1b. 9.59
1c. 11.58
2a. 9.8 in.
2b. 12 km
3. 9.5 ft

Objective 1

1. 4.123
3. 10.100

Objective 2

5. 2.2 cm

Objective 3

7. 10.3 ft
9. 8 ft

5.9 Problem Solving: Equations Containing Decimals

Key Terms

1. division property of equality
2. addition property of equality

Now Try

1a. $w = -13.9$
1b. $x = 19.6$
2a. $x = 3.05$
2b. $t = -8.5$
3a. $b = -1.6$
3b. $x = 0.53$
4. 32 min

Objective 1

1. $n = -4.6$
3. $h = 7.7$

Objective 2

5. $r = 2.5$

Objective 3

7. $y = -15$
9. $x = 29.4$

Objective 4

11. 30 mg

5.10 Geometry Applications: Circles, Cylinders, and Surface Area

Key Terms

1. radius
2. circumference
3. circle
4. π (pi)
5. surface area
6. diameter

Now Try

1a. 86 m

1b. 29.5 in.

2a. 138.2 yd

2b. 23.2 m

3a. 43.0 m^2

3b. 1519.8 yd^2

4. 127.2 ft^2

5. \$12.56

6. \$1.96

7a. 490.6 cm^3

7b. 678.2 cm^3

8. $V = 7956$ mm^3; $S = 2670$ mm^2

9. $V \approx 113.0$ ft^3; $S \approx 138.2$ ft^2

Objective 1

1. 4 ft
3. $6\frac{1}{4}$ yd

Objective 2

5. 28.3 yd

Objective 3

7. 22.3 yd^2
9. 1061.3 m^2

Objective 4

11. 6358.5 cm^3

Objective 5

13. 558 in.^{2}
15. 85.7 in.^{2}

Objective 6

17. 183.6 ft^2

Chapter 6 RATIO, PROPORTION, AND LINE/ANGLE/TRIANGLE RELATIONSHIPS

6.1 Ratios

Key Terms

1. ratio
2. numerator; denominator

Now Try

1a. $\frac{17}{8}$

1b. $\frac{8}{5}$

2a. $\frac{2}{3}$

2b. $\frac{2}{1}$

2c. $\frac{2}{3}$

3. $\frac{3}{8}$

4a. $\frac{19}{9}$

4b. $\frac{22}{17}$

5. $\frac{5}{7}$

Objective 1

1. $\frac{25}{19}$

3. $\frac{1}{4}$

Objective 2

5. $\frac{9}{2}$

Objective 3

7. $\frac{2}{7}$

9. $\frac{5}{6}$

6.2 Rates

Key Terms

1. unit rate
2. cost per unit
3. rate

Now Try

1a. $\frac{\text{1 dollar}}{\text{5 pages}}$

1b. $\frac{\text{15 strokes}}{\text{1 minute}}$

1c. $\frac{\text{33 strawberries}}{\text{2 cakes}}$

2a. 28 miles/gallon

2b. $1.26/pound

2c. $145/day

3. 5 pints at $1.70/pint

4a. eight-pack

4b. Brand T

Objective 1

1. $\frac{\text{7 pills}}{\text{1 patient}}$

3. $\frac{\text{32 pages}}{\text{1 chapter}}$

Objective 2

5. $225/pound

Objective 3

7. 24 ounces for $2.08
9. 5 cans for $2.75

6.3 Proportions

Key Terms

1. proportion
2. cross products
3. ratio

Now Try

1a. $\frac{24}{17}=\frac{72}{51}$

1b. $\frac{\$10}{7\text{ cans}}=\frac{\$60}{42\text{ cans}}$

2a. $\frac{9}{7}\neq\frac{4}{3}$, false

2b. $\frac{1}{3}=\frac{1}{3}$, true

3a. $306=306$, true

3b. $32\neq 35$, false

4a. 32

4b. 10.67

5a. $2\frac{3}{5}$

5b. 6.4

Objective 1

1. $\frac{50}{8}=\frac{75}{12}$

3. $\frac{3}{33}=\frac{12}{132}$

Objective 2

5. $\frac{6}{5}=\frac{6}{5}$; true

Objective 3

7. 36

9. 21

6.4 Problem Solving with Proportions

Key Terms

1. ratio
2. rate

Now Try

1. $\frac{23\text{ hr}}{4\text{ apt}}=\frac{x\text{ hr}}{16\text{ apt}}$; $x=92$ hr

2. $\frac{4}{5}=\frac{x}{540}$; $x=432$ people

Objective 1

1. \$108

3. approximately 333 deer

6.5 Geometry: Lines and Angles

Key Terms

1. ray
2. perpendicular lines
3. obtuse angle
4. point
5. angle
6. line
7. acute angle
8. degrees
9. parallel lines
10. line segment
11. intersecting lines
12. right angle
13. vertical angles
14. congruent angles
15. supplementary angles
16. complementary angles
17. alternate interior angles
18. corresponding angles
19. straight angle

Now Try

1a. line segment, $\overline{EF}$
1b. ray, $\overrightarrow{CD}$
1c. line, $\overleftrightarrow{RS}$

2a. intersecting
2b. parallel
3. $\angle VSW$ or $\angle WSV$

4a. straight
4b. obtuse
4c. acute

4d. right
5a. intersecting
5b. perpendicular

6. $\angle SRT$ and $\angle TRU$; $\angle URV$ and $\angle VRW$
7a. 18°

7b. 78°

8. $\angle OQP$ and $\angle PQN$; $\angle RST$ and $\angle BMC$; $\angle OQP$ and $\angle BMC$, $\angle PQN$ and $\angle RST$

9a. 98°
9b. 12°

10. $\angle RPS \cong \angle QPT$ and $\angle QPR \cong \angle TPS$

11. $\angle MKN$ and $\angle QKP$; $\angle MKQ$ and $\angle NKP$
12a. 33°

12b. 105°
12c. 42°
12d. 42°

13a. $\angle 5$ and $\angle 6$; $\angle 1$ and $\angle 2$; $\angle 7$ and $\angle 8$; $\angle 3$ and $\angle 4$

13b. $\angle 3$ and $\angle 6$; $\angle 7$ and $\angle 2$

14. $m\angle 1 = 55°$; $m\angle 2 = 55°$; $m\angle 3 = 125°$; $m\angle 4 = 125°$
$m\angle 5 = 125°$; $m\angle 6 = 125°$; $m\angle 7 = 55°$; $m\angle 8 = 55°$

Objective 1

1. ray, $\overrightarrow{CB}$
3. line segment, $\overline{KL}$

Objective 2

5. parallel

Objective 3

7. $\angle COD$
9. $\angle MON$

Objective 4

11. obtuse

Objective 5

13. perpendicular
15. intersecting

Objective 6

17. 164°

Objective 7

19. $\angle CAD \cong \angle BAE$
21. 73° and 107°

Objective 8

23. corresponding angles: $\angle 1$ and $\angle 5$; $\angle 2$ and $\angle 6$; $\angle 3$ and $\angle 7$; $\angle 4$ and $\angle 8$
    alternate interior angles: $\angle 4$ and $\angle 6$; $\angle 3$ and $\angle 5$
    $m\angle 1 = 37°$; $m\angle 2 = 143°$; $m\angle 3 = 37°$; $m\angle 4 = 143°$
    $m\angle 5 = 37°$; $m\angle 6 = 143°$; $m\angle 7 = 37°$; $m\angle 8 = 143°$

6.6 Geometry Applications: Congruent and Similar Triangles

Key Terms

1. congruent triangles
2. congruent figures
3. similar triangles
4. similar figures

Now Try

1. $\angle 1$ and $\angle 5$; $\angle 2$ and $\angle 4$; $\angle 3$ and $\angle 6$; $\overline{ST}$ and $\overline{VX}$; $\overline{RS}$ and $\overline{WV}$; $\overline{RT}$ and $\overline{WX}$

2a. SAS
2b. ASA
2c. SSS

3. $\overline{AB}$ and $\overline{PQ}$; $\overline{AC}$ and $\overline{PR}$; $\overline{BC}$ and $\overline{QR}$;
   $\angle A$ and $\angle P$; $\angle B$ and $\angle Q$; $\angle C$ and $\angle R$

4. 12.75 yd
5. 24
6. 21 ft

Objective 1

1. $\angle 1$ and $\angle 4$; $\angle 2$ and $\angle 5$; $\angle 3$ and $\angle 6$; $\overline{DF}$ and $\overline{XZ}$; $\overline{DE}$ and $\overline{XY}$; $\overline{EF}$ and $\overline{YZ}$

Objective 2

3. ASA

Objective 3

5. $\overline{PN}$ and $\overline{SR}$; $\overline{NM}$ and $\overline{RQ}$; $\overline{MP}$ and $\overline{QS}$;
   $\angle P$ and $\angle S$; $\angle N$ and $\angle R$; $\angle M$ and $\angle Q$

7. $\overline{HK}$ and $\overline{RS}$; $\overline{GH}$ and $\overline{TR}$; $\overline{GK}$ and $\overline{TS}$;
   $\angle H$ and $\angle R$; $\angle G$ and $\angle T$; $\angle K$ and $\angle S$

Objective 4

9. $m = 90$; $r = 21$

Objective 5

11. 36 m
13. 10 m

Chapter 7 PERCENTS

7.1 The Basics of Percent

Key Terms

1. ratio
2. percent
3. decimals

Now Try

1a. 10%
1b. 18%
1c. 85%
2a. 0.23
2b. 0.08
2c. 0.153
2d. 5.00
2e. 3.02
3a. 0.48
3b. 2.40 or 2.4
3c. 0.025
3d. 0.008
4a. 43%
4b. 230%
4c. 75.1%
5a. $\frac{3}{10}$
5b. $\frac{39}{50}$
5c. $1\frac{3}{4}$
6a. $\frac{109}{250}$
6b. $\frac{2}{9}$
7a. 15%
7b. $93\frac{3}{4}\%$
7c. $5\frac{5}{9}\%$ exact, or 5.6% rounded
8a. $6.15
8b. $23\frac{1}{2}$ days
9a. $3.65
9b. $9\frac{1}{2}$ pages

Objective 1

1. 68%
3. 45%

Objective 2

5. 3.10

Objective 3

7. 20%
9. 493%

Objective 4

11. $\frac{11}{60}$

Objective 5

13. 94%
15. 85.3%

Objective 6

17. $520

7.2 The Percent Proportion

Key Terms

1. whole
2. part
3. percent proportion

Now Try

1a. 59
1b. 7.75
1c. unknown
2a. 650
2b. unknown
2c. 3280
3a. unknown
3b. 950
3c. 2650
4. 2470
5. 4%
6. 150 cars
7a. 220%
7b. $94

Objective 1

1. 250%; unknown; $50
3. $7\frac{3}{4}\%$; $895; unknown

Objective 2

5. 141.0%

7.3 The Percent Equation

Key Terms

1. percent
2. percent equation

Now Try

1a. $50
1b. 15 hr
1c. 25 pounds
2a. $93.20
2b. 69.3 miles
3a. $5.49
3b. 0.307 mile
4a. $275.90
4b. 108 miles
5a. 220%
5b. 33%
6a. 300 tons
6b. 78

Objective 1

1. 25 minutes
3. $2000

Objective 2

5. 492 televisions

Objective 3

7. 75%
9. 400 magazines

7.4 Problem Solving with Percent

Key Terms

1. percent equation
2. percent of increase or decrease
3. percent proportion

Now Try

1. 66 points
2. 294 drivers
3. 23.3%
4. 165%
5. 77 micrograms
6. 125%
7. 28.3%

Objective 1

1. 782 members
3. 17%

Objective 2

5. 28.6%

7.5 Consumer Applications: Sales Tax, Tips, Discounts, and Simple Interest

Key Terms

1. interest rate
2. interest
3. interest formula
4. simple interest
5. principal
6. sales tax
7. discount
8. tax rate

Now Try

1. Tax: $6.23; Total: $95.23
2. 12%

3a. Est: $3; Exact: $3.56

3b. Est: $4; Exact: $4.75

4. $9.98
5. $4; $84
6. $124; $744
7. $29.75; $869.75

Objective 1

1. $22.75; $372.75
3. 5%

Objective 2

5. $9.00; $9.58; $12.00; $12.77

Objective 3

7. $30; $170
9. $36.18

Objective 4

11. $195; $975

Chapter 8 MEASUREMENT

8.1 Problem Solving with U.S. Measurement Units

Key Terms

1. metric system
2. unit fractions
3. U.S. measurement units

Now Try

1a. 1 lb

1b. 4 qt

2a. 210 inches

2b. 56 ounces

2c. 1.25 or $1\frac{1}{4}$ minutes

2d. $6.3\overline{3}$ or $6\frac{1}{3}$ hours

3a. 432 inches

3b. $\frac{1}{3}$ ft

4a. 21,120 feet

4b. $1\frac{1}{2}$ tons

4c. 13 quarts

5a. $\frac{3}{4}$ gal

5b. 7200 minutes

6a. $6.36 per pound

6b. 3.5 or $3\frac{1}{2}$ tons

Objective 1

1. 2000
3. 8

Objective 2

5. 5 gallons

Objective 3

7. 3.5 or $3\frac{1}{2}$ gallons
9. 3.75 or $3\frac{3}{4}$ pounds

Objective 4

11. $26\frac{1}{4}$ qt

8.2 The Metric System—Length

Key Terms

1. prefix	2. metric conversion line	3. meter

Now Try

1a. mm	1b. km	1c. cm
2a. 5.4 km	2b. 76 mm	3a. 675 mm
3b. 0.9865 km	3c. 431 cm	4a. 23,000 mm
4b. 0.042 m	4c. 0.95 km	

Objective 1

1. m
3. km

Objective 2

5. 0.45 km

Objective 3

7. 19.4 mm
9. 0.000035 cm

8.3 The Metric System—Capacity and Weight (Mass)

Key Terms

1. gram	2. liter

Now Try

1a. 5 mL	1b. 100 L	2a. 0.973 L
2b. 3850 mL	3a. 48 kg	3b. 590 g
3c. 3 mg	4a. 3720 mg	4b. 0.084 kg
5a. 125 mL	5b. 1.19 kg	5c. 400 m

Objective 1

1. mL
3. mL

Objective 2

5. 836,000 L

Objective 3

7. mg
9. g

Objective 4

11. 760 g

Objective 5

13. L
15. cm

8.4 Problem Solving with Metric Measurement

Key Terms

1. gram
2. meter
3. liter

Now Try

1. 1.54 kg
2. 200 g
3. 5600 mL

Objective 1

1. $12.99
3. 12 pills

8.5 Metric–U.S. Measurement Conversions and Temperature

Key Terms

1. Celsius
2. Fahrenheit

Now Try

1. 42.3 ft

2a. 21.3 lb

2b. 75.8 L

3a. 65°C

3b. 4°C

4. 50°C

5. 302°F

Objective 1

1. 468.5 km
3. 737.1 g

Objective 2

5. 37°C

Objective 3

7. 17°C
9. 204°C

Chapter 9 GRAPHS AND GRAPHING

9.1 Problem Solving with Tables and Pictographs

Key Terms

1. pictograph
2. table

Now Try

1a. 93%
1b. Southwest and United
1c. 80%
2a. $64.50
2b. $77.50
3a. 45 million
3b. 45 million

Objective 1

1. 177 calories
3. 486 calories

Objective 2

5. 3 million people

9.2 Reading and Constructing Circle Graphs

Key Terms

1. circle graph
2. protractor

1. $\frac{3}{58}$
2. $\frac{5}{9}$
3. \$142,500

4a. mysteries: 108°; biographies: 54°; cookbooks: 36°; romance novels: 90°; science: 54°; business: 18°

4b.

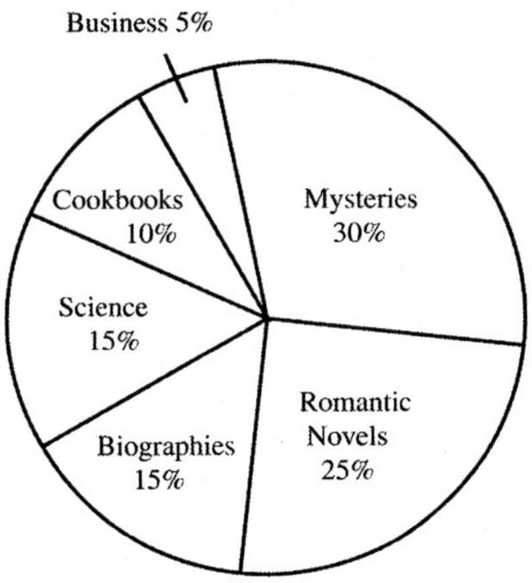

Objective 1

1. \$10,400
3. \$400

Objective 2

5. $\frac{1000}{1800}$ or $\frac{5}{9}$

Objective 3

7. parts: 5%; hand tools: 20%; bench tools: 25%; brass fittings: 35%; hardware: 15%

9.

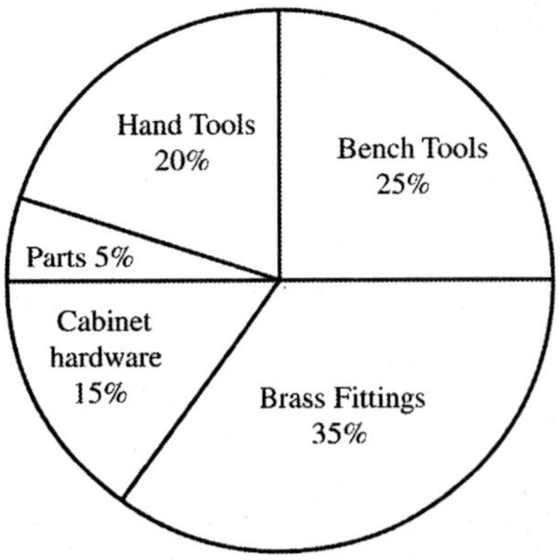

9.3 Bar Graphs and Line Graphs

Key Terms

1. line graph
2. double-bar graph
3. bar graph
4. comparison line graph

Now Try

1. 1800 students

2a. 250 female seniors

2b. 500 male freshmen

3a. $60

3b. December

4a. $1,500,000

4b. $4,000,000

4c. $1,000,000; 33%

Objective 1

1. 1600 students
3. 400 students

Objective 2

5. sophomores

Objective 3

7. September
9. $30

Objective 4

11. $2,000,000

9.4 The Rectangular Coordinate System

Key Terms

1. ordered pair
2. horizontal axis; x-axis
3. vertical axis; y-axis
4. paired data
5. coordinate system
6. quadrants
7. origin
8. coordinates

Now Try

1a. and 1b.

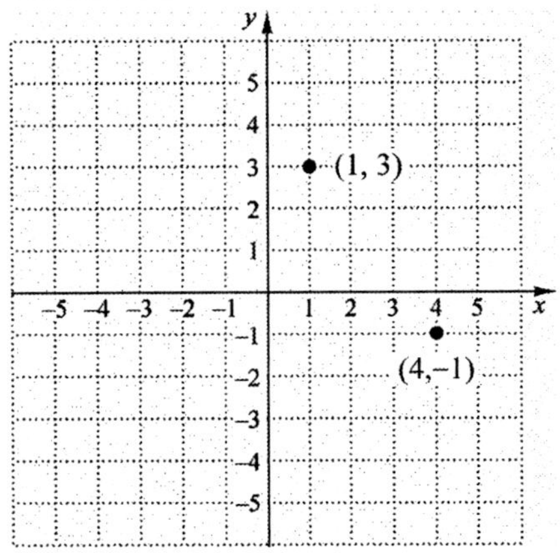

2a. and 2b.

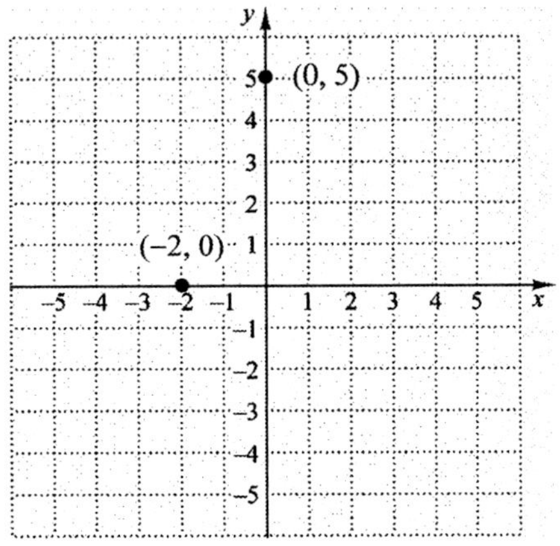

3a. (–4, –4)

3b. (1, –3)

3c. (2, 4)

3d. (0,–5)

4a. Quadrant I

4b. Quadrant II

4c. not in any quadrant

Objective 1

1.

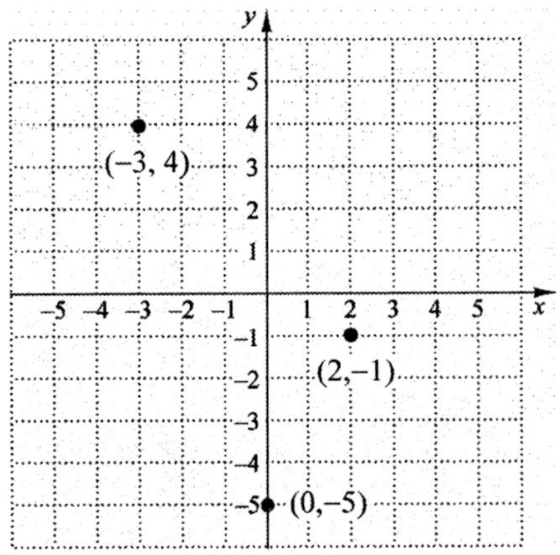

Objective 2

3. Quadrant IV
5. none

9.5 Introduction to Graphing Linear Equations

Key Terms

1. graph a linear equation
2. slope

Now Try

1.

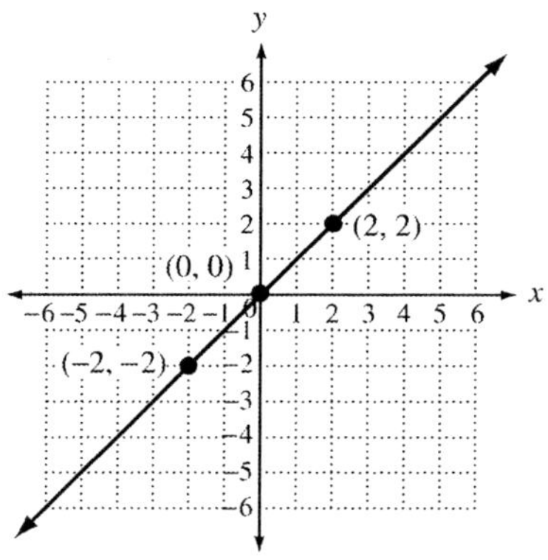

2.

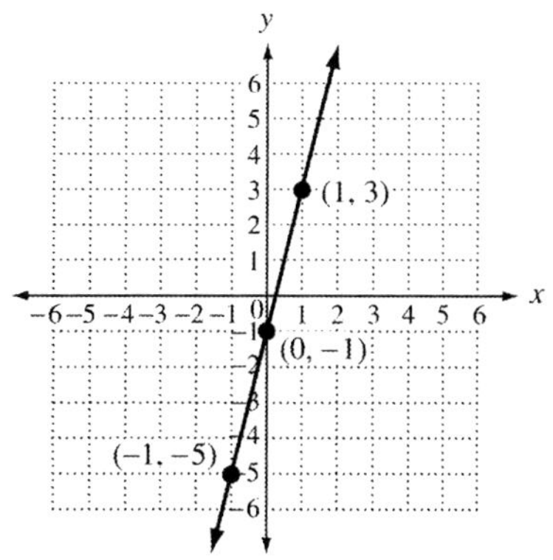

3.

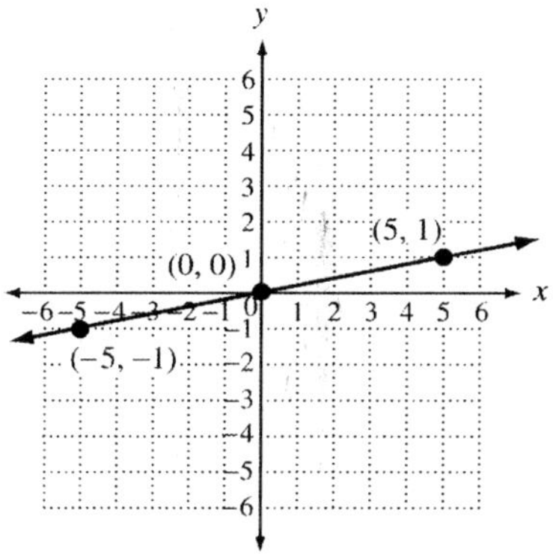

4.

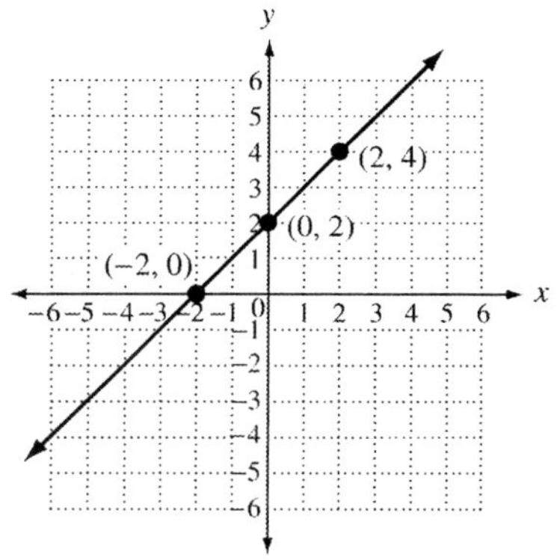

5. positive; increases

Objective 1

1.

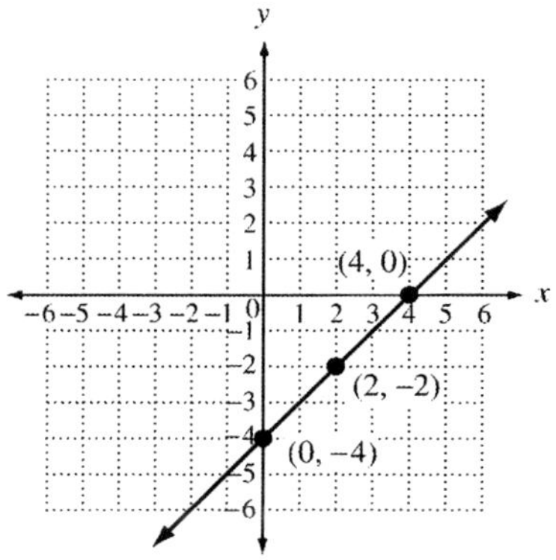

Objective 2

3. positive

Chapter 10 REAL NUMBERS, EQUATIONS, AND INEQUALITIES

10.1 Real Numbers and Expressions

Key Terms

1. whole numbers
2. integers
3. natural numbers
4. real numbers
5. irrational number
6. rational number

Now Try

1. 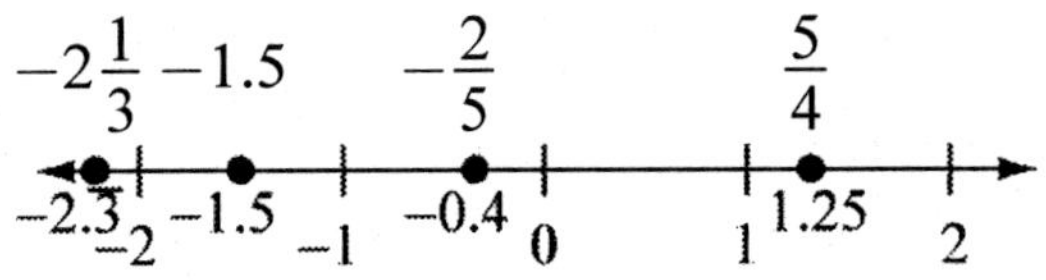

2a. Rational

2b. Rational

2c. Irrational

2d. Irrational

2e. Rational

2f. Rational

3a. True

3b. False

3c. True

3d. True

4a. $\frac{2}{7} < \frac{2}{5}$

4b. $-3 > -4$

4c. $-5 < 0$

4d. $11 > 8$

5. 124

6a. $-56y - 16$

6b. $-27a + 36$

7a. $-11x - 3$

7b. $15r + 2$

8a. $27y + 24$

8b. $-9 + 18r$

8c. $18a^2 - 20a + 6$

Objective 1

1. Rational; repeating decimal
3. Irrational; decimal value is nonrepeating, nonterminating

Objective 2

5. False

Objective 3

7. $\frac{2}{3} < \frac{3}{4}$
9. $\frac{3}{7} \geq \frac{3}{8}$

Objective 4

11. 48

Objective 5

13. $-d + 7$
15. $47n + 8$

10.2 More on Solving Linear Equations

Key Terms

1. identity
2. distributive property

Now Try

1. $x = -8$
2. $a = \frac{7}{3}$
3. $t = 6$
4. All real numbers
5. No solution
6. $x = 2$
7. $t = -9$

Objective 1

1. –1
3. $-\frac{5}{2}$

Objective 2

5. no solution

Objective 3

7. 2
9. –3

10.3 Formulas and Solving for a Specified Variable

Key Terms

1. formula
2. perimeter
3. area

Now Try

1a. $W = 4$

1b. $y_1 = -7$

2. $t = 1.5$ years

3. $b = \frac{2A}{h}$

4a. $y_1 = mx_1 - mx_2 + y_2$ or $y_1 = m(x_1 - x_2) + y_2$

4b. $n = \frac{S}{180} + 2$ or $n = \frac{S + 360}{180}$

Objective 1

1. $a = 36$
3. 12 ft

Objective 2

5. $r = \frac{A - P}{At}$, or $r = \frac{1}{t} - \frac{P}{At}$

10.4 Solving Linear Inequalities

Key Terms

1. linear inequality
2. inequalities

Now Try

1a. 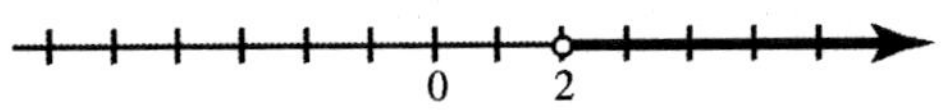

1b. 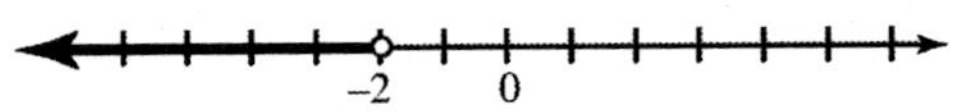

2. 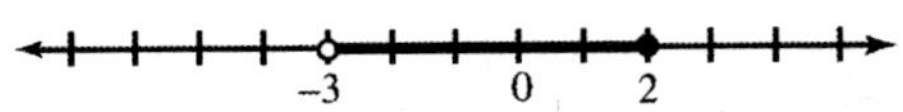

3. 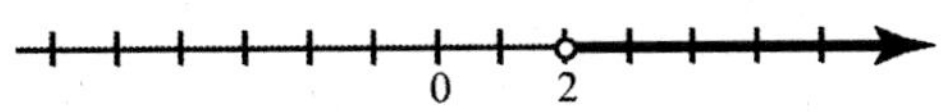

4. 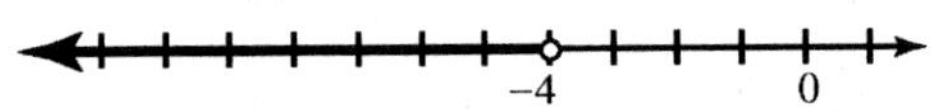

5. 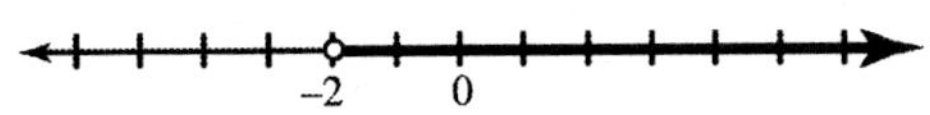

6. 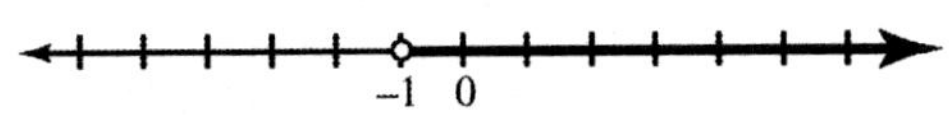

7. at least 55 students

Objective 1

1. 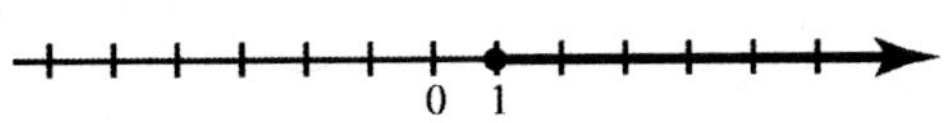

3. 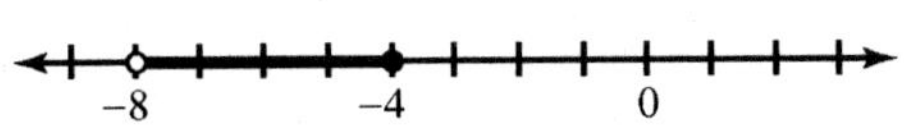

Objective 2

5. 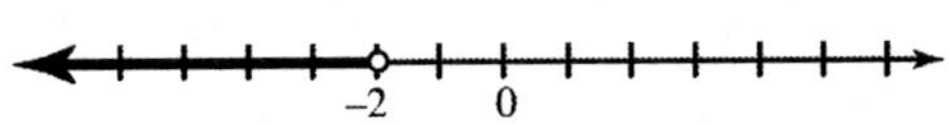

Objective 3

7. 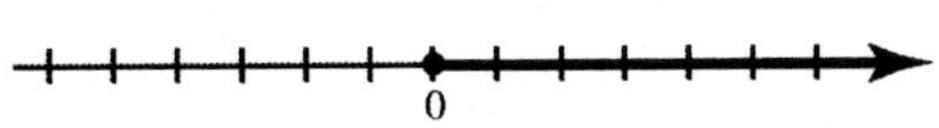

9. 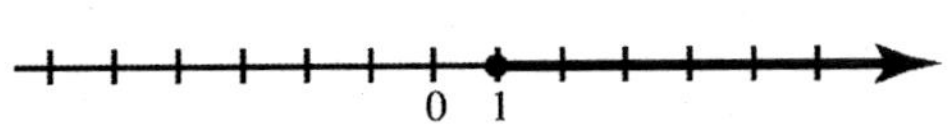

Objective 4

11. 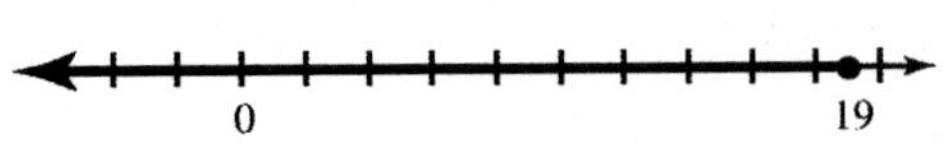

Objective 5

13. $38.84

15. 65 calories

Chapter 11 GRAPHS OF LINEAR EQUATIONS AND INEQUALITIES IN TWO VARIABLES

11.1 Linear Equations in Two Variables; The Rectangular Coordinate System

Key Terms

1. line graph
2. linear equation in two variables
3. bar graph
4. coordinates
5. x-axis
6. y-axis
7. ordered pair
8. rectangular (Cartesian) coordinate system
9. quadrants
10. origin
11. plot
12. scatter diagram
13. table of values
14. plane

Now Try

1a. 2002

1b. 1400

1c. Enrollment decreased by about 200 students from 2004 to 2005.

2a. 2000-2001, 2002-2003, 2004-2005

2b. 2001-2002

2c. 300 degrees

3a. yes

3b. no

4a. (5, 13)

4b. (–2, –15)

5a. (1, –4), (5, 12), (2, 0), (3, 4)

x	y
1	−4
5	12
2	0
3	4

5b. (6, –5), (6, 0), (6, 3)

x	y
6	−5
6	0
6	3

6.

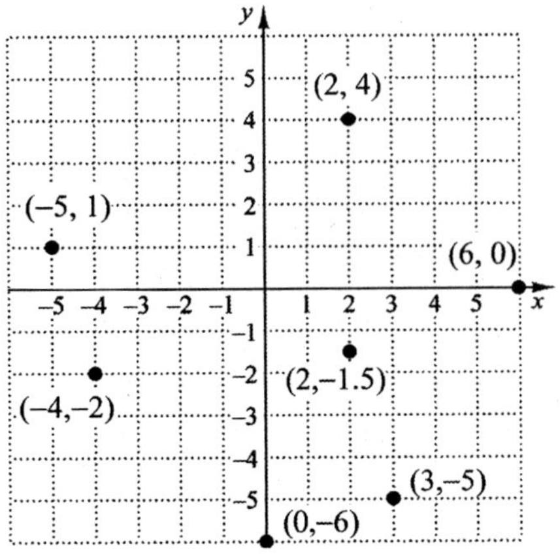

7a.

x (months)	y (balance due)
2	10,153.48
5	10,565.20
11	11,388.64

7b.

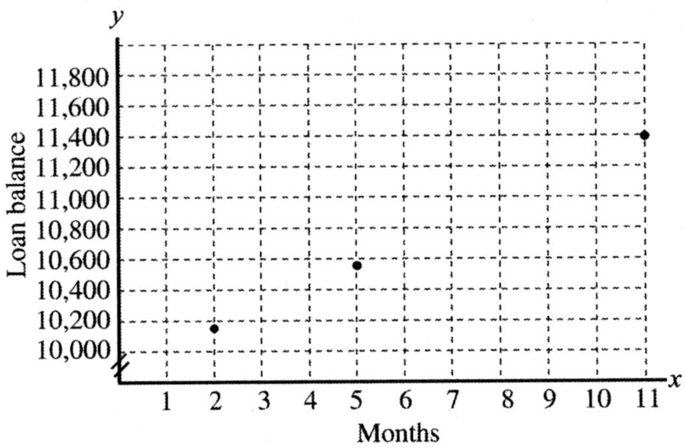

Objective 1

1. 2002, 2003, 2005   3. 325 MBA degrees

Objective 2

5. $\left(0, \frac{1}{3}\right)$

Objective 3

7. not a solution   9. a solution

Objective 4

11. (a) $(-4, -5)$; (b) $(2, 7)$; (c) $\left(-\frac{3}{2}, 0\right)$; (d) $(-2, -1)$; (e) $(-5, -7)$

Objective 5

13. $(-4, 4), (0, 4), (6, 4)$

Objective 6

15. 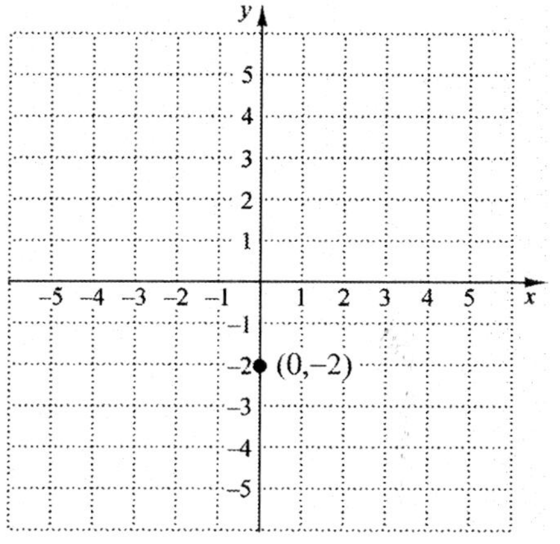

17. 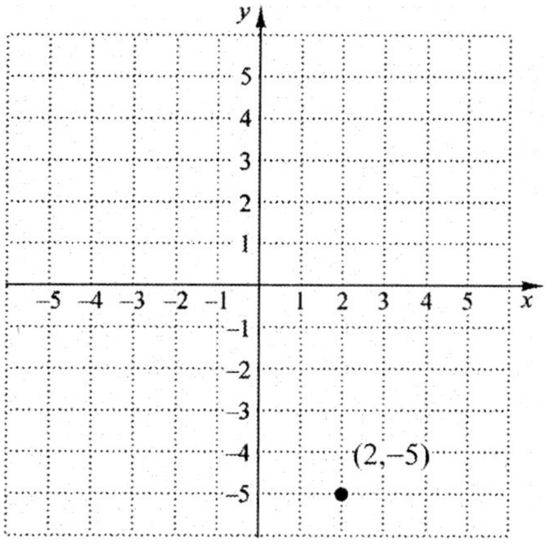

11.2 Graphing Linear Equations in Two Variables

Key Terms

1. y-intercept
2. x-intercept
3. graphing
4. graph

Now Try

1. 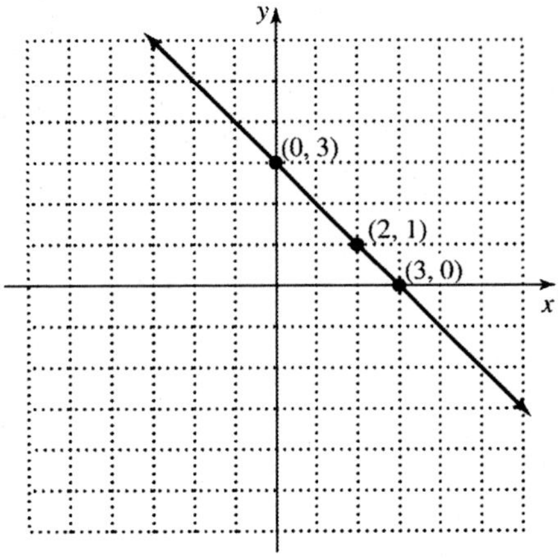

2. 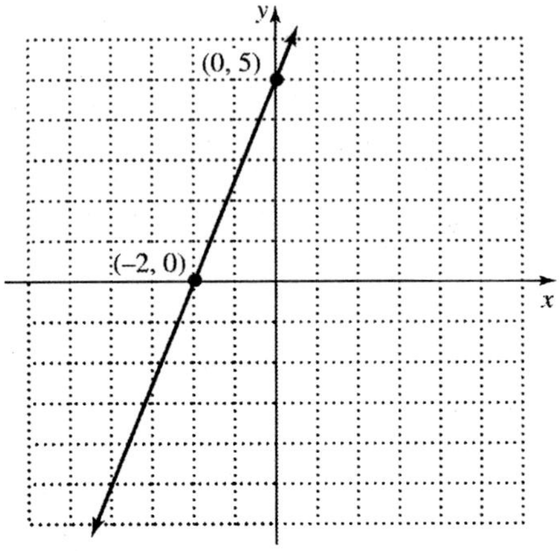

3. 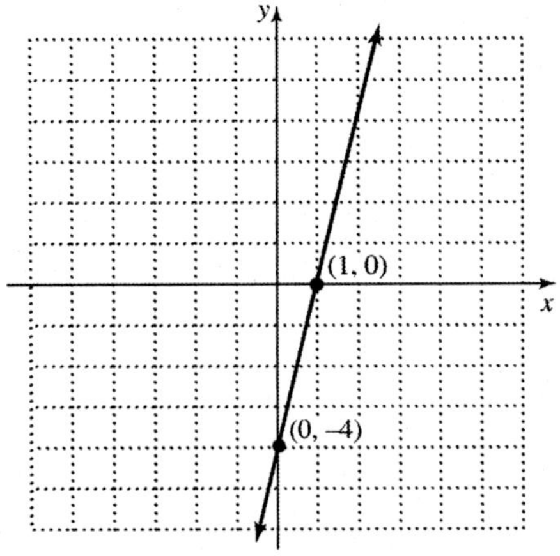

4. 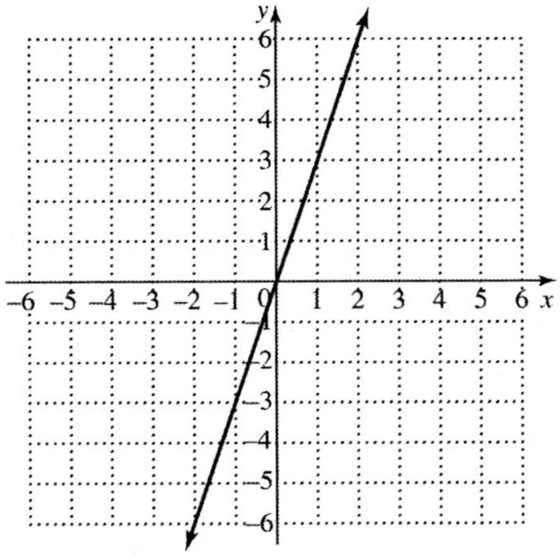

5. 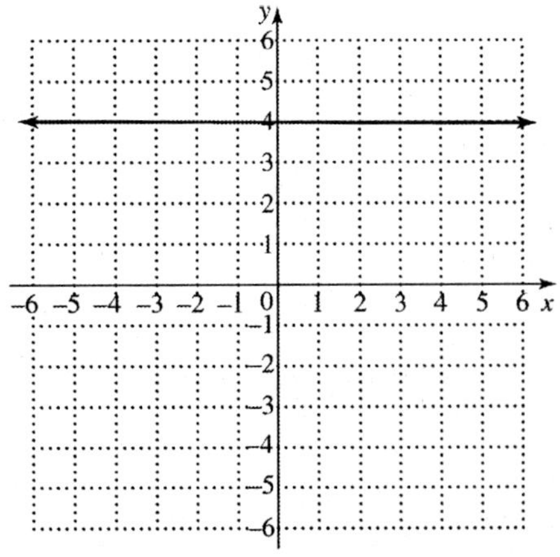

6. 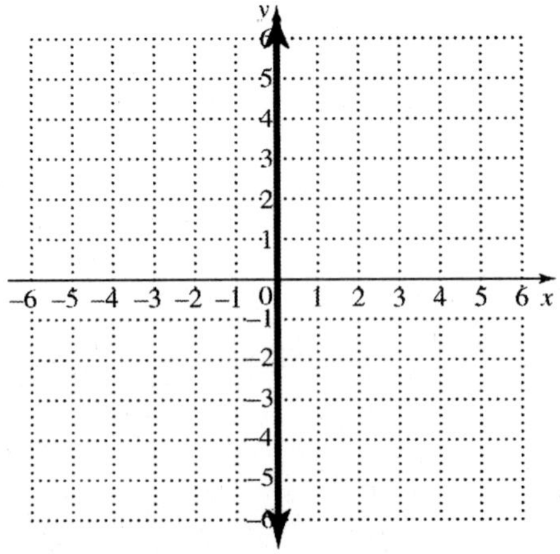

7a. 0 calculators, \$45
5000 calculators, \$42
20,000 calculators, \$33
45,000 calculators, \$18

7b. (0, \$45), (5, \$42), (20, \$33), (45, \$18)

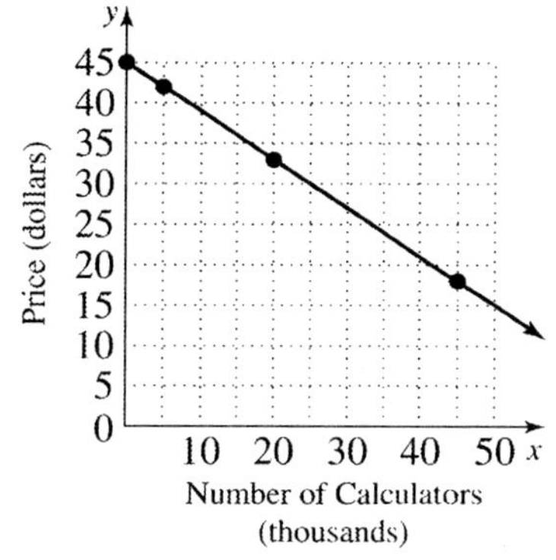

7c. 30,000 calculators, \$27

Objective 1

1. $(0,-2),\left(\frac{2}{3},0\right),(2,4)$

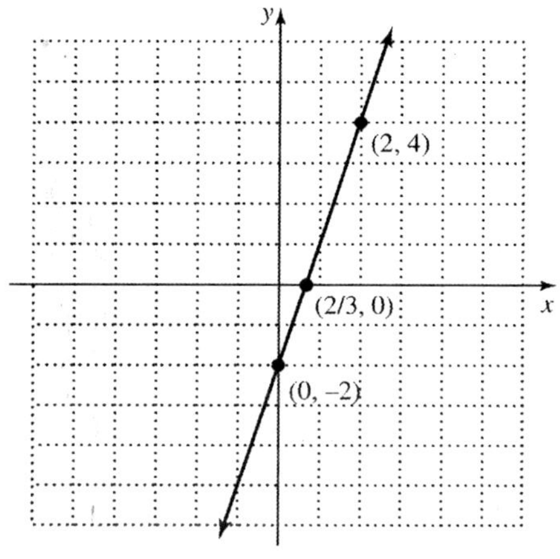

3. $\left(0,-\frac{1}{2}\right),(1,0),(-3,-2)$

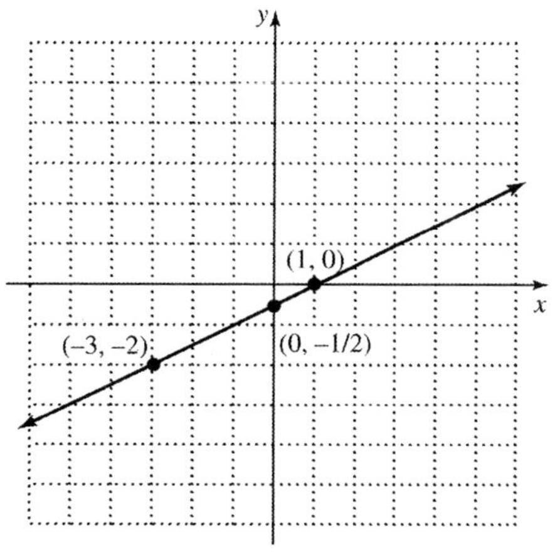

Objective 2

5.

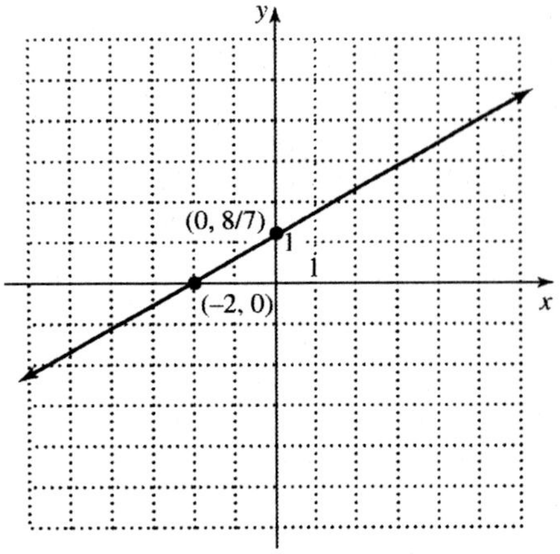

Objective 3

7. $x + y = 0$

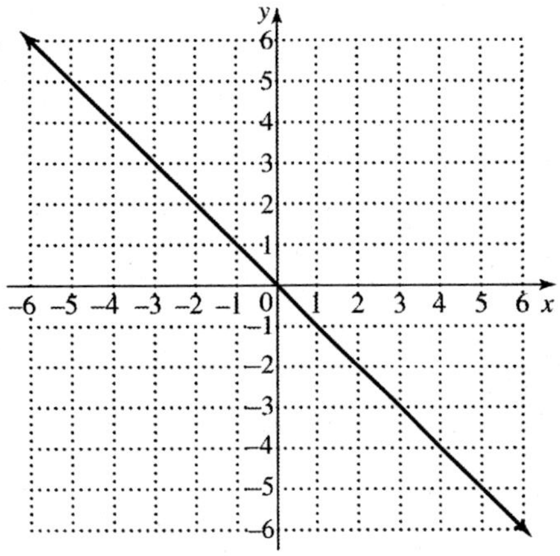

Objective 4

9. $x - 1 = 0$

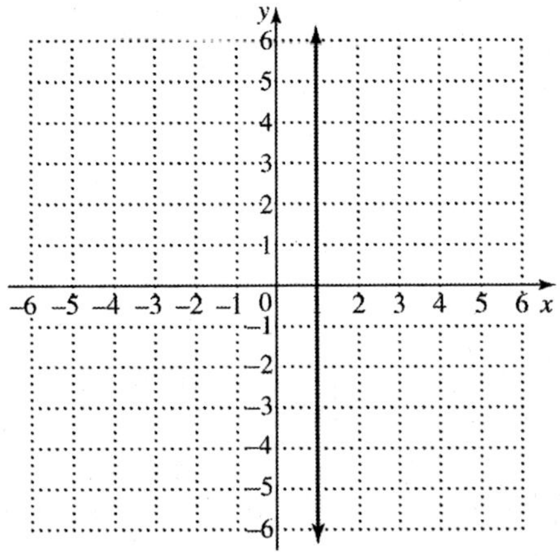

Objective 5

11. 2004, 4.9 million;
    2005, 5.53 million;
    2006, 6.16 million;
    2007, 6.79 million

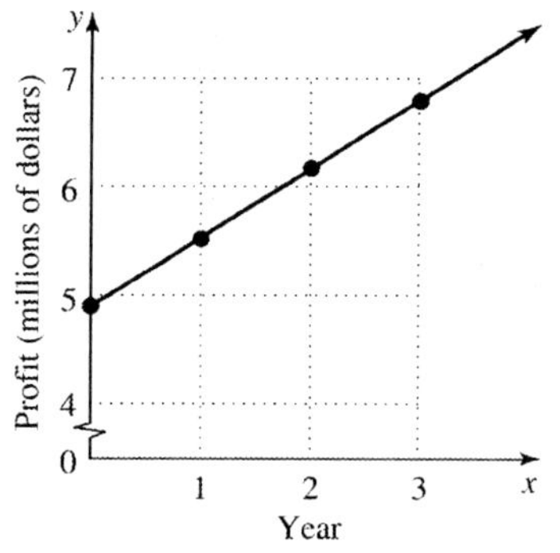

13. 48, 20°C;
    54, 22°C;
    60, 24°C;
    66, 26°C

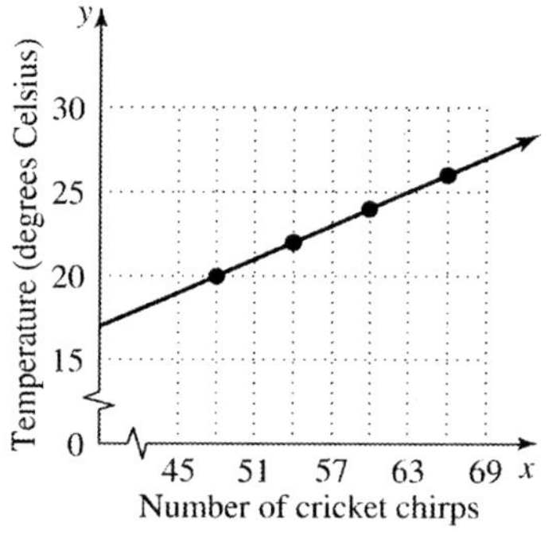

11.3 Slope of a Line

Key Terms

1. perpendicular lines
2. slope
3. rise
4. parallel lines
5. run

Now Try

1. $\frac{4}{1}$, or 4

2a. $-\frac{16}{9}$

2b. $\frac{9}{17}$

3. 0

4. undefined slope

5a. $\frac{7}{4}$

5b. $-\frac{3}{2}$

5c. $-\frac{1}{8}$

6a. perpendicular

6b. parallel

6c. neither

6d. neither

Objective 1

1. –2
3. 0

Objective 2

5. $\frac{2}{3}$

Objective 3

7. 1; 1; parallel
9. -3; $\frac{1}{3}$; perpendicular

11.4 Writing and Graphing Equations of Lines

Key Terms

1. point-slope form
2. standard form
3. slope-intercept form

Now Try

1a. slope: –12; y-intercept: (0, 6)

1b. slope: –1; y-intercept (0, –9)

1c. slope: 23; y-intercept (0, 0)

1d. slope: $-\frac{1}{7}$; y-intercept: $\left(0, -\frac{7}{5}\right)$

2. $y = \frac{7}{9}x + 8$

3a. 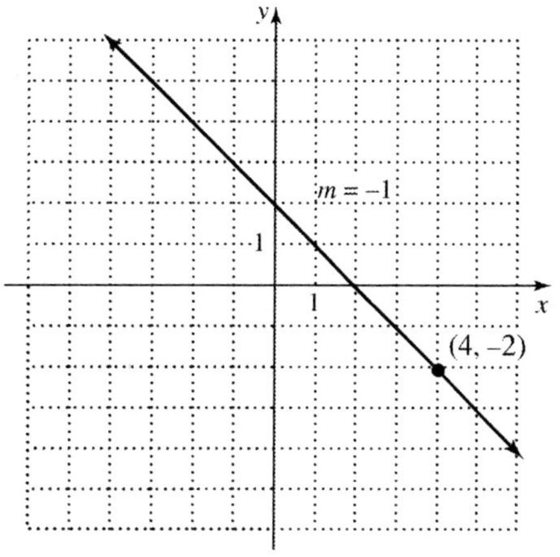

3b. 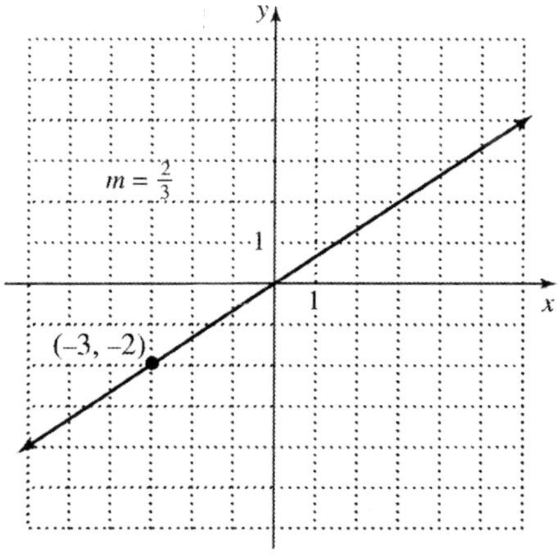

4. 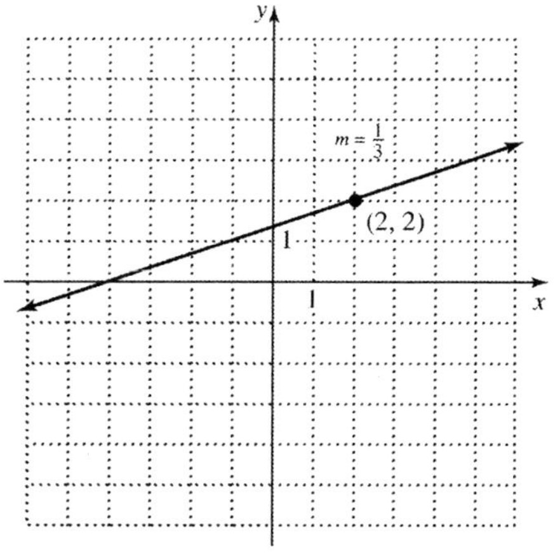

5. $y = 6x + 10$

6a. $y = \frac{1}{4}x - 9$

6b. $y = -\frac{2}{3}x + 15$

7. $y = -\frac{3}{4}x + \frac{81}{4}$; $3x + 4y = 81$

8.

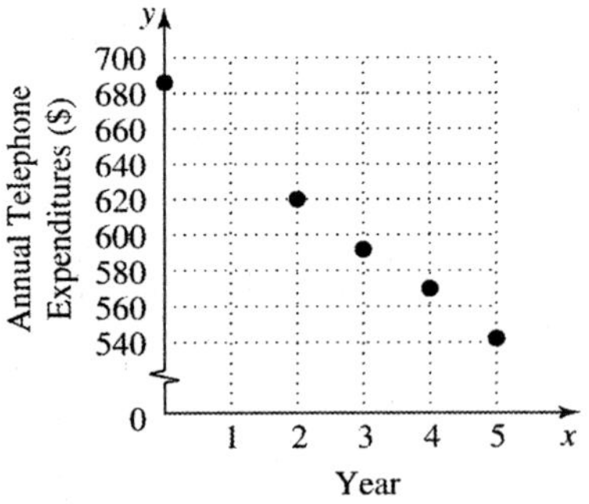

$y = -\frac{144}{5}x + 686$

Objective 1

1. slope: $\frac{3}{2}$; y-intercept: $\left(0, -\frac{2}{3}\right)$

3. $y = -3x + 3$

Objective 2

5.

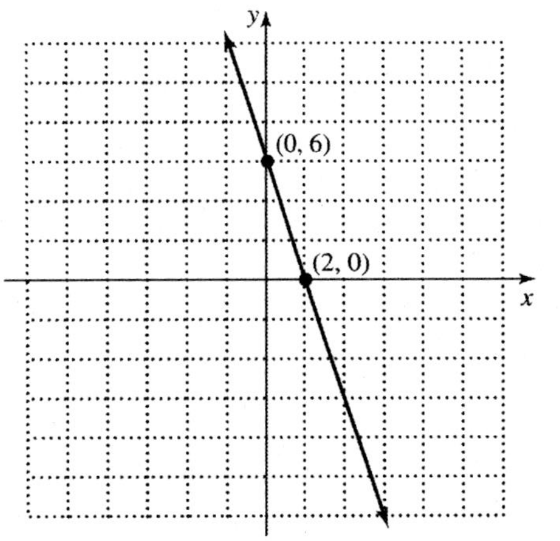

Objective 3

7. $y = -\frac{3}{5}x + \frac{11}{5}$

9. $y = -\frac{3}{2}x + 5$

Objective 4

11. $3x - 2y = 0$

Objective 5

13.

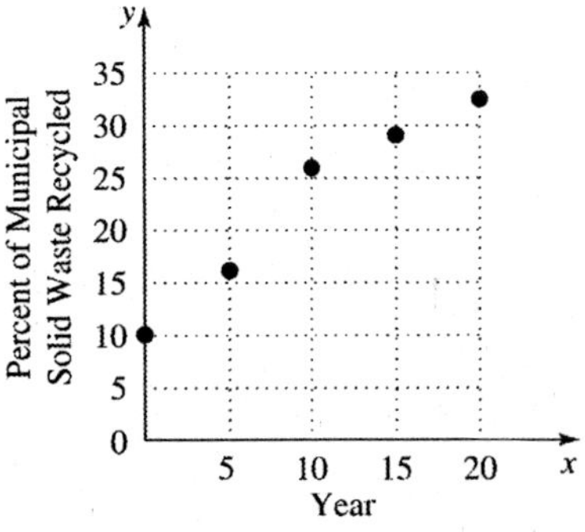

$y = \frac{28}{25}x + 10.1$

11.5 Graphing Linear Inequalities in Two Variables

Key Terms

1. boundary line
2. linear inequality in two variables

Now Try

1.

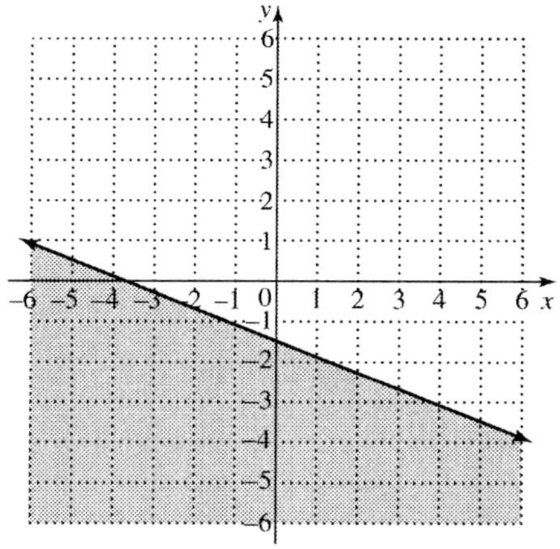

2.

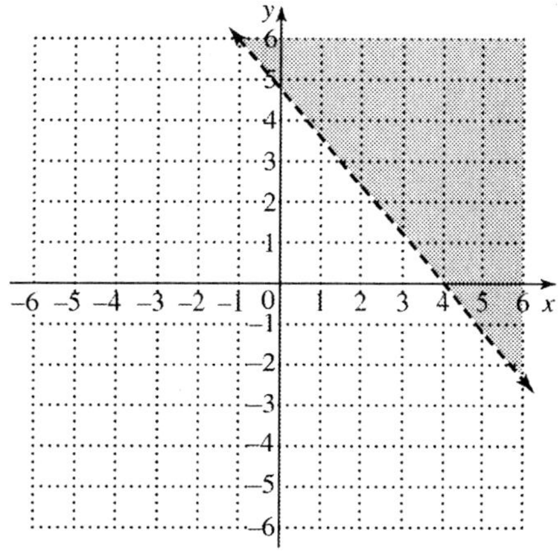

3.

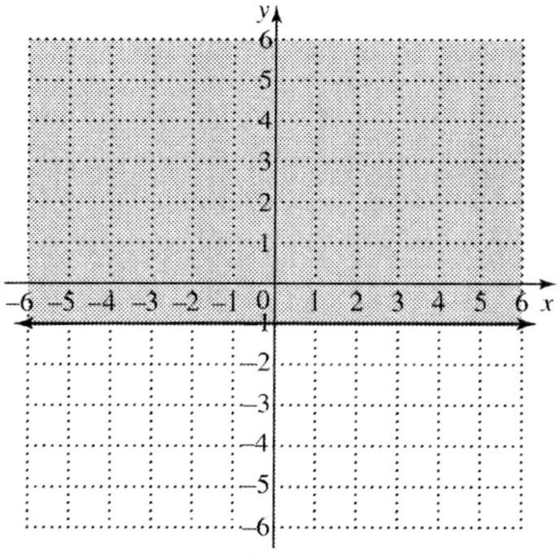

4.

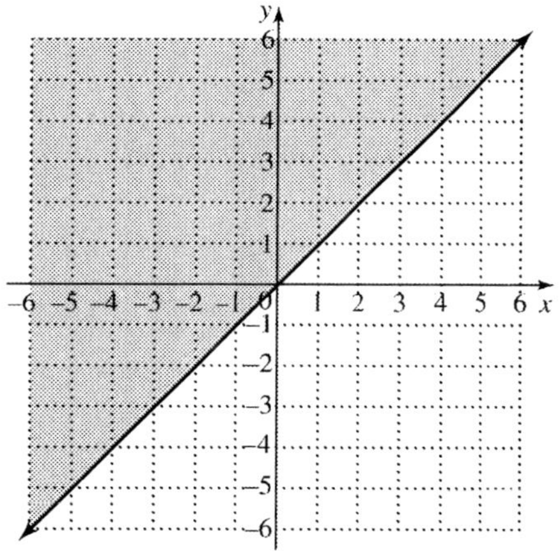

Objective 1

1.

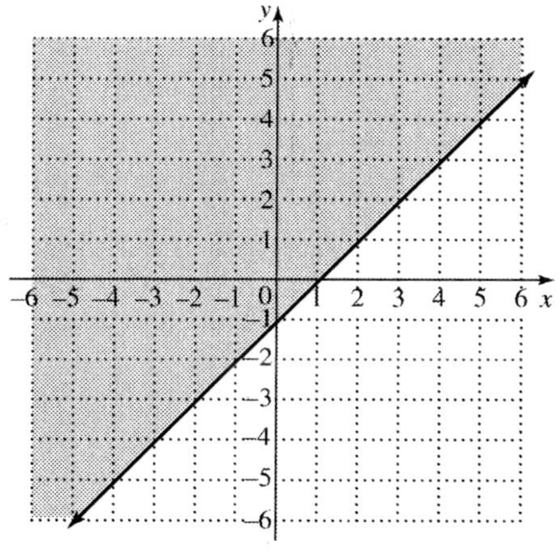

3.

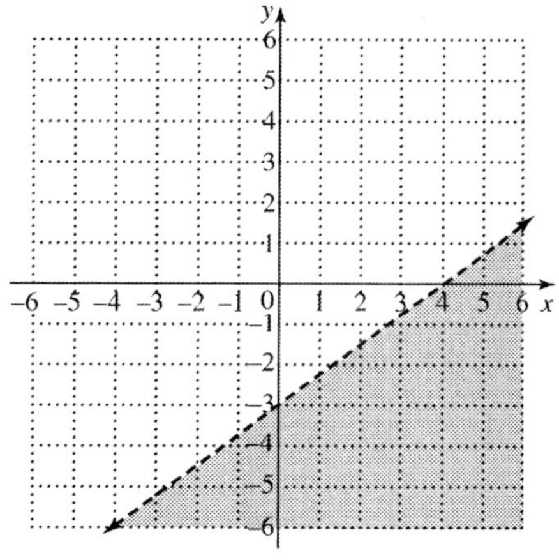

Objective 2

5.

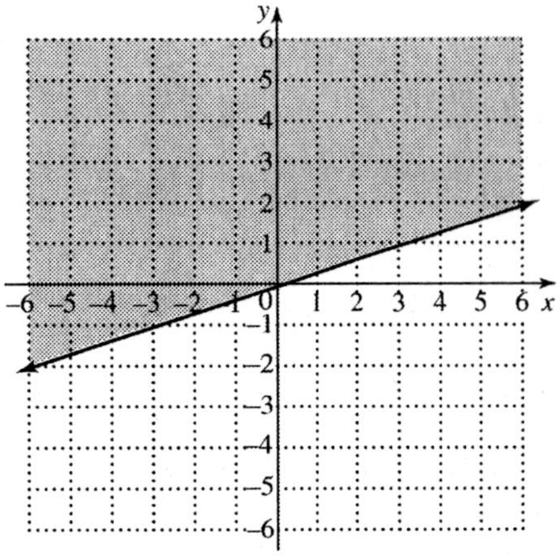

Chapter 12 SYSTEMS OF LINEAR EQUATIONS AND INEQUALITIES

12.1 Solving Systems of Linear Equations by Graphing

Key Terms

1. independent equations
2. consistent system
3. solution set of the system
4. solution of the system
5. dependent equations
6. inconsistent system
7. system of linear equations

Now Try

1a. yes

1b. no

2.

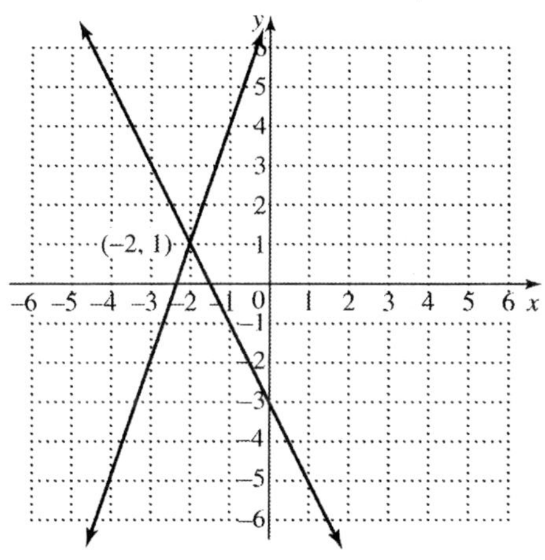

3a. $\{(x, y) | 4x - 2y = 8\}$

3b. $\varnothing$

4a. parallel lines; no solution

4b. same line; infinitely many solutions

4c. neither parallel nor same line; exactly one solution

Objective 1

1. no
3. no

Objective 2

5. 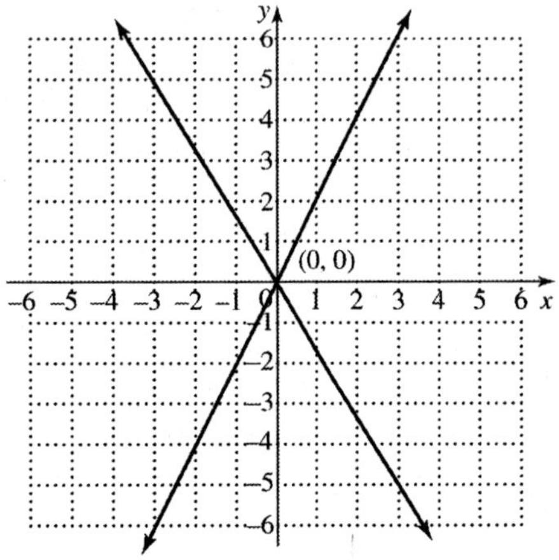

Objective 3

7. no solution

Objective 4

9. (a) neither (b) intersecting lines (c) one solution

11. (a) dependent (b) one line (c) infinitely many solutions

12.2 Solving Systems of Linear Equations by Substitution

Key Terms

1. ordered pair
2. substitution
3. dependent system
4. inconsistent system

Now Try

1. $\{(1, 6)\}$
2. $\{(9, -4)\}$
3. $\{(8, -2)\}$
4. $\varnothing$
5. $\{(x, y) | 5x + 4y = 20\}$
6. $\{(-2, 5)\}$
7. $\{(3, -5)\}$

Objective 1

1. $(2, 4)$
3. $(4, -9)$

Objective 2

5. $\{(x, y) \mid x - 2y = -6\}$

Objective 3

7. $(-9, -11)$
9. $\{(x, y) \mid 0.3x + 0.4y = 0.5\}$

12.3 Solving Systems of Linear Equations by Elimination

Key Terms

1. elimination method
2. addition property of equality
3. substitution

Now Try

1. $\{(8, 3)\}$
2. $\{(6, 5)\}$
3. $\{(-4, 9)\}$
4. $\left\{\left(\frac{9}{17}, \frac{11}{17}\right)\right\}$

5a. $\varnothing$

5b. $\{(x, y) \mid 9x - 7y = 5\}$

Objective 1

1. $(8, 3)$
3. $(5, 0)$

Objective 2

5. $\left(\frac{1}{2}, 1\right)$

Objective 3

7. $(2, -4)$
9. $(3, -2)$

Objective 4

11. $\{(x, y) \mid -x - 2y = 3\}$

12.4 Applications of Linear Systems

Key Terms

1. $d = rt$
2. system of linear equations

Now Try

1. 56 cm, 26 cm
2. 1500 general admission tickets; 750 reserved seats
3. water: 9 oz; 80% solution: 3 oz
4. Enid: 44 mph; Jerry: 16 mph
5. plane speed: 265 mph; wind speed: 35 mph

Objective 1

1. 32, 18
3. length: 14 ft; width: 11 ft

Objective 2

5. 30 \$5 bills; 60 \$10 bills

Objective 3

7. \$6 coffee: 100 lbs; \$12 coffee: 50 lb
9. \$1.60 candy: 20 lb; \$2.50 candy: 10 lb

Objective 4

11. plane A: 400 mph; plane B: 360 mph

12.5 Solving Systems of Linear Inequalities

Key Terms

1. solution set of a system of linear inequalities
2. system of linear inequalities

Now Try

1.

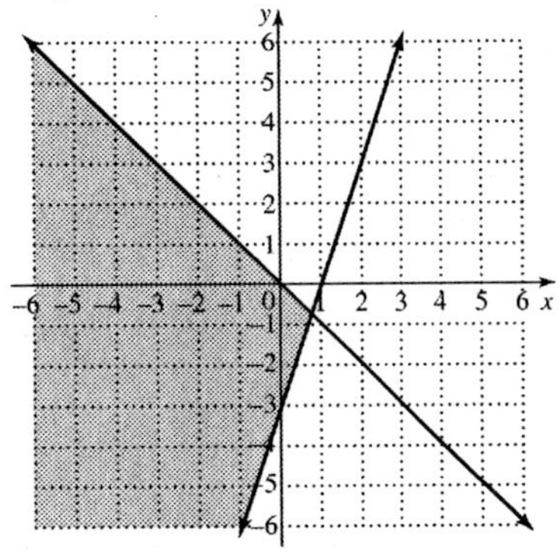

2.

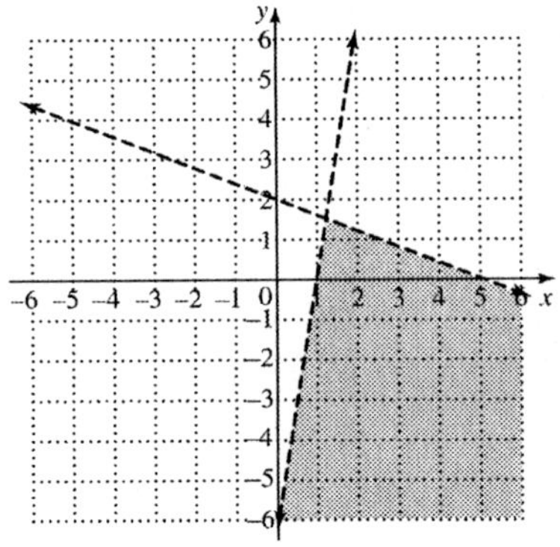

3.

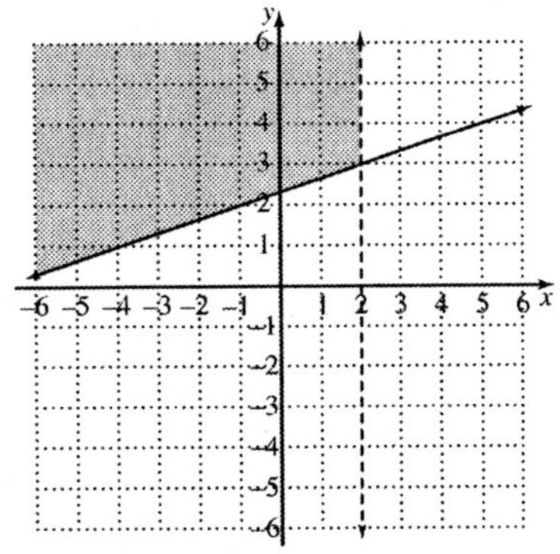

Objective 1

1.

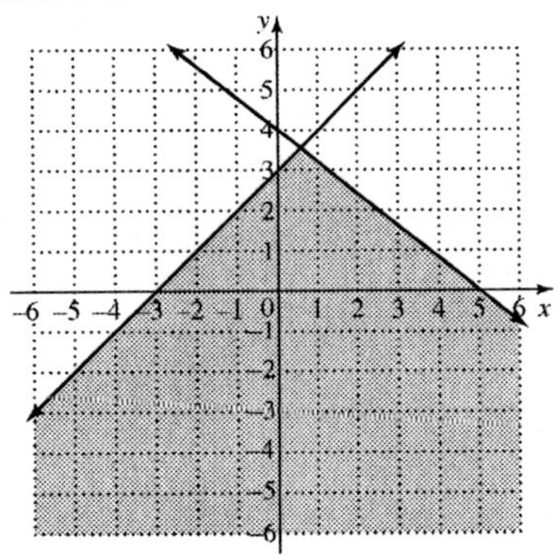

3.

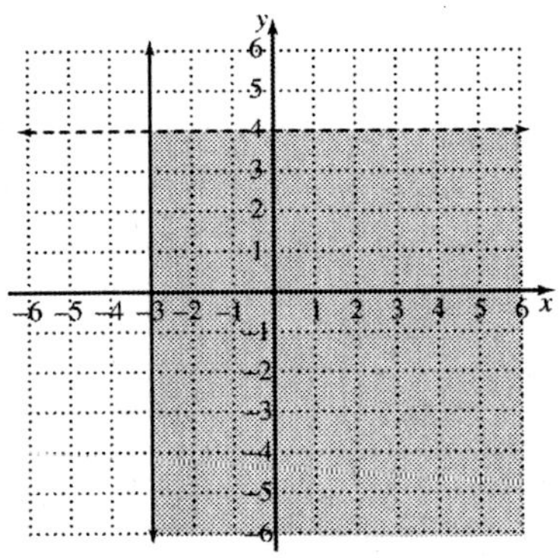

Chapter 13 EXPONENTS AND POLYNOMIALS

13.1 Adding and Subtracting Polynomials

Key Terms

1. degree of a term
2. descending powers
3. term
4. trinomial
5. polynomial
6. monomial
7. degree of a polynomial
8. binomial
9. like terms

Now Try

1a. $5x^4$

1b. $7pq$

1c. $-8x^7$

1d. $\frac{8}{5}x$

1e. $15m^3 + 29m^2$

1f. Cannot be simplified

2a. $8x^3 + 4x^2 + 6$; degree 3; trinomial

2b. $4x^5$; degree 5; monomial

3a. 1285

3b. 1357

4a. $5x^3 - 5x^2 + 4x$

4b. $9x^4 + 7x^2 + 7x - 6$

5a. $4x^3 + x + 12$

5b. $7x^3 + 8x^2 - 14x - 1$

6a. $2x + 2$

6b. $-11x^3 - 7x + 1$

7. $9x^3 + x^2 - 11x + 7$

8. $7 + 8x - 4x^2$

9a. $13a + ab$

9b. $x^2y + 4xy$

Objective 1

1. $-3z^3$
3. $8c^3 - 8c^2 - 6c + 6$

Objective 2

5. $-5.7d^8 - 1.1d^5 + 3.2d^3 - d^2$; degree 8; none of these

Objective 3

7. a. −51; b. 74
9. a. 29; b. 39

Objective 4

11. $4x^2 + 6x - 18$

Objective 5

13. $-8w^3 + 21w^2 - 15$
15. $3x^3 + 7x^2 + 2$

Objective 6

17. $-5ab - 6ac$

13.2 The Product Rule and Power Rules for Exponents

Key Terms

1. power
2. exponential expression
3. base

Now Try

1. 4^5; 1024

2a. base: 2; exponent: 6; value: 64

2b. base:2; exponent: 6; value: –64

2c. base: –2; exponent: 6; value 64

3a. 9^{13}

3b. $(-6)^6$

3c. x^6

3d. m^{27}

3e. 500

3f. 108

4. $18x^{11}$

5a. 4^{15}

5b. 7^8

5c. x^{30}

6a. $64a^3b^3$

6b. $7p^3q^3$

6c. $200x^8y^6$

6d. -8^{35}

7a. $\frac{125}{216}$

7b. $\frac{p^6}{q^6}$

7c. $\frac{1}{1024}$

8a. $\frac{5^5}{2^3}$, or $\frac{3125}{8}$

8b. $6^{11}x^{11}$

8c. $432x^{11}y^{12}$

8d. $-x^{27}y^{13}$

9a. $28x^5$

9b. $15m^7$

Objective 1

1. $\frac{1}{243}$

3. −6561; base: 3; exponent: 8

Objective 2

5. $8c^{15}$

Objective 3

7. 7^{12}

9. $(-3)^{21}$

Objective 4

11. $-0.008a^{12}b^3$

Objective 5

13. $-\frac{8x^3}{125}$

15. $-\frac{128a^7}{b^{14}}$

Objective 6

17. $32a^9b^{14}c^5$

Objective 7

19. $36x^5$

21. $28q^{11}$

13.3 Multiplying Polynomials

Key Terms

1. inner product
2. FOIL
3. outer product

Now Try

1a. $32x^4 + 64x^3$

1b. $-35m^8 + 42m^7 - 28m^6 + 7m^5$

2. $4x^7 - 2x^5 + 37x^4 - 18x^2 + 9x$

3. $28x^4 - 33x^3 + 51x^2 + 17x - 15$

4. $x^2 + 3x - 54$

5. $16xy + 72y - 14x - 63$

6a. $15k^2 + 44kn + 32n^2$

6b. $45p^2 + 11pq - 4q^2$

6c. $35x^5 - 195x^4 - 90x^3$

Objective 1

1. $35z^4 + 14z$

3. $-6y^5 - 9y^4 + 12y^3 - 33y^2$

Objective 2

5. $6m^5 + 4m^4 - 5m^3 + 2m^2 - 4m$

Objective 3

7. $20a^2 + 11ab - 3b^2$

9. $-6m^2 - mn + 12n^2$

13.4 Special Products

Key Terms

1. binomial
2. conjugate

Now Try

1. $x^2 + 12x + 36$
2a. $64a^2 - 48ab + 9b^2$
2b. $4a^2 + 36ak + 81k^2$
2c. $9a^2 - 66ab + 121b^2$
2d. $9p^2 + p + \frac{1}{36}$
2e. $45x^3 - 210x^2 + 245x$
3a. $x^2 - 81$
3b. $\frac{25}{36} - a^2$
3c. $x^3 - 144x$
4a. $64p^2 - 25$
4b. $121x^2 - y^2$
4c. $x^2 - \frac{9}{25}$
4d. $4p^5 - 144p$
5a. $x^3 + 18x^2 + 108x + 216$
5b. $81x^4 - 540x^3 + 1350x^2 - 1500x + 625$
5c. $-4x^4 - 36x^3 - 108x^2 - 108x$

Objective 1

1. $49 + 14x + x^2$
3. $16y^2 - 5.6y + 0.49$

Objective 2

5. $64k^2 - 25p^2$

Objective 3

7. $a^3 - 9a^2 + 27a - 27$
9. $256s^4 + 768s^3t + 864s^2t^2 + 432st^3 + 81t^4$

13.5 Integer Exponents and the Quotient Rule

Key Terms

1. power rule for exponents
2. base; exponent
3. product rule for exponents

Now Try

1a. 1
1b. 1
1c. −1
1d. 1
1e. 88
1f. 1
2a. $\frac{1}{64}$
2b. $\frac{1}{27}$
2c. 25
2d. $\frac{49}{36}$
2e. $\frac{8}{27}$
2f. $\frac{1}{8}$
2g. $\frac{11}{p^4}$
2h. x^8
2i. $\frac{b^4}{a^5}$
3a. $\frac{125}{36}$
3b. $\frac{y}{x^7}$
3c. $\frac{qr^5}{4p^3}$
3d. $\frac{625b^4}{a^4}$
4a. 9
4b. $\frac{1}{16}$
4c. 64
4d. z^{10}
4e. $\frac{a^3}{25}$
4f. $(a-b)^2$
4g. $\frac{36b^7}{a^8}$
5a. 6
5b. $3125b^5$
5c. $\frac{243}{32p^{20}}$
5d. $\frac{x^{12}y^{15}}{8000}$
5e. $56x$

Objective 1

1. −1
3. 0

Objective 2

5. $\frac{1}{m^{18}n^9}$

Objective 3

7. $\frac{k^4m^5}{2}$
9. $\frac{p^8}{3^5m^3}$ or $\frac{p^8}{243m^3}$

Objective 4

11. $a^{16}b^{22}$

13.6 Dividing a Polynomial by a Monomial

Key Terms

1. dividend
2. quotient
3. divisor

Now Try

1. $x^2 - 3x$
2. $3n^2 - 4n - \frac{2}{n}$
3. $\frac{7z^4}{2} - 4z^3 - \frac{5}{z} - \frac{3}{z^2}$
4. $-2a^3b^2 - 4a^2b + 3$

Objective 1

1. $2a^3 - 3a$
3. $-13m^2 + 4m - \frac{5}{m^2}$

13.7 Dividing a Polynomial by a Polynomial

Key Terms

1. divisor
2. quotient
3. dividend

Now Try

1. $2x^2 - 2x - 2 + \frac{-5}{5x - 1}$
2. $x^2 + 10x + 100$
3. $3x^2 + 5x + 5 + \frac{8x + 29}{x^2 - 4}$
4. $x^2 + \frac{1}{8}x + \frac{5}{8} + \frac{6}{8x - 8}$
5. $L = 2r^2 - r + 5$ units

Objective 1

1. $-3x + 4$
3. $a^2 + 1$

Objective 2

5. $3t + 4$ units

13.8 An Application of Exponents: Scientific Notation

Key Terms

1. scientific notation
2. power rule
3. quotient rule

Now Try

1a. 6.8×10^{8}

1b. 4.771×10^{10}

1c. 9.991×10^{0}

1d. 4.63×10^{-2}

1e. -8.48×10^{-4}

2a. 835,000

2b. 27,960,000

2c. –0.000164

3a. 2.7×10^{8}, or 270,000,000

3b. 3×10^{-8}, or 0.00000003

4. 5.45×10^{3} kg/m^{3}

5. 9×10^{23} grains of sand

Objective 1

1. 2.3651×10^{4}

3. -2.208×10^{-4}

Objective 2

5. 0.0064

Objective 3

7. 2.53×10^{2}

9. 4.86×10^{19} atoms

Chapter 14 FACTORING AND APPLICATIONS

14.1 Factors; The Greatest Common Factor

Key Terms

1. factoring
2. factored form
3. greatest common factor
4. factor

Now Try

1a. 6 1b. 8 1c. 1

2a. $6x^4$ 2b. q^3 3a. $8x(x^4+3)$

3b. $4y^2(5y^2-3y+1)$ 3c. $n^7(n+1)$ 3d. $8x^3(8x^2-5x+1)$

4. $-9pq(pq^2+3p^3q-1)$ 5a. $(y+8)(y+4)$

5b. $(z+5)(z^2-11)$ 6a. $(2+y)(x+7)$ 6b. $(9+t)(4x+1)$

6c. $(x-7)(4x+5y)$ 6d. $(x+7)(x^2-2)$ 7a. $(8x-3y)(7x+4)$

7b. $(9x-5)(y-2)$

Objective 1

1. 28
3. 6

Objective 2

5. $k^2m^4n^4$

Objective 3

7. $10x(2x+4xy-7y^2)$
9. $-13x^8(-2+x^4-4x^2)$

Objective 4

11. $(2x+y)(x-7y)$

14.2 Factoring Trinomials

Key Terms

1. factoring
2. greatest common factor
3. prime polynomial

Now Try

1. $(x+3)(x+8)$
2. $(y-7)(y-5)$
3. $(p+9)(p-3)$
4. $(a-17)(a+2)$

5a. prime

5b. prime

6. $(p-7q)(p+2q)$
7. $7x^4(x-5)(x-2)$

Objective 1

1. prime
3. $(x-11)(x+3)$

Objective 2

5. $2ab(a-3b)(a-2b)$

14.3 Factoring Trinomials by Grouping

Key Terms

1. coefficient
2. trinomial

Now Try

1. $(5x+2)(x+3)$

2a. $(7x-5)(2x+1)$

2b. $(3m-7)(m+2)$

2c. $(5x+3y)(2x-y)$

3. $3x^3(5x-3)(2x+7)$

Objective 1

1. $(4b+3)(2b+3)$
3. $(5c-7t)(2c-3t)$

14.4 Factoring Trinomials By Using the FOIL Method

Key Terms

1. inner product
2. FOIL
3. outer product

Now Try

1. $(3x+2)(x+5)$
2. $(3x+1)(5x+7)$
3. $(4x-1)(5x-2)$
4. $(4x+7)(2x-3)$
5. $(6x-5y)(4x+3y)$

6a. $2y(5y+6)(y+1)$

6b. $-6a(3a-5)(a-2)$

Objective 1

1. $(2q+1)(4q+3)$
3. $2(2c-d)(c+4d)$

14.5 Special Factoring Techniques

Key Terms

1. difference
2. perfect square trinomial

Now Try

1a. $(z+6)(z-6)$

1b. $(r+s)(r-s)$

1c. prime

1d. prime

2a. $(2x+9)(2x-9)$

2b. $(5t+7)(5t-7)$

2c. $(8r+3s)(8r-3s)$

3a. $10(3x+7)(3x-7)$

3b. $(x^2+8)(x^2-8)$

3c. $(p^2+16)(p+4)(p-4)$

4. $(p+8)^2$

5a. $(x-12)^2$

5b. $(8m+3)^2$

5c. prime

5d. $5x(2x+5)^2$

Objective 1

1. $(x-7)(x+7)$
3. prime

Objective 2

5. $(3j+2)^2$

14.6 Solving Quadratic Equations by Factoring

Key Terms

1. standard form
2. quadratic equation

Now Try

1a. $\left\{-12, \frac{7}{4}\right\}$

1b. $\left\{0, \frac{11}{6}\right\}$

2a. $\{-3, 8\}$

2b. $\{-8, 9\}$

3. $\left\{\frac{4}{5}, 3\right\}$

4a. $\left\{-\frac{3}{10}, \frac{3}{10}\right\}$

4b. $\{0, 11\}$

4c. $\left\{-\frac{1}{4}, 6\right\}$

5a. $\{8\}$

5b. $\left\{-\frac{9}{4}\right\}$

6. $\{-5, 0, 5\}$

7. $\left\{\frac{2}{5}, 2, 9\right\}$

Objective 1

1. $\left\{-\frac{5}{2}, 4\right\}$

3. $\left\{-4, \frac{3}{5}\right\}$

Objective 2

5. $\{-9, 0, 1\}$

14.7 Applications of Quadratic Equations

Key Terms

1. legs
2. hypotenuse

Now Try

1. width: 3 m, length: 5 m
2. −4, −3 or 3, 4
3. 8, 10
4. 16 ft
5. 1 sec
6. 110 items

Objective 1

1. width: 8 in., length: 24 in.
3. height: 4 ft, width: 6 ft

Objective 2

5. 6, 8

Objective 3

7. 45 m, 60 m, 75 m
9. 20 mi

Objective 4

11. 40 items or 110 items

Chapter 15 RATIONAL EXPRESSIONS AND APPLICATIONS

15.1 The Fundamental Property of Rational Expressions

Key Terms

1. rational expression
2. lowest terms

Now Try

1a. $\frac{3}{2}$ 1b. $\frac{7}{5}$ 1c. undefined

1d. 0 2a. $y \neq \frac{1}{7}$ 2b. $m \neq 5,\ m \neq -4$

2c. never undefined 3a. $\frac{8}{21}$ 3b. $\frac{3}{k^3}$

4a. $\frac{7}{9}$ 4b. $y+2$ 4c. $\frac{m+6}{2m+3}$

5. –1 6a. –1 6b. $-\frac{10x+9}{4}$

6c. in lowest terms 7. $\frac{-(10x-7)}{4x-3}, \frac{-10x+7}{4x-3}, \frac{10x-7}{-(4x-3)}, \frac{10x-7}{-4x+3}$

Objective 1

1. a. $-\frac{11}{9}$; b. –4
3. a. $-\frac{1}{6}$; b. $-\frac{7}{2}$

Objective 2

5. none

Objective 3

7. $\frac{-5b}{8c}$
9. $\frac{9(x+3)}{2}$

Objective 4

11. $\frac{-(2p-1)}{-(1-4p)}; \frac{1-2p}{4p-1}; \frac{-(2p-1)}{4p-1}; \frac{1-2p}{-(1-4p)}$

15.2 Multiplying and Dividing Rational Expressions

Key Terms

1. reciprocal
2. rational expression
3. lowest terms

Now Try

1a. $\frac{4}{15}$

1b. $\frac{4}{3x}$

2. $\frac{s^2}{6(r-s)}$

3. $\frac{35}{x}$

4a. $\frac{10b}{11a^3}$

4b. $\frac{w^2+6w-7}{w^2-6w}$

5a. $\frac{15}{2}$

5b. $\frac{y-2}{6(y+2)}$

6. $\frac{2}{25}$

7. $\frac{8x}{(x-3)^2}$

8. $\frac{-(m-8)}{5m(m+9)}$

Objective 1

1. $\frac{10m^3n}{3}$

3. $\frac{x+4}{2x-8}$

Objective 2

5. $\frac{1}{4r^2+2r+3}$

Objective 3

7. $-\frac{1}{2}$

9. $\frac{-3(6k-1)}{3k-1}$

15.3 Least Common Denominators

Key Terms

1. equivalent expressions
2. least common denominator

Now Try

1a. 72

1b. $24x$

2. $120a^4$

3a. $9w(w-2)$

3b. $(b+4)(b+1)(b-4)^2$

3c. $p-14$ or $14-p$

4a. $\frac{65}{30}$

4b. $\frac{27n^2}{66n}$

5a. $\frac{76}{24c-20}$

5b. $\frac{3(z+2)}{z(z-7)(z+2)}$ or $\frac{3z+6}{z^3-5z-14z}$

Objective 1

1. $108b^4$

3. $w(w+3)(w-3)(w-2)$

Objective 2

5. $30r$

15.4 Adding and Subtracting Rational Expressions

Key Terms

1. greatest common factor
2. least common multiple

Now Try

1a. $\frac{4}{5}$

1b. $2x$

2a. $\frac{137}{315}$

2b. $\frac{46}{63y}$

3. $\frac{4}{x-10}$

4. $\frac{7x^2+11x+8}{(x+2)(x+1)(x-4)}$

5. $\frac{x-y}{4x-3y}$

6. $\frac{7x-7}{x-3}$

7. $\frac{5x+63}{x(x-9)}$

8. 7

9. $\frac{8x^2+37x+15}{(x+5)(x-5)^2}$

10. $\frac{p+25}{(p+1)(p-1)(p-5)}$

Objective 1

1. $\frac{4}{w^2}$

3. $\frac{1}{x-2}$

Objective 2

5. $\frac{7z^2-z-6}{(z+2)(z-2)^2}$

Objective 3

7. $\frac{8z}{(z-2)(z+2)}$ or $\frac{8z}{z^2-4}$

9. $\frac{2m^2-m+2}{(m-2)(m+2)^2}$

15.5 Complex Fractions

Key Terms

1. complex fraction
2. LCD

Now Try

1a. $\frac{35}{51}$

1b. $\frac{90}{x}$

2. $\frac{bc^2}{a^2}$

3. $\frac{-9x+65}{2x-5}$

4a. $\frac{56}{99}$

4b. $\frac{12}{x}$

5. $\frac{5n-9}{3(6n+5)}$

6a. $\frac{5x+6}{x+15}$

6b. $\frac{x-1}{x-3}$

6c. $\frac{x+2}{3x-5}$

Objective 1

1. $\frac{7m^2}{2n^3}$

3. $\frac{9s+12}{6s^2+2s}$ or $\frac{3(3s+4)}{2s(3s+1)}$

Objective 2

5. $\frac{(x-2)^2}{x(x+2)}$

15.6 Solving Equations with Rational Expressions

Key Terms

1. extraneous solution
2. proposed solution

Now Try

1a. expression; $\frac{1}{2}x$

1b. equation; $\{14\}$

2. $\{18\}$

3. $\{8\}$

4. $\varnothing$

5. $\{6\}$

6. $\{0\}$

7. $\{-4\}$

8. $\{2\}$

9a. $b = aq + c$

9b. $y = \frac{x - wz}{w}$, or $\frac{x}{w} - z$

10. $x = \frac{yz}{y + z}$

Objective 1

1. equation; $\{-2\}$

3. expression; $\frac{41x}{15}$

Objective 2

5. $\{-4, 16\}$

Objective 3

7. $f = \frac{d_0 d_1}{d_0 + d_1}$

9. $q = \frac{2pf - Ab}{Ab}$ or $\frac{2pf}{Ab} - 1$

15.7 Applications of Rational Expressions

Key Terms

1. numerator
2. denominator
3. reciprocal

Now Try

1. 3
2a. 765 mi
2b. 4.1 hr (rounded)
2c. 243.2 mph (rounded)
3. 50 miles per hour
4. 3 miles per hour
5. $\frac{2}{5}$ hour

Objective 1

1. $-\frac{2}{3}$ or 1
3. $\frac{3}{5}$

Objective 2

5. 24 miles per hour

Objective 3

7. $2\frac{2}{5}$ hr
9. 72 hr

15.8 Variation

Key Terms

1. constant of variation
2. direct variation
3. inverse variation

Now Try

1. 168
2. 5
3. 50 pounds

Objective 1

1. 12
3. 15

Objective 2

5. 2.25

Chapter 16 ROOTS AND RADICALS

16.1 Evaluating Roots

Key Terms

1. radicand
2. perfect square
3. index (order)
4. square root
5. radical expression
6. principal square root
7. irrational number
8. radical
9. cube root

Now Try

1. 9, –9
2a. 13
2b. 41
2c. $\frac{3}{7}$
2d. 0.8
3a. 19
3b. 37
3c. n^2+5
4a. irrational
4b. rational
4c. not a real number
5a. 3.742
5b. 7.141
5c. –26.038
6a. $c=17$
6b. $a=\sqrt{52}\approx 7.211$
7. 35 miles
8a. 5
8b. –5
8c. 7
9a. 5
9b. –5
9c. none
9d. –5
9e. –5

Objective 1

1. 25, –25
3. $\frac{30}{7}$

Objective 2

5. not a real number

Objective 3

7. 5.657
9. 14.491

Objective 4

11. 9

Objective 5

13. –4
15. –1

16.2 Multiplying, Dividing, and Simplifying Radicals

Key Terms

1. radical
2. perfect cube
3. radicand

Now Try

1a. $\sqrt{91}$
1b. $\sqrt{42}$
1c. $\sqrt{15c}$

2a. $2\sqrt{3}$
2b. $7\sqrt{2}$
2c. $4\sqrt{5}$

2d. cannot be simplified
3a. $15\sqrt{2}$
3b. $6\sqrt{2}$

3c. $60\sqrt{2}$
4a. $\frac{13}{3}$
4b. 9

4c. $\frac{\sqrt{7}}{8}$
5. $3\sqrt{10}$
6. $\frac{\sqrt{7}}{6}$

7a. x^{10}
7b. $7x^4$
7c. $6x^2\sqrt{2}$

7d. $r^9\sqrt{r}$
7e. $\frac{\sqrt{11}}{y^2}$
8a. $3\sqrt[3]{2}$

8b. $3\sqrt[4]{2}$
8c. $\frac{2}{7}$
9a. m^3

9b. $5x^2$
9c. $2a\sqrt[4]{5a}$
9d. $\frac{x^7}{6}$

Objective 1

1. $\sqrt{65}$
3. $\sqrt{21x}$

Objective 2

5. $11\sqrt{3}$

Objective 3

7. $\frac{5}{9}$
9. $\frac{2}{25}$

Objective 4

11. $4x^2y^2\sqrt{2y}$

Objective 5

13. $-2\sqrt[5]{2}$
15. $\frac{5}{4}$

16.3 Adding and Subtracting Radicals

Key Terms

1. index
2. unlike radicals
3. like radicals

Now Try

1a. $10\sqrt{21}$
1b. $-6\sqrt{17}$
1c. $6\sqrt{14}$
1d. $2\sqrt{30}$
1e. cannot be combined
2a. $8\sqrt{6}$
2b. $\sqrt{2}$
2c. $27\sqrt{6}$
3a. $10\sqrt{5}$
3b. $5\sqrt{7k}$
3c. $32x\sqrt{3}$
3d. $11m\sqrt[3]{2}$

Objective 1

1. $5\sqrt{2}+\sqrt{3}$
3. $-5\sqrt{5}$

Objective 2

5. $27\sqrt{2}$

Objective 3

7. $4\sqrt{35}$
9. $39w\sqrt{6}$

16.4 Rationalizing the Denominator

Key Terms

1. product rule
2. rationalizing the denominator
3. quotient rule

Now Try

1a. $2\sqrt{7}$

1b. $\dfrac{\sqrt{3}}{3}$

2. $\dfrac{\sqrt{21}}{6}$

3. $\dfrac{\sqrt{10}}{15}$

4a. $\dfrac{5\sqrt{ab}}{b}$

4b. $\dfrac{m\sqrt{55n}}{5}$

5a. $\dfrac{\sqrt[3]{45}}{3}$

5b. $\dfrac{\sqrt[3]{33}}{3}$

5c. $\dfrac{\sqrt[3]{10x^2}}{2x}$

Objective 1

1. $\dfrac{3\sqrt{10}}{2}$

3. $\dfrac{3}{5}$

Objective 2

5. $\dfrac{m^2\sqrt{k}}{k^2}$

Objective 3

7. $\dfrac{\sqrt[3]{18}}{3}$

9. $\dfrac{\sqrt[3]{35x^2}}{7x}$

16.5 More Simplifying and Operations with Radicals

Key Terms

1. rationalize the denominator
2. conjugate

Now Try

1a. $\sqrt{10}$

1b. $-69-2\sqrt{30}$

1c. $-2+2\sqrt{21}$

2a. $36-10\sqrt{11}$

2b. $114+80\sqrt{2}$

2c. $144-24\sqrt{m}+m$

3a. 117

3b. $b-20$

4a. $\frac{10(7-\sqrt{10})}{39}$

4b. $\frac{-8\sqrt{6}-21}{3}$

4c. $\frac{7(5+\sqrt{b})}{25-b}$

5. $\frac{\sqrt{5}+5}{9}$

Objective 1

1. $4\sqrt{14}-63$

3. -38

Objective 2

5. $\frac{\sqrt{3}+2\sqrt{6}+\sqrt{2}+4}{-7}$

Objective 3

7. $\frac{1+\sqrt{3}}{3}$

9. $27\sqrt{3}+5$

16.6 Solving Equations with Radicals

Key Terms

1. extraneous solution
2. radical equation

Now Try

1. $\{11\}$
2. $\{3\}$
3. $\varnothing$
4. $\varnothing$
5. $\{7\}$
6. $\{-2, -1\}$
7. $\{5\}$
8. 6804 m^2

Objective 1

1. $\left\{\frac{8}{3}\right\}$
3. $\{2\}$

Objective 2

5. $\varnothing$

Objective 3

7. $\{8\}$
9. $\{3\}$

Objective 4

11. 499 sq cm

Chapter 17 QUADRATIC EQUATIONS

17.1 Solving Quadratic Equations by the Square Root Property

Key Terms

1. quadratic equation
2. zero-factor property

Now Try

1a. $\{-7, -1\}$

1b. $\{-10, 10\}$

2a. $\{-9, 9\}$

2b. $\left\{-\sqrt{23}, \sqrt{23}\right\}$

2c. $\left\{-5\sqrt{3}, 5\sqrt{3}\right\}$, or $\left\{\pm 5\sqrt{3}\right\}$

2d. $\varnothing$

3a. $\{-6, 16\}$

3b. $\left\{-3 \pm \sqrt{13}\right\}$

4. $\left\{\dfrac{3 \pm 4\sqrt{2}}{7}\right\}$

5. $\varnothing$

6. About 16.9 in.

Objective 1

1. $\{-4, -2\}$

3. $\{-7, 5\}$

Objective 2

5. $\left\{-7\sqrt{2}, 7\sqrt{2}\right\}$

Objective 3

7. $\{-6, 2\}$

9. $\left\{\dfrac{1}{5}, \dfrac{4}{5}\right\}$

Objective 4

11. 2 in.

17.2 Solving Quadratic Equations by Completing the Square

Key Terms

1. perfect square trinomial
2. square root property
3. completing the square

Now Try

1a. 25; $(x+5)^2$

1b. 121; $(x-11)^2$

2. $-5 \pm 4\sqrt{2}$

3. $\{-8+\sqrt{73}, -8-\sqrt{73}\}$

4. $\left\{\frac{5}{4}, \frac{11}{4}\right\}$

5. $\left\{\frac{7}{6}, 2\right\}$

6. $\varnothing$

7. $\{-3+\sqrt{30}, -3-\sqrt{30}\}$

8. 3 sec

Objective 1

1. $\{-4-2\sqrt{3},\ -4+2\sqrt{3}\}$

3. $\{-9, 7\}$

Objective 2

5. $\varnothing$

Objective 3

7. $\varnothing$

9. $\{-2-\sqrt{2},\ -2+\sqrt{2}\}$

Objective 4

11. 2 months, 4 months

17.3 Solving Quadratic Equations by the Quadratic Formula

Key Terms

1. standard form
2. constant
3. quadratic formula

Now Try

1a. $a = 12, b = -10, c = 7$

1b. $a = -4, b = -110, c = 8$

1c. $a = 15, b = 0, c = -8$

1d. $a = -8, b = -3, c = 0$

1e. $a = 30, b = 11, c = -39$

2. $\left\{-2, \frac{7}{3}\right\}$

3. $\{3+\sqrt{5}, 3-\sqrt{5}\}$

4. $\left\{\frac{11}{2}\right\}$

5. $\varnothing$

Objective 1

1. $a = 10,\ b = 4,\ c = 0$

3. $a = 1,\ b = 3,\ c = 9$

Objective 2

5. $\varnothing$

Objective 3

7. $\{2\}$

9. $\left\{\frac{9}{7}\right\}$

Objective 4

11. $\varnothing$

17.4 Graphing Quadratic Equations

Key Terms

1. line of symmetry
2. vertex
3. axis
4. parabola

Now Try

1.

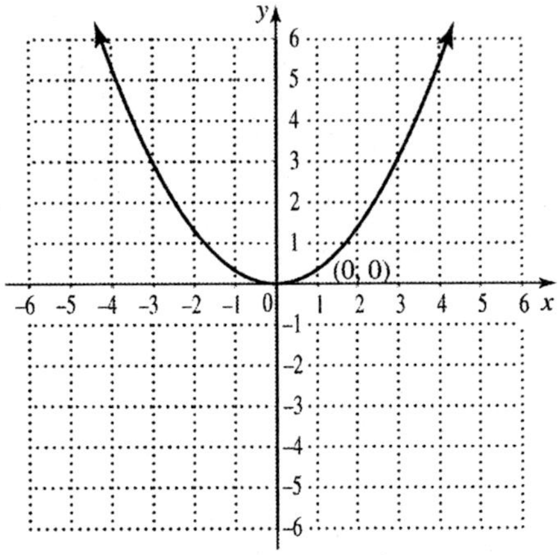

2.

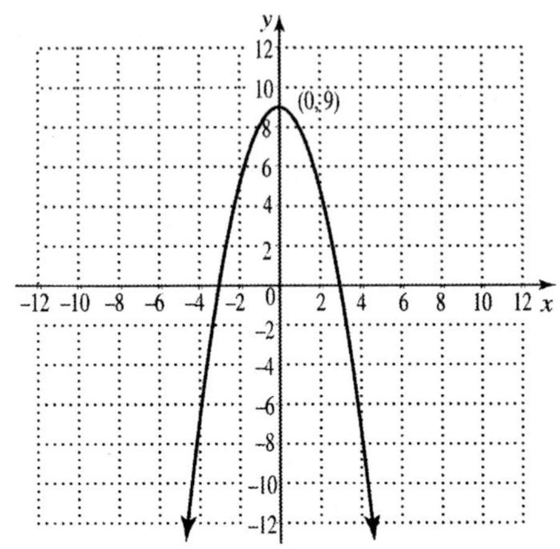

3.

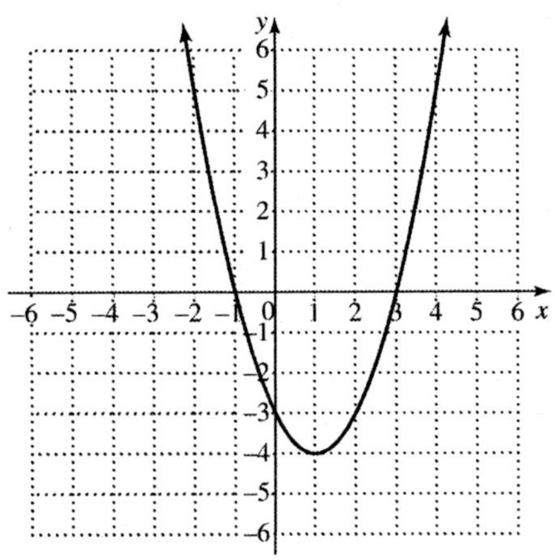

4.

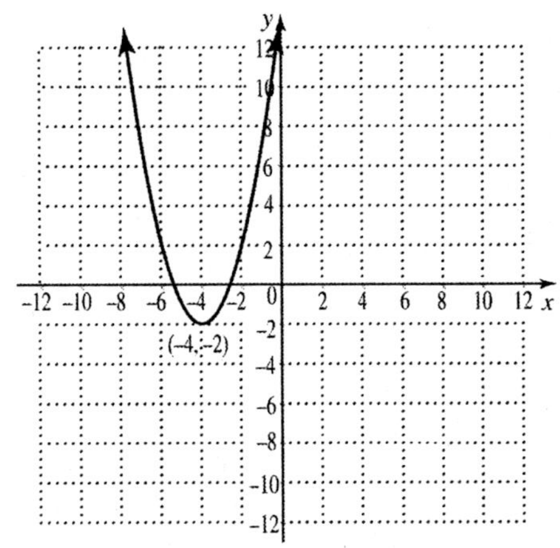

Objective 1; Objective 2

1.

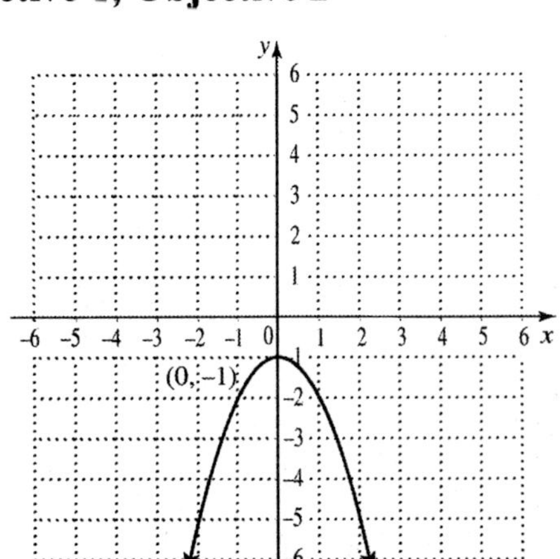

Vertex: $(0, -1)$

3.

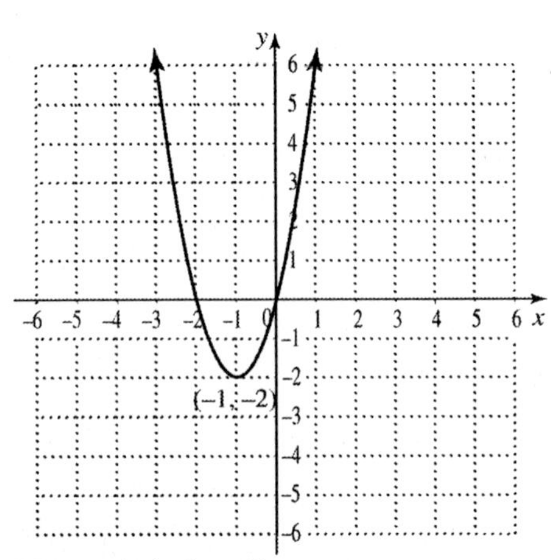

Vertex: $(-1, -2)$

5. 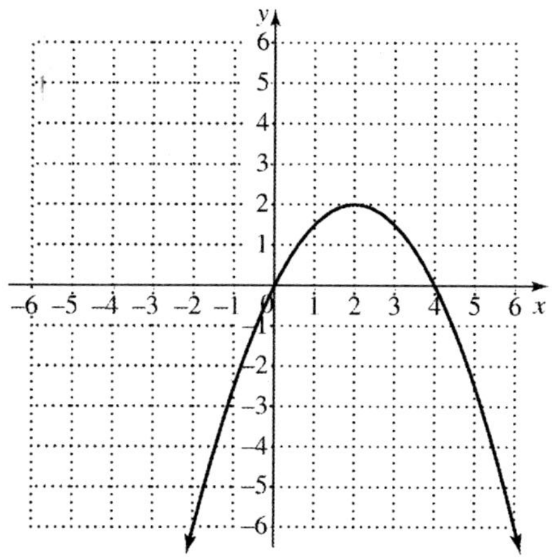

Vertex: (2, 2)

17.5 Introduction to Functions

Key Terms

1. relation
2. range
3. function
4. components
5. domain

Now Try

1a. domain: {6, 8, 10, 12}; range: {7, 9, 11, 13}

1b. domain: {1}; range: {11, 12, 13, 14}

2a. function

2b. not a function

3a. not a function

3b. function

3c. not a function

3d. function

4a. 5

4b. –10

4c. –15

5a. {(2003, 719 million), (2004, 817 million), (2005, 1018 million), (2006, 1093 million), (2007, 1262 million)}; yes

5b. Domain: {2003, 2004, 2005, 2006, 2007}
Range: {719 million, 817 million, 1018 million, 1093 million, 1262 million }

5c. $w(2003) = 719$ million; $w(2006) = 1093$ million

5d. 2005

Objective 1

1. domain: $\{-3, 0, 2, 5\}$; range: $\{-8, -4, -1, 2, 7\}$

3. domain: $\{-3, -2, -1, 0, 1\}$; range: $\{-5, 0, 5\}$

Objective 2

5. not a function

Objective 3

7. not a function
9. function

Objective 4

11. (a) 6; (b) 2; (c) 18

Objective 5

13. {(2008, 596 thousand), (2009, 625 thousand), (2010, 872 thousand), (2011, 795 thousand), (2012, 625 thousand)}; function

15. $r(2009) = \$625$ thousand; $r(2011) = \$795$ thousand

Chapter R WHOLE NUMBERS REVIEW

R.1 Adding Whole Numbers

Key Terms

1. commutative property of addition
2. addends
3. addition
4. associative property of addition
5. regrouping
6. sum (total)

Now Try

1a. $7 + 2 = 9$ and $2 + 7 = 9$

1b. $5 + 9 = 14$ and $9 + 5 = 14$

1c. $3 + 3 = 6$

2. 29
3. 788
4. 65
5. 11,179
6. 33 miles
7. 44 miles
8. incorrect; 1948

Objective 1

1. 12

Objective 2

3. 31

Objective 3

5. 987

Objective 4

7. 1041
9. 10,415

Objective 5

11. 44 miles

Objective 6

13. incorrect; 8612

R.2 Subtracting Whole Numbers

Key Terms

1. minuend
2. regrouping
3. subtrahend
4. difference

Now Try

1a. 13 – 8 = 5 or 13 – 5 = 8

1b. 12 – 9 = 3 or 12 – 3 = 9

2a. 9 = 7 + 2 or 9 = 2 + 7

2b. 320 = 190 + 130 or 320 = 130 + 190

3. minuend: 87; subtrahend: 12; difference: 75

4a. 43

4b. 111

4c. 6210

5a. correct

5b. incorrect;222

5c. incorrect; 1302

6. 39

7. 488

8. 2788

9. 5003

10a. correct

10b. incorrect; 378

11. 95 passengers

Objective 1

1. 187 − 38 = 149; 187 − 149 = 38

3. 785 + 426 = 1211

Objective 2

5. minuend: 35; subtrahend: 9; difference: 24

Objective 3

7. 6012

Objective 4

9. incorrect; 153

Objective 5

11. 198

13. 6703

Objective 6

15. $263

R.3 Multiplying Whole Numbers

Key Terms

1. commutative property of multiplication
2. factors
3. associative property of multiplication
4. multiple
5. product
6. partial products
7. chain multiplication problem

Now Try

1a. 42
1b. 0
1c. 16
2. 40
3a. 258
3b. 3969
4a. 460
4b. 70,800
4c. 519,000
5a. 17,800
5b. 8400
6a. 2880
6b. 28,116
6c. 8745
7a. 89,358
7b. 32,388,875
8. $5372

Objective 1

1. factors: 9, 12; 108
3. 72

Objective 2

5. 84

Objective 3

7. 216
9. 180,054

Objective 4

11. 20,000

Objective 5

13. 4416
15. 1,644,648

Objective 6

17. $9728

R.4 Dividing Whole Numbers

Key Terms

1. remainder
2. dividend
3. quotient
4. divisor
5. short division

Now Try

1a. $4\overline{)28}$ with quotient 7, or $\frac{28}{4}=7$

1b. $72 \div 9 = 8$ or $9\overline{)72}$ with quotient 8

1c. $32 \div 8 = 4$ or $\frac{32}{8} = 4$

2a. dividend: 63; divisor: 9; quotient: 7

2b. dividend: 56; divisor: 8; quotient: 7

2c. dividend: 54; divisor: 6; quotient: 9

3a. 0

3b. 0

3c. 0

3d. 0

4a. $4 \cdot 9 = 36$ or $9 \cdot 4 = 36$

4b. $3 \cdot 6 = 18$ or $6 \cdot 3 = 18$

4c. $8 \cdot 5 = 40$ or $5 \cdot 8 = 40$

5a. undefined

5b. undefined

5c. undefined

6a. 1

6b. 1

6c. 1

7. 23

8. 74 R 1

9. 261 R 5

10a. correct

10b. incorrect; 78 R1

11a. Yes

11b. No

12a. Yes

12b. No

13a. Yes

13b. Yes

13c. No

14a. Yes

14b. No

Objective 1

1. $3\overline{)15}$ with quotient 5; $\frac{15}{3} = 5$

Objective 2

3. dividend: 63; divisor: 7; quotient: 9
5. dividend: 44; divisor: 11; quotient: 4

Objective 3

7. 0

Objective 4

9. undefined
11. undefined

Objective 5

13. 1

Objective 6

15. 72 R 3

Objective 7

17. incorrect; 1522 R 5   19. incorrect; 6050

Objective 8

21. 2: yes; 3: no; 5: no; 10: no

R.5 Long Division

Key Terms

1. dividend
2. remainder
3. divisor
4. quotient
5. long division

Now Try

1. 73
2. 38 R 55
3. 209 R20

4a. 5
4b. 47
4c. 606

5a. 135
5b. 2006
6. incorrect; 37 R62

Objective 1

1. 77
3. 654 R22

Objective 2

5. 13

Objective 3

7. correct
9. incorrect; 460 R3